Advances in Concentrating Solar Thermal Research and Technology

Related titles

Performance and Durability Assessment: Optical Materials for Solar Thermal Systems
(ISBN 978-0-08-044401-7)

Solar Energy Engineering 2e
(ISBN 978-0-12-397270-5)

Concentrating Solar Power Technology
(ISBN 978-1-84569-769-3)

Woodhead Publishing Series in Energy

Advances in Concentrating Solar Thermal Research and Technology

Edited by

Manuel J. Blanco

Lourdes Ramirez Santigosa

ELSEVIER

AMSTERDAM • BOSTON • HEIDELBERG
LONDON • NEW YORK • OXFORD • PARIS • SAN DIEGO
SAN FRANCISCO • SINGAPORE • SYDNEY • TOKYO
Woodhead Publishing is an imprint of Elsevier

WP WOODHEAD PUBLISHING

Woodhead Publishing is an imprint of Elsevier
The Officers' Mess Business Centre, Royston Road, Duxford, CB22 4QH, United Kingdom
50 Hampshire Street, 5th Floor, Cambridge, MA 02139, United States
The Boulevard, Langford Lane, Kidlington, OX5 1GB, United Kingdom

ISBN: 978-0-08-100516-3 (print)
ISBN: 978-0-08-100517-0 (online)

Library of Congress Cataloging-in-Publication Data
A catalog record for this book is available from the Library of Congress

British Library Cataloguing-in-Publication Data
A catalogue record for this book is available from the British Library

For information on all Woodhead Publishing publications visit our website at https://www.elsevier.com/

Publisher: Joe Hayton
Acquisition Editor: Sarah Hughes
Editorial Project Manager: Anna Valutkevich
Production Project Manager: Debasish Ghosh
Designer: Christian J. Bilbow

Typeset by TNQ Books and Journals

Contents

List of contributors

R. Bader The Australian National University, Canberra, ACT, Australia

S.M. Besarati University of South Florida, Tampa, FL, United States

M.J. Blanco CSIRO, Newcastle, NSW, Australia

R. Buck Deutsches Zentrum für Luft-und Raumfahrt (DLR), Stuttgart, Germany

E.F. Camacho Universidad de Sevilla, Sevilla, Spain

D. Canavarro Largo Marques de Marialva, Evora, Portugal; University of Evora, Evora, Portugal

J. Chaves Light Prescriptions Innovators, Madrid, Spain

M. Collares-Pereira Largo Marques de Marialva, Evora, Portugal; University of Evora, Evora, Portugal

A. Fernández-García CIEMAT-Plataforma solar de Almería, Tabernas (Almería), Spain

A.J. Gallego Universidad de Sevilla, Sevilla, Spain

S. Giuliano Deutsches Zentrum für Luft-und Raumfahrt (DLR), Stuttgart, Germany

J. González-Aguilar IMDEA Energía Institute, Móstoles, Spain

D.Y. Goswami University of South Florida, Tampa, FL, United States

Z. Guan The University of Queensland, Brisbane, Queensland, Australia

L.L. Guerreiro University of Evora, Evora, Portugal

H. Gurgenci The University of Queensland, Brisbane, Queensland, Australia

C.K. Ho Sandia National Laboratories, Albuquerque, New Mexico, United States

K. Hooman The University of Queensland, Brisbane, Queensland, Australia

L. Irwin ManTech International Corporation, Arlington, Virginia, United States

W. Lipiński The Australian National University, Canberra, ACT, Australia

L. Martínez-Arcos CIEMAT-Plataforma solar de Almería, Tabernas (Almería), Spain

S. Miller CSIRO, Newcastle, NSW, Australia

A. Morales Centro de Investigaciones Energéticas, Medioambientales y Tecnológicas (CIEMAT), Madrid, Spain

K. Nithyanandam University of California, Los Angeles (UCLA), Los Angeles, California, United States

C. Pfefferkorn United States Nuclear Regulatory Commission, Washington, D.C., United States

R. Pitchumani Virginia Tech, Blacksburg, Virginia, United States

L. Ramírez CIEMAT, Madrid, Spain

M. Romero IMDEA Energía Institute, Móstoles, Spain

C. Sansom Cranfield University, Bedfordshire, United Kingdom

G. San Vicente Centro de Investigaciones Energéticas, Medioambientales y Tecnológicas (CIEMAT), Madrid, Spain

M.A. Silva-Pérez University of Sevilla, Seville, Spain

J. Stekli United States Department of Energy, Washington, D.C., United States

F. Sutter DLR German Aerospace Center, Institute of Solar Research, Plataforma Solar de Almería, Tabernas (Almería), Spain

R. Uhlig Deutsches Zentrum für Luft-und Raumfahrt (DLR), Stuttgart, Germany

L. Valenzuela CIEMAT, Plataforma Solar de Almería, Tabernas (Almería), Spain

J.M. Vindel CIEMAT, Madrid, Spain

E. Zarza Moya CIEMAT, Plataforma Solar de Almería, Tabernas (Almería), Spain

Editors' biographies

Prof. Manuel Blanco has more than 30 years of experience as a solar researcher and engineer and has contributed to advancing the state of the art of Concentrating Solar Thermal (CST) technologies. He is an expert on Thermodynamics, Heat Transfer, Optics, and Modeling of Energy Systems.

He is a Full Professor and holder of the European Research Area (ERA) Chair in Solar Thermal Energy of the Cyprus Institute's Energy, Environment and Water Research Center (EEWRC). He is also the Chair of the Executive Committee of SolarPACES, the Technology Collaboration Program of the International Energy Agency responsible for promoting the development and deployment of solar thermal power technologies and the solar-assisted manufacturing of fuels and chemicals.

Until August, 2016, he has been Science Leader of the Solar Energy Systems Research Group of CSIRO—Australia's National Research Agency, as well as the Director of the Australian Solar Thermal Research Initiative (ASTRI). He is the former Director of the Solar Thermal Energy Department of the National Renewable Energy Center of Spain (CENER), Chair and Full Professor of the Engineering Department of the University of Texas at Brownsville, and Director of the Plataforma Solar de Almería of CIEMAT, which is the largest European solar research infrastructure.

Manuel Blanco is the author of numerous of scientific publications and participates in a large number of national and international expert groups and committees in a senior capacity.

Dr. Lourdes Ramirez Santigosa is an expert on data analysis, mathematical modeling, measurements processes, and atmospheric physics. In 1990 she began working in the characterization of solar radiation for energy applications and has played for more than 12 years as a coordinator of the Solar Radiation Group at CIEMAT (Research Center for Energy, Environment and Technology). She worked at the National Renewable Energy Center (CENER) leading the Measurement and Characterization Service of the Solar Thermal Energy Department.

Throughout her career, she has worked on projects related to energy applications of solar radiation, taking part in more than 30 projects related to modeling solar radiation from satellite images, site selection of solar plants, studies of solar energy potential at regional or national level, tests characterization of components and solar systems, sensor calibration, series generation for systems simulation, and in general with the space-temporal characterization of solar radiation. She has published almost 30 original research papers, several books, and monographies as well as over 60 participations in international conferences.

Lourdes Ramirez has been appointed representative of CIEMAT in several international expert committees such as the International Electro-technical Committee, the International Energy Agency or the Implementation Group of Atmosphere (GMES), having been invited to research stays at Ecole des Mines de Paris (France) and at CSIRO (Australia). She is currently Senior Research at CIEMAT and also collaborates with the International University of Andalusia and with University Carlos III of Madrid teaching the Characterization of Solar Radiation for Energy Application and has supervised more than 10 PhD and Master theses.

Acknowledgment

First and foremost, we would like to express our appreciation and gratitude to:

- All the experts that have generously contributed to the book. Even though they are all working intensively advancing the frontiers of Concentrating Solar Thermal technologies, they have found the time to deliver excellent contributions.
- The colleagues that have provided us with reviews, feedback, and encouragement.
- The people from the publisher that have been involved in the concept, production and final publication of the book. They have been fantastic in their support and have always dealt with a positive and professional attitude with the many issues that always pop up on these endeavours.
- Our home organizations CIEMAT (Dr. Lourdes Ramirez), The Cyprus Institute (Prof. Manuel J. Blanco), and CSIRO (former organization of Prof. Manuel Blanco when he started working on the book).
- All our loves ones, which has supported as in this endeavour and have suffer the effect of our dedication to the work.

Manuel J. Blanco and Lourdes Ramirez Santigosa

Part One

Introduction

Introduction to concentrating solar thermal (CST) technologies

1

M.J. Blanco, S. Miller
CSIRO, Newcastle, NSW, Australia

1.1 The sun as an energy source

The sun is the most important energy source available to us. Outside the Earth's atmosphere, the average power of the solar radiation perpendicular to the main direction of the sun rays is of the order of 1.36 kW/m^2. This quantity, which is traditionally called the solar constant, is not a constant and varies inversely proportional to the square of the distance from Earth to the sun. It fluctuates about 6.9% during the year.

Incoming radiation from the sun passes through the Earth's atmosphere, whereby it is partially absorbed or scattered by components of the atmosphere, such as aerosols, gases, and particles. About 30% of solar radiation is reflected without being absorbed by the atmospheric components or surface of Earth, while the amount of scattering depends on the atmospheric conditions. The scattered solar radiation reaching the surface of the Earth is called diffuse radiation and may be experienced as glare on an overcast day. The solar radiation reaching the Earth's surface without scattering is called direct radiation and is able to cast a shadow.

In quantitative terms, the amount of solar energy reaching the Earth's surface, "ground-level," is about 885 million terawatt hours (TWh, or 3.06×10^{24} joules) per year [1]. In 2013, the annual anthropogenic energy consumption was 13,541 Mtoe (million tons of oil equivalent) [2], (0.157 TWh or 5.67×10^{20} joules) per year, just a fraction of a percentage point of the solar energy reaching the Earth's surface.

The physical impact of the sun and solar radiation on Earth affects, among other things, the currents and tides, the weather, and the biosphere. Thus, the sun is not only a direct renewable and practically inexhaustible source of energy but also the ultimate origin of most other renewable energy resources; wind energy, ocean energy, as well as biomass. It is worth noting that biomass was the original energy material source for fossils fuels, coal, shale, oil, and gas, albeit a million years in the making. This leaves geothermal and nuclear energy as the only sources of energy available on Earth that are independent of the sun.

The intensity of the direct solar radiation reaching the surface of the Earth is not geographically uniform. An approximate upper limit is 1 kW/m^2, although in many instances and regions of the world it is much less than this. The regions of the world that receive the most amount of direct solar radiation are typically located within the latitudes of $\pm 40°$, in what is sometimes referred to as the "sunbelt." In these regions the

Advances in Concentrating Solar Thermal Research and Technology
http://dx.doi.org/10.1016/B978-0-08-100516-3.00001-0

Table 1.1 Advantages and disadvantages of the sun as an energy source

Advantages	Disadvantages
Abundant, clean, and renewable	Relatively low surface density
Geographically distributed and consequently free of geopolitical tensions	Intermittent
High exergy content	Variable

annual energy per unit received by a surface that is kept at all instances perpendicular to the main direction of the sun rays is usually equal or larger than 1800 kWh/m^2/year.

The spectral distribution of the direct solar radiation just outside the Earth atmosphere is equivalent to that of a blackbody at 5777 K, ranging from near ultraviolet through to near infrared, with most photons and energy in the visible spectrum. The maximum percentage of heat that can be transformed into work for a given ambient temperature is known as the exergy of the heat. According to the second law of thermodynamics this exergy increases with the temperature at which the heat is delivered and decreases with increasing ambient temperatures.

The fact that solar radiation outside the Earth's atmosphere can be considered as heat delivered from a thermal reservoir at 5777 K implies that more than 94% of this heat can be transformed into work. At ground level, because of the interactions with the atmosphere, the spectral distribution of the direct solar radiation can no longer be considered equivalent to that of a blackbody. However, its exergy content is still very high—between 84% and 96% according to most authors [3].

In addition, of being intermittent due to the day and night cycles, the solar radiation on Earth, at ground level, is also variable. The amount of direct solar radiation reaching the ground depends on the atmospheric conditions in general and on cloud cover in particular. This variability, however, is significantly lower than the variability of other renewable energy sources, such as wind, and decreases with increased time intervals.

Table 1.1 summarizes the main advantages and disadvantages of the solar radiation as an energy source, as discussed.

All solar energy systems are designed to maximize the advantages provided by the sun as an energy source and to minimize the disadvantages.

1.2 Defining characteristics of CST technologies

Concentrating solar thermal (CST) technologies collect and concentrate radiation from the sun to transform it into high-temperature thermal energy. This thermal energy can later be used for a plethora of high-temperature thermal applications, such as heating and cooling, process heat, material processing, electricity production, or chemical processes.

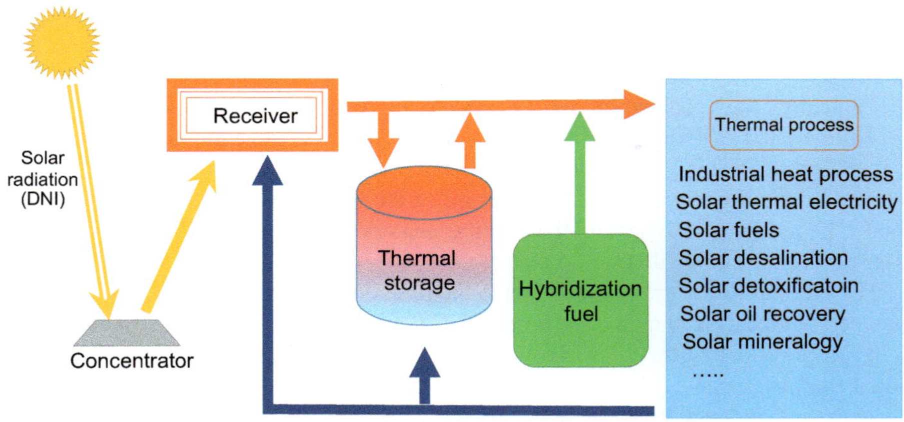

Figure 1.1 General scheme of a concentrating solar thermal (CST) system.

In most CST technologies, once the solar radiation is transformed into thermal energy, instead of using it when the transformation takes place, the thermal energy can be stored as such and used when it makes the most sense to do it, such as to maximize economic returns.

The high-temperature thermal energy derived from the solar radiation collected and concentrated in a CST system can also be hybridized; that is, mixed with thermal energies derived from other heat sources such as biomass or fossil fuels. The hybridization with biomass is particularly interesting, because it will produce a renewable energy system capable of delivering thermal energy to run industrial process or produce electricity 24 h a day, all days of the year.

Thus, the concentration of solar radiation, its intermediate transformation into high-temperature thermal energy, the possibility of storing this thermal energy to dispatch it when needed, and the possibility of hybridizing it with thermal energy from other heat sources are the defining characteristics of CST technologies. Those that set them apart from the rest of renewable energy technologies (Fig. 1.1).

1.3 Thermal efficiency and the need for concentration

Due to the relatively low surface density of direct solar radiation, to deliver a large amount of power a CST system needs a large area of solar radiation collecting surfaces. The efficient transformation of solar radiation into thermal energy however, is best done, in receivers with small surface area, since thermal losses are proportional to the area. Because of this, all CST systems collect and concentrate solar radiation using surfaces that are quite different from the receiver surfaces where the transformation of concentrated solar radiation into thermal energy takes place and much less expensive per unit area.

Traditionally, in a CST system, reflective surfaces are used to collect and concentrate the direct solar radiation upon the receiver. However, refractive surfaces can also be used.

Typically, the receiver of a CST system consists of an absorber material, or absorber coated material, upon which the concentrated solar radiation impinges and is transformed in thermal energy. A large fraction of this thermal energy is transformed into useful energy, for example, in the form of the increase in enthalpy of a working fluid. The rest is used to heat up the absorber material to its operating temperature and after that dissipated as thermal losses.

The thermal losses are proportional to the receiver area and temperature difference between the receiver area and the ambient. Conductive and convective losses are linearly proportional to this temperature difference, while radiative losses are proportional the difference of the fourth power of these temperatures. Because of this, as the operating temperature increases, radiative losses become more and more dominant. This is the reason why, for low- to mid-temperature CST systems, the system designers make a significant effort to keep convection and conduction losses at bay, using for instant vacuum technologies to reduce these losses to a minimum, while for very high-temperature CST systems where the dominant losses are radiative these techniques are seldom used, since they are ineffective in reducing radiative losses.

Thus, in order to increase the thermal efficiency, that is, the ratio of the useful energy delivered by the receiver to the concentrated solar radiation impinging upon it, the system designer can:

- increase the concentration, that is, the ratio between the total area of the solar collecting surfaces and the total area of the surface of the receiver;
- decrease the operating temperature of the receiver, thereby, reducing the temperature difference with the ambient;
- use specially engineered materials and techniques to increase the absorption of the solar radiation in the receiver and to minimize its thermal losses, such as spectrally selective absorber coatings, evacuated tubes, and materials and meta-materials with the appropriate thermal properties.

To achieve high thermal efficiencies while operating at high temperature almost inevitably requires achieving high concentration ratios (Figs. 1.2 and 1.3).

1.4 Limits of concentration

How much direct solar radiation can be concentrated is limited by the second law of thermodynamics. The limit is related to the fact that as seen from the surface of Earth the sun subtends a finite solid angle. This implies, that direct solar radiation is not completely parallel, but distributed over a cone of directions around the direction from the observer to the center of the solar disk defined by a half-angle of 4.58 mrad.

In terms of the refractive indices at the entrance and exit of the concentrator, n_{in} and n_{out}, its acceptance angle, θ, and the angular spread of the concentrated radiation

Figure 1.2 Solar field of SENERtrough® cylindrical-parabolic collectors. Arcosol and Termesol thermosolar plants, owned by Torresol Energy. (Source: SENER/Torresol Energy).

incident upon the receiver, ϕ (see Fig. 1.4), the limit of concentration for point focus concentrators (3D concentration) is given by the following equation:

$$C_{g3D} = \frac{n_{out}(\sin \phi)^2}{n_{in}(\sin \theta)^2} \tag{1.1}$$

For linear concentrator (2D concentration) the limit of concentration is the square root of the 3D limit. Particularizing Eq. (1.1) for the concentration of solar radiation on the surface of the Earth, assuming $n_{in} = n_{out}$, $\theta = \theta_{sun}$, and $\phi = \pi/2$ yields a maximum 3D concentration ratio of about 47,500. Consequently, the maximum 2D concentration is of the order of 218.

Figure 1.3 Gemasolar thermosolar plant, owned by Torresol Energy and developed by SENER. (Source: SENER/Torresol Energy).

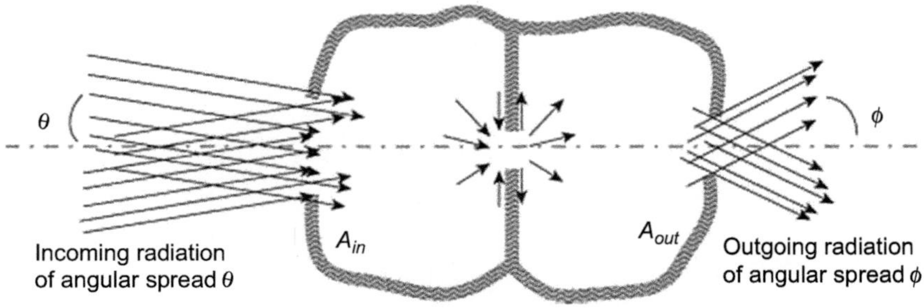

Figure 1.4 An arbitrary concentrator accepting radiation with a half-angle θ over area A_{in} and sending it over area A_{out} with half-angle ϕ.
From Lovegrove K, Pye J. Fundamental principles of concentrating solar power (CSP) systems. In: Lovegrove K, Stein W, editors. Concentrating solar power technology. Woodhead Publishing; 2012 [chapter 2].

Traditional image-forming concentrators, such as parabolic troughs or parabolic dishes fall short of the concentration limits for linear and point focusing concentrators, respectively. The concentration of the best possible parabolic dish concentrator is of the order of 25% of the maximum 3D concentration limit. To approach the limits of concentration with actual physical devices, one has to use new types of concentrators that do not form images and are adequately known as nonimaging concentrators.

1.5 Optimum operating temperature to maximize light-to-work conversion efficiency

The second law of thermodynamics also limits the conversion of thermal energy into work. The maximum efficiency at which heat taken from a heat reservoir at temperature T_{hot} can be transformed into work by a heat engine in contact also with a cold reservoir at temperature T_{cold} is given by Carnot's law:

$$\eta_{Carnot} = 1 - \frac{T_{cold}}{T_{hot}} \tag{1.2}$$

According to this law, the efficiency of the conversion of thermal energy into work increases with the temperature at which the thermal energy is delivered to the heat engine.

In a CST system designed to produce work, the overall light-to-work efficiency is the product of two efficiencies:

- the thermal efficiency with which the direct solar radiation from the sun is transformed into useful thermal energy;
- the efficiency with which the useful thermal energy is subsequently transformed into work by a heat engine.

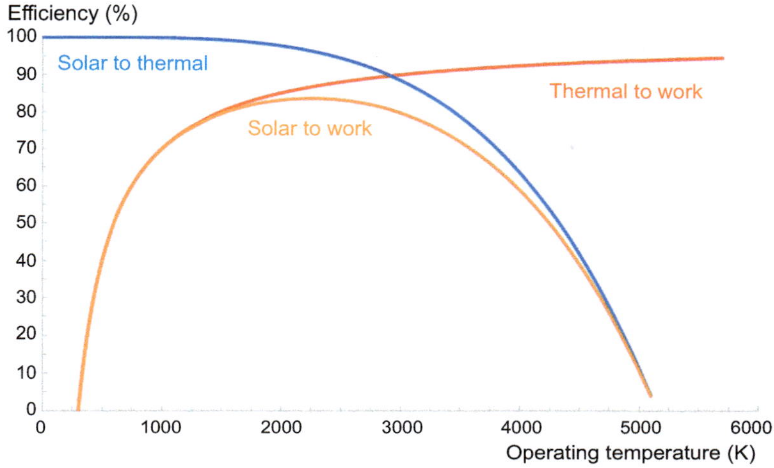

Figure 1.5 Solar-to-thermal, thermal-to-work, and overall solar-to-work efficiencies in a CST system.

While the efficiency at which the direct solar radiation is transformed into useful thermal energy decreases with the operating temperature, the efficiency at which the useful thermal energy is transformed into work increases with it.

The overall light-to-work efficiency is, therefore, the product of two functions with opposite tendencies with regard to the variation of the operating temperature of the receiver and as such it must have an optimum; that is, a value of the operating temperature of the receiver for which the light-to-work efficiency is a maximum (Fig. 1.5).

Since the higher the concentration the higher the temperature at which the receiver can operate with a given amount of thermal losses, the optimum operating temperature increases with concentration.

1.6 Main commercially available solar concentrating technologies

Currently, there are four main commercially available technologies to concentrate the direct solar radiation into a receiver, where this radiation will be transformed in thermal energy. Two of these solar concentrating technologies are linear-focusing technologies and the other two are point-focusing technologies. All of them are imaging technologies, but can be combined with non-imaging secondary reflectors to further increase the concentration they can achieve.

1.6.1 Line focus solar concentrators

Line focus solar concentrators have reflectors that concentrate solar radiation onto a linear receiver. The two dominant concentrator technologies are parabolic troughs and linear Fresnels. A linear Fresnel concentrator approximates a parabolic trough

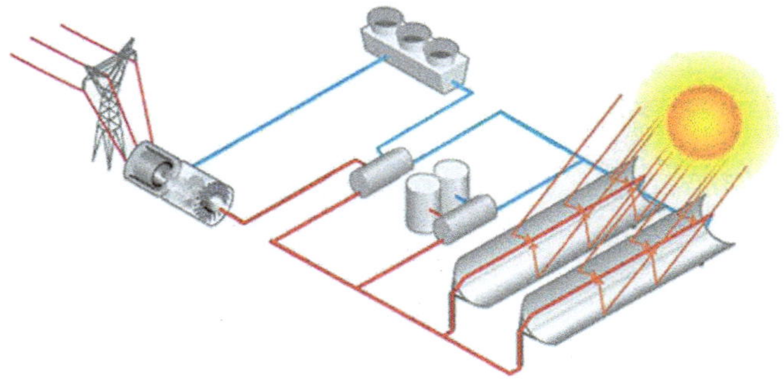

Figure 1.6 Parabolic trough concentrator with moving receiver and thermal energy storage tanks. Used with permission from European Solar Thermal Electricity Association (ESTELA). www.estelasolar.org.

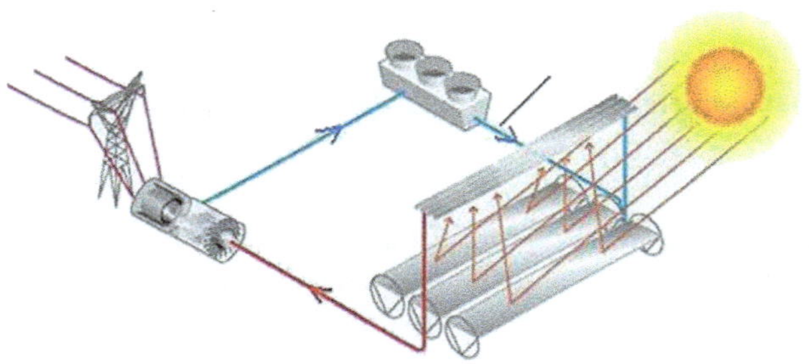

Figure 1.7 Linear Fresnel concentrator with fixed receiver and no thermal energy storage. Used with permission from European Solar Thermal Electricity Association (ESTELA). www.estelasolar.org.

by having independent reflectors rather than a more continuous reflector surface. While there may be some loss of optical efficiency from having independent reflectors, there may be other advantages that contribute to the overall collector efficiency or cost effectiveness (Figs. 1.6 and 1.7).

1.6.1.1 Parabolic trough

A conventional parabolic trough solar concentrator comprises a parabolic reflector, with:

- reflector support structure;
- pylons with joint attachments to allow single-axis solar tracking;
- pier foundations;
- a receiver fixed to the reflector support structure;
- pipework to convey the heat transfer fluid to the receiver and to its storage or utilization point.

The troughs are laid out in parallel rows as a solar field with spacing between the rows to minimize shading of the reflectors, while allowing sufficient access for maintenance and minimizing the pipework and parasitic pumping energy for the heat transfer fluid. The heat transfer fluid normally enters at one end of the trough and leaves at the other, although it might also have tube-in-tube arrangement to allow entry and exit at one end (Fig. 1.8).

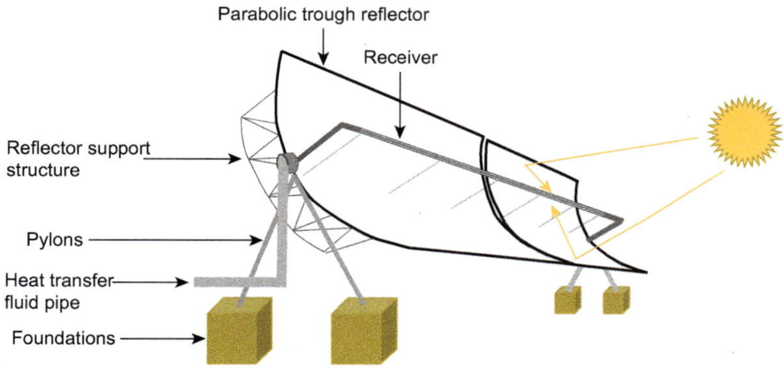

Figure 1.8 Schematic of a parabolic trough collector.

The troughs are normally installed to maximize the annual energy output, with the collector aligned from solar north to south, allowing tracking from east to west. The troughs may also be installed to maximize the energy output at solar noon in winter, with the collector aligned from east to west.

Commercially available trough solar concentrators achieve concentration ratios within the 50 to 80 range.

1.6.1.2 Linear Fresnel

A conventional linear Fresnel solar concentrator comprises reflectors that may be flat or slightly parabolic, with:

- reflector support structure;
- frame with joint attachments to allow single-axis solar tracking;
- pier foundations;
- a receiver fixed above, but independent of the reflector support structure;
- pipework to convey the heat transfer fluid to the receiver and to its storage or utilization point.

The individual reflectors are laid out parallel to the ground such that each reflector has a different focal length to its receiver. The spacing of the reflectors is close to minimize the discontinuity in the reflective area or aperture. The width of the reflectors is optimized to allow access to the reflectors for maintenance, while not being too large to complicate the support structure or tracking (Fig. 1.9).

While a reflector may be placed so that it can be associated with more than one receiver, access to the receiver for maintenance also needs to be considered. The alignment of linear Fresnel systems is the same as for parabolic troughs.

Commercially available linear Fresnel solar concentrators achieve concentration ratios within the 30 to 70 range.

1.6.2 Point-focus solar concentrators

Point-focus systems have reflectors that concentrate solar radiation onto a central receiver that is effectively a point compared to the reflector. The two dominant

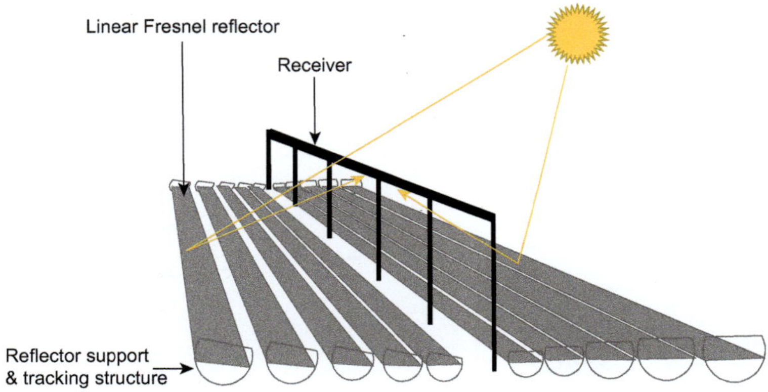

Figure 1.9 Schematic of a linear Fresnel collector.

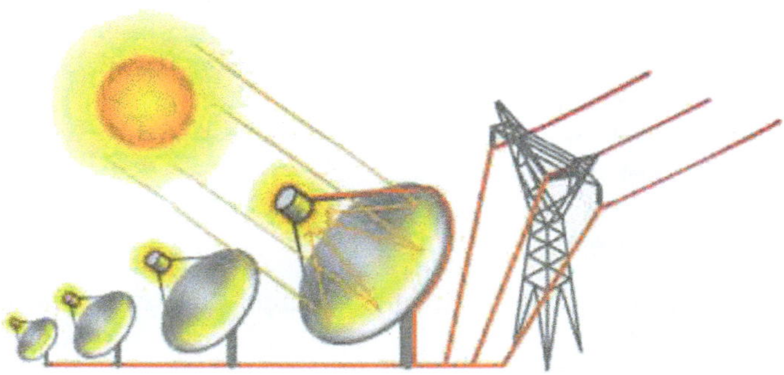

Figure 1.10 Parabolic dish concentrators with moving receiver and no thermal energy storage. Used with permission from European Solar Thermal Electricity Association (ESTELA). www.estelasolar.org.

technologies are parabolic dishes and central receiver systems, known as solar towers. The solar tower has independent reflector facets, known as heliostats, rather than a more continuous reflector surface. While there may be some loss of optical efficiency from having heliostats, there may be other advantages that contribute to the overall collector efficiency or cost effectiveness (Figs. 1.10 and 1.11).

1.6.2.1 Parabolic dish

A conventional parabolic dish collector comprises a dish parabolic reflector, with:

- reflector support structure;
- pylons with joint attachments to allow double-axis solar tracking;
- pier foundations;
- a receiver, fixed to the reflector support structure;
- pipework to convey the heat transfer fluid to the receiver and to its storage or utilization point.

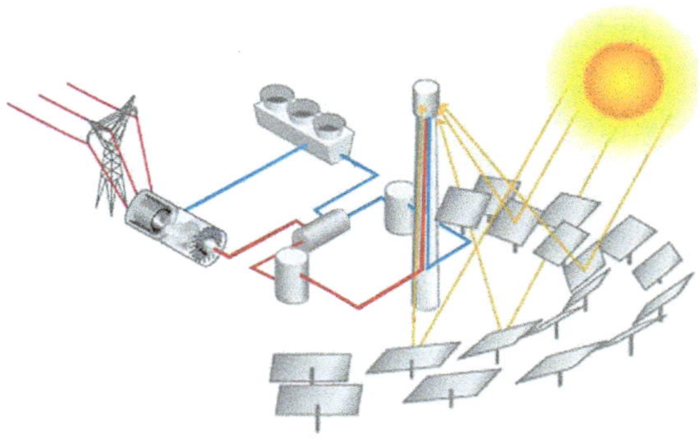

Figure 1.11 Heliostat field-central receiver concentrator with fixed receiver and thermal energy storage. Used with permission from European Solar Thermal Electricity Association (ESTELA). www.estelasolar.org.

The parabolic reflectors may be continuous or comprise of discrete elements conforming to a parabolic shape. The receiver is fixed to the reflector support structure so that both the dish and receiver track the sun. The size of the receiver needs to be optimized to minimize the shadow it, and its support structure, might create on the reflector. The mass of the receiver needs to be optimized to minimize the mass that needs to track the sun. The most common consideration is for a parabolic dish to have a Stirling engine placed at the receiver. Alternatively, the receiver might have a heat transfer fluid to drive an independent process or heat engine (Fig. 1.12).

The dishes are laid out as a solar field with spacing to minimize collisions and shading of the collectors, while allowing sufficient access for maintenance and minimizing the pipework and parasitic pumping energy for the heat transfer fluid.

Commercially available dish concentrators achieve concentration ratios of more than 2000.

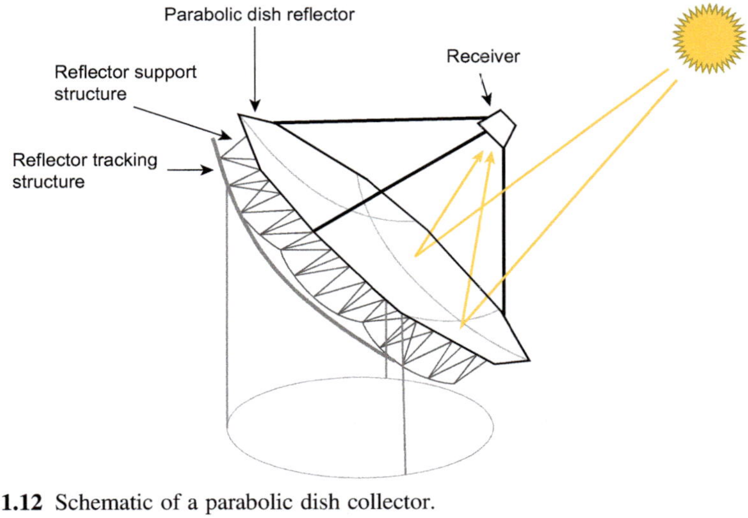

Figure 1.12 Schematic of a parabolic dish collector.

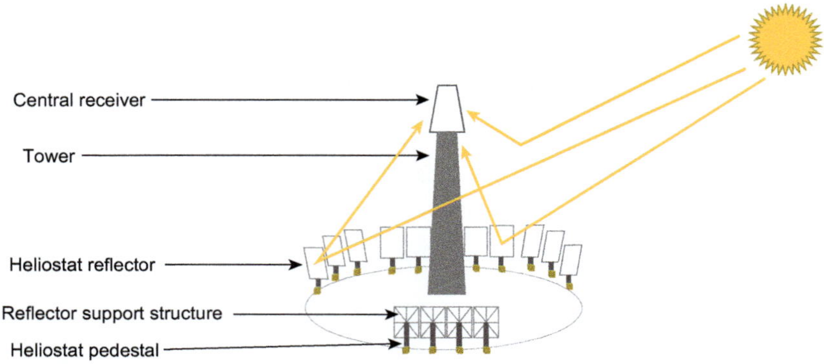

Figure 1.13 Schematic of a heliostat field: central receiver system or solar tower.

1.6.2.2 Heliostat field-central receiver

A conventional heliostat field-central receiver solar concentrator, comprises heliostat reflectors, that may be flat or slightly parabolic, with:

- reflector support structure;
- pedestals with joint attachments to allow double-axis solar tracking;
- pier foundations;
- a receiver mounted on a tower;
- pipework to convey the heat transfer fluid to the receiver and to its storage or utilization point.

The heliostat reflectors are placed in a solar field surrounding the tower. The receiver and tower need to be optimized to minimize the shadow they might create on the solar field. The solar field needs to be optimized in terms of the heliostat size, closeness to minimize the discontinuity in the reflective area, or aperture, spacing to minimize collisions, while also allowing sufficient access for maintenance. The system layout needs to be optimized to minimize the pipework and parasitic pumping energy for the heat transfer fluid (Fig. 1.13).

Commercially available heliostat field, central receiver solar concentrators, achieve concentration ratios within the 500 to 800 range.

1.7 Industry and market trends

1.7.1 Solar thermal electricity

The current state-of-the art of CST technologies for electricity production are:

- parabolic trough system using synthetic oil as the heat transfer fluid, with a two-tank molten salt storage system;
- linear Fresnel system using water as the heat transfer fluid, for direct steam generation without storage;

- central receiver tower system using molten salt as the heat transfer fluid, with a two-tank molten salt storage system.

These three CST commercial technologies for electricity production typically use a Rankine cycle to transform the useful thermal energy generated in the receiver or receivers into electricity. For parabolic dish systems, the most demonstrated concept uses a Stirling engine, although these systems are not fully commercially available.

Parabolic troughs are linear focusing technologies, with a concentration ratio of 50–80. The trough reflects the sun's rays onto a linear receiver fixed at the focus on the parabola, and tracks the sun in one direction—normally from east to west. The receiver usually consists of an absorber metal pipe insulated inside an evacuated glass tube. A synthetic oil is generally used as the heat transfer fluid, which can be heated to temperatures of around 400°C. In heat exchangers, the hot heat transfer fluid passes countercurrent to the working fluid of the power block, which for steam means the water passes through a preheater, boiler, and superheater. The resultant superheated steam temperature is at the lower end of the subcritical Rankine cycle efficiency scale, of say 34%.

The Rankine cycle is also used in thermal power plants involving biomass, fossil fuel, and nuclear energy with the thermal efficiency of the cycle being dependent upon the temperature and pressure of the steam (Table 1.2). The Rankine cycle involves water being heated to create pressurized steam that is fed into an expander, such as a turbine. Expansion of the steam causes the turbine to rotate, converting thermal energy into mechanical energy that can drive an alternator or generator, to create electricity. As the steam expands, it also cools and the temperature of the steam leaving the turbine is dependent upon the extent of the expansion pressure drop. Although the steam leaving the turbine may have cooled sufficiently to create aerosols of water, this saturated steam normally needs to be cooled further to be pumped as a liquid. The saturated steam is normally condensed using a cooling tower that rejects heat to a water source, or to the ambient air, or a hybrid air and water system. The water is then pumped back to the heat source to generate steam again to close the loop on the working fluid.

Table 1.2 Steam conditions and Rankine cycle efficiency

Steam	Temperature (°C)	Pressure (MPa)	Thermal efficiency (%)
Subcritical	565/565	16.5	34–38
Supercritical	565/585	>24.8	38–41
Ultra-supercritical	593/621 and above	>24.8	41–42
Advanced ultra-supercritical	677 and above	>34.5	>42

Adapted from Electric Power Research Institute. Australian power generation technology report; 2015.

More than 90% of current commercial CST power plants are based on parabolic trough technology, less than 5% based on linear Fresnel, and about 5% based on central receiver tower. Thus, the most proven CST power plants use parabolic troughs, although the rapidly emerging alternative is a central receiver tower, due to the higher temperature and higher Rankine cycle efficiency.

Linear Fresnel systems are linear focusing technologies, with a concentration ratio of 30–70. The reflectors focus the sun's rays onto a fixed linear receiver by tracking the sun in one direction—normally from east to west. The receiver usually consists of absorber metal pipes in an insulated cavity with a glass window. The heat transfer fluid is normally water allowing direct steam generation at higher temperatures than is possible with synthetic oil in parabolic troughs. The higher temperatures of around 450°C with steam requires the receiver and pipework from the solar field to have a higher pressure rating. The resultant superheated steam temperature is at the lower end of the subcritical Rankine cycle efficiency scale, of say 36%.

Central receiver towers are point-focusing technologies with a concentration ratio of about 500–800. A field of many mirrors, called heliostats, reflects the sun's rays onto a central receiver located on a tower. Each heliostat has two-axis-tracking to direct the sun's rays toward the central receiver, which is normally an array of absorber tubes, referred to as an external receiver. The heat transfer fluid in the absorber tubes could be molten salt, which can be heated to temperatures of around 560°C, or steam. Other heat transfer fluids are under consideration, as well as cavity receivers, which partially insulate the absorber tubes to reduce heat losses. The resultant superheated steam temperature is at the higher end of the subcritical Rankine cycle efficiency scale, of say 38%.

The overall conversion efficiency of solar energy into electricity is dependent upon each step in the conversion process, including the thermal efficiency of the power block. The thermal efficiency of a Rankine cycle depends upon the temperature and pressure of the steam (Table 1.2). To increase the temperature of a CST system, the concentration ratio needs to be increased, and it follows that point-focus systems achieve this more readily than line-focus systems.

Alternative power blocks to the Rankine cycle include the Brayton and Stirling cycles. The Brayton cycle uses a gas as the working fluid, which needs to be compressed rather than pumped through the cycle to the turbine. If the cycle is open, air can be used as the working fluid, with gases such as helium, argon, and supercritical carbon dioxide being preferred for a closed cycle. The Stirling cycle also uses a gas as the working fluid, but involving a reciprocating engine to compress and convert the expanding gas into mechanical energy.

The IEA 2014 roadmap reported 4 GW of installed CST power plants worldwide. Following a detailed analysis of the main markets and project pipelines, the roadmap reported an anticipated deployment of 11 GW of CSP plants by 2020. The IEA also proposed that "achieving this roadmap's vision of 1000 GW of installed CSP capacity by 2050 would avoid the emissions of up to 2.1 gigatonnes (Gt) of carbon dioxide (CO_2) annually."

The present CST industry exists due to establishment of the initial trough plants at Kramer Junction in California USA, in the 1980s and deployment of mainly trough but also Fresnel and solar tower plants in Spain from 2005 to 2012.

The 1980s trough plants saw the construction on nine solar energy generating systems (SEGS) totaling 354 MW and in 2015 were all still operating. These plants ranged in net output from 13.8 to 80 MW and progressed commercially trough development in terms of collector, receiver, thermal oil used as the heat transfer fluid, and trialing thermal storage of this thermal oil. These plants were at three different locations (Daggett, Kramer Junction, and Harper Lake) and the deployment ceased when the energy crisis ended and legislative support ended. In their more than 30 years of operation, the Kramer Junction SEGS plants in particular have demonstrated that:

- CST plants are very reliable;
- they operate beyond their originally estimated operational life;
- they improve their performance and reduce their operational cost with time, as the team that operates and maintains them gets more experienced in carrying out these activities.

The 30 MW SEGS VI plant at Kramer Junction is the most characteristic of these SEGS plant. The details of the plant are in Table 1.3 and the plant flowsheet is in Fig. 1.14.

The industry was revived in 2006 with feed-in tariffs legislated in Spain to promote deployment of renewable energy technologies. Most of these plants:

- were trough technology, although some used linear Fresnel and tower technology;
- incorporated thermal energy storage, based on nitrate molten salts, of 4–15 h at nominal plant capacity;
- had a high proportion (60%) of local content, that is, goods and services from local companies.

Troughs were the preferred technology due to the proven experience of the SEGS plants in California providing a low project risk for the finance sector, although CST

Table 1.3 SEGS VI plant details

Location	225 km northeast of Los Angeles, California 35°00′ N; 117°33′ W
Insolation	2725 kWh/m^2/year (7.6 kWh/m^2/day)
Solar field	Luz LS-2 troughs
	Aperture area 188,000 m^2
	UVAC receivers
	Therminol VP-1 heat transfer fluid
Thermal storage	Nil
Power cycle	30.0 MWe steam turbine capacity
	Wet cooling system
	~19% peak plant efficiency (~10% annual)

Adapted from Patnode A. Simulation and performance evaluation of parabolic trough solar power plants [Master of Science thesis (Mechanical Engineering)]. University of Wisconsin-Madison; 2006.

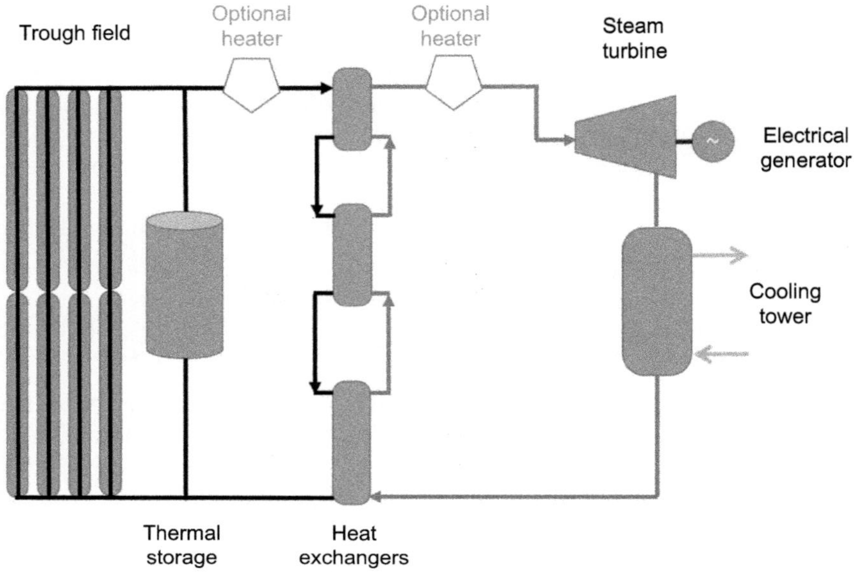

Figure 1.14 SEGS VI plant flowsheet.
Adapted from Status report on solar thermal power plants: experience, prospects and recommendations to overcome market barriers of parabolic trough collector power plant technologies. s.l. Pilkington Solar International; 1996.

power plants using linear Fresnel and solar tower technologies were also constructed. There was widespread integration of thermal energy storage, based on the use of hot and cold tanks of molten nitrate salts. The local content included design, operation, and maintenance from being a modern and industrialized country with a long tradition of CST research and active commitment to the development of CST technologies since mid-1970s. The 50 MW Andasol I plant in Granada is typical of the trough plants; details of the Andasol I plant are in Table 1.4 and the plant flowsheet is in Fig. 1.15. The 20 MW Gemasolar Thermasol plant in Andalucía is representative of the tower plants installed in Spain. A key difference is the Gemasolar Thermasol plant is the first CST plant to operate continuously, firstly overnight and then for over 30 days.

The best market potential for CST power plants is in regions with direct solar insolation greater than 1800 kWh/m^2/year which occur mainly within latitudes of $\pm 40°$, which includes the Americas, Australia, China, India, Middle East, North Africa, and South Africa. In 2015, the emerging markets were predominantly in Morocco, South Africa, China, and Chile where the insolation coexisted with increasing demand and policy drivers for renewable energy and/or stable grid supply.

1.7.2 Industrial process heat

The opportunities for industrial process heat present themselves in niches. While the trend to improve the performance of CST power plants is to increase the operating temperature and thereby the concentration ratio, the opportunities for industrial process heat tend to be at lower temperatures. In this section, we consider the higher

Table 1.4 Andasol I plant details

Location	10 km east of Guadix, Spain (37°13′ N; 3°04′ W)
Insolation	2136 kWh/m^2/year (5.85 kWh/m^2/day)
Solar field	Skal_ET troughs (148.5 m × 6 m wide, 156 loops of four collectors, 17.2 m row spacing)
	Aperture area 510,120 m^2
	UVAC receivers
	Therminol VP-1 heat transfer fluid
Thermal storage	7-h full-load storage
	Dual tank system (14 m height × 37 m diameter, T_{hot} 386°C, T_{cold} 292°C)
	Solar salt storage medium (60% $NaNO_3$ + 40% KNO_3, 28,500 tons, 1010 MWt)
Power cycle	49.9 MWe steam turbine capacity
	Wet cooling system
	~28% peak plant efficiency (~15% annual)
Costs	US $380 million or €310 million for construction (2009)
	13 c/kWh of electricity produced

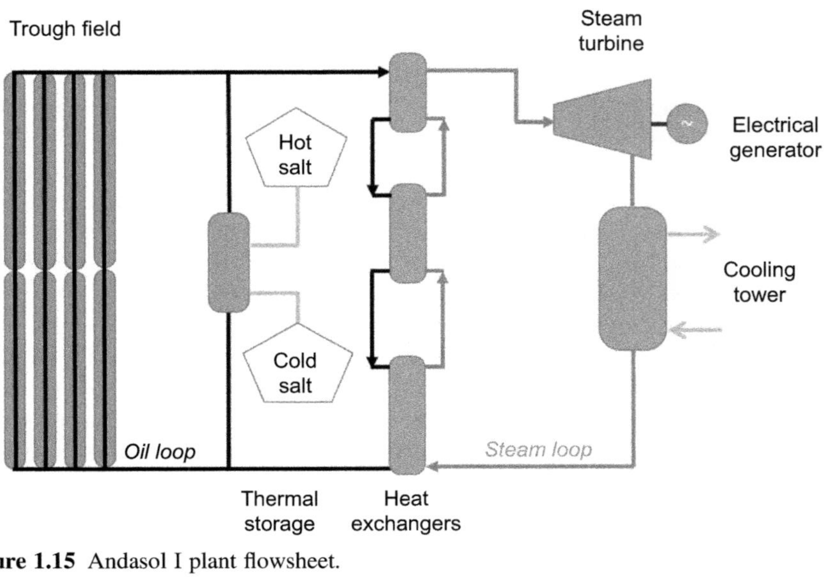

Figure 1.15 Andasol I plant flowsheet.

temperature industrial process heat in the solar chemistry which focuses on solar fuels and mineral processing.

The International Energy Agency (IEA) has a technology collaboration program (TCP) on solar heating cooling (SHC) that focuses on low temperature solar thermal applications for hot water, space heating for buildings, pool heating, and drying processes. Due to end-use demand, these processes consume more energy than transport or electricity, but in most instances can be readily achieved by non-CST technologies. These technologies are scalable and suited to distributed energy systems as the transport of thermal energy is an important aspect of the overall annualized efficiency. These non-concentrating technologies also benefit from the high efficiencies of thermal energy storage.

The industrial process heat applications that fall between the non-CST and high-temperature solar chemistry are typically from 150 to 450°C. At the lower end of this temperature scale, CST collectors may be preferred due to the amount of heat required. These applications may include direct or indirect:

- boiling, drying, pasteurizing, rendering, sterilizing, and washing related to food and beverages;
- bleaching, drying, dyeing, and washing related to textiles, leather, clothing, and footwear;
- boiling, distilling, and reaction processes related to chemical production;
- bleaching and drying related to pulp and paper manufacture or recycle;
- boiling, drying, and preheating related to desalination, sludge drying, and water treatment;
- delivery of steam- or hot-pressurized water for driving absorption chillers, enhanced oil recovery or soil remediation, and hydrothermal processing, such as dewatering or the Bayer process.

Absorption chillers are commercial cooling equipment that can be predominantly driven by thermal energy to provide cooling air-conditioning for commercial buildings, hospitals, and hotels. A two-stage absorption chiller requires a thermal input of about 180°C that can be easily met using a small aperture trough or linear Fresnel collector system and may be mounted on the roof of the building being serviced. The heat transfer in the solar collector can be oil or water based, but needs to match the requirements of the absorption chiller. Absorption chillers operate best with stable thermal input, so should include several hours of thermal storage. The distance of the absorption chiller from the collector, or chiller from its load, needs to be considered to minimize the thermal losses and parasitic energy losses related to pumping. To achieve the best outcome there needs to be an integrated design of the collectors, storage, chiller, and building. There are deployments around the world of solar-driven absorption chillers.

Thermal desalination uses heat to remove salts from feed water to improve its quality. The feed water may be freshwater with salinity up to 1500 ppm, brackish water with salinity up to 10,000 ppm, or seawater with salinity above 10,000 ppm. While potable water, for drinking, is often the desired water quality, desalination may also be used to provide industrial process water that is not as high a quality as potable water. There are a range of thermal desalination processes ranging from high-temperature multistage flash (MSF) and multieffect evaporation (MEE) distillation to low-temperature membrane distillation and humidifying–dehumidifying stills. The high-temperature processes can also be high-throughput systems, and powered by CST collectors. The low-temperature process can be powered by non-CST collectors or as bottoming cycles, using the waste heat from other processes.

Enhanced oil recovery (EOR) uses steam to reduce the viscosity of oil to increase the rate and extent of oil recovery from oil sands. This steam can be generated using concentrating solar collectors, and this is used during the day to displace a fossil fuel normally used to create the steam. There are deployments in California of enhanced oil recovery: a 29 MW central receiver tower system at the Coalinga Oil Fields and an enclosed trough demonstration at the McKittrick Oil Field.

Hydrothermal processing uses steam- or hot-pressurized water to induce chemical reactions such as the digestion of bauxite, known as the Bayer process, or hydrothermal dewatering of lignite materials, such as biomass or brown coal. The Bayer process uses a solution of sodium hydroxide at temperature of up to typically 240°C, with a pressure of about 35 bar, to dissolve the aluminum from the ore. These temperatures are readily achievable with concentrating solar collectors. In 2016 it was announced that Australian researchers are looking to develop the technologies and process knowledge to enable the progressive integration of low-temperature CST into the existing Bayer process.

1.7.3 Solar chemistry and material processing

Solar chemistry covers solar-driven thermochemical processes involving reactions that convert solar energy into chemical energy by driving endothermic reactions. The reactions required to contribute to the production of fuels and materials are typically dissociation reactions that occur at high temperatures and therefore require high concentrations of solar radiation and point focus collector systems. While dissociation reactions may also be used to provide products, or intermediaries for products, as a form of long-term energy storage, these dissociation reactions may also be for more short-term thermochemical energy storage for electricity production or industrial process heat. In this section, the reactions are assumed to require direct use of the solar energy as indirect use is an industrial process heat application.

The endothermic reactions of most interest in solar chemistry are:

- water dissociation or thermolysis, which requires temperatures above 2200°C;
- reduction of carbon dioxide, which requires temperatures above 1950°C;
- hydrogen sulfide dissociation at temperatures around 1200°C;
- methane reforming, albeit with steam (wet reforming) or carbon dioxide (dry reforming), or both steam and carbon dioxide (mixed reforming), at temperatures around 820°C;
- water splitting, involving thermochemical cycles, where the temperatures of interest range from 600 to 900°C;
- reduction of metal oxides where the temperatures of interest range from 600 to 900°C;
- sulfur trioxide dissociation at temperatures around 730°C;
- ammonia dissociation at temperatures around 480°C;
- methanol dissociation at temperatures around 360°C.

While solar-driven thermochemical processes for reduction of water or carbon dioxide to produce net carbon emissions-free fuels have long-term attractiveness, the thermodynamics and high temperatures pose a range of mid-term challenges.

Ammonia as a fuel with its dissociation as a reversible thermochemical energy storage system has been of significant interest. Despite the simplicity of the ammonia dissociation reaction, with just ammonia, nitrogen, and hydrogen involved, the storage of hydrogen poses a significant challenge. The production of ammonia by the Haber process involves the reaction of nitrogen (N_2) with hydrogen (H_2), where the hydrogen often comes from fossil fuels, such as steam reforming of methane.

Methane reforming is also a relatively simple reaction where methane is partially oxidized by steam (H_2O) and/or carbon dioxide (CO_2) to produce a mixture of hydrogen (H_2) and carbon monoxide (CO), known as syngas. However, methane reforming does not occur independently of the water–gas shift reaction ($CO + H_2O \leftrightharpoons CO_2 + H_2$), which is an equilibrium that affects the extent of methane reforming. The equilibrium can be influenced by using excessive amounts of water or steam to promote the production of hydrogen, or by using membrane reactors to separate the products for conversion to continue in an integral reactor. Syngas can also be produced by reforming of petroleum gas or gasification of carbonaceous feedstocks such as coal and biomass. Syngas can then be used to produce bulk chemical commodities such as methanol (CH_3OH), dimethyl ether (DME, CH_3OCH_3) or synthetic liquid fuels ($C_nH_{2n\pm2}$).

The reduction of metal oxides can be used to process mineral ores into metals, particularly if the metal is volatile and becomes gaseous, and then condensed or as thermochemical looping cycles for short-term energy storage, particularly if the metal remains a solid.

1.8 Research priorities, strategies, and trends

The major research programs around the world differ in priorities due to the maturity of the CST industry, policies to prevent climate change and adopt renewable energy technologies, as well as local innovation culture. Compared with other global challenges, the development of CST technologies has been predominantly led by large national research laboratories with support from the university sector. This has been due to the scale of the facilities needed to perform prototype testing. For example, a new photovoltaic solar cell can be tested at a square centimeter, but a laboratory-scale CST test may need a megawatt of thermal energy. Since 2000s there has been an increase in CST research by the university sector, which has been facilitated and supported by the Solar Facilities for the European Research Area (SFERA) program in Europe and by the Multi-University Research Initiative (MURI) in the United States, despite there being quite different research strategies.

The European research strategy is driven by industry through the European Solar Thermal Electricity Association (ESTELA), while the US research strategy is driven by the Department of Energy (US DOE).

The two main goals for the ESTELA Strategic Energy Technology plan are:

- to contribute to achieve the EU targets for 2020 and beyond by implementing large-scale demonstration projects to be carried out by industry and aimed at increasing the competitiveness of the solar thermal electricity sector;

- to enhance market penetration and to consolidate the European industry global leadership throughout medium-term research activities aimed at reducing generation and operation costs in solar thermal electricity generation plants.

The three technology objectives that followed as part of The European Solar Thermo-Electric Industry Initiative (STEII) in 2009 were to:

- increase efficiency and reduce generation, operation, and maintenance costs;
- improve dispatchability;
- improve environmental profile.

Details of these objectives were presented in ESTELA's "Strategic Research Agenda 2020–2025" in 2013. The first objective on efficiency and cost contains cross-cutting research issues that are independent of the CST technology as well as specific technology issues. The second objective on dispatchability considers hybridization, storage, and forecasting tools. The third objective on environmental profile considers minimizing the impact of heat transfer fluids and reduction in water consumption, including desalination.

The aim of US DOE SunShot Initiative is to reduce the total installed cost of solar energy systems to 6 cents per kilowatt-hour (kWh) by 2020. The SunShot Vision Study (2012) predicted this price would allow solar technologies to satisfying about 14% of US electricity demand by 2030 and 27% by 2050. This vision also stated the projected technology and cost improvements to existing and emerging CSP technologies would be achieved through

- reducing solar field costs by reducing the cost of the collector support structure, reflectors, and receivers;
- identifying heat transfer fluids that operate at higher temperatures;
- reducing the cost and improving the operating temperature range of thermal energy storage;
- adopting cooling technologies that minimize water use and cost;
- developing advance power block technologies; while
- ensuring reliability and availability of materials, manufacturing and supply chains.

The US DOE has had a range of research programs on aspects of these issues, often with specific target metrics for the subsystem research derived from systems analysis. In the last few years, US DOE has begun to focus on the subsystems integrated into complete systems.

The much smaller Australian Solar Thermal Research Initiative (ASTRI) has focused on the key challenges of CST power technologies. The overall goal is to significantly reduce the levelized cost of electricity over 8 years from 2012, with four integrated nodes of research:

- *Reducing capital expenditure*—reducing the cost of building solar thermal power plants.
- *Increasing the capacity factor*—increasing operation hours to sell more electricity to the grid.
- *Improving efficiency*—producing more solar thermal power without increasing cost.
- *Adding product value*—with a dual motivation on [1] reducing the operating and maintenance costs of solar thermal power plants as well as [2] developing niche solar fuel technologies within levelized cost of fuel and greenhouse gas reduction boundaries.

This strategic research commenced at early technology readiness levels, taking a systems approach to subsystem evaluation, with the intent of progressing into subsystem demonstration(s). Other Australian research has been targeted at subsystem development at high temperatures for high efficiency, or niche concept demonstrations. In 2015, ARENA funded a specific round of projects on "Integration of renewable energy for industrial process (excluding electricity generation)," whereby the two successful projects involving CST were:

- integrating CST energy into the Bayer alumina process;
- a robotic vision system for automatic inspection and evaluation of solar plants.

Countries such as China, India, and Chile, and regions such as Africa and the Middle East all have major initiatives to deploy CST technologies, as well as both near- and long-term research programs to support the local industry. This makes their programs similar to the ESTELA approach.

This book on advances in concentrating solar power research and technology provides an overview beyond the state-of-the-art, with a focus on advanced CST concepts that are emerging as incremental or step changes in CST technology.

The book begins by considering the mirrors to collect and concentrate solar radiation, then linear collection systems along with their heat transfer fluid and receivers. We then move into point collection systems, their heat transfer fluid and receivers. Thermal energy storage and the use of thermal energy are considered independently of the collection system. The initial focus on using power cycles is to convert thermal energy into electricity, but then alternative end uses in industrial process are heat and direct promotion of solar chemical reactions. Finally, we consider the need for control of CST plants in variable solar conditions, along with solar forecasting and now casting and the long-term solar variability for bankability assessments.

The chapters and descriptions are:

- "Advanced mirror concepts for concentrating solar thermal systems," which outlines significant advances in anti-soiling coatings, highly reflective mirror materials, and high-temperature mirrors for use as secondary concentrators.
- "Improved design for linear Fresnel reflector systems" considers the benefits and challenges of achieving higher concentration values than troughs.
- "A new generation of absorber tubes for concentrating solar thermal (CST) systems" discusses the durability cost and maintenance challenges and solutions for heat collector elements (HCE) or trough receivers.
- "Innovative working fluids for parabolic trough collectors" describes the state-of-the-art of working fluids for troughs to analyze three alternative working fluids in terms of thermohydraulic and cost issues.
- "A new generation of solid-particle and other high-performance receiver designs for concentrating solar thermal (CST) central tower systems" reviews novel high-temperature receiver designs but focuses on several particle receivers concepts.
- "Next generation of liquid metal and other high-performance receiver designs for concentrating solar thermal (CST) central tower systems" revives the concept of sodium as a heat transfer fluid and includes consideration of it boiling. It also discusses the use of mixed liquid metals.
- "Supercritical CO_2 and other advanced power cycles for concentrating solar thermal (CST) systems" reviews the potential power cycles that can be employed for solar thermal power plants in the near future.

- "Advances in dry cooling for concentrating solar thermal (CST) power plants" looks at the performances losses from dry cooling when electricity demand and prices are highest, and how these loses can be minimized.
- "High temperature latent heat storage for concentrating solar thermal (CST) systems."
- "Thermal energy storage concepts for direct steam generation (DSG) solar plants" summarizes recent research from the use of the existing commercial systems with optimized power blocks, to three-part storage systems that combine the use of sensible and latent heat storage.
- "Advanced control strategies to maximize ROI and the value of the concentrating solar thermal (CST) plant to the grid" describes two examples of how advanced control techniques can help to optimize the operation of solar plants and thus maximize the return of investment (ROI).
- "Linear Fresnel reflector (LFR) plants using superheated steam, molten salts, and other heat transfer fluids" proposes the need for advanced optics and molten salt heat transfer fluids to improve this technology's market potential.
- "Central tower systems using the Brayton cycle" summarizes past developments and discusses the current status of solar-hybrid gas-turbine system (SHGT) technology.
- "Solar power towers using supercritical CO_2 and supercritical steam cycles and decoupled combined cycles" presents power cycle concepts poised to become the next generation of commercial CST plants, although in some cases significant technology developments are required.
- "Thermochemical energy storage for concentrating solar thermal (CST) systems" is discussed in terms of its value proposition when high-energy densities and isothermal energy outputs are required, noting integration poses significant operation challenges.
- "Solar thermal processing" provides an overview of applications other than electricity generation, with focus on *solar fuels* and *solar material processing and production.* At present these high-temperature direct solar thermal processing applications have progressed to the proof of concept.

References

[1] International Energy Agency (IEA). Technology roadmap. Solar thermal electricity. 2014.
[2] International Energy Agency (IEA). Key world energy statistics. 2015.
[3] Zamfirescu C, Dincer I. How much exergy one can obtain from incident solar radiation? Journal of Applied Physics 2009;105:044911.
[4] Lovegrove K, Pye J. Fundamental principles of concentrating solar power (CSP) systems. In: Lovegrove K, Stein W, editors. Concentrating solar power technology. Woodhead Publishing; 2012 [chapter 2].
[5] Electric Power Research Institute. Australian power generation technology report. 2015.
[6] Patnode A. Simulation and performance evaluation of parabolic trough solar power plants [Master of Science thesis (Mechanical Engineering)]. University of Wisconsin-Madison; 2006.
[7] Status report on solar thermal power plants: experience, prospects and recommendations to overcome market barriers of parabolic trough collector power plant technologies. s.l. Pilkington Solar International; 1996.

Part Two

Advances in the collection and concentration of sunlight

Advanced mirror concepts for concentrating solar thermal systems

2

A. Fernández-García [1], F. Sutter [2], L. Martínez-Arcos [1],
L. Valenzuela [1], C. Sansom [3]
[1]CIEMAT-Plataforma solar de Almería, Tabernas (Almería), Spain; [2]DLR German Aerospace Center, Institute of Solar Research, Plataforma Solar de Almería, Tabernas (Almería), Spain; [3]Cranfield University, Bedfordshire, United Kingdom

Nomenclature

Acronyms	
CESA	*Central Electrosolar de Almería*
CIEMAT	Centro de Investigaciones Energéticas, Medioambientales y Tecnológicas (Spanish research center)
CPC	Compound parabolic concentrators
CPV	Concentrating photovoltaic
CST	Concentrating solar thermal
CSP	Concentrating solar power
DLR	Deutsches Zentrum für Luft- und Raumfahrt (German aerospace research center)
ESTELA	European Solar Thermal Electricity Association
IEA	International Energy Agency
ISE	Institute for Solar Energy systems
O&M	Operation and maintenance
PMMA	Polymethylmethacrylate
PSA	*Plataforma Solar de Almería*
PSI	Paul Scherrer Institute (Swiss research institute)
PV	Photovoltaic
ST	Solar tower

Continued

Advances in Concentrating Solar Thermal Research and Technology
http://dx.doi.org/10.1016/B978-0-08-100516-3.00002-2

Symbols	
γLV	Liquid–vapor interfacial energy
γSL	Solid–liquid interfacial energy
γSV	Solid–vapor interfacial energy
θc	Contact angle

2.1 Introduction

Mirrors are of primary importance in concentrating solar thermal (CST) technology because they are the first component to be reached by solar beams in the energy conversion process and because the large mirror surfaces covered by the solar field involve a significant cost, both for the initial investment and during the operation and maintenance (O&M) of the plant. Consequently, all stakeholders playing an important role in this technology are paying special attention to the evolution and advances in solar mirrors (also known as reflectors).

According to the materials roadmap published by the European Commission, the target performance in terms of reflectance for silvered reflectors in the following decades (2020–2030) is set at 95–96%, considering a reference starting point of 94% in 2010. In relation to economic requests, it is intended to achieve a reduction of 25% in the reflector cost compared to 2010 values [1]. The solar reflector performance is expected to be maintained with low degradation levels during its lifetime, which is projected to be from 10 to 30 years under severe outdoor environments [2] or even more than 30 years [3].

The materials roadmap mentioned in the preceding paragraph, published by the European Commission in 2011, focused on the materials R&D and the related product development for the next years on the following aspects [1]:

- applied research to further develop low-lead solutions toward "zero lead" mirrors with long-term corrosion protection against weathering;
- develop advanced mirror protective coatings with anti-soiling function and high abrasive and erosion resistance;
- improve resistance of nonglass mirrors to surface degradation in different climatic conditions and under abrasion and erosion;
- develop improved accelerated aging tests;
- applied research for improved low-iron glass with reduced transmission losses; method for recycling process, method for treatment of raw materials to reduce the iron content;
- develop reflector materials with high-temperature stability suitable for secondary reflectors [Fresnel collectors, solar towers (STs), and so on];
- development of accelerated aging testing taking into account specifications of different applications and loads (primary, secondary, climate variability, abrasion, erosion, and so on).

In the same context, one of the key technological milestones stated by the International Energy Agency (IEA) for 2018 in its roadmap is the development of lightweight, low-cost reflector optics [4]. According to this document, this action should be driven by the CST industry with support from research institutions.

Finally, it is also important to mention that the strategic research agenda (2020–2025) elaborated by the European Solar Thermal Electricity Association (ESTELA) also reports on the baselines for the short- to long-term research for solar thermal electricity technology. Concerning reflectors, several research priorities are identified in this document [5]:

- develop lightweight and durable reflective surfaces;
- enhance anti-soiling properties of reflectors to reduce water consumption;
- increase the reflectance of glass-based reflectors by increasing the transmission of glass.

This chapter presents a description of the main advanced mirror concepts developed in the last 7 years, currently under study, or expected to be analyzed in the coming years by the research institutes and the industrial sector, according to the previous recommendations given by the reference institutions.

2.2 Anti-soiling coatings

Anti-soiling coatings are currently receiving major attention by reflector manufacturers, researchers, and plant developers because of their potential to achieve a significant reduction of the soiling accumulation on the reflector surfaces and, consequently, on the cleaning activities and water consumption of CST plants. At the same time that these coatings accomplish their main goal, that is, preventing the dust particles from settling or sticking on the mirror surface, they should not have a detrimental effect on the optical properties of the reflector. Additionally, the coating and reflector combination must be durable and reliable while experiencing high temperatures, sandstorms, and higher than normal exposure to both ultraviolet (UV), visible, and infrared radiation, and must prove their ability to withstand any cleaning procedures [6].

Cuddihy was the first in postulating the following six characteristics of low-soiling surfaces [7]:

- hardness (less susceptible to embedding particles or being damaged by them);
- smoothness (less likely to trap particles);
- hydrophobic (less attractive to ionic species, adsorption of solids, and retention of water);
- low surface energy (lower chemical reactions);
- nonstickiness (chemically clean of sticky materials, surface, and bulk);
- cleanliness (chemically clean of water-soluble salts, which are likely to link other soiling agents, and first-period elements, surface, and bulk).

Also, some of the anti-soiling coatings that have been studied present a charge on the surface to repel charged dust particles. In general, the anti-soiling effect is based on hydrophobic or the hydrophilic properties, which are, respectively, water repellent or water attracting. When a liquid and a solid are in contact, forming an interface, the contact angle is defined as the angle formed by the intersection of the liquid–solid interface and the liquid–vapor interface (geometrically acquired by applying a tangent line

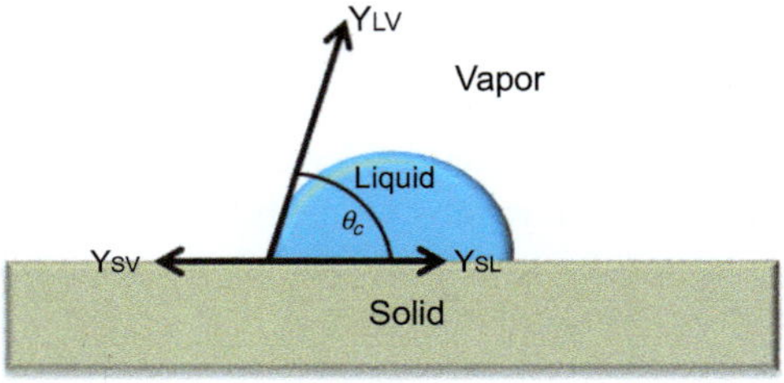

Figure 2.1 Contact angle, Θ_c, definition.

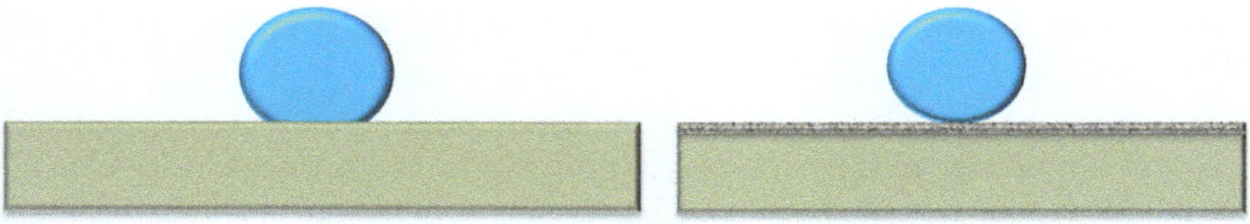

Figure 2.2 Schematic of a hydrophobic surface (left) and a super-hydrophobic surface (right).

from the contact point along the liquid–vapor interface in the droplet profile). This angle depends of the interfacial energy of the three interfaces, γ_{LV}, γ_{SV}, and γ_{SL} (Fig. 2.1).

Water repellent hydrophobic coatings have a low surface energy or polarity and a contact angle between the surfaces and water above 90° (see Fig. 2.2 left), being super-hydrophobic when this angle is above 120° (see Fig. 2.2 right). Materials used for hydrophobic coatings are mainly fluoropolymers [8]. This behavior promotes water droplets rolling over the surface, and so removing soiling particles during this kinetic movement, which achieves the self-cleaning (SC) or easy-to-clean effect [6]. The super-hydrophobic behavior is also known as the "lotus effect" due to the similarity with the phenomena occurring in the lotus plant.

On the other hand, water attracting hydrophilic coatings have high surface energies, leading—assuming the presence of enough water—to a closed water film and low contact angles. With contact angles of less than 10° (super-hydrophilic) dirt particle transportation off the surface is guaranteed under terms of a certain surface inclination [8]. The extreme situation is a totally wet surface (see Fig. 2.3).

Materials such as silicon oxides (SiOx) and titanium oxides (TiOx) are used for such coatings [8]. In particular, TiO_2 has photocatalytic properties that have been known about for many years [9]. Photons with energy greater than the band gap of TiO_2 create electron–hole pairs that in the presence of air can speed up or catalyze the conversion of organic matter to carbon dioxide and water. Only very thin layers of TiO_2 are needed and as a consequence they can have a minimal impact on the reflectance properties of solar mirrors [10].

Figure 2.3 Scheme of a super-hydrophilic surface.

Several studies about anti-soiling coatings were realized for applications in concentrating photovoltaic (CPV) systems. An anti-soiling layer consisting of a WO_3 photocatalyst was coated on a polymethylmethacrylate (PMMA) substrate, which is a primary material used for the manufacturing of Fresnel lenses for CPV modules. The anti-soiling layer was prepared using modified WO_3 and partially hydrolyzed tetraethyl orthosilicate in the photocatalytic surface layer. The three layers were deposited by the spin-coating method. The samples with and without the anti-soiling coating were exposed to a sand impingement. The electrostatic potential of the sample was measured with a digital static meter and the mass of the sample was measured with an electronic balance. The mass of adherent sand was more than 0.010 g for the sample without the coating and about 0.005 g for the sample with the coating. The electrostatic potential of the sample without the coating increased with increasing amount of incident sand, reaching a maximum value of 0.25 kV. On the other hand, the electrostatic potential of the sample with the coating was suppressed to 0.10 kV. The presence of electrostatic charges on the surface of the samples was a main factor for the adhesion of sand, and it could be suppressed by the anti-soiling photocatalytic layer [11].

Asahi Kasei Corporation, a Japanese chemical company, developed a new SC coating for photovoltaic (PV) applications. The performance of PV modules can be increased due to the two main properties of the coating film: anti-reflectiveness and SC effect. The coating solution was prepared by mixing of hybrid polymer emulsion and water-dispersed metal-oxide nanoparticles. Its solid content was around 5 wt% at pH = 4. There was a small amount of alcohol in the solution. For this purpose, six polycrystalline silicon PV modules of the same manufacturer and technology, three with coated surface and three without it, were tested under outdoor conditions for one year in the laboratory of PV systems at the University of Málaga, southern Spain. In this study the objective was to evaluate the losses caused by the decrease in transmittance of the module surface due to dust accumulation. The results showed that daily energy soiling losses were greater for modules without SC coating. Modules with coating film had an average daily energy soiling loss of 2.5%, whereas for the uncoated modules this value was 3.3%. The coated modules always showed lower losses due to soiling [12].

The effect of special coatings to enhance the reflection and SC properties of the front glass on PV modules was investigated in [13]. Prototype-coated glass samples with three different types of coatings were examined: an antireflection (AR) coating, a SC coating, and a multilayer (ML) coating consisting of both AR and SC. The glass samples were exposed to atmospheric conditions for 3 weeks. The results showed the transmittance decrease of the three coatings: 2.63% for regular glass, 1.75% for AR, 1.30% for SC, and 0.85% for ML. The decrease in transmittance was least when ML coating is used. The measurements showed that the coatings improve the SC properties of the glass samples [13].

The only solar mirror with anti-soiling coating currently marketed is duraGLARE, by the company Flabeg [14]. The objectives of this coating are to improve the solar output of the entire solar field by repelling the dust particles blown on the mirrors (dust repellent properties) and to reduce costs for cleaning and maintenance activities (easy-to-clean properties). This coating reaches the highest optical transparency and therefore no major objectionable influence to the initial reflectance of the solar mirror is to be measured. In addition to dust repellent properties, the coating means that the mirrors are cleaned very easily by rinsing with water only. As the dust adhesion of the few remaining particles is so low, the water can wash off the dust from the complete mirror and restore the full initial reflectance values. These properties were tested and proved by a huge number of measurements, laboratory tests, and outdoor exposure campaigns. In addition to several durability tests (UV, several temperature changes, thermal cycling, humidity, dry heat, copper acetic acid and neutral salt spray, chemical resistance, and sand storm tests), outdoor tests have been performed at the Plataforma Solar de Almería (PSA) since 2011, in real-field environments, with significantly positive results. The specular reflectance (at 660 nm) of coated and standard mirrors, located side by side with a 45° installation angle, was measured regularly with a D&S reflectometer [15]. The data were compared and evaluated. These data showed clearly the additional performance of the coated anti-soiling mirrors. The duraGLARE-coated mirrors showed at least 45% less soiling than standard mirrors [14]. According to the manufacturer, this product reduces the settling of dust on the mirror surface, which it is expected to increase the average reflectance of the solar field by 1−2% [16].

The European Commission has funded in January 2016 a 4-year project named WASCOP[1] whose main goal is to develop a revolutionary innovation in water management of concentrating solar power (CSP) plants—providing a flexible integrated solution comprising different innovative technologies and optimized strategies for the cooling of the power block and the cleaning of the solar field optical surfaces [17,18]. Among the different tools proposed, anti-soiling coatings for reflectors (developed by the Spanish company Rioglass and the Spanish not-for-profit foundation Tekniker) are included as a key action to save water in the cleaning activities by reducing the amount of dust deposited on the reflector surfaces.

2.3 High-reflective mirror materials

Raising the reflectance of the solar mirror permits a reduction in the solar field size, while maintaining the electricity output and the investment cost for collectors or heliostats. Even small improvements of 0.5 ppt in reflectance will have a high impact on the annual revenues of the power plant.

High reflectance can be achieved with silver as the reflecting surface and thin transparent front coats. Silvered-polymer films achieve solar-weighted hemispherical reflectance values around 94%, which is still below the state-of-the-art 4 mm silvered-glass mirror reflectance (they achieve around 94.7%). Silvered-polymer films have the benefit of being flexible, allowing one to construct any kind of collector geometry. Several research studies have been conducted with polymer film reflectors for concentrating solar applications [19,20]. However, in terms of specularity and durability, glass-based mirrors are superior.

The trend is therefore going toward thin silvered-glass mirrors. The reduction of the glass thickness from 4 mm to 1 mm boosts the reflectance around 1 ppt. Commercially available thin-glass mirrors (around 1 mm glass thickness) achieve around

[1] This project has received funding from the European Union's Horizon 2020 research and innovation program under grant agreement No. 654479, project WASCOP.

Figure 2.4 Examples of concentrating technologies with silvered thin-glass reflectors at PSA: parabolic dish collector (left) and small-sized parabolic-trough collector (right). PSA.

95.7% solar-weighted hemispherical reflectance. The durability of this type of materials has already been widely checked [21,22]. Further improvements of the reflectance can be achieved by further reducing the front glass thickness. Prototype mirrors of more than 97% solar-weighted hemispherical reflectance have been produced with 100 μm ultra-thin glass. The use of thin-glass mirrors also decreases the material weight, allowing the reduction of collector cost due to the requirement for less support material, simpler foundations, cheaper motors, and reduced transport costs.

Thin-glass mirrors need to be supported by a backing structure to provide stiffness against wind loads. Several substrate materials are being investigated for this purpose (concrete, composite materials, aluminum, steel, and glass). Thin-glass mirrors <1 mm are flat and flexible, and the substrate material needs to provide the shape accurately. Examples of collectors with silvered thin-glass reflectors are presented in Fig. 2.4. The bonding between the mirror and backing material (e.g., using adhesives) needs to be durable and must not introduce waviness or additional shape errors. Therefore, special care must be taken with the gluing process.

RAISELIFE[2] is a 4-year research project just approved by the European Commission (starting on the April 1, 2016) which focuses on extending the in-service lifetime of five key materials for CSP technologies: (1) protective and anti-soiling coatings of primary reflectors, (2) high-reflective surfaces for heliostats, (3) high-temperature secondary reflectors, (4) receiver coatings for STs and line-focus collectors, (5) corrosion-resistant high-temperature metals and coatings for steam and molten salts [23]. As mentioned, one of the functional materials under investigation is a novel high-reflectance lightweight heliostat to be developed by BrightSource Industries Israel Ltd. In addition to exceeding the reflectance targets

[2] This project has received funding from the European Union's Horizon 2020 research and innovation program under grant agreement No 686,008, project RAISELIFE.

of the materials roadmap (described in [2]), this new mirror presents the following advantages:

- reduced heliostat cost by weight reduction of support materials;
- reduced transport costs due to a significant volume reduction;
- more flexible optical design due to greater bendability for curved mirrors;
- increased abrasion resistance.

2.4 High-temperature mirrors for secondary concentrators

Secondary concentrators are employed in solar thermal concentrating technologies to redirect solar radiation reflected by the primary concentrators to the focal point or line. The main goal of these components is to increase the concentrated solar flux density and hence decrease thermal radiation losses by reducing the receiver size. Also, these components are used to increase the optical efficiency of the system by improving the flux distribution homogeneity or reducing the amount of solar flux missing the receiver due to scattering (spillage).

Compound parabolic concentrators (CPCs) are typically used as secondary concentrators, either for concentrating solar radiation on line-focusing systems (named 2D-CPC) or point-focusing systems (named 3D-CPC) [24]. Normally, 2D-CPC are used in linear Fresnel collectors and 3D-CPC are employed in STs or "beam-down" tower systems, which allow the placing of solar receiver/reactors on the ground rather than on the tower, with the concentrated irradiation entering from the top [25,26].

Typically, 2D-CPC systems are not actively cooled and, hence, they must keep their functionality during operation under high temperatures (up to 350°C) [27]. 3D-CPC systems are submitted to even higher temperatures, mainly due to the amount of concentrated solar radiation that is being absorbed by the mirror, and to a lesser degree due to convective heat exchange with its surroundings. Consequently, they are normally cooled (down to around 85°C, mainly with water), due to the high radiative flux density achieved [27]. This solution may comprise some technical risks and cause a plant shutdown in case of any failure of the cooling system. By contrast, the development of a mirror for 3D-CPC systems without cooling involves a significant technical challenge.

The secondary concentrators of the REFOS [28,29] and SOLGATE research projects [30] were installed in the Central Electrosolar de Almería 1 (CESA-1) central tower at the PSA at about 60 m height (see Fig. 2.5 left). The receiver system consisted of three pressurized absorbers, with the corresponding secondary concentrators, named SecNT1, SecNT2, and SecNT3 (see Fig. 2.5 right). The three absorbers and concentrators were connected in series so the circulating fluid temperature rose from SecNT1 to SecNT3. The secondary concentrators used in these projects were water-cooled systems, with a maximum expected temperature of about 85°C. The concentrators were composed of several pieces of silvered thin-glass reflectors (with a reflectance of about

Figure 2.5 CESA-1 tower (left) and REFOS secondary concentrator unit composed of three secondary concentrators (right) at PSA.
PSA.

90%), glued to a 20-mm-thick aluminum structure with thermally conductive adhesive [28]. A photographic investigation was performed on the secondary concentrators for an independent determination of its optical behavior, using a method based on the inversion of the ray paths.

According to the results obtained, tests with high-power levels showed that the cooling of the front edges of the aluminum plates was insufficient. At these places, the elevated temperatures caused some glass mirrors cracks [28] (see Fig. 2.6).

Solar reflectors for secondary concentrators are permanently exposed to environmental conditions, high radiation fluxes and elevated temperatures that potentially cause stress and degradation over time. Research work to analyze the durability of

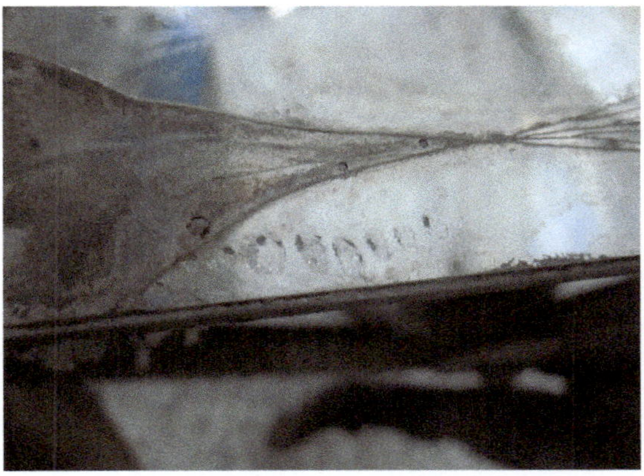

Figure 2.6 Mirror cracks appeared during the testing of the REFOS secondary concentrator, close to the edges.
PSA.

solar reflectors for secondary concentrators by simulating these conditions was performed under the SFERA-I European project, with the participation of three research institutes, CIEMAT-PSA (Spain), DLR (Germany), and PSI (Switzerland) [27]. A complete set of accelerated aging tests for the reflector material used in secondary concentrators (both cooled and uncooled) was specifically designed and performed on nine different material types (one first-surface aluminum reflector, three thin silvered-glass reflectors with different edge protections, three thin silvered-glass reflectors with different edge protections glued to an aluminum back structure, one laminated silvered-glass reflector, and one thick silvered-glass reflector) from five manufacturers. The performed accelerated aging tests covered salt spray tests, constant temperature, temperature cycling, humidity, and temperature combined with humidity tests. The comparison with naturally aged secondary concentrators revealed that the simulated degradation under accelerated conditions performed in this work did reproduce the most frequent degradation patterns suffered in real operating conditions. According to the results obtained, aluminum reflectors and thin silvered-glass reflectors glued to an aluminum structure showed minimum reflectance losses and structural degradation under the operation conditions of cooled 3D secondary concentrators. The following critical aspects to avoid reflector degradation were identified for cooled systems: to select a suitable adhesive material to glue the thin silvered-glass reflector to the support aluminum structure, to properly protect reflector edges, to design a suitable cooling system, and to avoid the combination of high radiation fluxes with mechanical stress. In addition, laminated silvered-glass reflectors were adequate for uncooled 2D secondary concentrators (typically for linear Fresnel collectors) [27].

As mentioned, one of the conclusions obtained in [27] was that a chemical reaction between the reflector and the glue occurred in some cases, mainly after the humidity and damp heat (high temperature and humidity) accelerated aging tests (see Fig. 2.7). Therefore, an activity has been included in the ongoing DETECSOL project (funded by the Spanish government) to select a suitable adhesive material and procedure to glue the thin silvered-glass reflector to the support aluminum structure [31].

Figure 2.7 Picture of a silvered thin-glass reflector glued to an aluminum back sheet, after the damp heat test, where corrosion appeared due to a chemical reaction provoked by the glue under the test conditions. PSA.

Experiments are being conducted with three different temperature-resistant adhesive materials and also three different seals to adequately protect the edges. The goal of this activity is to recommend a suitable combination for further secondary concentrator designs that minimize the edge corrosion and avoid any degradation mechanisms due to the interaction of the adhesive with the reflector.

The research center Fraunhofer ISE (Institute for Solar Energy Systems), from Germany, investigated several approaches to avoid the silver corrosion in high-temperature reflectors for secondary concentrators of linear Fresnel collectors, where temperatures of up to 300°C can be reached, involving a crucial challenge for the reflective coating [32].[3] To investigate it, a silver layer was deposited together with barrier and adhesion layers by sputtering to form a ML stack on glass. The best-known form of silver degradation is "tarnishing" due to reaction with H_2S from the atmosphere. This was suppressed by suitable barrier layers on the air side. A further degradation mechanism is pinhole corrosion, which can occur in two main forms: few large defects or many small pinholes. An essential effect for the mechanism of this corrosion is the tendency of silver layers to agglomerate. This agglomeration process can be suppressed by coating the silver layer with a cover layer. However, this protective function is affected by local imperfections, like dust particles, resulting in large defects. As the barrier property is disturbed at such spots, oxygen can enter and oxidize the interface between the silver and the cover layer, which reduces the interfacial energy. Agglomeration starts and may lead to a local delamination, hence the defect grows. This process can be retarded by increasing the interface energy by suitable intermediate layers. Thermal stress in individual layers was determined by measuring the curvature of thin, coated substrates at elevated temperatures. By varying the substrate material, it was possible to estimate the elasticity module and the thermal expansion coefficient for the very thin layers. As expected, low coefficients of expansion were obtained for the dielectric barrier layers but silver is characterized by a large coefficient of expansion, which leads to high thermal stresses on glass. Nevertheless, Fraunhofer ISE succeeded in modifying the deposition process and reducing the expansion coefficients of the resulting silver layer to a third of the original value. Combinations of these approaches resulted in reflective coatings that were stable at 250°C. At 350°C, the solar reflectance decreased by 7% over 5 months for a modified layer that showed pinhole corrosion. By contrast, the type with large defects degraded by only 1.5% after modification.

Finally, as mentioned in Section 2.3, one of the research activities that are being addressed in the European project RAISELIFE[4] [23] is to develop a secondary mirror specifically designed for CST applications, which maintains its optical and mechanical properties for operating temperatures of up to 350°C. This activity is also being developed by the research center Fraunhofer ISE, thanks to their significant expertise on this topic.

[3] This work is supported by the German Federal Ministry for the Environment, Nature Conservation and Reactor Safety (BMU).

[4] This project has received funding from the European Union's Horizon 2020 research and innovation program under grant agreement No 686,008, project RAISELIFE.

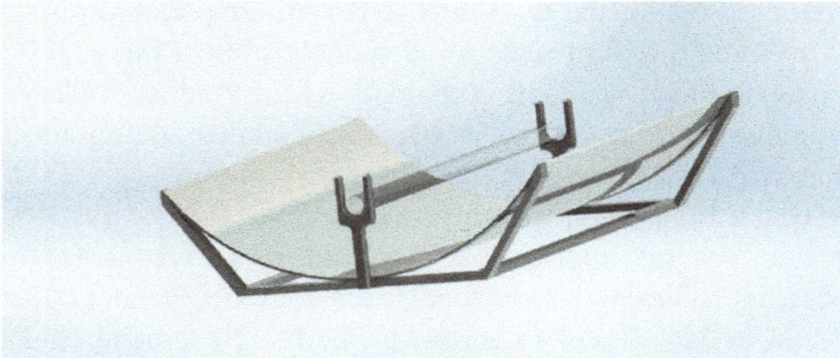

Figure 2.8 MEXSOL prototype.

2.5 Low-cost mirrors based on stainless steel

Stainless steel is not commonly used as a reflective metal for CST mirrors because it presents quite low reflectance. However, this material may represent an alternative solution for processes whose energy demand is low and where implementation costs are more important than efficiency. Another advantage of this material is its high market availability.

The solar-weighted hemispherical reflectance attained by stainless steel[5] was reported to be 0.572, with a maximum hemispherical spectral value of 0.680 at 1100 nm [33]. Different reflectance spectra were reported as a function of the polishing treatment. In particular, the reflectance of mechanically polished stainless steel containing 18% Cr and 10% Ni was found to be 8–10% lower than with the same material polished electrochemically at 80°C in a solution of 60% H_3PO_4, 20% H_2SO_4, and 20% H_2O [34].

There exist proposals of solar dryers with stainless steel as the reflective material for agro-industrial applications [35]. A low-cost parabolic-trough collector incorporating a polished stainless steel reflector, named MEXSOL, was developed to be coupled to a dried hybrid energy system [36] (see Fig. 2.8). It was calculated that the solar system may produce sufficient thermal energy to cover the energy demands of processes such as chilli pepper dehydration. In this work, the optimum polishing grade for AISI 304 stainless steel samples was analyzed by testing four types of polishing sheets to obtain several roughness levels. Hardness, reflectance, and durability of seven samples, composed of the best suited stainless steel reflectors, were tested and compared with aluminum reflectors. According to the conclusions of this work, very promising results were achieved because the average hemispherical reflectance value was 0.70 [36].

2.6 Conclusions

To conclude, Table 2.1 includes a summary of the potential benefits and challenges of the different advanced mirror concepts described in this chapter, as well as the future research that needs to be performed by the research institutes.

[5] The stainless steel type was not specified in the publication.

Table 2.1 Summary of the advanced mirror concepts

Advanced concept	Benefits	Challenges	Research needs
Anti-soiling coating	Reduction of the amount of soil or dust particles accumulated on the reflector surfaces and, consequently, to save effort and cost of the cleaning tasks, and to save water	• Minimum interference on the initial reflectance • Suitable durability in outdoor conditions • Low cost	Assure the coating efficiency over the expected lifetime using accelerated aging and real operating conditions
High-reflective mirrors	Raised reflectance, which permits a decrease in the solar field size while keeping the power output and investment cost, and flexibility in the collector design	• Appropriate solutions for the back-support structure • Suitable durability in outdoor conditions • Low cost	Study of appropriate adhesive materials to attach the mirror to the back-support structure and assure the durability of ultra-thin glass reflectors
High-temperature mirrors for secondary concentrators	Increased concentrated solar flux density and decreased thermal radiation losses by reducing the receiver size	• Acceptable stability and reduced thermal stress under high radiation flux and temperatures • Minimum interference with the adhesive material	Analysis of the influence of the high radiation flux and thermal stress on the mirrors and selection of suitable adhesive materials
Low-cost mirrors based on stainless steel	High market availability, flexibility in the collector design and reduced costs	• High reflectance • Suitable durability • Low cost	Study of the efficiency of different approaches to increase the reflectance and durability

References

[1] European Commission. Commission staff working paper SEC(2011) 1609 final. 2011. Materials roadmap enabling low-carbon energy technologies.
[2] Kennedy CE, Terwilliger K, Milbourne M. Development and testing of solar reflectors. Tech. Rep. No. NREL/CP-520–36582. Golden: NREL; 2004.
[3] Pitchumani R. Concentrating solar power program. SunShot Grand Challenge and Peer Review 2014. US Department of Energy (DOE); Anaheim, CA.
[4] Technology roadmap: solar thermal electricity. Paris: International Energy Agency (IEA); 2014.
[5] Solar thermal electricity. Strategic research agenda 2020–2025. European Solar Thermal Electricity Association (ESTELA); 2012.
[6] Sarver T, Al-Qaraghuli A, Kazmerski LL. A comprehensive review of the impact of dust on the use of solar energy: history, investigations, results, literature and mitigation approaches. Renewable and Sustainable Energy Reviews 2013;22:698–733.
[7] Cuddihy EF. Theoretical considerations of soil retention. Solar Energy Materials 1980;3: 21–33.
[8] Lorenz T, Klimm E, Weiss KA. Soiling and anti-soiling coatings on surfaces of solar thermal systems—featuring an economic feasibility analysis. Energy Procedia 2014;48: 749–56.
[9] Hashimoto K, Irie H, Fukishima A. TiO_2 photocatalysis: a historical overview and future prospects. Japanese Journal of Applied Physics 2005;44(12):8269–85.
[10] Atkinson C, Sansom CL, Almond HJ, Shaw CP. Coatings for concentrating solar systems—a review. Renewable and Sustainable Energy Reviews 2015;45:113–22.
[11] Sueto T, Ota Y, Nishioka K. Suppression of dust adhesion on a concentrator photovoltaic module using an anti-soiling photocatalytic coating. Solar Energy 2013;97:414–7.
[12] Piliougine M, Canete C, Moreno R, Carretero J, Hirose J, Ogawa S, Sidrach-de-Cardona M. Comparative analysis of energy produced by photovoltaic modules with anti-soiling coated surface in arid climates. Applied Energy 2013;112:626–34.
[13] Appels R, Lefevre B, Herteleer B, Goverde H, Beerten A, Paesen R, De Medts K, Driesen J, Poortmans J. Effect of soiling on photovoltaic modules. Solar Energy 2013;96: 283–91.
[14] Schwarberg F, Schiller M. Enhanced solar mirrors with anti-soiling coating. SolarPACES 2012. In: 18th International conference on solar power and chemical energy systems. Marrakech (Morocco). September, 11–14; 2012.
[15] http://www.devicesandservices.com/prod02.htm.
[16] http://www.flabeg-fe.com/en/solar-mirrors/duraglare.html.
[17] http://cordis.europa.eu/project/rcn/199297_en.html.
[18] http://wascop.eu.
[19] DiGrazia MJ, Gee R, Jorgensen GJ, Bingham C. Service life prediction for ReflecTech® mirror film. In: WREF 2012, World renewable energy forum; 2012.
[20] Almanza R, Hernández P, Martínez I, Mazari M. Development and mean life of aluminium first-surface mirrors for solar energy applications. Solar Energy Materials and Solar Cells 2009;93:1647–51.
[21] Kennedy CE, Terwilliger K. Optical durability of candidate solar reflectors. Journal of Solar Energy Engineering Trans ASME 2005;127(2):262–9.
[22] Sutter F, Fernández-García A, Wette J, Heller P. Comparison and evaluation of accelerated aging tests for reflectors. Energy Procedia 2014;49:1718–27.

[23] http://cordis.europa.eu/project/rcn/200815_en.html.
[24] Winston R, Minano JC, Benitez P. Nonimaging optics. Amsterdam: Elsevier Academic press; 2005.
[25] Yogev A, Kribus A, Epstein M, Kogan A. Solar Tower Reflector Systems: a new approach for high temperature solar plants. International Journal of Hydrogen Energy 1998;23: 239–45.
[26] Segal A, Epstein M. The optics of the solar tower reflector. Solar Energy 2000;69:229–41.
[27] Fernández-García A, Cantos-Soto ME, Röger M, Wieckert C, Hutter C, Martínez-Arcos L. Durability of solar reflector materials for secondary concentrators used in CSP systems. Solar Energy Materials and Solar Cells 2014;130:51–63.
[28] Buck R, Bräuning T, Denk T, Pfänder M, Schwarzbözl P, Tellez F. Solar-hybrid gas turbine-based power tower systems (REFOS). Journal of Solar Energy Engineering 2002; 124(1):2–9.
[29] Schmitz M, Schwarzbözl P, Buck R, Pitz-Paal R. Assessment of the potential improvement due to multiple apertures in central receiver systems with secondary concentrators. Solar Energy 2006;80:111–20.
[30] SOLGATE. Solar hybrid gas turbine electric power system. Final Publishable Report. Contract ENK5-CT-2000–00333. European Commission; 2006.
[31] http://www.idi.mineco.gob.es/stfls/eSede/Ficheros/2015/Anexo_I_Ayudas_Concedidas_Proyectos_IDi_Retos_2014.pdf.
[32] Dallmer-Zerbe K, Georg A, Graf W, Klimm E, Kühne M, Platzer W. High-temperature corrosion of coatings for secondary reflectors. Fraunhofer ISE Annual Report. 2011.
[33] Echazú R, Cadena C, Saravia L. Estudio de materiales reflectivos para concentradores solares. AVERMA; 2000. p. 08–11.
[34] Okicć M, Roušar I, Táborský Z, Roháček K. Reflectivity and surface composition of electrochemically and mechanically polished stainless steel. Materials Chemistry and Physics 1987;17(3):301–9.
[35] Pangavhane DR, Sawhney RL. Review of research and development work on solar dryers for grape drying. Energy Conversion and Management 2002;43:45–61.
[36] García-Ortiz Y, Yáñez-Mendiola J, Valenzuela L. Cylindrical parabolic collectors material from low cost (stainless steel) applied to dried hybrid system. Dyna 2016;91(1):1–7.

Improved design for linear Fresnel reflector systems

3

M. Collares-Pereira [1,2], D. Canavarro [1,2], J. Chaves [3]
[1]Largo Marques de Marialva, Evora, Portugal; [2]University of Evora, Evora, Portugal;
[3]Light Prescriptions Innovators, Madrid, Spain

3.1 Introduction (motivation)[1]

3.1.1 Low energy cost

This chapter presents a state-of-the-art of the present challenges for cost reduction of concentrated solar power (CSP) systems and several ways to overcome the existing difficulties. In particular, a new look over linear Fresnel reflector (LFR) systems is presented from the point of view of optics and how new optical designs can help to achieve higher performances and better cost-effectiveness factors.

In fact, delivering (thermal or electrical) energy at the lowest possible cost is an ever-present concern of the solar collector designer. To achieve this goal there are several paths to be followed: For instance, (1) increase collector efficiency, (2) increase system efficiency, (3) reduce collector costs, (4) reduce system costs, and (5) reduce O&M (operation and maintenance) costs.

Collector efficiency can be achieved by improving optical and thermal efficiency at the desired operating temperature. System efficiency can be improved in many different ways, by increasing thermodynamic conversion efficiency as in the case of heat into electricity and by reducing heat losses and parasitic losses (for instance, pumping power). Collector cost reduction may come through increased scale in production, simpler manufacturing techniques, new materials, new design that results in reduced weight, and other mechanical variables. System cost can likewise be reduced in many different ways: in particular, through the reduction of the type and number of components (this includes pipe length, heat transfer fluid (HTF) volume and pumps, heat exchangers, and so on). Finally, O&M costs can also be reduced in many different ways, in particular, through the adoption of new operating and control strategies.

Perhaps the most powerful handle in practically all of these avenues for cost reduction is the concept of concentration. There are several reasons for this.

[1] This chapter is written with thermal energy and thermal energy conversion into electricity in mind; even though the optical considerations made here could also apply to direct electricity conversion through the use of PV cells (CPV), LFR optics does not provide solar concentration values high enough to have a practical application in CPV, an area where the word high (HCPV) is added to really represent concentration values 10 to 20 times higher and readily achievable with 3D optics (including Fresnel lenses).

Advances in Concentrating Solar Thermal Research and Technology
http://dx.doi.org/10.1016/B978-0-08-100516-3.00003-4

First and foremost is the fact that collector losses are directly proportional to receiver area. Thus, the smaller it is the better. Since concentration is the ratio of aperture area to receiver area, the larger the concentration, the lower the thermal losses will be on a unit aperture area basis. This results in higher collector efficiency at higher operating temperatures and, in turn, higher thermodynamic conversion efficiency from heat to electricity.

However, another effect is the following: for the same receiver (the usual 70 mm receiver evacuated tube) an increase in concentration means a larger primary aperture which, in turn, for the same collector field power with several parallel rows of troughs, would result in less rows, less pipes, less parasitic and thermal losses, less components, less O&M, and so on.

3.1.2 Concentration

Concentration is deeply and inversely related to the acceptance angle 2θ of the concentrator, and there is a limit [1] to the concentration that can be achieved for any given half acceptance angle θ which, for 2D or linear optics, is given by

$$C_{max} = A_{apert}/A_{receiver} = n/\sin(\theta) \quad (3.1)$$

where n is the index of refraction at the receiver aperture.

Practical concentrators that achieve this limit area said to belong to the category of nonimaging optics (NIO). Conventional focusing optics, like that of the parabolic troughs widely used today, falls way short of this limit. For a parabola of aperture "a" (see Fig. 3.1) and a tubular receiver of radius "r" the concentration [2] is given by

$$C_{par} = a/2\pi r = n/\sin\theta * \sin\varphi/\pi \quad (3.2)$$

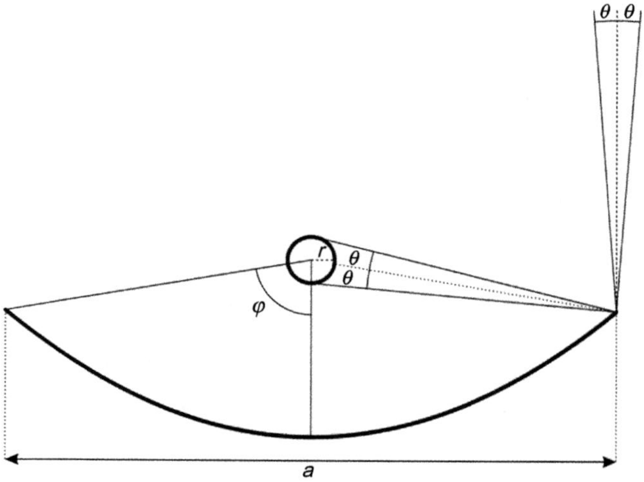

Figure 3.1 Parabolic trough concentrator designed for a tubular receiver and half acceptance angle of θ. The size of the receiver is exaggerated for better viewing.

with $n = 1$ and φ representing the (rim) angle with the optical axis that describes the extension of the parabolic arc.

The half acceptance angle θ is usually chosen to correspond to between two and three times the half angular width of the sun (0.27 degrees). With the choice of parameters $\theta = 2.5*$ (0.27 degrees) $= 0.675$, $\varphi \approx 90$ degrees, r = 70 mm [3], C = 27X (see Eq. 3.2), one gets a = 5.93 m. Most parabolic trough collectors on the market for solar thermal electricity (STE) production have values corresponding to these figures.

To increase concentration with PT (Parabolic Trough) technology requires reducing the angle θ (at the extra cost of more precise manufacturing and tracking accuracy) and that is the way being proposed by some manufacturers. For instance the so called Ultimate-Trough (about 8 m aperture width) solar concentrator [4].

A further possibility would be to seek ways to bring the focusing parabola closer to the concentration limit of Eq. (3.1) through the use of second-stage concentrators and without sacrificing the acceptance angle value, as explained in [5,6]. The result is that of primaries with more than 11-m aperture width, but that extra size on a tracking trough does not look practical at all.

Fortunately, there is a completely different alternative and that comes from LFR technology, the subject of this chapter.

3.1.3 Etendue

Before going into LFR concentrators design, a further comment about the concept of "etendue" [7] is in order.

Any optical device has an entrance and an exit aperture. Geometrical optics handles light rays. All that enters the device may or may not exit it. An efficient device is one that manages to have all entering rays, exiting it as well. The precise way to talk about this, in geometrical optics, is to talk about "etendue." The most efficient device is the one that conserves "etendue," that is, no light rays are lost.

Simply put, "etendue" is a geometrical quantity [7] that measures the amount of "room" available for light to pass through. "Etendue" dU is defined (Fig. 3.2) as follows:

$$dU = dA * \cos\theta * d\Omega. \tag{3.3}$$

Spatial "room": dA * cos θ (light is crossing dA in a direction θ).
Angular "room" (the solid angle): $d\Omega$.

In geometrical optics, energy flux is simply a constant multiplying "etendue" in optical systems with no losses or diffusion.

3.1.4 CAP Concentration acceptance product [7]

Another concept worth recalling, before going into LFR concentrators design, is that of CAP.

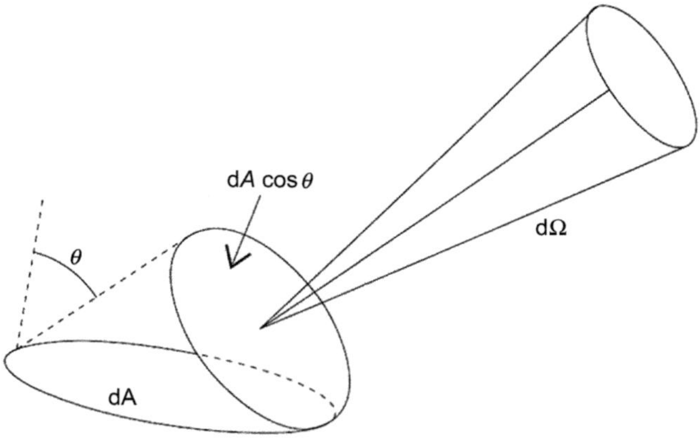

Figure 3.2 Light rays cross section dA at an angle θ as a bundle within the solid angle dΩ [7].

CAP for any given concentrator is given as

$$\mathrm{CAP} = \mathrm{C} * \sin\theta \tag{3.4}$$

For $n = 1$, CAP = 1 when $\mathrm{C} = \mathrm{C_{max}}$, given by Eq. (3.1). Otherwise it is always <1. Its value gives a measure of how far a given optic is from the limit. CAP of parabolic troughs (Eq. (3.2)) is typically below 0.3.

3.1.5 Summary: one recipe for low-cost energy delivery

In summary, one recipe for lower-cost energy delivery is to seek higher concentration and "etendue" conservation. This means that light rays need to be properly handled through the optics so that as little as possible are lost and, at the same time, to seek NIO solutions capable of delivering much higher concentration values. That and a high CAP value are also important, bringing into play the relevance of the acceptance angle being as large as possible.

It should be noted that reaching the limit of Eq. (3.1) requires "etendue" conservation. However, "etendue" conservation per se does not imply that the highest concentration will be reached.

3.2 Advanced linear Fresnel reflector concentrators

3.2.1 Conventional LFR

A possible way to concentrate sunlight is using a Fresnel reflector: a large number of small mirrors (called heliostats) that mimic the behavior of a large concentrator, replacing it. These heliostats move to track the apparent motion of the sun, keeping its light concentrated onto the receiver.

Fig. 3.3 shows a classic LFR configuration for a flat receiver.

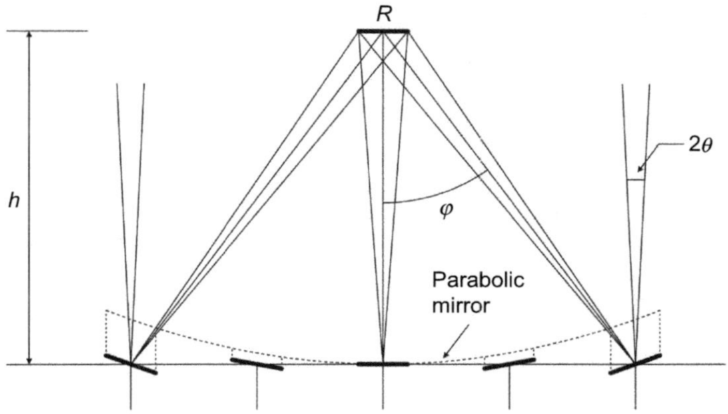

Figure 3.3 LFR for a flat receiver R, 2θ acceptance angle, and rim angle ϕ.

This optical configuration results from a parabolic trough (see *dashed line* in Fig. 3.3) to which the flat mirrors would be tangent and would end up being placed on a horizontal plane.

The conventional LFR is, thus, also very far from the limits of concentration, as established by Eq. (3.1).

It is possible to show [7] that the highest concentration achievable in the case of Fig. 3.4 is C_{max} * 0.45 when $\phi = 40.4$.[2]

Fig. 3.4 (a) and (b) helps to understand why the limit is really far away for conventional LFR: the "etendue" of the incoming radiation is not conserved on its way to the receiver, through shading and blocking from one mirror to the next and eventual loss to the ground from the space between the mirrors.

Given these limitations, the claim made earlier about the fact that LFRs can be a solution to the problem of practical high-concentration devices requires further elaboration.

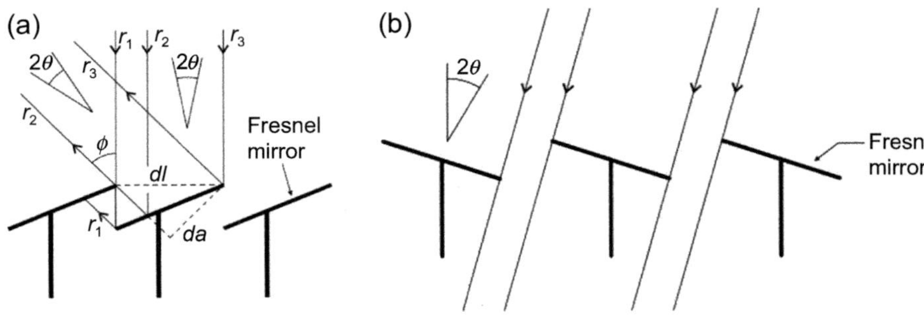

Figure 3.4 (a) Blocking and shading [9]. (b) "Etendue" lost to the ground.

[2] With LFR it is hard to have large rim angles; in the case of the parabola a rim angle approaching 90 degrees gives the highest concentration; with LFR approaching 90 degrees places the flat mirrors at very large distances and R would be very large for any acceptance angle value.

3.2.2 Advanced concepts

3.2.2.1 "Etendue" conservation

In order to approach the limit in (1) the first try goes toward taking proper care of the incoming "etendue."

The first contribution comes from [8] with the idea that instead of just one receiver a multiple receiver concept might be considered, as shown in Fig. 3.5. This concept was baptized as compact LFR, or CLFR for short. Compact signifies that less spacing can now be left between mirrors for ground loss reduction in larger primaries.

This concept adds a degree of freedom. That of sending the reflected radiation to one or the other receiver, thereby controlling blocking.

The second concept comes from the idea of placing the mirrors on an "etendue"-conserving curve [9,10], shown in Fig. 3.6.

The "etendue"-conserving curve uses an extra degree of freedom, that of raising/lowering the height of each mirror with respect to its neighbors and thereby further reducing (even eliminating) shading and blocking.

3.2.2.2 Toward maximal concentration

Some LFRs on the market [11,12] already incorporate a further step in the right direction: the use of second-stage concentration to enhance the overall concentration of the device. Fig. 3.7 shows two examples.

The collector in Fig. 3.7(b) is designed for a primary width of 16.45-m counting the spacing between mirrors and a total mirror width of 12 m. This figure should already be compared with about 6-m aperture width of the conventional parabolic troughs. It already clearly goes in the right direction of reducing the number of rows, components,

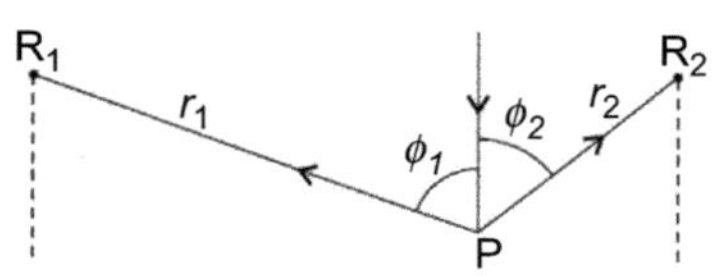

Figure 3.5 Multiple receiver LFR concept, CLFR.

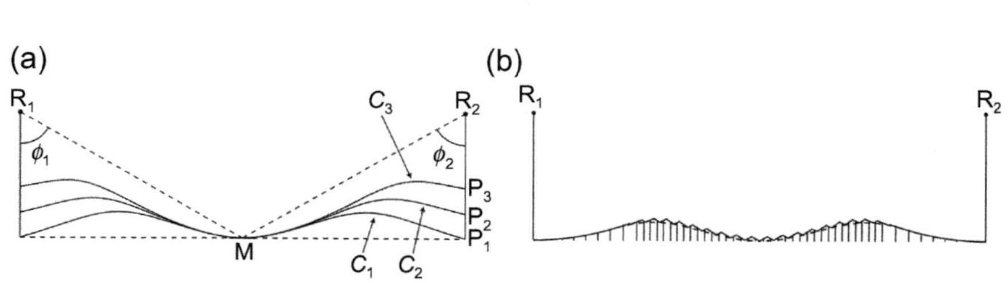

Figure 3.6 (a) "Etendue" conserving curves in a CLFR concept [9]. (b) A practical case [9].

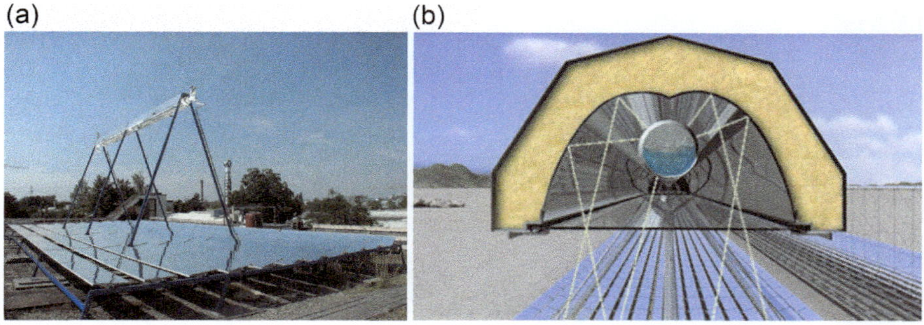

Figure 3.7 (a) LFR with second-stage CPC-type reflector and a 70-mm receiver evacuated tube [11]. (b) LFR with second-stage CPC-type reflector and a non-evacuated 70-mm tube [12].

and so on, referred to in the preceding paragraphs as quite important for cost reduction. Because mirrors are relatively small and lie low, LFR solutions need to cover less ground for the same peak power, when compared with parabolic trough fields and they possess another simplifying and important feature: the receiver is fixed, that is, only the primary mirrors track.

However, in both cases (a) and (b), this conventional Fresnel primary/second stage–type combination does not yet reach the maximum concentration possible and does not conserve "etendue." These optics also do not attempt at solving or reducing the gap problem; that is, there is a large distance between the receiver tube and the outer glass tube in and the same distance between the start of the CPC-type concentrator cusp and the atmospheric receiver (see Fig. 3.7 (b)); a number of rays get lost because they go through the gap without reaching the receiver (see also [2]).

To come closer to the limit, there are many different possible ways to follow the recipe presented earlier. In what follows a few of these are shown.

The receiver may be atmospheric or in an evacuated environment. For very high temperatures (above 400°C up to 600°C), as of 2016, there are no selective coatings for atmospheric receivers (tubular or not) with guaranteed long-term stability/durability, and therefore the following discussion concerns only evacuated tube technology. Even so the reader is referred to [9,14] for a detailed solution.

All solutions shown in the following paragraphs are designed for the same evacuated tubular receiver, a sort of standard in the market today, the 70-mm receiver [3,4] and thus results can be intercompared at once.

Beside the already-mentioned CPC-type concentrator [1,2] there are several other second stage–type concentrators that can be combined with a primary concentration stage. Examples are compound elliptical concentrator (CEC) [1,2,7], tailored-edge ray concentrator (TERC) [7,15], simultaneous multiple surface (SMS) [7,16], and aplanatic optics [7,19].

In theory it can be shown that TERCs provide the highest possible concentration, since they are explicitly designed for the limit in Eq. (3.1). However, they must be severely truncated since by definition they extend from the secondary to the primary, and the resulting full shadow would make them useless.

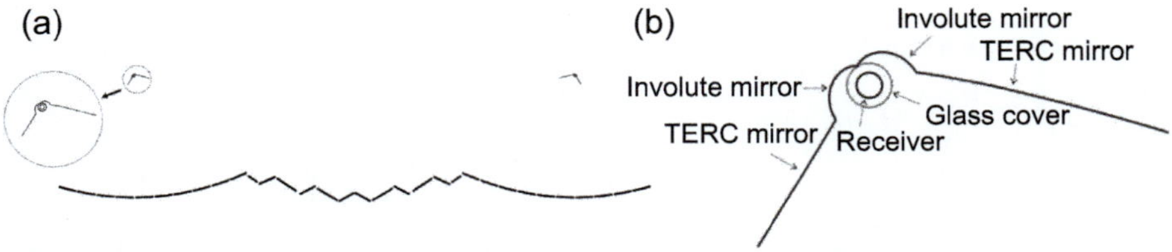

Figure 3.8 The CLFR "etendue-matched" concentrator. (a) Overview of the concentrator. (b) Details of the secondary mirror adapted to an evacuated tubular receiver.

Fig. 3.8 [17] shows a solution for evacuated tubular receivers, using a TERC-type secondary and the primary mirrors placed on an "etendue"-conserving curve [17,18]. It is also a CLFR solution. The "upper" part of the second-stage concentrator is the usual involute-type mirror [1,2,7].

The overall concentration is C = 50X for a half acceptance angle $\theta = 0.47$ degree.

The primary has $\approx$31 m, a highly desirable feature, when talking about system costs, as explained. Tower height is 6.5 m, mirror lengths are 0.52 m, 0.86 m, and 1.15 m, and maximum height on the "etendue"-conserving curve is 1.3 m, referenced to the lowest point on the same curve.

Another possible solution [19] is presented in Fig. 3.9. It is an SMS single receiver solution with a very large primary (20 m) designed for the 70 mm evacuated tube and

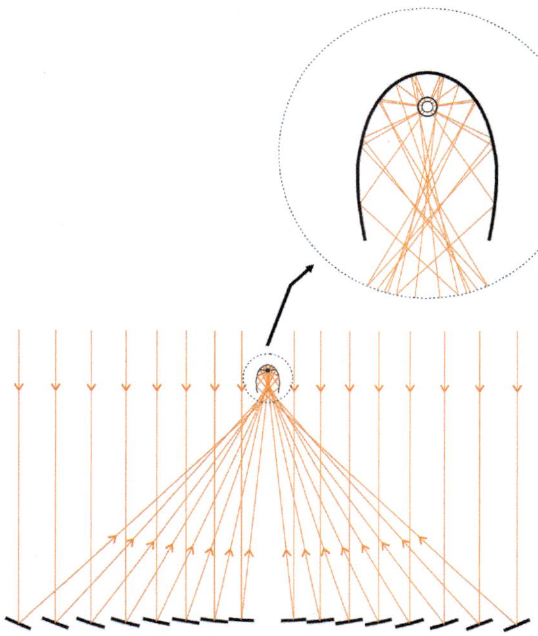

Figure 3.9 The Fresnel XX SMS concentrator.

to solve the gap problem (losses in gap) at the same time reducing the losses through the fact in the other solutions some rays still need to traverse several times the outer glass envelope. This SMS solution strives for having the largest number of rays going only once through the glass envelope. The "etendue" is conserved through the SMS solution development condition, of simultaneous primary and secondary optimization. C achieved is 73.71X and $\theta = 0.44$ degree for a CAP of 0.57 [13].

However, it should be noted that the SMS solution of Fig. 3.9 has a second-stage concentrator with a size that is typically larger than other types, as, for instance CECs [6]. A larger second stage means larger shadow over the primary for off-axis incidence angles. The last configuration [20], shown next, proposes a second stage of the CEC type and some of the other features of the preceding solution.

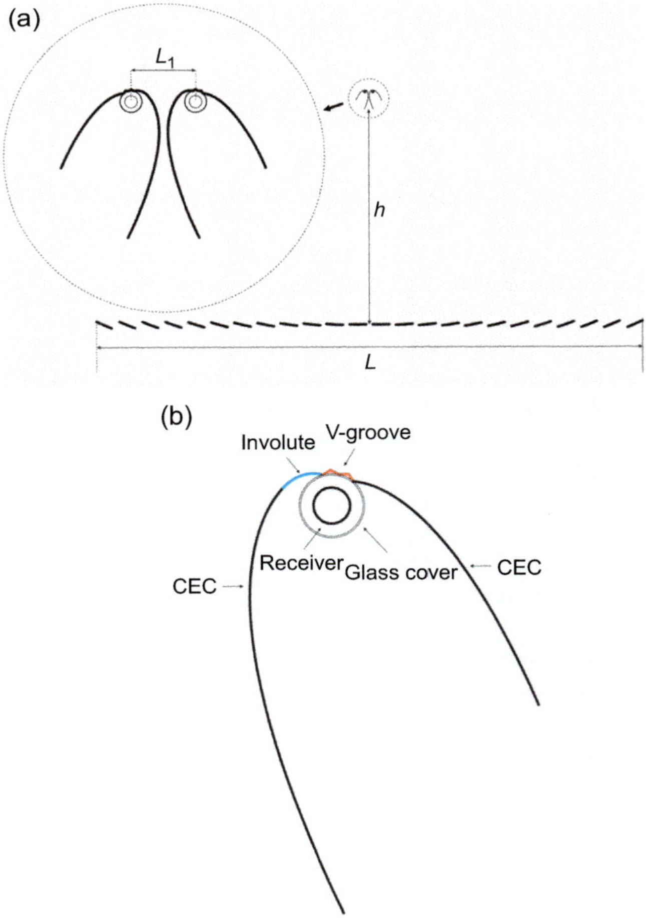

Figure 3.10 (a) Cross section of the asymmetric dual macrofocal CEC LFR concentrator. (b) Details of the second stage showing asymmetric CEC, the involute part, and the V-groove addition, for zero gap losses.

It is a multiple (2) receiver solution, as the TERC solution of Fig. 3.8, but the two receivers are placed on a single tower. The idea is to join two asymmetric solutions, as shown in Fig. 3.10(a) and (b).

The second stage is composed of two macrofocal CECs [7], asymmetric, and containing a V-groove section that is designed to eliminate the possibility of rays escaping through the gap between receiver and outer glass tube. The total primary width is now 26 m. The configuration shown was designed for a slightly lower concentration (45X) but with a much higher full acceptance angle $2\theta = 2 * 0.75$ degrees to relax tolerances for tracking accuracy, manufacture, and installation. Yet its CAP = 0.58, a very high value indicating an optical behavior approaching the limits, as commented earlier. Compared to the CAP of the parabolic trough and for the same acceptance angle, one can see that this configuration achieves about double the concentration.

The performance of some of these advanced concepts are presented in Chapter 15 - "Linear Fresnel reflector (LFR) plants using superheated steam, molten salts, and other heat transfer fluids".

3.3 Conclusion

What these advanced concepts accomplish is to provide much higher concentration values (more than the double of what parabolic troughs provide). This means less thermal losses at higher temperatures, themselves associated with larger efficiency conversion from heat to electricity and a number of other advantages like substantially larger apertures, thus a substantial reduction of rows in a collector field, less HTF, less components, less parasitic losses and thermal system losses, and less O&M.

The advanced LFR solutions shown correspond to very large primaries, larger than 20 m. The last one (Fig. 3.10) with 26 m, implies, when compared with 6 m aperture PTs in collector field, to a potential reduction of the number or rows by a factor larger than 4. This is a substantial reduction of the number of pipes, valves, bents, HTF fluid volume, thermal and parasitic pumping power losses, and so on, also with benefits on O&M tasks and costs.

On top of this, one should add the inherent advantages of LFR systems, more compact (tighter ground cover) and with fixed receivers, a substantial simplification, since it eliminates the need for flexible hosings or swivel-type joints, present in all PT fields. Other ones are the lowest cost associated with flat mirrors, simple tracking mechanisms, better response to high winds.

However, LFRs have, in comparison with parabolic troughs, an extra penalty from effects arising from cosine of incidence angle. The question is: can the advantages of the advanced LFR concepts, compensate for this? The answer is yes, and is given in Chapter 15.

As a final remark, LFRs have a great potential for cost reduction but some inherent problems to solve as well. These problems are mainly related, as mentioned earlier, the optical losses due to the cosine effect. In this sense, future developments will be strongly connected with more advanced designs and configurations, seeking maximum concentration, and maximum efficiency in order to compensate the aforementioned losses.

References

[1] Winston R, Miñano JC, Benítez P, Shatz N, Bortz JC. Nonimaging optics. Amsterdam: Elsevier Academic Press; 2005.
[2] Rabl A. Active solar collectors and their applications. Oxford: Oxford University; 1985.
[3] Schott Solar. <http://www.schott.com/csp/english/schott-solar-ptr-70-receivers.html>
[4] Flabeg—Ultimate Trough. <http://www.flabeg-fe.com/en/engineering/ultimate-trough.html>
[5] Canavarro D, Chaves J, Collares-Pereira M. New second-stage concentrators (XX SMS) for parabolic primaries; Comparison with conventional parabolic troughs concentrators. Solar Energy June 2013;92:98–105.
[6] Canavarro D, Chaves J, Collares-Pereira M. A novel compound elliptical-type concentrator for parabolic primaries with tubular receiver. [Submitted to Solar Energy].
[7] Chaves J. Introduction to nonimaging optics. Second ed. CRC Press, Taylor and Francis Group; 2016.
[8] Mills D, Morrison G. Compact linear Fresnel reflector solar thermal powerplants. Solar Energy 2000;68:263–83.
[9] Chaves J, Collares-Pereira M. Etendue-matched two stage concentrators with multiple receivers. Solar Energy 2010;84:196–207.
[10] Chaves J, Collares-Pereira M. Primary concentrator with adjusted etendue combined with secondaries associated to multiple receivers and with convection reduction, International Patent No. PCT/PT2009/000026.
[11] Industrial Solar. <http://www.industrial-solar.de/content/>
[12] Novatec Solar. <http://www.novatecsolar.com/>
[13] Canavarro D, et al. Simultaneous multiple surface method for linear Fresnel concentrators with tubular receiver. Solar Energy December 2014;110:105–16.
[14] Horta P, Collares-Pereira M, Canavarro D, Guerreiro L. Modeling thermal losses in a new CLFR "Etendue Matched" non-evacuated collector cavity. In: Proceedings 17th International SolarPACES symposium, September, 20–23, Granada (Spain); 2011.
[15] Gordon JM, Ries H. Tailored edge-ray concentrators as ideal second stages for Fresnel reflectors. Applied Optics 1993;32:2243.
[16] Miñano JC. High efficiency nonimaging optics. 2003. United States Patent 6.6639.733 B2.
[17] Chaves J, Canavarro D, Collares-Pereira M. Compact TERC LFR solar concentrator designed for direct molten salt operation as heat transfer fluid in evacuated tubes. [Patent pending].
[18] Gordon J. Aplanatic optics for solar concentration. Optics Express 2010;18(S1):A41–52.
[19] Canavarro D, et al. Infinitesimal etendue and simultaneous multiple surface (SMS) concentrators for fixed receiver troughs. Solar Energy November 2013;97:493–504.
[20] Chaves J, Canavarro D, Collares-Pereira M. Dual asymmetric macrofocal CEC LFR solar concentrator. [patent pending].

Part Three

Advances in the thermal conversion of concentrated sunlight

A new generation of absorber tubes for concentrating solar thermal (CST) systems

4

A. Morales, G. San Vicente
Centro de Investigaciones Energéticas, Medioambientales y Tecnológicas (CIEMAT), Madrid, Spain

4.1 Introduction

Solar receiver tubes are among the most important components in a solar thermal electricity (STE) plant, and they are subject to considerable technical and scientific developments with continuous improvements to increase optical and thermal properties and durability.

Main components of a heat collecting element or solar receiver are shown in Fig. 4.1, and they are:

- metallic pipe;
- glass pipe;
- selective absorber;
- antireflective coating (ARC);
- getters;
- glass–metal welding;
- bellows.

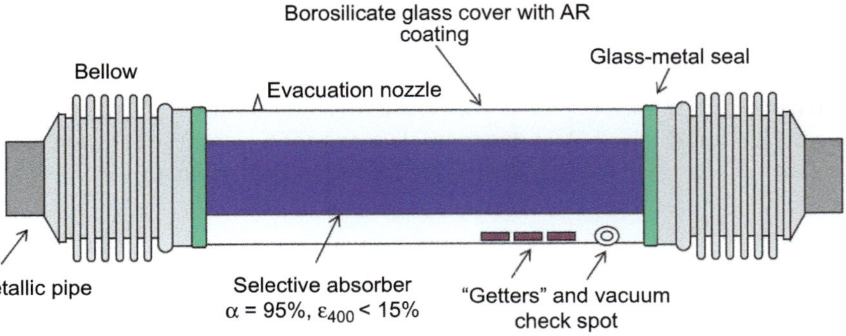

Figure 4.1 Receiver tube main parts.
Courtesy of Flagsol.

Advances in Concentrating Solar Thermal Research and Technology
http://dx.doi.org/10.1016/B978-0-08-100516-3.00004-6

The glass cover is connected to the steel pipe using metallic expansion bellows to compensate different linear thermal expansion of glass and steel, when solar receiver is working at nominal temperature. The glass to metal welding is used to connect the glass cover and the flexible bellows, and it is shielded from concentrated solar radiation to avoid the thermal and mechanical stress that could affect welding performance and durability. The heat thermal fluid (HTF) is pumped into the metallic pipe to collect thermal energy absorbed by the selective absorber and vacuum is established in the annulus to reduce thermal losses.

Parabolic trough technology needs to increase performance to achieve a solar field system with improved efficiency and lower operation and maintenance (O&M) costs. The performance of the STE plant is mainly related to the optical and thermal properties of the collector system and to the maximum operation temperature. The commonly used oil-based HTFs, however, are restricted to operation temperatures of 400°C, and therefore limit the overall plant efficiency. The key innovation to overcome this limitation is the use of high-temperature HTF, such as molten salts or direct steam generation (DSG), which enables operation temperatures up to 550°C. On the other hand, project execution and solar field construction costs can be significantly reduced, with a potential reduction of levelized cost of electricity (LCOE), at higher operational temperatures in a molten salt power plant with thermal storage unit, of above 20% compared with a nonstorage standard HTF plant and 15% to an oil-based HTF plant with thermal storage [12].

Main two challenges in parabolic trough receivers are cost-reducing measures (including manufacturing, installation, and maintenance) and durability.

Cost reduction can be afforded in several ways: using cheaper materials without affecting performance and durability; increasing tube length from actual 4 to 6 m or even longer sizes and with new simpler and innovative designs such as re-evacuable tubes or dynamic vacuum systems. Longer pipes present a higher useful surface, less costly welding between pipes, larger collectors that require fewer servomotors and controls, and so on.

Main degradation problem observed in receiver tubes, operating in commercial solar power plants, is vacuum loss in the inner annulus produced by hydrogen diffusion from oil-based HTF through metal pipe wall. Getters have been extensively used to try to reduce hydrogen partial pressure in vacuum annulus, but the 25 years expected durability is not achieved. Moving from oil-based HTF systems to molten salts or DSG will reduce vacuum degradation problems and will help to reduce receiver tube cost.

4.2 Glass cover

Glass cover is used in heat collector elements (HCEs) to reduce convective thermal losses and to preserve selective absorber from outdoor exposition. It is a glass pipe, concentric to metallic absorber pipe, and it needs to have high solar transmittance and a low thermal expansion coefficient to provide good thermal shock resistance, close to the thermal expansion coefficient of metallic absorber pipe at operating temperature, in order to reduce mismatch between both linear expansion coefficients.

Vacuum is stablished in annulus between glass and metallic pipes to reduce thermal losses by conduction–convection and to avoid thermal oxidation of the selective absorber deposited on metallic pipe. This annulus is sealed with a glass–metal welding, and a metallic bellow is used to compensate the mismatch expansion coefficient between glass and metal pipes.

4.2.1 Glass composition

The glass material used as glass cover in HCE has to be borosilicate glass. This glass is characterized by the presence of substantial amounts of silica (SiO_2) and boric oxide ($B_2O_3 > 8\%$) as glass network formers. In glasses used in industrial applications, B_2O_3 content is 12–13% and the SiO_2 content is over 80% that provides high chemical durability and low thermal expansion (3.3×10^{-6}/K). In order to increase expansion coefficient to a value closer to stainless steel, alkaline earths and alumina are added. In addition to about 75% SiO_2 and 8–12% B_2O_3, these glasses contain up to 5% alkaline earths and alumina [18]. To this subtype of slightly softer glasses (as compared with nonalkaline-earth borosilicate glass), which have thermal expansions between 4.0×10^{-6}/K and 5.0×10^{-6}/K, belong the chemically high-resistant varieties used in receiver tubes. High-borate borosilicate glasses containing 15–25% B_2O_3, 65–70% SiO_2, and smaller amounts of alkalis and Al_2O_3 as additional components, have low softening points and low thermal expansion.

Borosilicate glasses have a solar transmittance from 0.91 to 0.93, good thermal shock resistance, and good outdoor durability. Most common borosilicate glasses used in solar receivers are:

- Borosilicate 5.1 (linear expansion coefficient of 5.1×10^{-6}/K) that presents alkaline earths and alumina in its composition. It has a solar transmittance of about 0.92 and good chemical and mechanical properties to be employed in this application.
- Normal borosilicate 3.3 (linear expansion coefficient of 3.3×10^{-6}/K) without alkaline earths and alumina in its composition. It has a solar transmittance of about 0.92, good chemical properties, and the lowest thermal expansion coefficient that provides the highest thermal shock resistance. Its linear thermal expansion coefficient is lower, so it has better thermal shock resistance than borosilicate 5.1 glass, but elongation mismatch between glass and metallic pipes are higher.

4.2.2 AR coating

The glass jackets of receiver tubes are coated by a film on both sides (inner and outer) to reduce the reflection losses in the glass, thereby increasing the optical efficiency of the receiver tube. This film, known as ARC has to satisfy two requeriments (related to coating thickness, and refractive index value) to obtain the destructive interference of light which is reflected at the glass–coating interface and at the coating–air interface. These values come from solving Fresnel's law equations for normal incidence and for a determined wavelength. The wavelength value selected is usually around 600 nm for solar applications, as this wavelength value is centered in the maximum irradiance zone of solar spectrum. In this way, the optimal ARC must have a thickness of about 150 nm and a refractive index value of 1.22.

The material most widely used as ARC on glass is silicon dioxide (SiO_2), and the low refractive index value is achieved by introducing porosity in the coating. This porous nature, necessary to increase the glass transmittance, is the weak spot of this material in terms of durability. These pores easily absorb water and other volatile organic components, increasing the refractive index of the coating and lowering the transmittance. This process, mostly reversible, is known as "breathing of the coating" [26]. Another consequence of the porous structure is the weak mechanical performance coming from the weak binding force between the silica particles and substrate, as well as between the particles [19].

The technology most widely used for preparing the ARC on glass receiver tubes is sol–gel dip-coating method. It allows to coat both sides of the tubes at the same time, and it is a low-cost technology and is easy to scale up. The porous silica coatings can be produced by two routes: from colloidal solutions (obtained by basic catalysis of metal alkoxides or commercial SiO_2 colloidal dispersion), which is the route used by Schott [36], or from polymeric solutions (obtained by acid catalysis of metal alkoxides) where the porous structure is achieved by adding a compound to the solution which is removed during a heat treatment, leaving pores in its place [23]. This is the route used by Archimede Solar. The first one produces coatings with excellent optical properties but with poor mechanical properties, and improvement methods have been developed to get better coating adhesion on the glass and better mechanical properties [16, 36]. The second route produces coatings with better mechanical properties, although these coatings can undergo the same process of water adsorption in the pores or quick soiling due to the porous structure.

At the end, the big challenge in the ARC is to solve the contradiction of the porous structure between low refractive index and high stability, and future trends and developments go toward this goal.

The modification of the porosity type in the silica coating can increase the durability of the ARC. For example, DSM has developed a porous silica ARC coating for photovoltaic (PV) glass covers, with close surface porosity and strong binding with the glass that increase the coating durability [7]. Other important development work is referred to develop multifunctional coatings that work not only as ARC but also as easy-to-clean or self-cleaning coating. In this way, they will increase the receiver efficiency and moreover they will minimize the costs associated to clean receivers in solar plants. The soiling of glass tube receivers in solar plants decreases significantly the optical efficiency. Moreover, the porous nature of the ARC makes it more liable to soiling than bare glass or mirrors. There are two strategies for obtaining these easy-to-clean or self-cleaning surfaces: applying a treatment or coating that makes the surface hydrophobic and applying a photocatalytic coating that produces hydrophilic surface [20]. The mechanisms that take place in both strategies are different but both give place to surface that avoid the soiling of the coatings by any cause. The key points of these multifunctional coatings are the stability of them and, overall, to producing anti-soiling behavior without decreasing the transmittance. Future researches have to focus on these restrictions.

4.3 Steel tube

HTF is pumped into the receiver metallic pipes along collector lines to be heated by solar concentrated radiation. Hence, metallic pipe needs to fulfill several requirements:

- Good thermal conductivity to transfer heat collected by the selective absorber deposited on its outer surface.
- Good mechanical properties to withstand high temperature and operation pressures.
- Good corrosion resistance regarding HTF employed in the system.

4.3.1 Steel composition and durability

Metals with high thermal conductivity, such as copper or aluminum, cannot be used for this application because of their poor mechanical properties at high temperatures and pressures required for this solar application, hence steel is commonly used.

Carbon and low-alloyed stainless steels have better thermal conductivity than high-alloyed austenitic stainless steel, and they are cheaper but these steels have lower corrosion resistance than austenitic stainless steel.

Carbon steel ASTM 335 grade P22 has been used in DISS facility in PSA, with DSG with water as HTF, and corrosion problems were observed due to erosion and cavitation produced by biphasic flow. Now, in this technology, producers are using AISI 316 austenitic stainless steel to guarantee mechanical robustness against high operating pressures and radial temperature gradients produced by biphasic flow. Receiver thickness increases dependently on the maximum operative pressure (100 bar), and it leads to a thicker tube (>4.5 mm) that directly reflects into an effective cost increase of the technology [5].

When synthetic oil and organics such as biphenyl/diphenyl oxide systems are used as HTF, austenitic AISI 321L stainless steel is the usual option due to its lower hydrogen permeation in comparison with other austenitic stainless steel. Hydrogen diffusion is the major drawback of this technology because of HTF thermal degradation, and high hydrogen partial pressures are achieved in vacuum annulus [21]. Typical wall thickness employed is 2 mm.

Molten salt receiver tubes commonly use a mixture of $NaNO_3$ (60 wt%)–KNO_3 (40 wt%) salts (solar salt) or a mixture of $NaNO_3$ (7 wt%)–KNO_3 (53 wt%)–$NaNO_2$ (40 wt%) salts (HITEC). There are several reports in the literature on corrosion issues of these molten salts with metallic alloys, and it was found that the nickel alloys with 15–20% chromium content performed the best corrosion resistance, whereas iron alloys with low or almost zero nickel content showed poor corrosion resistance at high temperatures [10, 15, 38]. Hence, AISI 321 stainless steel is commonly used in molten-salt receiver tubes.

Other HTF materials, such as superheated steam [14] or cooled gasses [24, 25], are now being studied in order to increase thermal efficiencies of solar power plant, and they open the possibility to reduce requirements and cost of the metallic tube used in receiver tubes.

4.3.2 Selective absorber

Selective absorber is the main responsible of optical and thermal efficiency of a receiver tube. It absorbs concentrated solar radiation that reaches the metallic pipe, and the collected energy is transferred to the HTF. Receiver tube thermal losses by conduction and convection are reduced with vacuum environment within metal and glass pipes and radiative losses are minimized by the low thermal emittance of the selective absorber.

Selective absorbers have to withstand temperatures between 30 and 50°C over nominal temperature depending on wall pipe thickness and pipe material. Absorber collects solar radiation and has to transfer heat to the HTF, hence increasing wall thickness and decreasing steel heat transfer coefficient lead to a higher absorber temperature [32].

Also the intensity distribution on the absorber surface has an impact on the stresses on the absorber system, because due to the concentrator system, usually one side of the absorber tube receives an overwhelming fraction of the total irradiation on the tube. For parabolic trough the outside part of the pipe receives just nonconcentrated radiation (900 W/m^2), whereas the side facing the reflector in the average receives about 35–40 kW/m^2.

Main degradation processes in selective absorbers can be summarized as follows:

- Thermal stress: Temperature differences within the high-temperature receivers, and even for homogeneous temperature fields, the nonmatching thermal expansion coefficients of adjacent layers, may lead to thermally induced stresses on the individual layers. This in turn may induce flaking or chipping off of layers. Cracks are developed and barrier layers are destroyed locally. The problem could be solved by introducing special adhesion layers between problematic materials.
- Oxidation: If vacuum is lost in the receiver tube, a strong degradation is produced due to oxidation in high-temperature absorbers. Metallic layers of the infrared (IR) mirror and/or the metallic particles in the cermet layer are subject to oxidation. It has to be considered that antireflective and barrier layers that protect absorber and IR reflector are quite thin layers, and they do not protect effectively the absorber system against degradation.
- Diffusion: Diffusion processes may take place between substrate and absorber systems or within the individual absorber layers. Hence, it is required to use thin and dense barrier layers between substrate and IR reflector and, usually, between IR reflector and absorbing layer. Barrier layers of Al_2O_3 and SiO_2 have been successfully used to reduce diffusion.

Commercial selective absorbers have been improved dramatically since the first black chrome absorber used in SEGS LS-2 solar collectors [8]. First developed sputtered cermet, $Mo-Al_2O_3$ coating, demonstrated good thermal performance in evacuated HCE configurations, but showed limited durability when exposed to air at parabolic trough operating temperatures. When vacuum is reduced and absorber is exposed to air at temperatures above 300°C, the Mo–Al_2O_3 cermet degrades and coats the glass tube with an opaque white precipitate similar to a fluorescent light [13]. This strong absorber degradation impacts strongly solar field performance. Air leakage at the glass to metal seal appears to be the primary cause of failure, resulting in a loss of vacuum and oxygen exposure of the cermet coating. Historically, HCE failure rates were about 3–4%/year [29]. Molybdenum was removed from cermet composition to avoid this problem.

As of 2016, sputtered cermets from various suppliers present better optical properties than original Mo/Al_2O_3, but thermal stability in air is still not available. Only Archimede Solar Energy sputtered cermet shows a relatively good stability in air at high temperature up to 450°C and good optical properties, with solar absorptance $(\alpha) > 0.95$ and $\varepsilon_{400°C} < 0.10$ [30].

Huiyin Group, in China, has developed a sputtered selective absorber cermet with good optical performance, and it has been successfully used in Fresnel systems without thermal degradation.

In order to avoid absorber thermal degradation in air at high temperature, it is necessary to carefully select both the IR reflector and absorber material. Commonly used IR reflectors, such as copper, silver, or molybdenum are not stable in air and, if vacuum were affected, IR reflector oxidizes and optical properties are degraded. β-Tungsten is the most promising metal to be used as IR reflector. It has a high thermal stability in air at high temperature, and it can be deposited by sputtering [2, 35].

Absorber cermets present a similar problem as IR reflectors. Metal particles are not stable in air, and they are oxidized at high temperature if air enters vacuum annulus. A quite promising absorber has been used for flat collectors and medium temperature applications produced by sol–gel deposition technology that uses a $CuMnO_4$ spinel mixed oxide [3]. This absorber is stable in air at temperatures higher than 1000°C, and it is almost transparent to IR radiation. This absorber has been optimized with a dense and a porous antireflective layers, and thermal emittance as low as 0.05 at 400°C can be obtained with a solar absorptance as high as 0.95 [22], using an aluminum IR reflector. Combining this spinel absorber with an IR reflector stable at high temperature, such as platinum or β-tungsten, it is possible to prepare a selective absorber stable in air at temperatures of 550°C.

4.4 Vacuum maintenance

Receiver tubes have vacuum in the annulus between metallic and glass pipes, usually lower than 10^{-3} mbar, to reduce conductive–convective thermal losses and to preserve selective absorber from air oxidation. Annulus space is sealed using a glass–metal welding that attaches borosilicate glass cover with a metallic ring that is finally welded to a bellow installed to compensate dilatation mismatch between glass and steel pipes.

Vacuum maintenance is one of the most critical issues in solar receivers' design, construction, and maintenance cost. Vacuum degradation leads to a dramatic increase in thermal losses due to conduction—convection; it can produce selective absorber degradation and it implies that tube needs to be replaced. Replacing pipes represents a high cost because it is necessary to disconnect its collector line in the solar field until a new receiver is procured and the pipe is replaced.

Most usual problems in vacuum maintenance, observed in plant operation, have been glass—metal welding failure and hydrogen diffusion from HTF to vacuum annulus, minimized with getter materials, hydrogen-permeable membranes, using a low partial pressure of an inert gas in vacuum gap or receiver pipes that can be evacuated in plant operation when getters are exhausted and hydrogen content is too high.

Assuming that there are no vacuum leakages in receiver tubes, vacuum maintenance can be affected by hydrogen diffusion from oil-based HTF through metal pipe wall to vacuum annulus, due to thermal oil degradation at high temperature.

There are three ways to reduce or avoid hydrogen diffusion problem:

- replacing oil-based systems by molten salts or DSG in future solar power plants;
- using hydrogen diffusion barriers on steel pipe to block hydrogen diffusion—as of 2005 such barriers tested had not produced desired effects [17];
- removing or reducing hydrogen content from HTF using filters with any material that adsorbs hydrogen or the simplest and probably also the most cost-effective approach that consists of venting the fluid system on a regular basis so that the gas does not migrate through the steel receiver wall [21].

Once hydrogen has entered in the vacuum annulus, heat losses are typically increased by a factor of six compared with the vacuum original situation. Hence, it is critical to avoid this scenario reducing hydrogen partial pressure in oil-based HTF or reducing or removing hydrogen in vacuum annulus.

4.4.1 Glass to metal seal

One of the problems related to constructing glass to metal seals lies in the fact that metals generally have a linear thermal expansion coefficient considerably greater than typical borosilicate glasses, and this leads to the establishment, in the constructed joint, of mechanical stresses that can easily lead, during thermal heating to which the joint is subjected both in the step of manufacturing and using the collector, to breaking the joint.

In order to solve the problem of mechanical stresses in the seal, there are several approaches:

Glass to metal joints are made with metals or alloys having a value of coefficient of linear thermal expansion very close to that of the glass used in the temperature range comprised between ambient temperature and the glass annealing one. Kovar has a linear thermal expansion coefficient value of 5.5×10^{-6}/K from 20 to 200°C and 4.9×10^{-6}/K at 400°C, and it is quite close to that of alkaline earths borosilicate glasses (Schott 8414). Kovar has extensively been employed to make glass to metal seals in Schott's HCE [9].

Kovar has now been replaced by cheaper metal alloys such as SS AISI 430 in Archimede's HCE [1, 31], which has an intermediate value between 3.3 borosilicate glass and AISI 321 SS.

Housekeeper method in which AISI 321 stainless steel is welded directly to borosilicate 3.3 glass tube, is the simplest and cheaper solution and, although it had problems in the past, it has successfully employed in Huiyin's receiver tubes [37].

In first solar power plants, main operation failure was vacuum loss due to glass—metal welding breakage. This degradation problem has been considerably reduced, but still some glass—metal welding breakage has been reported in commercial power plants.

Another factor observed in HCE initial generations, which produced receiver tubes failure, is concentrated solar radiation that reaches directly the bellow, especially during low-sun-angle operating conditions [28], and glass—metal welding that induces thermal stresses in the welding producing glass—metal welding failures. Hence, efficient internal and external shields need to be employed to completely cover bellows and glass—metal welding from concentrated solar radiation.

4.4.2 Getters

Since the construction of the first solar trough systems, it has been a common practice among manufacturers of receiver tubes to incorporate so-called "hydrogen getters" inside the vacuum space of each HCE. These materials absorb any traces of hydrogen gas that might be formed over time as a result of thermal degradation of the HTF. The getter consists of a material that adsorbs hydrogen to form a hydride and thereby, removes hydrogen from the annular volume. The limitation of this method is the finite capacity of the getter for hydrogen. Once the getter saturates, it cannot adsorb additional hydrogen allowing the concentration of hydrogen in the annulus to increase.

Observations, however, indicate that the long-term exposure of the organic HTF to 400°C temperatures can lead to hydrogen pressures that exceed the gas-absorbing capacity of the hydrogen getters, thereby causing the significant heat losses that are currently observed in several solar trough plants. Solar receiver producers [33] claims for an effective partial pressure of dissolved hydrogen lower than 30 Pa and thermal oils, after several hundreds of hours under operation in real conditions, have values really higher. Getters are quite expensive materials, so the solution would not be to increase the amount of them.

4.4.3 Hydrogen-permeable membranes

Another method to remove hydrogen from the annulus is to locate a hydrogen-permeable membrane as a barrier between the annular volume and ambient air. The membrane is most commonly a thin layer of palladium that is selectively permeable to hydrogen. At elevated temperatures, hydrogen permeates through the membrane from the annulus to ambient air where it reacts with oxygen to form water. This method works in principle, but practical implementation results in failure of the palladium membrane due to corrosion of the membrane or hydrogen embrittlement when operating at the design temperature, so it is not used in new generation receiver tubes.

4.4.4 Low partial pressure inert gases

Hydrogen is a quite dangerous gas in vacuum because it has significant thermal conductivity due to its low molecular weight and correspondingly high molecular velocity. Hence, presence of low partial pressures of hydrogen gas in the annulus significantly decreases the thermal performance of the HCE.

Moderating the movements of hydrogen with a heavy mass noble gas (e.g., xenon) the heat loss can be reduced to a level 30–40% above the vacuum ($p < 10^{-3}$ mbar).

This strategy, to introduce a low partial pressure of a noble gas inside vacuum annulus, has not been employed yet in commercial pipes because thermal losses are higher than conventional evacuated pipes. In 2010s, new [33] PTR70 Premium Receivers incorporate a noble gas capsule that is activated with a laser, breaking it, reducing significantly thermal losses in the receiver [34] when a significant partial pressure of hydrogen gas is detected.

4.4.5 Re-evacuable pipes

Probably, the most obvious solution to solve vacuum maintenance problem is to design solar receivers that can be evacuated during operation on demand. These receivers can be continuously evacuated in plant operation in a dynamic system or re-evacuated when hydrogen partial pressure is high enough.

A new absorber tube has been developed by ARIES INGENIERIA company, under HITECO project from the EU 7th Framework Programme. It presents a continuous open chamber in which the dynamic vacuum is held and internal steel tube and external glass tube are totally independent [4]. Vacuum level and gas composition inside the chamber can be maintained continuously with minimal thermal losses. ARIES claims that vacuum does not have to be continuously created; on the contrary, the operation would have a very large periodicity with little energy consumption—months—achieving minimal maintenance by the use of conventional pumping systems [11]. Fig. 4.2 shows the diagram of HITECO receiver.

Huiyin Group has developed a serviceable HCE [37] that provides capability to evacuate vacuum tubes during operation in a solar plant and regenerate getters located in a "getter box." The design is shown in Fig. 4.3. It presents a reusable nozzle to connect a vacuum pump to the vacuum tube under operation in a power plant when hydrogen pressure increases. During exhausting procedure, it is possible to regenerate getters, which are located in an external getter box, with an induction heater. It is known that getters release adsorbed gases at high temperature.

4.5 Bellows

The bellow is the airtight connection of the absorber and the glass envelope. Depending on the specific receiver design, the bellow might also have to support the weight of the glass envelope. At room temperature, a typical absorber tube has a length of 4060 mm. At operating temperature of 400°C the absorber tube of the receiver

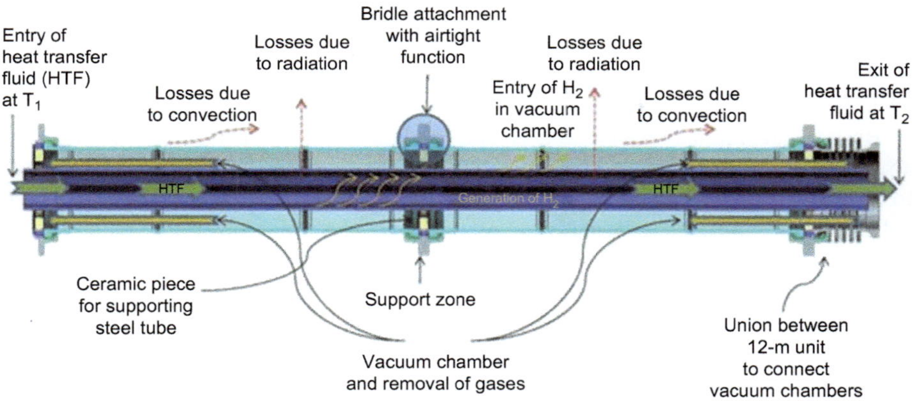

Figure 4.2 Diagram showing new HITECO's receiver tube.
The research leading to these results is founded from the European Union 7th Framework Programme (FP7/2007-2013) under grant agreements n° 256830 (HITECO).

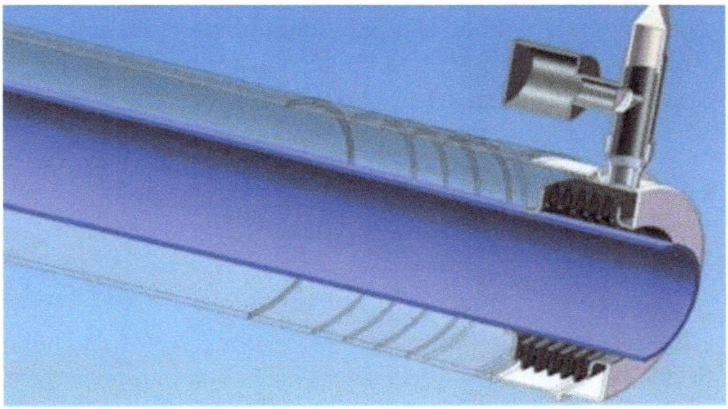

Figure 4.3 Re-evacuation system in Huiyin's receiver tube.
Courtesy of Huiyin.

expands about 28 mm at 400°C [27], while the glass envelope expands only 5 or 8 mm if borosilicate 3.3 or 5.1 are used. As the receiver temperature is significantly reduced during the night, the bellow experiences expansion and compression every day that the plant is under operation. Considering a solar receiver life time of 25−30 years, the receiver will suffer about 10,000 expansion−compression cycles from 400 to 100°C.

Main degradation problems observed in metallic bellows is due to the process known as intergranular stress corrosion cracking (ISSC). This corrosion process is produced when three conditions are present [6]:

- a susceptible material;
- an aggressive medium;
- stresses on the material produced during welding procedure or plant operation.

Typical susceptible materials are some austenitic stainless steel, such as "H" grades ($C > 0.04\%$) of 304 and 316 stainless steels, in which chromium carbide precipitation occurs, at temperatures between 530 and 590°C, producing chromium depletion at grain boundaries.

Bellows are not in contact with HTF and one bellow part is inside vacuum chamber, so it is protected against any aggressive medium. The other bellow surface is exposed to air and it is necessary to avoid the formation of an aggressive environment. Both solar receiver tube ends are usually covered with an insulating material to reduce thermal losses and it is critical to select a suitable insulating material without chlorides traces and very low humidity content to preserve receiver from aggressive mediums than could lead to ISCC.

Residual stresses can be introduced by cold deformation and forming, welding, heat treatment, machining, and grinding.

In ISCC, usually, most of the surface remains unaffected but with fine cracks penetrating into the material. In the microstructure, these cracks can have an intergranular morphology. ISCC is classified as a catastrophic form of corrosion, as the detection of such fine cracks can be very difficult, and it can produce vacuum loss and bellow degradation with complete receiver damage.

Recommended materials for bellows are stainless steels stabilized with titanium (Grade 321) or niobium (Grade 347) that are less susceptible to suffer ISCC.

Bellow design has been optimized to obtain maximum active length at operating temperature and now HCE producers present values close to 97% at 400°C.

4.6 Conclusion

HCEs are key elements in the development and performance of parabolic trough collectors. They are the responsible to collect solar energy and to transfer heat collected to the HTF. Main challenges to be met in the future are to increase durability and to reduce cost and maintenance. Regarding durability, vacuum maintenance is the major challenge because thermal losses and selective absorber stability are directly dependent on vacuum level. Vacuum degradation mainly occurs by glass to metal seal failures due to mismatch between glass and metal linear expansion coefficients, that can be solved with an adequate selection of metal and glass composition, or hydrogen produced by synthetic oil thermal degradation that diffuses through stainless steel wall and reaches vacuum annulus. Hydrogen diffusion problem cannot be avoided only with getters or selective-permeable membranes and in this way, dynamic vacuum systems or re-evacuable pipes seem to be the best options for oil systems. The necessity to increase operation temperature to improve efficiency implies replacing synthetic oil with other HTFs with higher thermal stability and thermal conductivity, such as molten salt, and thus, hydrogen diffusion problem can be avoided.

Selective absorber thermal stability, both in vacuum and air, has to be increased to raise operation temperature in solar thermal plants to improve efficiency and minimize damage if vacuum is lost. IR reflectors with higher thermal stability in air, such as tungsten, and more thermally stable spinel and cermet absorbing layer materials are now being used, with promising results.

ARCs have to improve mechanical properties to withstand erosion and abrasion phenomena produced by sand particles in air and cleaning procedure in normal plant maintenance. Multifunctional coatings that work as ARC and are easy-to-clean or self-cleaning, are now being developed and soiling of the coatings is strongly reduced.

Finally, bellow design has been optimized to obtain maximum active length at operating temperature, with values close to 97% at 400°C and efficient internal and external shields need to be employed to completely cover bellows and glass–metal welding from concentrated solar radiation.

References

[1] Angelantoni G. Archimede's social presentation. 2008. Available from: http://www.ice.gov.it/sedi/umbria/energia/angelantoni/Archimede%20Company%20Profile.pdf.

[2] Antonaia A, Esposito S, Addonizio ML, Guglielmo A. Solar selective absorber based on double nitride composite material and process for its preparation. 2011. EP 2721353 A1.

[3] Bayón R, San Vicente G, Maffiotte C, Morales A. Preparation of selective absorbers based on CuMn spinels by dip-coating method. Renewable Energy 2008;33(2008):348–53.

[4] Cachafeiro H. New concepts advance solar thermoelectric energy. EuroPhotonics 2012. Available from: http://www.photonics.com/Article.aspx?AID=52546.

[5] Chiarappa T. Performance of direct steam generator solar receiver: laboratory vs real plant. Energy Procedia May 2015;69:328–39.

[6] Cramer Jr SD, Covino Jr BS. Handbook "corrosion: fundamentals, testing, and protection, vol. 13A. ASM Handbook; 2003, ISBN 978-0-87170-705-5.

[7] DSM ARC. Available from: https://www.dsm.com/corporate/markets-products/markets/energy/anti-reflective-coating.html.

[8] Dudley VE, Kolb GJ, Mahoney A, Mancini TR, Matthews CW, Sloan M, et al. SEGS LS-2 solar collector: test results. SAND94–1884. Sandia National Laboratories; December 1994.

[9] Fark C. Setting the benchmark in receiver technology. In: Delhi International Renewable Energy Conference (DIREC) 2010 October 27–29, 2010; 2010.

[10] HITEC heat transfer salt. Report of Coastal Chemical Co. LLC, Houston, TX. Available from: http://stoppingclimatechange.com/MSR%20-%20HITEC%20Heat%20Transfer%20Salt.pdf.

[11] HITECO. HITECO (new solar collector concept for high temperature operation in CSP applications). 2013. Final Report Summary EU-FP7-ENERGY reference: 256830.

[12] Kearney AT. ESTELA industry roadmap report. 2010. Available from: https://www.atkearney.de/documents/856314/1214532/BIP_Solar_Thermal_Electricity_2025.pdf/3535c6f9-ad45-442a-b0f6-0630b21a05cb.

[13] Kennedy CE, Price H. Progress in development of high-temperature solar-selective coating. In: Proceedings of ISEC2005 International Solar Energy Conference August 6–12, 2005 (2005), Orlando, Florida USA; 2005.

[14] Koretz B, Afremov L, Chernin O, Rosin C. Molten salt thermal energy storage for direct steam tower systems. In: SolarPACES 2011, 20–23 September 2011, Granada, Spain; 2011.

[15] Kruizenga AM, Gill DD, Laford M, Mcconohy G. Corrosion of high temperature alloys in solar salt at 400, 500 and 680°C. 2013. Sandia Report 2013. Sand 2013–8256.

[16] Kryzak M, Helsch G, Heide G, Frischat GH. Method of making a glass body with a phosphorous- and porous SiO_2-containing coating, glass body made thereby and solution for making same. 2003. US Patent number US20030170465.

[17] Kuckelkorn T, Graf W. Radiation-selective absorber coating with an adherent oxide layer and method of making same. 2005. US7793653 Patent.

[18] Martienssen W, Warlimont H. Springer handbook of condensed matter and materials data. Springer Science & Business Media; September 21, 2006.

[19] Mehner A, Dong J, Prenzel T, Datchary W. Mechanical and chemical properties of thick hybrid sol–gel silica coatings from acid and base catalyzed sols. Journal of Sol-Gel Science and Technology 2010;54:355–62.

[20] Midtdal K, Jelle BP. Self-cleaning glazing products: a state-of-the-art review and future research pathway. Solar Energy Materials & Solar Cells 2013;109:126–41.

[21] Moens L, Blake DM. Mechanism of hydrogen formation in solar parabolic trough receivers. In: 14th Biennial CSP SolarPACES Symposium, 4–7 March 2008, Las Vegas, Nevada; 2008.

[22] Morales A, San Vicente G, Germán N. Air stable selective absorber spinel prepared by dip-coating method. 2016 [To be published].

[23] Morales A. Sol-gel process for the preparation of porous coatings, using precursor solutions prepared by polymeric reactions. 2002. European Patent number EP 1329433.

[24] Muñoz J, Zarza E, Díez LE, Martínez-Val JM, López C, Gavela R, et al. The new technology of gas cooled trough collectors. In: SolarPACES 2011, 20–23 September 2011 Granada, Spain; 2011.

[25] Muñoz-Anton J, Biencinto M, Zarza E, Díez LE. Theoretical basis and experimental facility for parabolic trough collectors at high temperature using gas as heat transfer fluid. Applied Energy December 15, 2014;135:373–81.

[26] Nielsen KH, Kittel T, Wondraczek K, Wondrackzek L. Optical breathing of nano-porous antireflective coatings through adsorption and desorption of water. Scientific Reports 2014; 4. Article number 6595.

[27] Pernpeintner J, Sutter F, Happicha CH, Lüpfert E, Schiricke B, Lichtenthäler N, et al. Durability testing of parabolic trough receivers –Overheating, thermal cycling, bellow fatigue and antireflex-coating abrasion. Energy Procedia 2015;69:1540–50.

[28] Price H, Lüpfert E, Kearney D, Zarza E, Cohen G, Mahoney R. Advances in parabolic trough solar power technology. Transactions of the ASME May 2002;124:109–25.

[29] Price H, Hale MJ, Mahoney AR, Gummo C, Fimbres R, Cipriani R. Developments in high temperature parabolic trough receiver technology. In: Proceedings of the 2004 International Solar Energy Conference, July 11–14, 2004 Portland, OR; 2004 (ISEC200465178).

[30] Raccurt O, Disdier A, Bourdon D, Donnola S, Stollo A, Gioconia A. Study of the stability of a selective solar absorber coating under air and high temperature conditions. Energy Procedia 2015;69:1551–7. SolarPACES 2014.

[31] Raggi C, Chiarappa T. A glass-to-metal joint for a solar receiver. 2013. EP 2626336 A1 European Patent.

[32] Roldán MI, Valenzuela L, Zarza E. Thermal analysis of solar receiver pipes with superheated steam. Applied Energy March 2013;103:73–84.

[33] Schott PTR®70 Premium Receivers brochure. Available from: http://www.schott.com/csp/english/download/schott_ptr70_premium_4th_generation_datasheet.pdf.

[34] Sohra O, Möllenhoffa M, Kuckelkorna T. Laser induced release of encapsulated noble gas in SCHOTT receiver. Energy Procedia December 2014;49:2070–9.

[35] Stelmakh V, Rinnerbauer V, Joannopoulos JD, Soljacic M, Celanovic I, Senkevich JJ. Evolution of sputtered tungsten coatings at high temperature. Journal of Vacuum Science & Technology A 2013;31:061505.

[36] Trapp D, Tratzky S. Method of making a wipe-resistant antireflection coating on a borosilicate glass body and coating solution for use in said method. 2010. US Patent number US7842339, 30-11-2010.

[37] Ven K. Receivers for parabolic troughs. CSP Today 2014; 2014. Available from: http://social.csptoday.com/sites/default/files/5aHUIYIN.pdf.

[38] Vignarooban K, Xinhai X, Arvay A, Hsu K, Kannan AM. Heat transfer fluids for concentrating solar power systems – a review. Applied Energy 2015;146(2015):383–96.

Innovative working fluids for parabolic trough collectors

5

E. Zarza Moya
CIEMAT, Plataforma Solar de Almería, Tabernas (Almería), Spain

5.1 Introduction

Sixty-three solar thermal electric (STE) plants with parabolic trough collectors (PTCs) were fully operational during mid-2015. Sixty-one of these sixty-three plants were using thermal oil as working fluid in the solar field, thus proving the commercial maturity of using thermal oil as working fluid in STE plants with PTCs. The use of thermal oil in the solar field is usually known as "*heat transfer fluid* (HTF) technology" because the thermal oil transfers the thermal energy delivered by the solar field to the thermal energy storage (TES) and to the steam generating system producing the steam for the plant power conversion system (PCS).

Thermal oil has been used as working fluid in PTCs since the 1970s, when first small demonstration projects with PTC were developed in USA and Europe. It was also used in the experimental distributed collector system (DCS) of the International Energy Agency (IEA) Small Solar Power Systems (SSPS) project implemented in Almería (Spain) early in the 1980s, composed of single-axis (ACUREX-3001) and two-axis (Heliomann-3/32) PTC solar fields. Thermal oil was then also used in nine SEGS (solar electricity generating systems) installed by the company LUZ Industries in California (USA) during the period 1985–1991. Subsequently, during the 2007–13, it was used in 45 STE plants implemented in Spain, and in the SOLANA and MOHAVE solar plants installed later in USA.

Thermal oils available in 2010s have better properties than those available during 1970s and 1980s. The maximum working temperature of the one used in the 1980s for the IEA SSPS project (i.e., Therminol-55 oil) was 300°C, and the first SEGS plant used another one, ESSO-500 oil, with a maximum working temperature of 310°C. Modern STE plants use thermal oils with a maximum working temperature of 398°C (Dowtherm A and Therminol VP-1 oils). This higher working temperature has allowed higher superheated steam temperature for the PCS and therefore higher plant efficiencies.

The lifetime of currently available thermal oils exceeds 30 years, if a suitable treatment system (the so-called *oil ullage system*) is installed in the plant and the maximum operating temperature is kept below the limit defined by the manufacturer. Another advantage of thermal oils is the existence of a complete set of thermal oils with different properties and costs. The selection of the best option for a specific project depends on the required working conditions. There are cheap and low vapor pressure oils for working temperatures lower than 350°C, while more expensive oils with higher

Advances in Concentrating Solar Thermal Research and Technology
http://dx.doi.org/10.1016/B978-0-08-100516-3.00005-8

vapor pressure are available for temperatures up to 420°C (e.g., Syltherm 800). The step from 398°C to 420°C in the maximum working temperature imposes a significant increase in the price. This is the reason why modern STE plants are using thermal oils with a working temperature limit of 398°C.

The main reasons why thermal oils have been used in solar fields with PTC since the 1970s are their affordable price, low vapor pressure, good thermal stability, and long lifetime if the working conditions recommended by the manufacturers are fulfilled. However, it has been clear from the very beginning that thermal oils are not the perfect working fluid for PTC, because they also have some important constraints. The three main constraints associated with them are:

- environmental contamination in case of leaks, because thermal oils are pollutant and the ground affected by a leak has to be decontaminated, which is expensive and increases the plant operation and maintenance (O&M) costs;
- fire hazards when there is a leak of hot oil because the fire point is usually below the solar field working temperatures—fires of different levels have been reported in many STE plants due to leaks of hot oil;
- limited working temperature—thermal oils currently available at affordable prices have a maximum working temperature of 398°C, which limits the temperature of the superheated steam delivered to the PCS to about 385–390°C; this limitation jeopardizes the overall plant efficiency.

Although thermal oils are neither user friendly nor environment friendly, globally they are currently the best available and fully proven option at commercial scale for large solar fields with PTC working at temperatures above 300°C. Other alternatives, like pressurized liquid water, can be a better solution for lower temperatures (<250°C) because the associated vapor pressure is affordable.

Taking into consideration the limitations and drawbacks of thermal oils when they are used as working fluid in PTC, several alternative working fluids have been investigated and important advancements have been achieved during the 1990s and 2000s, so that thermal oils could be replaced by other fluids in future STE plants with PTC. Three different working fluids have been investigated so far:

- liquid-water/steam (the so-called direct steam generation, DSG, technology);
- pressurized gases, like CO_2, N_2 or air;
- molten salts, composed of ternary or binary salt mixtures (potassium and sodium nitrates mainly).

These three alternative working fluids have advantages and disadvantages when they are compared to thermal oils traditionally used in PTC. Table 5.1 shows a summary of main advantages and disadvantages, and it clearly depicts that none of these fluids is a perfect candidate, because they all have several drawbacks when they are compared to thermal oils. All the advantages and disadvantages listed in Table 5.1 are discussed and analyzed in detail in the relevant sections of this chapter devoted to each working fluid.

The global efficiency of an STE plant mainly depends on the efficiency of both the PCS and the solar field. Concerning the working fluid there are two main parameters influencing the plant efficiency: the pressure drop in the solar field circuit and the heat

Table 5.1 **Alternative working fluids versus thermal oils**

Fluid	Advantages	Disadvantages
Direct steam generation (DSG)	• Simpler plant configuration • Higher steam temperature • No fire hazard • No pollutant	• Lack of a cost-effective thermal storage system • Complexity of solar field control • Higher pressure in solar field piping
Pressurized gases	• Higher steam temperature • Cheaper thermal storage • No fire hazard • No pollutant	• Lower heat transfer coefficient • Complexity of solar field control • Higher pressure in solar field piping • Higher pumping power
Molten salts	• Higher steam temperature • Cheaper thermal storage • No fire hazard • No pollutant	• Freezing hazard • Complex solar field design • Higher electricity self-consumption

transfer coefficient in the receiver tubes. The pressure drop has a direct impact on the plant electricity consumption because the pumping power demanded by the solar field is the main fraction of the overall electricity consumption of the plant. Concerning the heat transfer coefficient, a high value of this coefficient is essential for the integrity of the receiver tubes because a too low coefficient can contribute to the bending and even breakage of the tubes due to a high temperature difference between the irradiated and the nonirradiated parts of the tubes. Taking into account the importance of the pressure drop in the solar field circuit and the heat transfer coefficient in the receiver tubes, these two parameters are analyzed in this chapter in the relevant section devoted to each of the three alternative working fluids considered.

5.2 Direct steam generation

In the DSG process liquid water is pumped to the solar field inlet to be preheated, evaporated, and converted into saturated or superheated steam (depending on the solar field design) as it circulates through the receiver tubes. Hence, collector rows in a DSG solar field are composed of three consecutive sections: water preheating section, water evaporating section, and steam superheating section. DSG was the first option considered to replace thermal oils. In fact, the first large commercial solar plant built in the 20th century used DSG. At the beginning of the 20th century the American engineer Frank

Shuman designed the solar plant that was inaugurated at Meady (Egypt) in 1912 with PTC evaporating water and producing saturated steam at 0.1 MPa [16]. The operation of this first DSG plant had already brought out the technical constraints associated with this technology. The next references found in the literature about the DSG process were published early during the 1970s and 1980s [25,27,35], with theoretical results obtained at the University of Tel-Aviv (Israel) and SERI (Solar Energy Research Institute, USA) about possible flow instabilities and control-related problems in DSG solar fields.

Later on, in 1988 the company LUZ International (the promoter of the SEGS plants in California, USA) came to the conclusion, after the implementation of the first SEGS plants, that in future, thermal oil had to be replaced by another working fluid in STE plants to improve the plant efficiency and to reduce costs, and they considered DSG the best option. Therefore, they launched an ambitious research and development (R&D) program, the so-called *Advanced Trough System* (*ATS*) project to fully evaluate this option and develop a new generation of STE plants based on DSG. However, this program was stopped by the bankruptcy of LUZ in 1991.

After the demise of LUZ, several experimental facilities and R&D projects were promoted in Europe to continue with the study and evaluation of the feasibility of the DSG process. The HIPRESS facility implemented by the German ZSW [22], the test facility installed by Belgo Instruments International (BII) in Sde Boker (Israel) [38], the GUDE project [13,14], the ARDISS project [1], the PRODISS project [12], the DISS project [39], the DISTOR project [2], and the DUKE project [11] are the best examples of R&D activities performed during the period 1990–2015 on DSG.

It must be pointed out here the great contribution of the European DISS project to the development of the DSG technology. The test facility implemented at the Plataforma Solar de Almería (PSA) within the DISS-phase I project [17] was the first life-size DSG test facility available in the world to study the DSG process under real solar conditions. The experimental results gathered in this facility during the test campaign performed within the DISS project clarified many of the open technical questions related to the feasibility of the DSG process. The advantages and disadvantages of the DGS process versus thermal oils summarized in Table 5.1 are further explained in Section 5.2.1.

5.2.1 Advantages and disadvantages of the DSG process versus thermal oil

Fig. 5.1 shows the simplified diagram of an STE plant with PTC and thermal oil as working fluid. The plant is composed of four main subsystems: solar field, TES, steam generator, and PCS. When DSG is used in the solar field, the steam generator is no longer needed because the steam for the PCS is directly provided by the solar field. The way how DSG simplifies the overall plant configuration becomes evident if the steam generator is removed from Fig. 5.1. The removal of the steam generator has a twofold benefit: the simplification of the plant configuration and a lower investment cost because the steam generating system is quite expensive.

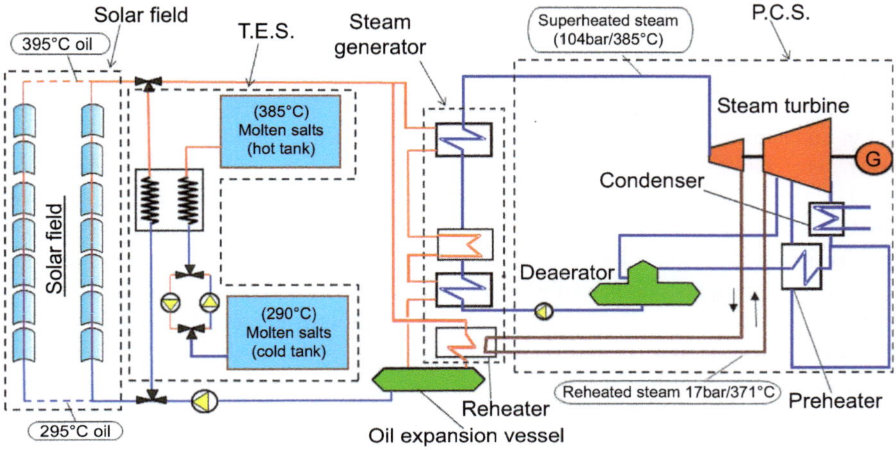

Figure 5.1 Simplified scheme of an STE plant with thermal oil as working fluid.

Another important benefit of DSG is the possibility to increase the temperature of the steam provided to the PCS, thus increasing the overall plant efficiency because of better thermodynamic parameters of the Rankine cycle. Due to the thermal stability limitation of thermal oils (~400°C), the maximum steam temperature that can be achieved is about 385°C, whereas DSG has no thermal limitation and steam at temperatures even higher than 500°C can be produced in the solar field.

The nonpollutant nature of water is another benefit of DSG, because water leaks have neither fire nor contamination hazards, while oil leaks are very dangerous for people and for the environment. Fires in commercial STE plants due to oil leaks in the solar field have been reported by the plant owners and even videos can be found in Internet (e.g., https://youtu.be/aGyvicfgZm0). Such a problem will never occur if thermal oil is replaced by DSG.

Concerning disadvantages of DSG versus thermal oil, the lack of a suitable and cost-effective TES is a significant barrier for the commercial deployment of DSG STE plants. Dispatchability is the main benefit of STE plant when they are compared with wind farms or photovoltaic (PV) plants. Since dispatchability is provided by the TES systems, the lack of a competitive TES for DSG is a significant constraint of this alternative working fluid. Solar plants using thermal oil can be provided with TES systems based on molten salts (see Fig. 5.1), so that the thermal energy is stored as sensible heat (i.e., temperature increase) of a binary mixture of molten salts, the so-called *solar salt* (i.e., 40% of potassium nitrate and 60% of sodium nitrate). This type of TES system had been implemented in 21 STE plants with PTC at the end of 2015. With a specific investment cost of about 35–40 €/kWh of capacity, TES systems using molten salts are cost-effective and they provide the STE plants with an excellent degree of dispatchability. However, this type of TES system is not suitable for solar plants with DSG, because the steam provided by the solar field will condense as it trespasses its thermal energy to a storage media, and condensation is a thermodynamic process taking place at constant temperature (phase-change process), and therefore the released heat cannot be stored as sensible heat, but as latent heat in a storage media changing its phase (from solid to liquid during the condensation of the steam, TES charging process). During the TES discharging process the storage media will release heat at constant temperature during its phase change from liquid to solid. This is the reason why DGS systems require TES systems based on phase-change materials (PCM). Although several small prototypes of TES systems based on PCM have been successfully tested [2,7,8,20], costs are still too high and a suitable design for large TES systems (i.e., $\geq$1 GWh) is not available yet.

Another constraint of DSG is the complexity of the solar field control, because uneven two-phase flow (i.e., liquid water + steam) conditions in parallel rows of PTC are likely to happen during solar radiation transients [36,37] due to the very different density of the fluid in the three consecutive sections composing each collector row (i.e., liquid water at the preheating section, saturated liquid water + steam in the evaporating section and steam at the superheating section). Since the density of the steam is much lower than that of the liquid water, the fluid speed inside the superheating steam receiver tubes is much higher than the speed at the preheating and evaporating sections, thus provoking a much higher pressure drop per unit length at the superheating section. Therefore, for instance, for a working pressure of 10 MPa the pressure drop per meter of receiver tube in the superheating section is about 10 times higher than the pressure drop per meter in the preheating section (see Section 5.2.2). This fact makes the superheating section to govern the overall pressure drop in the collector, so that any change in the length of the steam superheating section has a significant influence in the overall pressure drop in the complete collector row. During a solar radiation transient due to clouds (see Fig. 5.2(b)), if a control valve is not available at the inlet of each row, the length of the steam superheating sections of those rows affected by the shadow of the clouds are reduced because the lower solar radiation impinging on the collector row requires longer preheating and evaporating sections to convert the same amount of liquid water into saturated steam, and the length available for steam superheating is shorter (see Fig. 5.2), thus significantly reducing the overall pressure drop in the complete row and enhancing the increase of the feed water flow through the rows affected by the cloud if there is not a control valve at the inlet of each row to introduce an additional pressure loss and keep stable the pressure at the row outlet.

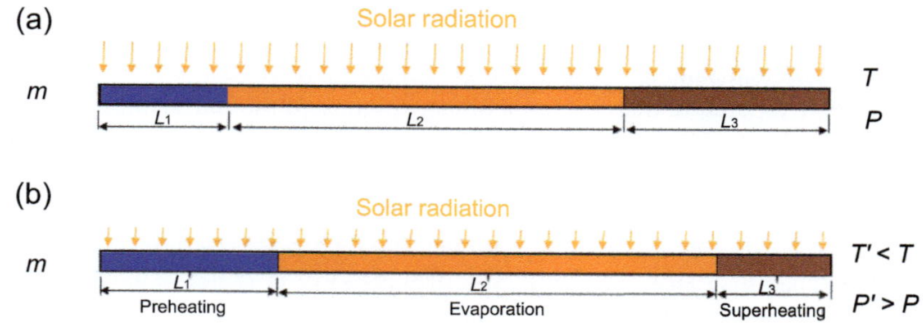

Figure 5.2 Simplified scheme of a DSG collector row under a solar radiation transient, (a) before the transient, (b) during the transient.

Therefore, control devices must be installed at each row to avoid flow instabilities. This problem does not happen in solar fields using thermal oil because it remains liquid all over the solar field and its density is very similar at both solar field inlet and outlet, so that cloud transients only affect the oil temperature at the row outlet, without disturbing the feed flow.

Although DSG imposes higher pressures in the solar field circuit (due to the higher vapor pressure of the water versus thermal oil), this is not really a great problem, because it can be overcome by the implementation of thicker steel pipes and mechanically more resistant fittings. Since this solution increases the investment cost just little, it is not really a great problem for the commercial use of DSG instead of thermal oil.

Another experimental result obtained at the PSA DISS test facility is that working pressures above 100 bar are not recommended for large DSG solar fields because water leaks at flanges and nonwelded connections will often be likely to occur due to the daily thermal cycles at the solar field (warming up in the morning and cooling down overnight).

5.2.2 Thermo-hydraulic aspects

Pressure drop in the solar field and heat transfer coefficients in the receiver tubes with DSG are analyzed in this section and operational limits are defined according to affordable pressure drop and thermal stress in the receivers.

When the design of a DSG solar field is optimized from a thermodynamic stand point, the number of PTC connected in series within each row is higher than in solar fields using thermal oil. For instance, the number of PTC to be connected in series within each row is usually four in STE plants using thermal oil if standard PTC composed of 12 parabolic trough modules with a parabola width of 5.7 m and a length of 11.9 m are used, while the optimum number of collectors in each row would be between six and eight if DSG is used. This fact must be taken into consideration when comparing the overall pressure drop with thermal oil and with DSG. However, the greater length of the collector rows is by far compensated by the lower fluid flow circulating through each row, because typical oil flow per row in commercial STE plants is of about 4–5 kg/s, while in a DSG collector row would be lower than 2 kg/s. This is the reason why the pumping power required for solar fields with thermal oil is significantly higher (even more than 20%, depending on the design) than for DSG solar fields.

Each row of a DSG solar field is composed of three consecutive sections: liquid water preheating, evaporating, and steam superheating. The third section does not exist in those DSG solar fields designed to produce saturated steam. The performance of these three sections concerning pressure drop and the heat transfer coefficients in the receiver tubes is very different. The specific pressure drop (i.e., Pa/m) increases from the preheating section to the steam superheating section, because the density of the fluid circulating inside the receiver tubes decreases and the velocity increases. Thus, for instance, the density of the saturated steam is about 10 times lower than the density of the liquid water for a working pressure of 10 MPa, and the corresponding fluid velocity changes in the same proportion. Since the specific pressure drop for a single-phase fluid (i.e., liquid or gas) in a cylindrical pipe is given by the following equation the pressure drop per unit length obtained when the fluid is steam is about 10 times higher than that obtained for liquid water because the pressure drop is proportional to the square of the velocity. This is the reason why the steam superheating section in a DSG solar field is the section governing the overall pressure drop in the solar field.

$$\Delta p = \frac{f}{2} \cdot \rho \cdot v^2 \cdot (1/d) \tag{5.1}$$

where Δp, specific pressure drop, Pa/m; f, Moody's friction factor, dimensionless; ρ, fluid density, kg/m^3; v, fluid velocity, m/s; d, pipe inner diameter, m.

Fig. 5.3 shows the scheme of one row of an optimized DSG solar field to convert liquid water at 150°C/10.7 MPa into superheated steam at 450°C/10 MPa. The row is composed of eight north–south oriented PTCs connected in series, each of them composed by 12 parabolic trough modules with a parabola width of 5.7 m and a length of 11.9 m. The direct normal solar irradiance considered for the simulation was 950 W/m^2, with an incidence angle of 15 degrees. The receiver tube diameters considered are $Ø_{ext}/Ø_{int}$ = 70/59 mm.

Fig. 5.3 shows that the pressure drop along the three consecutive sections of the row (i.e., water preheating, water evaporating, and steam superheating) is 0.03, 0.37, and 0.3 MPa respectively, in spite of the length of each section (350, 600, and 250 m, respectively). It is evident that the average specific pressure drop in the steam superheating section is 1200 Pa/m, while the specific pressure drop in the preheating and evaporating sections are 86 Pa/m and 617 Pa/m, respectively. It must be remarked here that an overall pressure drop lower than 1 MPa in the PTC rows is a good value

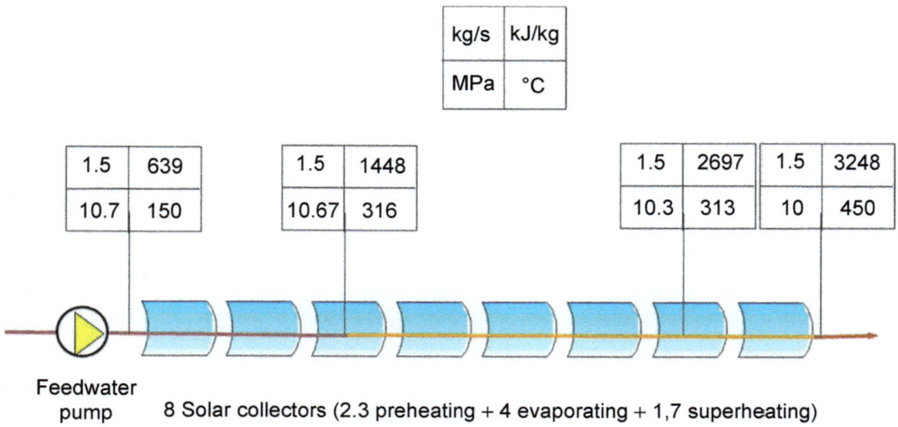

Figure 5.3 Simplified scheme of an optimized DSG PTC row to produce superheating steam at 450°C/10 MPa.

for DSG because the associated pumping power is affordable and lower than that of a solar field with the same nominal thermal power and using thermal oil.

One of the open technical questions concerning DSG that were clarified by the R&D activities performed during the 1990s was the pressure drop associated with the two-phase flow (i.e., liquid water and saturated steam) existing in the solar field evaporating section, which has a great contribution to the overall pressure drop in the solar field because the typical length of the evaporating section at each row is usually about 50% of the total row length including the three consecutive sections (i.e., liquid water preheating, evaporating, and steam superheating). It has been experimentally proven that the correlations proposed by Friedel [10] and Chisholm [6] provide accurate values of the pressure drop in straight pipes and elbows of the evaporating section, respectively. Although long evaporating sections are needed in DSG solar fields because of the high evaporation latent heat of the water, the total pressure drop in the steam superheating section, which has a quite shorter length, is similar because the fluid velocity is higher.

Due to the high pressure drop associated with the piping of the steam circuit, because of the high speed of the steam (the diameter of the pipes would be too big otherwise), it is necessary to reduce to a minimum the length of the steam header connecting the solar field outlet with the PCS inlet. This is the main reason why the effect of scaling up the size of a DSG solar field is not always favorable, and DSG solar fields for STE plants with a unit power higher than 50 MWe does not seem to be a good option. In fact, two 50-MWe DSG plants are more cost-effective than a single 100-MWe plant [9].

One of the conclusions obtained concerning working pressure for DSG solar fields is that working pressures lower than 5 MPa are not recommended for solar fields composed of large collectors (e.g., EuroTrough, and TCP-150 collectors) because the pressure loss jeopardizes the plant efficiency due to high pumping losses; 7.5 MPa seems to be a good compromise between low working pressures with high pumping losses and high working pressures suffering from often water leaks in the flanges and nonwelded connections of the piping due to thermal cycling. However, working pressures lower than 5 MPa are feasible for smaller collectors used in industrial process heat applications. Even 3 MPa is a feasible working pressure for small PTC with DSG. Hence, the recommended maximum working pressure depends on the receiver tubes diameter and row length.

The heat transfer coefficient in the inner side of the receiver tubes in a DSG solar field is another major technical issue, because the receiver tubes could be damaged if they are not refrigerated properly by the fluid circulating inside. Experimental studies developed at the Plataforma Solar de Almería during the project DISS [40] showed that the heat transfer coefficients in the water preheating section are high enough to guarantee a good refrigeration of the receiver tubes. There are different two-phase flow patterns possible in the evaporating section, and Fig. 5.4 shows in a very simplified way the two-phase flow pattern map for a cylindrical horizontal pipe. As shown in this figure, four main flow patterns are possible depending on the superficial liquid and steam velocities, V_l and V_g, respectively, and the convection heat transfer coefficient inside the pipe strongly depends on the flow pattern. The superficial steam and liquid

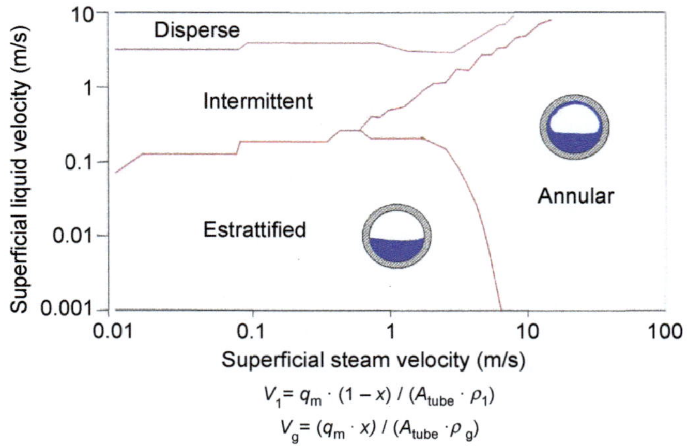

Figure 5.4 Simplified two-phase flow pattern map for a cylindrical horizontal pipe.

velocities depend on the total water mass flow, q_m, the steam quality, x, the liquid and steam densities, ρ_l and ρ_g, and the inner transversal cross-sectional area of the pipe, A_{tube}. The intermittent, disperse, and annular flow patterns provide a good heat transfer coefficient because the inner wall of the receiver tubes are wetted by liquid water. The stratified flow pattern is more dangerous because the liquid water remains at the bottom of the pipe while the saturated steam is above the liquid water, and there is no turbulence between liquid water and steam inside the receiver tube.

Due to the liquid water stratification, the heat transfer coefficient at the bottom of the steel receiver tubes is much higher than at the upper part of the tube and a great temperature difference there could be between the lower and upper parts of the receiver tube. This temperature difference is less dangerous when the solar collectors are looking upward (this happens around solar noon when the collector rotation axis is oriented north–south, Fig. 5.5(b)) because the concentrated solar flux heats the lower part of the receiver tube, which is refrigerated by the liquid water. Under extreme working conditions (low water flow and low level of the stratified liquid water inside the pipe, h_{liquid}) the temperature difference between the lower and upper part of the receiver tube can be high enough to bend the steel tube and thus breaking the glass cover. This problem even happens in solar fields with thermal oil when the oil flow is too low and the convection heat transfer coefficient is not high enough to provide a good cooling of the receiver tube. In the case of water stratification, the probability to have this problem in the solar collectors is much higher, and therefore stratification must be avoided by keeping a feed flow high enough.

At the steam superheating section there is a single-phase flow: steam, and the convection heat transfer coefficient is the same all around the inner wall of the receiver tubes. However, the steam flow must be high enough to provide the receiver tubes with a proper cooling when the incident solar flux is high (around solar noon) and more heat has to be extracted from the receiver tube by the steam circulating inside. Since the feed flow is somehow a function of the solar radiation level (i.e., the higher

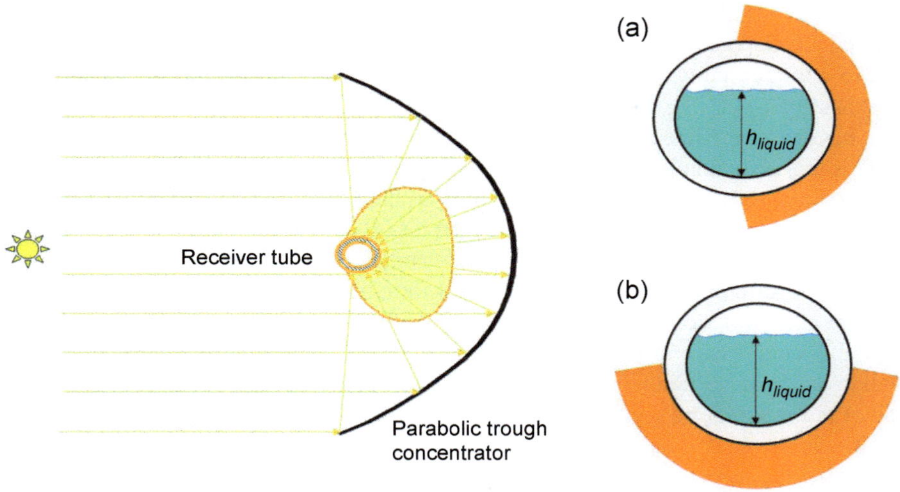

Figure 5.5 Effect of water stratification inside the receiver with DSG when collector rotation axis is oriented North-South, (a) at Sunrise and Sunset times, (b) at solar noon.

the solar flux onto the receiver tubes is, the higher the feed water flow required to keep the steam temperature constant at the outlet of the solar field) a proper design of the DSG solar field to assure a suitable feed flow when solar radiation is high is enough to guarantee the required convection heat transfer coefficient at the steam superheating section and also avoid stratification in the evaporating section. It has been proven that convection heat transfer coefficients equal or higher than 800 $W/m^2 \cdot K$ are good enough to avoid the bending of the steel receiver tube.

What has been explained concerning pressure drop and heat transfer coefficients in collector rows with DSG clearly shows that the main challenge in the design process of a DSG solar field is to achieve a good equilibrium between pressure drop and heat transfer coefficients. This equilibrium depends on the feed water flow selected for the design conditions, because it has to be high enough to avoid stratification and low heat transfer coefficients inside the receiver tubes, while keeping the overall pressure drop at a reasonable limit. Experimental results obtained at the PSA have shown that this equilibrium can be achieved with feed water flows within a wide range and the designer only has to be conscious of the necessity of this equilibrium.

5.2.3 State of the art of direct steam generation in parabolic trough collectors

The technical feasibility of the DSG process has been proven experimentally, and the knowledge required to design commercial solar fields using this technology exists nowadays. The DISS test facility of the PSA has been operated for more than 9000 h with encouraging results at three pressure levels (3, 6, and 10 MPa) and a maximum superheated steam temperature of 400°C [40]. This facility was upgraded in 2013-4 within the framework of the project DUKE to produce superheated steam at 10 MPa/500°C. Abengoa Solar has also successfully operated their experimental DSG parabolic trough facility at Sanlucar La Mayor (Sevilla). However, due to confidentiality issues no detailed experimental results have been published by Abengoa Solar.

It has been experimentally proved that the best configuration for collector rows with DSG is that shown in Fig. 5.6, with the end of the evaporating section separated from the beginning of the superheating section by a liquid-water/steam separator. With this configuration, the feed water flow at the inlet is higher than the steam flow required at the outlet, thus assuring an excess of liquid water at the end of the evaporating section. This excess of liquid water is recirculated to the inlet of the collector row and mixed with the feed water coming from the solar field feed pump. The control requirements for the system shown in Fig. 5.6 are very small because the system adapts itself very well to solar radiation transients. Thus, for instance, the amount of liquid water at the end of the evaporating section decreases and the steam flow entering the superheating section increases if solar radiation increases, thus compensating the higher solar flux impinging the receiver tubes in the steam superheating section. The small amount of water injected at the inlet of the last collector in the steam superheating section allows for a very accurate and fast steam temperature control.

The first commercial DSG plant was installed in Thailand and put into operation in 2010 [19]. It is the plant Thai Solar One (TSE-1), with a nominal power of 5 MWe. The DSG solar field of this plant produces superheated steam at 340°C/3.4 MPa.

Production of saturated steam is easier than superheated steam, because the solar field can be operated with a feed water flow at solar field inlet higher than the saturated steam flow to be delivered by the solar field, and the excess of liquid water is then separated from the saturated steam at the solar field outlet, so that solar radiation variations will only modify the amount of saturated steam and liquid water at the solar field outlet, and the steam pressure and temperature will therefore change only a little. With an excess of liquid water into the preheating and evaporating sections the control system required for saturated steam production is simpler than for superheated steam.

Ball joints for DSG systems up to 10 MPa/500°C have been successfully tested at PSA and technology for compact water/steam separators is also available. However, TES with phase-change materials have been developed and tested for small capacities only (≤5 MWh).

The use of DSG PTCs to feed industrial heat processes (IHP) is nowadays a very interesting application for this technology because this type of commercial applications usually requires saturated steam at low or medium working pressures, and the state-of-the-art of the DSG technology is already suitable to undertake these

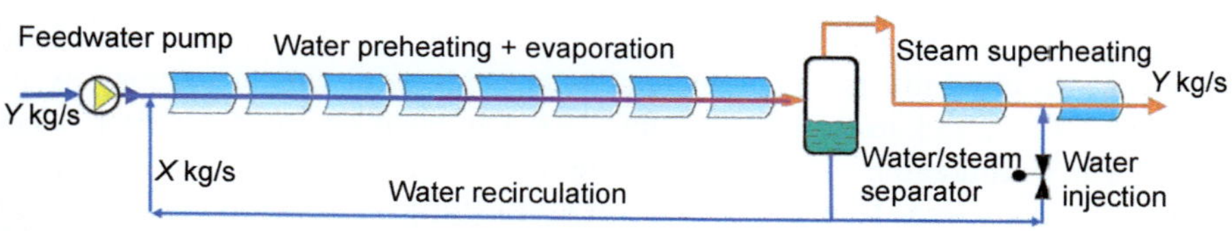

Figure 5.6 The best configuration for a DSG collector row.

applications. Since the solar field size required for IHP applications is usually much smaller than for STE plants, the pressure drop in long steam headers does not exist in these small- or medium-size applications, while the production of saturated steam simplifies the solar field control, and the capacity of the thermal storage systems required is usually small, thus making feasible the use of the PCM technology already tested successfully for capacities of about 5 MWh.

The DSG technology, including TES with PCM, is therefore already available for small IHP applications demanding saturated steam. For large DGS solar fields producing superheated steam, further R&D activities are needed in order to optimize the solar field control. The feasibility of replacing control valves by orifice plates to avoid flow instabilities during solar radiation transients must be investigated, together with the development of commercial compact liquid-water/steam separators for different mass flows in order to decouple the pre-heating and evaporating sections from the steam superheating section.

Another interesting R&D topic related to DSG is the investigation of control systems and operation strategies for superheated steam production without water/steam separators inserted in the solar field (the so-called *once-through* solar field configuration).

R&D activities are also needed to develop TES systems with PCM suitable for large storage capacities ($\geq$1 GWh). Availability of this type of TES is essential for the commercial deployment of large STE plants using the DSG technology.

5.3 Molten salts

As depicted in Table 5.1, the replacement of thermal oil by molten salts has several benefits. The first one is the possibility to work at higher temperature in the solar field (550–560°C) thus overcoming the thermal limit imposed by current thermal oils. This higher temperature at the solar field not only increases the efficiency of the PCS thermodynamic cycle, but it also enhances the integration of the TES into the plant because the heat exchangers (HXs) acting as interface between the TES and the solar field are no longer needed. Additionally, the investment cost of the TES is significantly cut down because the inventory of storage media is reduced by 64% due to the higher temperature difference between the cold and hot tanks (from 290°C to 550°C) instead of 90°C (from 290°C to 380°C) with thermal oil. Fig. 5.7 shows the simplified scheme of an STE plant using molten salt in the solar field. A direct comparison with the plant scheme for STE plants using thermal oil (see Fig. 5.1) clearly shows the simplification of the interface between the solar field and the TES.

Molten salts have high density, high heat capacity, high thermal stability, and very low vapor pressure even at elevated temperatures, what are important features for a heat transfer fluid. However, working temperatures higher than 400°C with current molten salts require the use of stainless steel in piping, vessels, and elements, because of corrosion issues when carbon steel is used for sodium and potassium nitrates at this temperature [15].

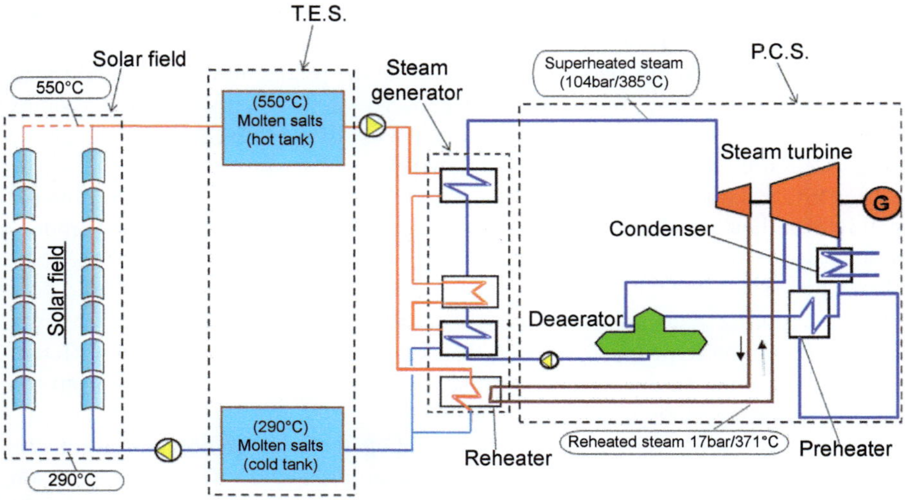

Figure 5.7 Simplified scheme of an STE plant with molten salt as working fluid.

Another benefit of molten salts is the avoidance of the fire and contamination hazards due to leaks of thermal oil in piping and vessels, because molten salts currently used (binary mixture of potassium nitrate and sodium nitrate) have been traditionally used by farmers as fertilizer. Furthermore, when molten salt is poured onto the ground it freezes immediately and remains as a thick solid film that can be easily recovered and reused, thus avoiding the high decontamination costs associated with oil leakages on ground.

Parasitic pumping losses are lower with molten salt because the volumetric flow and the pressure drop are lower than with thermal oil. This is another benefit of using molten salts in the solar field.

Nevertheless, the use of molten salts in STE plants with PTCs also has some problems. The most outstanding problem is the significant freezing hazard due to the high melting point of the current salts. The salt used in commercial STE plants is not an eutectic mixture of potassium and sodium nitrate, and therefore the freezing and melting processes take place in the temperature range from 220°C (melting) to 240°C (freezing). This high freezing temperature introduces a significant hazard of salt plug formation in the pipes when the plant is not operating and the ambient temperature is low. It has to be taken into consideration that in a 50 MWe STE plant with 1 GWh thermal storage there are more than 80 km of receiver tubes and interconnection piping in the solar field. The existence of a fluid with a freezing point higher than 225°C in such a complex and large circuit is rather risky.

Salt crystallization inside the pipes, valves, or any other element installed in the salt circuit (e.g., flow meters and pressure sensors) can lead to a very dramatic situation concerning not only the availability of the salt system, but also concerning its integrity. Once a salt plug is formed, the first problem is the difficulty to find its location. This search for the location of the plug can take a lot of time, and it could even be an impossible task when the circuit layout is very complex (e.g., a high number of valves, pipe fittings, and accessories). If the search is successful, the second problem to face is the melting of the salt plug without damaging the element at that place. Salt-specific volume increases by 4–6% during the melting process, and this volume increase will produce an extremely high overpressure if melting does not start at either of the ends of the plug. If melting starts inside the salt plug, the overpressure caused by the volume increase is likely to burst the pipe or valve casing. The design and implementation of an efficient heating system is the only way to prevent salt from freezing and to avoid these problems.

Two different electricity-based concepts are used in solar plants for heating pipelines and components in salts systems: impedance heating and mineral insulated heat tracing. Impedance heating directly heats the pipeline by flowing electrical current through the pipeline wall (Joule effect) by direct connection to a low voltage, high current source from a dual-winding power transformer. This concept heats uniformly the pipe and it is best suited for long straight runs of pipe without components, and it is critical that piping and instruments installed in an impedance system are not grounded. Even though high currents are flowing through the pipes, there is no safety hazard for personnel because the voltage is low. Impedance heating is the option used nowadays for the receiver pipes of PTCs using molten salt as working fluid, because the glass cover surrounding the inner steel pipe makes the use of heat tracing unfeasible.

Mineral-insulated electrical heat tracing (EHT) is nowadays the most common option in molten salt applications to maintain the temperature of pipelines (not the receiver tubes) and components above the salt freezing point, typically at 290°C.This option requires a more careful design and installation than an impedance system, but it represents no risk for the instrumentation or personnel. The electrical consumption of impedance systems in the steady state at 290°C is higher than in heat tracing systems, mainly due to higher electrical losses in the feed cables and power transformer [26].

Other options to avoid salt freezing inside the piping and vessels are under study (skin effect, microwaves, magnetic fields, and so on) but the most developed and installed options are the EHT or impedance systems. The existence of these systems increases the electricity self-consumption of the plant to an extent that in cold winter months with cloudy days the net electricity production of the STE plant could be negligible because most of the electricity produced by the plant would be consumed by the anti-freezing system.

The need for an anti-freezing system in all the solar field piping, instruments, and receiver tubes makes the solar field design more complex and expensive. Fig. 5.8 shows an example of heat tracing system (a) and impedance system (b).

The freezing point of the eutectic mixture of potassium and sodium nitrates (221°C) is a bit lower than the mixture used nowadays with the name of *solar salt* (240°C), but the cost would be higher because the percentage of potassium nitrate in the eutectic mixture (46%) is higher than in the *solar salt* (40%), and potassium nitrate is more expensive than sodium nitrate. The reason why the composition of the so-called "solar salt" is different from the eutectic mixture is only the cost.

(a) (b)

Valve provided with heat-tracing system in a molten-salt circuit

Connection of an impedance system to the receiver tubes of a PTC

Figure 5.8 Detail of a valve with electrical heat-tracing system (a) and receiver tube with impedance system (b) to avoid salt freezing.

Another drawback of *solar salt* is its incompatibility with graphite components used in the sealing and packing of valves and instruments, because it dissolves graphite and it ages in contact with oxygen.

At present there are several salts available for PTCs, with different melting points and maximum working temperature. Table 5.2 shows the components, melting point, and maximum working temperature of the three more common salts commercially available for solar applications [28].

Table 5.2 Main salts currently available for STE plants

Salt	Components	Melting point (°C)	Maximum working temperature (°C)
Hitec XL	Sodium nitrate Potassium nitrate Calcium nitrate	140	500
Hitec	Sodium nitrate Potassium nitrate Sodium nitrite	142	538
Solar salt	Sodium nitrate Potassium nitrate	240	593

One of the first detailed studies of a molten salt heat transfer fluid in a parabolic trough solar field was published in 2003 [18]. A techno-economic assessment of the potential electricity cost reduction that could be achieved with the replacement of thermal oil by two different molten salts (i.e., the so-called *solar salt* and Hitec XL) in PTCs was carried out in that study for a 50 MWe (net power) STE plant provided with a 6-h two-tank TES system. The same salt was assumed for both the solar field and the TES. The main conclusion of that study was that in the most favorable case an LEC (levelized electricity cost) reduction of US $10/MWh could be achieved with a solar field outlet temperature of 450°C, and US $15/MWh for a temperature of 500°C.

5.3.1 Thermo-hydraulic aspects

Thermo-hydraulic aspects are not expected to be a problem if molten salts are used as working fluid in PTCs. The density of molten salts is more than two times higher than the density of thermal oils. Hence, for instance, the densities of solar salt and VP-1 thermal oil at 350°C are 1868 and 761 kg/m^3, respectively. This high density reduces the pressure loss in the solar field and the corresponding pumping loses, while keeping a good heat transfer coefficient inside the receiver tubes. In fact, one of the benefits of using molten salt is the low pressure loss in the solar field, which significantly reduces the electricity consumption for pumping, thus increasing the overall plant efficiency [18].

The temperature rise in a solar field with molten salt would be of about 200°C (assuming an output temperature of 500°C), while with thermal oil it is 100°C only. The typical temperature rise in a single collector with molten salt would therefore be two times higher than with oil. In commercial STE plants the solar field is composed of parallel rows with four 150-m collectors or six 100-m collectors connected in series in each row. If six 100-m collectors connected in series is assumed, the typical temperature rise in a single collector with oil is 17°C, while it would be 34°C with *solar salt*. Table 5.3 shows the pressure drop, the pumping power required, the heat transfer coefficient and the net thermal gain in a Eurotrough-100 collector working with *solar salt* and with VP-1 oil under the same solar conditions (DNI = 900 W/m^2 with an incidence angle of 20 degrees) and ambient temperature (20°C). It is shown in the table that the mass flow and pressure loss is lower with molten salt, and both effects combined—low mass flow and low pressure loss—lead to relatively low pumping power, which is seven times lower than with oil. The thermal gain with molten salt is a bit lower due to higher thermal loss in the collector. However, the higher thermal loss is more than compensated by the much higher efficiency of the PCS.

Data given in Table 5.3 clearly depict that no thermo-hydraulic problem is associated with the use of molten salt. The low pressure loss and pumping power in the solar field if molten salt is used makes feasible the use of collector rows with more PTCs connected in series. A total row length of 800 m with a mass flow per row of about 8 kg/s is fully feasible concerning pressure loss and pumping power.

Table 5.3 Mass flow, temperatures, pressures, and pumping power using solar salt and thermal oil VP-1 in an ET-100 parabolic trough collector, with receiver tubes of ϕ_{ext} = 70 mm and wall thickness of 2.5 mm, and direct normal irradiance of 900 W/m^2 with an incidence angle of 20 degrees

Mass flow and inlet temperature	Outlet temperature (°C)	Density (kg/m^3)	Pressure loss (Pa)	Pumping power (kW)	Convection heat transfer coefficient at inlet–outlet (W/m^2·K)	Thermal gain (kW)
Solar salt 6.0 kg/s T_{in} = 300°C	334	1890	3.01×10^4	0.096	2770–3060	302
VP-1 7.5 kg/s T_{in} = 300°C	317	804	7.69×10^4	0.717	5110–5276	303

5.3.2 State of the art of molten salt as heat transfer fluid in parabolic trough collectors

At the end of 2015 there were 22 commercial STE plants using molten salt in the thermal storage system, and only one plant was using molten salt as working fluid in the solar field. This plant is the 5 MWe Archimede solar plant, owned by ENEL and built at Priolo Gargallo near Syracuse in Sicily, Italy. The solar field of this plant has inlet–outlet temperatures of 290°C/550°C, and it is composed of nine parallel loops with six 100-m and 590 m^2 PTCs in each loop. Although this plant was inaugurated in July 2010, no public information has been issued with O&M data (e.g., gross/net electricity production, solar field performance and investment, and O&M costs). The actual performance of this plant is unknown, and this lack of information could be somehow discouraging because when a first-of-its-kind plant is successfully built and operated, information is rapidly disseminated in congresses and specialized magazines.

Also in Italy and since July 2013, a demo plant with PTCs using *solar salt* is in operation at Massa Martana (Italy), close to the receiver tubes manufacturing plant of the company Archimede Solar Energy (ASE). This plant has no PCS, and it is provided with a steam generating system to evacuate the thermal energy delivered by the solar field, which is composed of a single loop with six 100-m and 590 m^2 PTCs connected in series. The information published by ASE at the end of 2015 did not cover a complete year and it did not include information about the internal electricity consumption, which would be very useful to evaluate the commercial feasibility of PTCs with molten salts.

Since the high melting point is the main disadvantage associated with molten salts when used in solar fields with PTCs, a great effort has been devoted during the past years to develop new salts with low melting point and good thermal stability. As a result of this R&D effort investigating new salt mixtures the development of a new salt fulfilling these objectives has been announced from time to time. However, the economic benefit of using salt with lower freezing point is not fully clear, because a techno-economic comparison between two 50 MWe STE plants with 7.5-h TES using two different salt mixtures as working fluid in the solar field: (1) a binary mixture of potassium and sodium nitrates with a freezing point of 238°C (the so-called *solar salt*) and (2) a ternary mixture of potassium, lithium, and sodium nitrates with a freezing point of 120°C showed that [32]:

- thermo-physical properties of both salts are very similar, the only significant difference is their freezing point;
- the cost of electricity produced by both plants is practically the same, because the lower thermal loss at the solar field using the ternary mixture is compensated by the higher cost of the ternary mixture and the HXs required between the solar field and the TES (solar salt is used in the TES of both plants due to its lower cost and therefore no HX is needed between the solar field and TES when the binary mixture is also used in the solar field);
- plant operation is easier in the case of *solar salt* at both the solar field and TES;
- it is extremely difficult to precisely determine the influence over a plant lifetime of the theoretical advantage of using a ternary low-melting mixture in the solar field during maintenance operations (e.g., pipelines draining and filling); experience gained at the 5 MWe Archimede solar plant, owned by ENEL, is essential to this extent, but such information is not public.

In this study, the temperatures considered for the cold/hot storage tanks were: 290°C/550°C when *solar salt* was assumed in the solar field, and 280°C/540°C when the ternary mixture was assumed in the solar field. These storage temperatures imply a temperature rise of about 260°C in the solar field for both cases. The results obtained by Sau [32] clearly show that development of new salt mixtures with low melting point will be less useful if the new salt is much more expensive than the solar salt. The R&D activities aimed at developing new salt mixtures must therefore have a twofold objective: development of new salt mixtures with lower melting point without increasing the cost too much. This constraint is probably the reason why the so-called *solar salt* and the commercial ternary mixture called Hitec XL are still considered the best options commercially available nowadays for STE plants, despite the new salt mixtures that are announced from time to time. An example of innovative salt mixture that has not achieved commercial success is the novel mixture developed by Ref. [28] with a low melting point (65°C) and high thermal stability (up to 560°C) The composition of this advanced salt mixture is:

Component	Weight percent (%)	Mole percent (%)
$LiNO_3$	8	15
$NaNO_3$	6	10
KNO_3	23	30
$CsNO_3$	44	30
$Ca(NO_3)_2$	19	15

Since the cost of raw materials for this salt mixture is considerably higher than the simple binary solar salt, it has not been adopted by the STE sector so far.

There are also many R&D groups investigating the possibility to improve the thermo-physical properties of salt mixtures using nanoparticles. It has been proven at laboratory scale that the heat capacity and thermal conductivity can be significantly improved by adding a small percentage ($\leq$1.0 wt.%) of silica or alumina nanoparticles [5].

As explained in Section 5.3.1 with simulation results, the main barrier at present to use molten salt in PTCs are not the thermo-hydraulic aspects, but the uncertainties associated with the O&M requirements over a plant lifetime of a large solar field piping filled with a fluid of high freezing point.

5.4 Compressed gases

The use of pressurized gases is another option to replace thermal oil in PTCs. They are safe and clean fluids from an environmental viewpoint because they do not pose the fire hazards and environmental constraints associated with thermal oils, and they are able to work at higher temperatures in the solar field without thermal stability problems. Additionally to the higher efficiency of the power block due to higher working temperatures at the solar field, higher temperature differences between solar field inlet and outlet would enhance and make more efficient the integration of TES systems based on latent heat (e.g., two-tank molten salt systems), increasing the storage capacity per volume and thus reducing the amount of storage medium. The potential benefits of using pressurized gases are therefore evident.

However, pressurized gasses also have some disadvantages when compared to thermal oils. The lower density of compressed gases reduces the heat transfer coefficients in the receiver tubes and increases the required pumping power. The only way to reduce the negative impact of a lower fluid density is increasing the working pressure at the solar field piping and associated components (i.e., HXs and vessels), because the required pumping power of gases to compensate circuit pressure drop is inversely proportional to the square of the pressure [24]. This is the reason why a high working pressure is needed to avoid excessive pumping consumptions in circuits with gases.

The evaluation of the technical and economic feasibility of pressurized gases as HTF was experimentally addressed by CIEMAT in the *gas-cooled solar collectors* project [29]. This project included the analysis of several working gases (helium, CO_2, N_2, and air) and the design, construction, and testing at PSA of the first-of-its kind experimental solar facility composed of two 50-m parabolic-trough collectors connected to a complete hydraulic circuit using pressurized gases. Fig. 5.9 shows a photograph and the diagram of the gas-cooled solar test facility installed at the PSA in 2008 to investigate the feasibility of pressurized gases in PTCs. The two solar collectors can be connected in series or in parallel, depending on the intended outlet temperature (in series for 525°C, or in parallel for 400°C), while the nominal inlet temperature was 225°C in both cases.

The solar field was connected to a two-tank molten salt TES in order to evaluate the performance of not only the solar collectors with pressurized gas, but also the overall performance of the system composed by the solar field and the TES. This facility delivered very interesting and useful results related to the feasibility of pressurized gases for PTCs [23,24,29]. A complete description of this facility can be found in Ref. [29].

The requirements in the selection of the proper gas involve thermo-hydraulic and economical characteristics, keeping in mind the feasibility of a commercial plant. This principle leads to the gases that have been extensively used in the industry for similar working conditions. The state-of-the-art drives to reduce the selection to He, CO_2, and N_2 (or air, with similar thermo-hydraulic properties to N_2). Table 5.4 shows the main thermo-physical properties of these gases. Considering thermal conductivity and specific heat, He seems to be a good option. However, He must be discarded because the small size of its atoms is a serious barrier to avoid leaks, and its high cost makes it unaffordable for large systems. The other two gases (CO_2 and N_2) are widely commercialized, but in the sense of working experience and thermo-hydraulic behavior, CO_2 shows a higher density at the same values of pressure and temperature. This provides lower pumping consumptions [24], increasing in this way the overall efficiency of the installation. This is the reason why CO_2 is considered

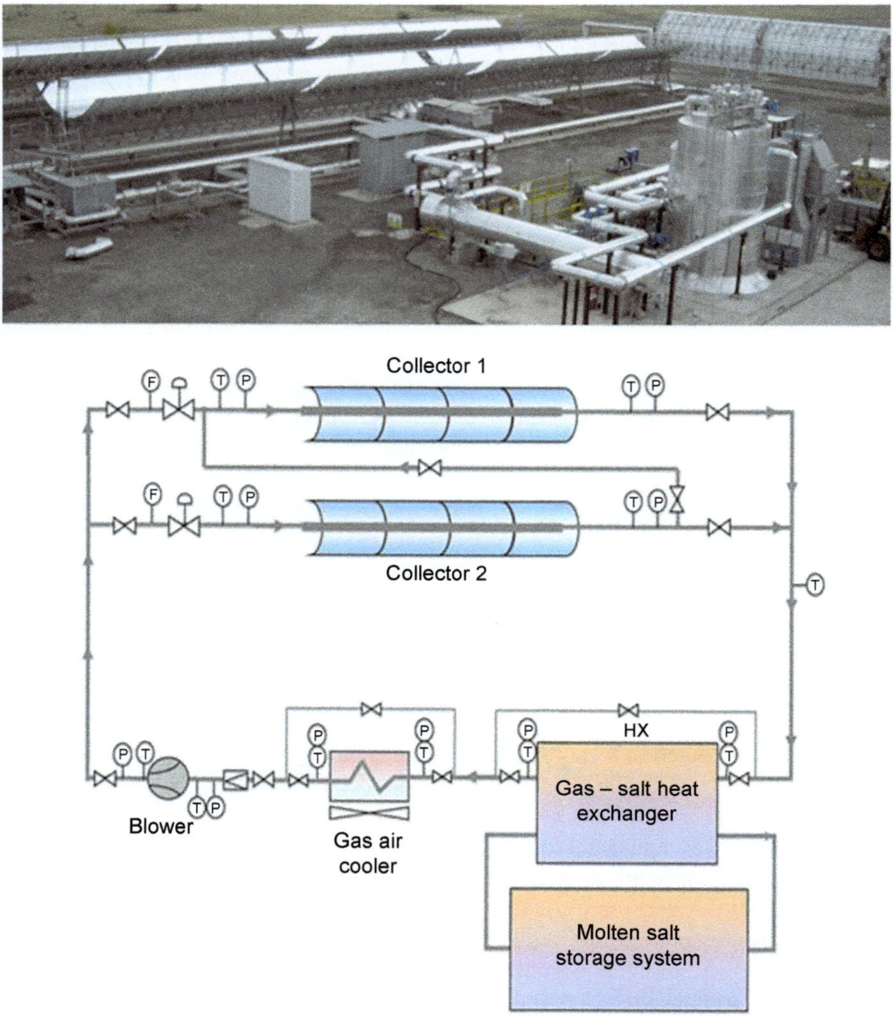

Figure 5.9 Scheme of the PSA test facility to evaluate the feasibility of pressurized gases in PTCs under real solar conditions.

nowadays the best candidate for using pressurized gas as heat transfer fluid, especially when working at supercritical conditions ($T >$ 31°C and $P >$ 7.4 MPa).

However, CO_2 also has some problems. In case of water presence, CO_2 could react to form carbonic acid (H_2CO_3), which is corrosive to carbon steel. Since carbon steel is widely used in solar plants, the use of CO_2 would imply that the water content should be carefully controlled to avoid formation of carbonic acid due to humidity condensation inside the piping and vessels during cold periods. Another major problem found at PSA when working with supercritical CO_2 at temperatures higher than 400°C was its incompatibility with the graphite sealing used in the ball joints installed at the end of

Table 5.4 **Gases of interest: thermal properties at 100 bar and 400°C [21,33,34]**

Gas	Density (kg/m^3)	Thermal conductivity (W/m °C)	Viscosity (kg/m s)	Specific heat (J/kg °C)	Prandtl (−)
Carbon dioxide	78.83	0.04968	$31.48 \cdot 10^{-6}$	1171	0.74
Helium	7.024	0.27840	$35.07 \cdot 10^{-6}$	5186	0.65
Nitrogen	47.96	0.05074	$32.58 \cdot 10^{-6}$	1114	0.71

the receiver tubes of PTCs to allow their thermal expansion and rotation [29]. The graphite sealings were damaged very soon at working temperatures higher than 400°C. However, no problem was found for temperatures lower than 400°C. Although the origin of this problem is not explained in the existing bibliography, it is probably due to the fact that CO_2 always behaves as a "nonpolar " solvent that selectively dissolves the lipids that are water-insoluble compounds such as vegetal oils, butter, fats, and hydrocarbons.

Although air does not have the strong drying requirements of CO_2, it would still require preventive measures in order to limit the water content in the circuit. On the contrary, N_2 presents less corrosion issues than either CO_2 or air and can be obtained on site in a simple way from compressed air by means of a nitrogen generator. Therefore, nitrogen could be the most feasible HTF among the proposed gases for PTCs.

The overall costs of a large solar field with nitrogen as HTF will be higher than that of a synthetic-oil field due to the increasing cost of distributed blowers and HXs. On the other hand, higher temperature differences in TES imply a lower molten salts inventory and therefore the cost of storage will be lower for the N_2 plant. Moreover, the cost of synthetic oil and the corresponding equipment (pumps, expansion vessels, conditioning system, and so on) will be replaced by a lower cost of nitrogen.

5.4.1 Thermo-hydraulic aspects

The two main thermo-hydraulic aspects that must be considered when evaluating the feasibility of pressurized gases as HTF in PTCs are: (1) pressure drop in the piping, and (2) convection heat transfer coefficient in the receiver tubes. As explained in Section 5.2.1, the pressure loss has a direct and great impact on the overall system efficiency due to the pumping power required to overcome the pressure loss, while the heat transfer coefficient affects the lifetime of the receiver tubes and their thermal efficiency.

It can be theoretically demonstrated [24] that if working pressure is increased 10 times, the pumping power decreases 100 times. Therefore, pumping consumption using pressurized gas can be kept at a reasonable level rising the working pressure. However, the length of the piping is also a major parameter influencing the overall solar field pressure loss, as depicted by (Eq. 5.1), and both parameters (working pressure and piping length) must be carefully defined during the design phase to keep the overall pressure loss in the solar field at an affordable level.

Table 5.5 shows for a mass flow of 1.1 kg/s of CO_2 and N_2 the pressure loss (ΔP), the convection heat transfer coefficient in the receiver tubes (h), the pumping power and the maximum temperature gradient in cross sections of the steel receiver tubes of a ET-100 PTC composed of eight 12.2 × 6.76 m modules connected in series and provided with ϕ70 mm steel receiver tubes made of stainless steel (thermal conductivity 18 W/m·K) with a wall thickness of 6.5 mm to withstand the high temperatures and pressures. Table 5.5 shows that pumping power is divided by 10 (approximately) if the working pressure is increased from 2.5 MPa to 7.5 MPa. However, even for high pressures the pumping power with CO_2 is much higher than for thermal oil. The pressure loss in a ET-100 PTC using Therminol VP-1 thermal oil with a mass flow of 5 kg/s and an inlet temperature of 325°C is 3.6×10^4 Pa, which demands a pumping power of 0.24 kW only (45% of the pumping power with CO_2 and 15.5% of the pumping power with N_2, both at 7.5 MPa). This great difference between thermal oil and pressurized CO_2 or N_2 concerning the required pumping power is even more important when the complete solar field piping is taken into account for pressure loss calculation. The influence of fittings and ball joints used at the interconnections between adjacent collectors is also very important when calculating the pressure loss, thus demanding an optimized design of the collector rows to keep the pumping power at a reasonable level when using pressurized gas as working fluid in the solar field.

The many simulation studies performed to find an optimum layout for parabolic trough solar fields with pressurized gas to keep the pressure loss at an affordable value have led to a modular design [3] composed by basic collector loops connected in parallel to a distribution system that transfers the thermal energy delivered by the modules to a central thermal storage or steam generating systems using molten salt as heat carrier and storage media. Fig. 5.10 shows the optimized collector loop recommended for solar fields using pressurized gases. With this approach, each collector loop will transfer its thermal energy to a molten salt circuit acting as a heat carrier between the modules and the central thermal storage system or the HX where the superheated steam is produced for the power block. Each collector module is composed of four 100-m PTCs (ET-100 type), one blower to pump the pressurized gas, three HXs to transfer the thermal energy from the module to the molten salt distribution circuit, and the connecting pipes between components. In nominal operation, the pressurized gas is heated through each collector from 310°C up to 525°C and then driven through gas/salt HXs, transferring the useful thermal power, 330 kW in each end HX and 660 kW in the central HX, to the molten salt distribution circuit. With this layout, the pressure drop is kept low enough not to penalize the overall plant efficiency. The use of a molten salt circuit to carry the heat from the modules to the central storage system avoids the extremely high pressure loss that would occur if the pressurized gas is circulated through the piping connecting the collectors to the thermal storage system.

The reason why the pressurized gas is heated from 310°C to 525°C in only one 100 m collector is the high pressure loss that would result if the temperature step is

Table 5.5 Temperatures, pressures, and maximum temperature gradient in cross sections using CO_2 in an ET-100 parabolic trough collector, with receiver tubes of ϕ_{ext} = 70 mm and a wall thickness of 6.5 mm, and direct normal radiation of 900 W/m^2

Mass flow and inlet temperature	Outlet temperature (°C)	Inlet pressure (MPa)	Pressure loss (Pa)	Pumping power (kW)	Convection heat transfer coefficient at inlet-outlet (W/m^2·K)	Max. ΔT in cross sections (°C)
CO_2 1.1 kg/s T_{in} = 225°C	507	2.5	12.9×10^4	5.24	980–1115	51
	500	5.0	6.3×10^4	1.26	1006–1122	50
	495	7.5	4.1×10^4	0.53	1034–1129	50
N_2 1.1 kg/s T_{in} = 225°C	505	2.5	21.1×10^4	13.86	1005–1064	50
	503	5.0	10.3×10^4	3.42	1016–1067	50
	502	7.5	6.9×10^4	1.55	1026–1071	50

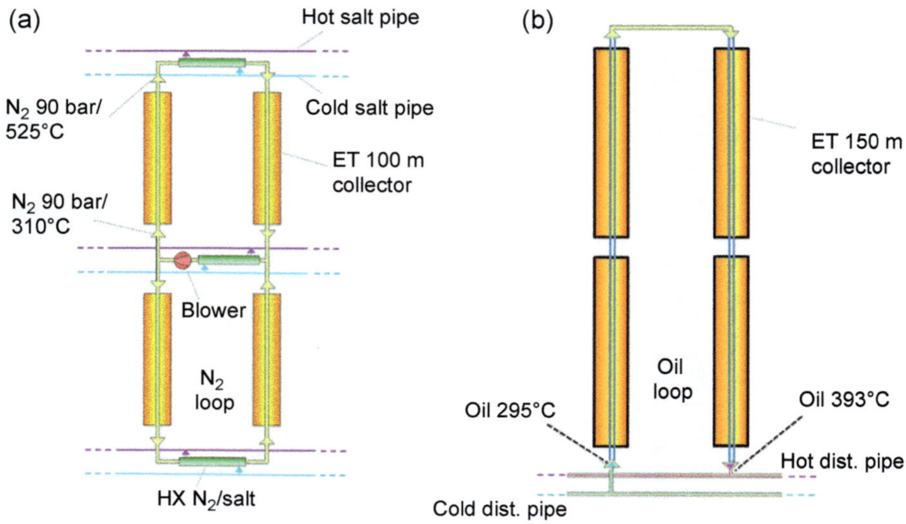

Figure 5.10 (a) Simplified scheme of an optimized collector module for solar fields using pressurized gases and Rodríguez, M.M., Márquez, J.M., Biencinto, M., Adler J.P., Díez L.E. First experimental results of a solar PTC facility using gas as the heat transfer fluid. In: SolarPACES 2009 Conference. Berlin, Germany. (b) a typical collector module of commercial STE plants using thermal oil (Biencinto et al., 2014).

achieved in a 150 m collector or even in two 100 m collectors connected in series. The gas flow and therefore the gas velocity would have to be increased if the collector length is increased, thus increasing the pressure loss significantly according to (Eq. 5.1). Hence, the configuration shown in Fig. 5.10(a) has a limited pressure loss, while keeping a reasonably good convection heat transfer coefficient at the inner wall of the receiver tubes.

5.4.2 State of the art of pressurized gases as heat transfer fluids in parabolic trough collectors

With the optimized solar field design composed of many modules (Fig. 5.10(a)) connected in parallel to a molten-salt circuit, the overall plant efficiency is similar to the efficiency of a traditional STE plant with thermal oil. Several simulation studies have been performed during the period 2009–15 to compare the efficiency of a traditional STE plant with thermal oil and an STE plant with pressurized gas. Due to the problem detected at PSA with the graphite sealing of ball joints when compressed CO_2 at temperatures above 400°C is used, the use of this gas is not foreseen at present for PTCs, and all these simulation studies have been performed assuming pressurized N_2 in the solar field.

A detailed performance model and annual yield comparison of parabolic trough STE plants with either pressurized nitrogen or synthetic oil as heat transfer fluid can be found in Biencinto et al. [4]. Two STE plants with a nominal power of

50 MWe, one using pressurized N_2 and another one using Therminol VP-1 were modeled in detail, and their performance was simulated during a whole year. A 6-h capacity thermal storage system with molten salts in a two-tank system was assumed for both plants. The net collection area (388.233 m^2) and the type of collector (ET-100) assumed for both plants were also the same. Pressure loss in the solar field was simulated with a detailed piping model of both plants. The main conclusions from this study are the following:

- The net electricity production of both plants was very similar (only 0.91% lower for the N_2 plant).
- The annual parasitic consumption in the case of N_2 is 4.6% higher than in the VP-1 plant.
- The higher parasitic consumption of the N_2 plant is compensated by the higher efficiency of its power block due to its higher nominal temperature.
- Higher yields can be expected for the N_2 plant in summer months.

In summary, that study proved that similar annual performances can be attained for parabolic plants using synthetic oil and compressed N_2. However, there are significant uncertainties concerning the investment cost for the N_2 plant, because the high number of blowers (177) and HXs (531) required in the solar field are likely to increase both the investment and the O&M costs. In order to improve this technology, future works should investigate new gases as heat transfer fluids with better thermo-hydraulic properties than N_2. Also development of commercial PTCs with larger aperture area and bigger receiver tubes could help to reduce the number of HXs per collector loop.

Since CO_2 has better thermo-hydraulic properties than N_2, the investigation of the incompatibility problem found with the graphite sealing of commercial ball joints is also an interesting topic for R&D.

The study of the technical feasibility of a mixture of water and N_2 seems also an interesting topic for R&D projects related to this technology, because the mixture of steam and N_2 have better thermo-hydraulic properties than N_2 alone. During cold periods overnight the water can condensate and start-up procedures would be required to avoid the damage of the blowers due to liquid water.

The pilot plant tested at PSA with pressurized CO_2 [29] showed a very user-friendly start-up and shutdown procedures, together with a good performance and easy control under solar radiation transients, so that this technology could be an interesting option for industrial heat applications demanding a moderated amount of heat, because an automatic operation procedure seems feasible.

5.5 Conclusions

Taking into account the explanations in Sections 5.2, 5.3, and 5.4 concerning the three options analyzed in this chapter (i.e., water and steam, molten salts, and pressurized gases) to evaluate their benefits and drawbacks when used as working fluids in PTCs to replace the synthetic oil currently used in commercial plants, it seems evident that none of these options is the perfect one in the short term. All of them have either important open questions that must be further investigated or technical challenges that

Table 5.6 Main technical challenges for DSG, molten salts, and compressed gases

DSG	Molten salts	Compressed gases
Latent-heat thermal storage systems with phase-change materials	New salt mixtures with lower melting point and good thermal conductivity	New gases mixtures (i.e., water and gas) with good thermos-hydraulic behavior
Mass-flow control devices cheaper than conventional control valves	Nonrotating devices to absorb thermal expansion of receiver tubes	Incompatibility between CO_2 and graphite

have not been met yet (Table 5.6). Therefore, it is foreseen that thermal oil will still be used in STE parabolic trough plants in the short term.

Nevertheless, the search for innovative working fluids and the investigation of the open questions mentioned in this chapter concerning DSG, molten salts, and pressurized gases must continue in parallel with the use of synthetic oil during the next years.

Regarding DSG process, the state-of-the-art of this technology already allows its application in industrial process heat applications demanding saturated steam within the pressure range 0.1–7.5 MPa if no thermal storage is required or its storage capacity is small. There are many industrial processes in which this technology can be coupled to providing heat in a clean and environment-friendly way. It is foreseen that development of this market niche will start as soon as public incentives are available. This type of commercial applications is very suitable for SME and would be a very useful way to reactivate local economies because most of the components can be manufactured locally.

Commercial development of small-size and compact liquid-water/steam separators to be inserted in the DSG rows of collectors to separate the end of the evaporating section and the beginning of the steam superheating section is also an interesting topic.

Development of mass flow control devices cheaper than conventional control valves is also an interesting topic because the control of the feed water flow at the inlet of each collector row seems to be necessary in order to assure a good control of the steam pressure and temperature at the outlet. At present, the high cost of conventional control valves is a barrier to be competitive with conventional solar fields using thermal oil, which do not require control valves at the inlet of each collector row.

Nowadays the most important R&D topic related to DSG is the development of a cost-effective thermal storage system based on latent heat in a temperature range suitable for PTCs ($T \leq 300°C$). The availability of such storage system would significantly enhance the commercial deployment of PTCs with DSG.

Concerning molten salts, a key R&D topic is the development of new salt mixtures with lower melting point and affordable cost, the lower the better, because molten salts currently available pose a significant crystallization risk in the receiver tubes and piping during cold overnight periods. Large commercial plants in routine operation with molten salts are not likely to be implemented in the short term with the current salts. This is the main reason why the investigation of novel salt mixtures is very important nowadays. As already depicted by Ref. [28], this R&D field is extremely large because the number of possible combinations with high-order mixtures is huge and it increases exponentially with the number of components. A eutectic mixture exhibits the lowest melting point of any similar mixture with the same components and it is quite sensitive to the weight percent of each component in the mixture. A small deviation may have a significant effect on the resulting melting point. If it is assumed that the amount of each component in a salt mixture can be controlled at 1% rate, then with a two-component system there are 101 possible combinations, with a three-component system there are 5151 combinations, and with a four-component system there are 176,851 possible combinations.

Another interesting topic concerning molten salts is the development of suitable devices to replace the ball joints used nowadays to allow thermal expansion of the receiver tubes in PTCs, because the designs used in thermal oil plants do not perform well with molten salts. It has been explained in Section 5.3 that sodium and potassium nitrates dissolve graphite, which is a rather usual raw material in the packing and sealing of ball joints and valves.

R&D activities aimed at improving the thermo-physical properties of molten salts, heat capacity and thermal conductivity mainly, are also very interesting nowadays. Results obtained at laboratory scale have proven that the addition of nanoparticles seems to be a good option to achieve these improvements.

Development of procedures to detect and eliminate salt plugs inside the piping without damaging the piping due to expansion of the salt during the melting process, is another interesting R&D topic that would enhance the commercial deployment of large solar fields with PTCs using molten salts.

For pressurized gasses, large commercial plants using CO_2 or N_2 do not seem cost-effective because of the high pumping power required. This technology seems cost-effective for small systems only, due to its easy control and operation. The use of other gases should be investigated to check the possibility to find options for Brayton cycles working at moderate temperature levels. Some studies have shown that this R&D field is very interesting [30,31].

As explained in Section 5.4.2, the study of the technical feasibility of a mixture of water and N_2 seems also an interesting R&D topic related to pressurized gases, because the mixture of steam and N_2 have better thermo-hydraulic properties than N_2 alone. During cold periods overnight the water can condensate and start-up procedures would be required to avoid the damage of the blowers due to liquid water.

References

[1] Ajona JI, Herrmann U, Sperdutto F, Farhina-Mendes J. Advanced receiver for direct solar steam (ARDISS). Final Report of Project ARDISS (E.U. JOULE II Programme contract No: JOU2-CT94-0311). May 1997.

[2] Bayón R, León J, Rojas E, Valenzuela L, Zarza E, León-Alonso J. Analysis of the experimental behavior of a 100 kWth latent heat storage system for direct steam generation in solar thermal power plants. Applied Thermal Engineering 2010;30:2643–51.

[3] Biencinto M, González L, Zarza E, Díez LE, Muñoz J, Martínez-Val JM. Modeling and simulation of a loop of parabolic troughs using nitrogen as working fluid. In: SolarPACES Conference, Marrakech, Morocco; 2012.

[4] Biencinto M, González L, Zarza E, Díez LE, Muñoz J. Performance model and annual yield comparison of parabolic-trough solar thermal power plants with either nitrogen or synthetic oil as heat transfer fluid. Energy Conversion and Management 2014;87:238–49.

[5] Chieruzzi M, Cerritelli GF, Miliozzi A, Kenny JM. Effect of nanoparticles on heat capacity of nanofluids based on molten salts as PCM for thermal energy storage. Nanoscale Research Letters 2013;8:448.

[6] Chisholms D. Two-phase flow in bends. International Journal of Multiphase Flow 1980;6: 363–7.

[7] Eck M, et al. Direct steam generation in parabolic troughs at 500°C- a German-Spanish project targeted on component development and system design. In: 14th international symposium on concentrated solar power and chemical energy technologies, Las Vegas, USA; March 3–7, 2008.

[8] Eck M, et al. Test and demonstration of the direct steam generation (DSG) at 500°C. In: Proceedings of the 15th CSP SolarPACES symposium, Berlin, Germany; 2009.

[9] Feldhoff JF, Schmitz K, Eck M, Laing D, Ortiz-Vives F, Schnatbaum-Lauman L, Schulte-Fischedick J. Comparative system analysis of parabolic trough power plants with DSG and oil using integrated thermal energy storage. In: Proceedings of the 17th CSP SolarPACES symposium, Granada, Spain; 2011.

[10] Friedel L. Improved friction pressure drop correlation for horizontal and vertical two-phase flow. 3R Internacional 1979;18(7):485–91.

[11] Feldhoff JF, Eickhoff M, Kellerb L, León-Alonso J, Meyer-Grünefeldtb M, Valenzuela L, Pernpeintnerd J, Hirsch T. Status and first results of the DUKE project – component qualification of new receivers and collectors. Energy Procedia 2014;49:1766–76.

[12] Geskes P, Eck M. Fluctuation flow patterns and pipewall temperatures in horizontal absorber pipes with direct steam generation under changing conditions. Journal de Physique IV. In: Pr3: proceedings of 9th international symposium on solar thermal concentrating technologies, held at Font-Romeu (France) on June 22–26, 1998, vol. 9. France: EDP Sciences; 1999. p. 450–95.

[13] Goebel O, Hennecke K. Investigation of thermohydraulics in a parabolic trough absorber tube with direct steam generation (DSG). In: Becker M, Böhmer M, editors. Proceedings of 8th international symposium on solar thermal concentrating technologies held at Cologne (Germany) on October 7–11, 1996. Heidelberg (Germany): C.F. Müller Verlag; 1997, ISBN 3-7880-7616-X. p. 787–814.

[14] Goebel O. Thermohydraulics of direct steam generation. Journal de Physique IV. In: Pr3: proceedings of 9th international symposium on solar thermal concentrating technologies, held at Font-Romeu (France) on June 22–26, 1998, vol. 9. France: EDP Sciences; 1999. p. 481–7.

[15] Goods S, Bradshaw R, Chavez J. Corrosion of stainless and carbon steels in molten mixtures of industrial nitrates. Technical report SAND 94-8211. 1994.

[16] Günther H. Technische Traeume (Technical dreams). Zurich: Rascher und Cie. Verlag; 1922.

[17] Hennecke K, Zarza E. Direct solar steam generation in parabolic troughs (DISS). Update on project status and future planning. Journal de Physique 1999;9(3):3-469–3-474.

[18] Kearney D, Herrmann U, Nava PO, Kelly B, Mahoney RM, Pacheco J, Cable R, Potrovitza N, Blakeand D, Price H. Assessment of a molten salt heat transfer fluid in a parabolic trough solar field. Journal of Solar Energy Engineering May 2003;125(2): 170–6.

[19] Khenisi A, Krüger D, Hirsch T, Hennecke K. Return of experience on transient behavior at the DSG solar thermal power plant in Kanchanaburi, Thailand. Energy Procedia May 2015;69:1603–12.
[20] Laing D, Bahl C, Bauer T, Lehmann D, Steinmann W-D. Thermal energy storage for direct steam generation. Solar Energy 2011;85:627–33.
[21] McCarty RD, Arp VD. A new wide range equation of state for helium. Advances in Cryogenic Engineering 1990;35:1465–75.
[22] Müller M, Lippke F. Lab-scale high-pressure water/steam test loop for research on two-phase flow phenomena in parabolic trough solar power plants with direct steam generation; thermodynamics and first operational results. In: Proceedings of 6th international symposium on solar thermal concentrating technologies, held at Mojacar (Spain) in September 1992. Madrid (Spain): CIEMAT; 1993, ISBN 8478341633. p. 695–712.
[23] Muñoz J, Zarza E, Díez LE, Martínez-Val JM, López C, Gavela R. The new technology of gas cooled trough collectors. In: SolarPACES Conference, Granada, Spain; 2011.
[24] Muñoz-Anton J, Biencinto M, Zarza E, Díez LE. Theoretical basis and experimental facility for parabolic trough collectors at high temperature using gas as heat transfer fluid. Applied Energy 2014;135(C):373–81.
[25] Murphy LM, May EK. Steam generation in line focus solar collectors: a comparative assessment of thermal performance, operation stability and cost issues. Technical report SERI/TR-632–1311. Golden (Colorado): EEUU, Solar Energy Research Institute; 1982.
[26] Pacheco JE, Kolb WJ. Comparison of an impedance heating system to mineral insulated heat trace for power tower applications. In: Proceedings of the 1997 ASME international solar energy conference, Washington, DC; 1997.
[27] Pederson RJ, May EK. Flow instability during direct steam generation in a line-focus solar-collector system. Technical report SERI/TR-632–1354. Golden (Colorado): EEUU, Solar Energy Research Institute; 1982.
[28] Raade J, Padowitz D. Development of Molten salt heat transfer fluid with low melting point and high thermal stability. Journal of Solar Energy Engineering July 2011;133(3).
[29] Rodríguez MM, Márquez JM, Biencinto M, Adler JP, Díez LE. First experimental results of a solar PTC facility using gas as the heat transfer fluid. In: SolarPACES Conference, Berlin, Germany; 2009.
[30] Rovira A, Muñoz J, Montes MJ, Martinez-Val JM. Optimization of Brayton cycles for low-to-moderate grade thermal energy sources. Energy 2013;55:403–16.
[31] Rovira A, Rubbia C, Valdés M, Martinez-Val JM. Thermodynamic cycles optimized for medium enthalpy units of concentrating solar power. Energy 2014;67:176–85.
[32] Sau S, Corsaro N, Crescenti T, Dóttavi C, Liberatore R, Licoccia S, Russo V, Tarquino P, Tizzoni AC. Techno-economic comparison between CSP plants presenting two different heat transfer fluids. Applied Energy 2016;168:96–109.
[33] Span R, Wagner W. A new equation of state for carbon dioxide covering the fluid region from the triple-point temperature to 1100 K at pressures up to 800 MPa. Journal of Physical and Chemical Reference Data 1996;25(6).
[34] Span R, Lemmon EW, Jacobsen RT, Wagner W, Yokozeki A. A reference equation of state for the thermodynamic properties of nitrogen for temperatures from 63.151 to 1000 K and pressures to 2200 MPa. Journal of Physical and Chemical Reference Data 2000;29(6).
[35] Taitel Y, Dukler AD. A model for predicting flow regime transition in horizontal and near horizontal gas-liquid flow. AIChe Journal 1976;22(1):47–55.

[36] Valenzuela L, Zarza E, Berenguel M, Camacho E. Control concepts for direct steam generation process in parabolic troughs. Solar Energy 2005;78(2):301–11. http://dx.doi.org/10.1016/j.solener.2004.05.008.

[37] Valenzuela L, Zarza E, Berenguel M, Camacho E. Direct steam generation in solar boilers. IEEE Control Systems Magazine 2004;24(2):15–29.

[38] Ven J. Direct steam generation using a water injection system. In: Faiman D, editor. Proceedings of the 5th sede boqer symposium on solar electric production. Beersheba (Israel): Ben-Gurion University of the Negev; 1993. p. 37–43.

[39] Zarza E, Valenzuela L, Eickhoff M, Weyers D-H, León J, Hennecke K, Eck M. The DISS project: direct steam generation in parabolic troughs. Operation and maintenance experience and update on project status. In: Proceeding of ASME forum 2001. Solar energy: the power to choose, held at Washington (USA); April 23–25, 2001 [Book of proceedings edited by ASME].

[40] Zarza E. Generación Directa de vapor con colectores solares cilindro parabólicos [Doctoral thesis presented at the University of Seville]. Proyecto Direct Solar Steam (DISS); 2003. Available on-line at: http://www.tesisenred.net/handle/10803/113784/browse?value=Zarza+Moya%2C+Eduardo&type=author.

A new generation of solid particle and other high-performance receiver designs for concentrating solar thermal (CST) central tower systems

6

C.K. Ho
Sandia National Laboratories, Albuquerque, New Mexico, United States

6.1 Introduction

6.1.1 Background

Higher efficiency power cycles are being pursued to reduce the levelized cost of energy from concentrating solar power-tower technologies [1]. These cycles, which include combined air-Brayton, supercritical-CO_2 (sCO_2) Brayton, and ultra-supercritical steam cycles, require higher temperatures than those previously achieved using central receivers. Current central receiver technologies employ either water/steam or molten nitrate salt as the heat transfer and/or working fluid in subcritical Rankine power cycles. The gross thermal-to-electric efficiency of these cycles in currently operating power-tower plants is typically between 30% and 40% at turbine inlet temperatures <600°C. At higher input temperatures, the thermal-to-electric efficiency of the power cycles increases following Carnot's theorem. However, at temperatures greater than 600°C, molten nitrate salt becomes chemically unstable, producing oxide ions that are highly corrosive [2], which results in significant mass loss [3].

6.1.2 Technical challenges and requirements

As discussed in Ho and Iverson [4], unique challenges associated with high-temperature receivers include the development and use of geometric designs (e.g., dimensions and configurations), materials, heat-transfer fluids, and processes that maximize solar irradiance and absorptance, minimize heat loss, and have high reliability at high temperatures over thousands of thermal cycles. Advantages of direct heating of the working fluid include reduced exergetic losses through intermediate heat exchange, while advantages of indirect heating include the ability to store the heat transfer media (e.g., molten salt and solid particles) for energy production during nonsolar hours.

Advances in Concentrating Solar Thermal Research and Technology
http://dx.doi.org/10.1016/B978-0-08-100516-3.00006-X

Ho and Iverson [4] show that a high concentration ratio on the receiver and reduced radiation losses are critical to maintain high thermal efficiencies at temperatures above 650°C. Reducing the convective heat loss is less significant, although it can yield a several percentage point increase in thermal efficiency at high temperatures (note that the convective heat loss in cavity receivers can be a factor of two or more greater than that in external receivers because of the larger absorber area [5]). Increasing the solar absorptance, α, and/or decreasing the thermal emittance, ε, can also increase the thermal efficiency.

6.1.3 Overview of chapter and introduction to next-generation receivers

This chapter provides an overview of next-generation high-temperature solar thermal receivers, including particle receivers and other high-performance receiver designs that increase solar absorptance while reducing radiative and convective losses. The particle receivers are categorized as either direct or indirect particle heating receivers. For each receiver type, the following are presented: operating principle, review of previous modeling and testing activities, expected outlet temperature and thermal efficiency, advantages, and remaining challenges.

6.2 Particle receivers[1]

Falling particle receivers are being investigated to enable higher operating temperatures (>700°C), inexpensive direct storage, and higher receiver efficiencies for concentrating solar power technologies, thermochemical reactions, and process heat [6–23]. Unlike conventional receivers that employ fluid flowing through tubular receivers, falling particle receivers use solid particles that are heated—either directly or indirectly—as they fall through a beam of concentrated direct solar radiation. Once heated, the particles may be stored in an insulated tank and used to heat a secondary working fluid (e.g., steam, CO_2, or air) for the power cycle (Fig. 6.1). Particle receivers have the potential to increase the maximum temperature of the heat-transfer media to over 1000°C. Thermal energy storage costs can be significantly reduced by directly storing heat at higher temperatures in a relatively inexpensive medium (i.e., sand-like particles). Because the solar energy is directly absorbed in the particles, the flux limitations associated with tubular central receivers (high stresses resulting from the containment of high temperature, high pressure fluids) are significantly relaxed. The falling particle receiver appears well suited for scalability ranging from 10 to 100 MW_e power-tower systems.

[1] This section is reprinted with permission from Ho CK. A review of high-temperature particle receivers for concentrating solar power. Applied Thermal Engineering [in press]. http://dx.doi.org/10.1016/j.applthermaleng.2016.04.103.

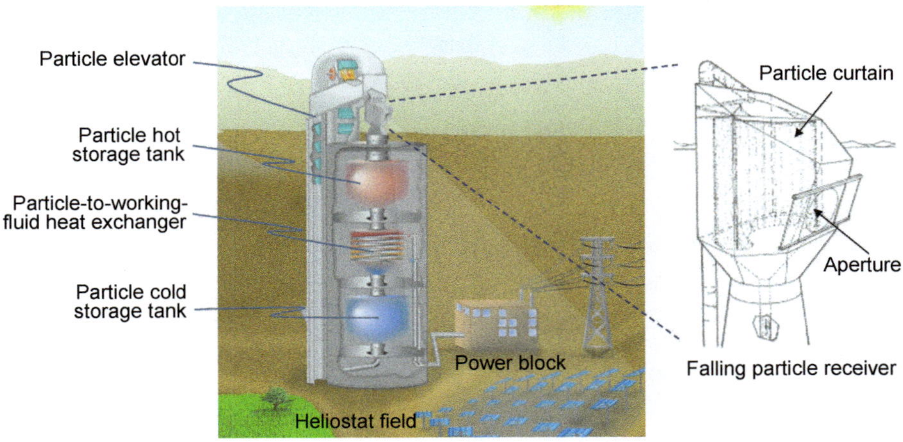

Figure 6.1 Falling particle receiver system with integrated storage and heat exchange.

Previous studies have considered alternative particle receiver designs including free-falling [18], centrifugal [20,21,24,25], flow in tubes with or without fluidization [15,22,23,26–29], multipass recirculation [9,17] north- or south-facing [6,11], and face-down configurations [30]. In general, these particle receivers can be categorized as either direct or indirect particle heating receivers. Direct particle heating receivers irradiate the particles directly as they fall through a receiver, while indirect particle heating receivers utilize tubes or other enclosures to convey and heat the particles.

6.2.1 Direct particle heating receivers

6.2.1.1 Free-falling particle receivers

The most basic form of a direct particle heating receiver consists of particles falling through a cavity receiver, where the particles are irradiated directly by concentrated sunlight. The particles are released through a slot at the base of a hopper above the receiver, producing a thin sheet (or curtain) of particles falling through the receiver (Fig. 6.2).

A number of assessments and studies have been performed on direct free-falling particle receivers since its inception in the 1980s [6–10,12–14,17–19,31–46]. In 2010, Tan et al. provided an overview of the prior research on free-falling particle receivers [19]. The majority of those studies focused on modeling the particle hydraulics and radiant heat transfer to falling particles. Various geometries and configurations of falling particle receivers have been considered, including north/south-facing cavity receivers as well as face-down cavity receivers with a surrounding heliostat field [8,30,39]. In 2008, Siegel et al. performed one of the first on-sun tests (in batch mode) of a simple free-falling particle receiver [18,47]. Those tests achieved about 50% thermal efficiency, and the maximum particle temperature increase was about 250°C. During 2015 and 2016, Ho et al. performed on-sun tests of a 1 MW_{th}

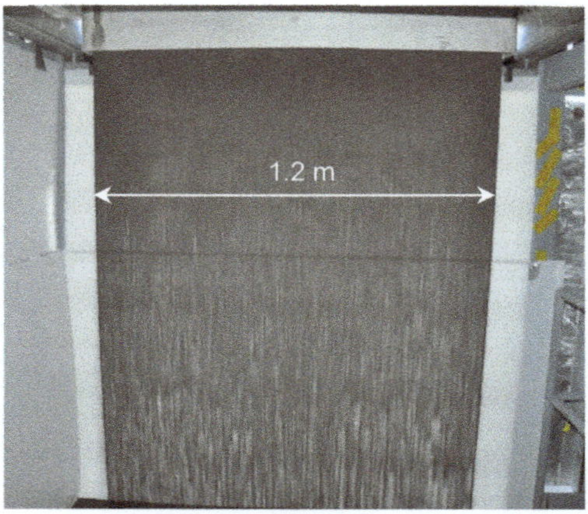

Figure 6.2 Falling particle curtain released through 1.2 m × 11.1 mm discharge slot aperture.

Figure 6.3 On-sun testing of a falling particle receiver at the National Solar Thermal Test Facility at Sandia National Laboratories, Albuquerque, NM.

continuously recirculating particle receiver with bulk particle outlet temperatures reaching over 700°C, and thermal efficiencies from about 50–80% [48,49] (Fig. 6.3). Results showed that the particle temperature rise and thermal efficiency were dependent on particle mass flow rate and irradiance. Higher particle mass flow rates yielded greater thermal efficiencies but lower particle temperature rise. As the particle mass flow rate increased (by increasing the particle discharge slot aperture size), the solids volume fraction increased and the particle curtain became more opaque. Thus, while more sunlight was intercepted and absorbed by the curtain for a greater thermal efficiency, additional shading and blocking reduced the bulk outlet temperature of the particles for a given irradiance. At higher irradiances of 1000 suns and higher, a greater amount of energy is absorbed by the particles for a given receiver size with relatively less heat loss than for lower irradiances. Technical challenges that were identified during the tests included nonuniform irradiance distributions on the particle curtain, variable mass flow rates, wind impacts, particle loss through the aperture, particle elevator reliability, and wear on the receiver walls from direct flux and high temperatures (>1000°C).

The heat gain and exit temperature of particles falling through concentrated sunlight depends on the particle mass flow and amount of time spent in the heated region of the receiver. Increasing this residence time is a critical aspect in achieving desired high temperatures. One way to increase the residence time is to recirculate the particles through the receiver multiple times, increasing in temperature over each successive drop [6,30,38]. Although particle recirculation is an attractive means to increase particle heating, additional particle elevators or conveyance systems would be required, which would increase complexity and cost. Previous studies have modeled recirculating particle flow through the receiver, but prototypes have not yet been demonstrated.

Kim et al. [13] performed tests of particles free-falling along a 3 m drop length to evaluate the influence of wind direction (induced by fans). They found that the most particles were lost through the aperture when the wind was parallel to the aperture and when the cavity depth was shallow. The least amount of particle loss occurred when the wind was oriented directly toward (normal to) the aperture. Air recirculation and air curtains have been proposed as a means to mitigate the impacts of wind on particle flow and to reduce convective losses [19,40,42,50–53]. Tan et al. [19,51–53] simulated the use of an aerowindow (transparent gas stream along the aperture) to mitigate heat loss and wind impacts in falling particle receivers (Fig. 6.4). Tan et al. [53] found that aerowindows could reduce the heat loss by up to 10% depending on external wind direction and speed. However, no tests or validation studies were performed, and few parametric analyses have been conducted to evaluate important air-recirculation parameters. Ho et al. [40,42] performed experimental and numerical studies that evaluated the impact of an air curtain on the performance of a falling particle receiver. Unheated experimental studies were performed to evaluate the impact of various factors (particle size, particle mass flow rate, particle release location, air-curtain flow rate, and external wind) on particle flow, stability, and loss through the aperture (Fig. 6.4). Numerical simulations were performed to evaluate the impact of

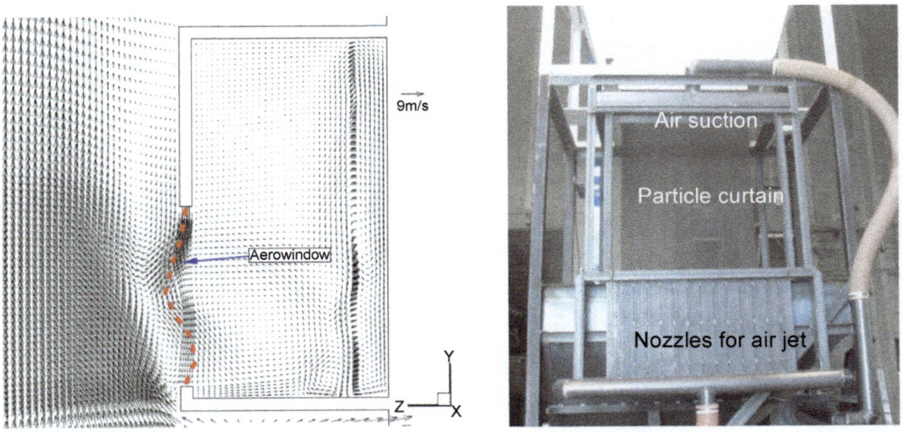

Figure 6.4 Left: Air curtain modeling for particle receivers [53]. Right: Experimental system to test air curtains for particle receivers [42].

an air curtain on the thermal efficiency of a falling particle receiver at different operating temperatures. Results showed that the air curtain reduced particle loss when particles were released near the aperture in the presence of external wind, but the presence of the air curtain did not generally improve the flow characteristics and loss of the particles for other scenarios. Larger particles and mass flow rates were also shown to reduce particle loss through the aperture. Numerical results showed that the presence of an air curtain could reduce the convective heat losses, but only at higher temperatures (>600°C) when buoyant hot air leaving the aperture was significant.

6.2.1.2 Obstructed particle receivers

Another method to increase the residence time of particles within the concentrated sunlight is to obstruct the flow with porous structures or an array of obstacles that mechanically impede their descent and slow the downward velocity while still allowing direct absorption of concentrated solar energy. Early concepts of obstructed flow designs were introduced by Sandia during the 1980s by using ceramic structures suspended from the back wall to decelerate the particles [32]. No analytical or experimental studies were published, however. King Saud University and the Georgia Institute of Technology investigated the use of interconnected porous structures (metallic or ceramic foam blocks) to slow the flow of particles [54].

Additional studies evaluated the use of a staggered array of porous mesh structures [9,55] to impede the flow of particles for increased residence time. In 2015, Ho et al. performed on-sun tests of a particle receiver consisting of a staggered array of stainless steel chevron-shaped mesh structures [48] (Fig. 6.5). Peak particle temperatures reached over 700°C near the center of the receiver, but the particle temperature increase near the sides was lower due to a nonuniform irradiance distribution. At a particle inlet temperature of about 440°C, the particle temperature increase was nearly 30°C per meter of drop length, and the thermal efficiency was about 60% for an average irradiance of 110 kW/m^2. At an average irradiance of 211 kW/m^2, the particle temperature increase was about 60°C per meter of drop length, and the thermal efficiency was about 65%. While the obstructed-flow design seemed to improve the particle heating and reduce the impacts of wind and particle loss through the aperture, there were problems with the stainless steel 316 mesh materials overheating, oxidizing,

Figure 6.5 Images of particle flow over a staggered array of chevron-shaped mesh structures [48].

and deteriorating as a result of direct irradiance from the concentrated sunlight and wear from the particles. New materials and operational strategies are being investigated to mitigate mesh deterioration.

Another obstructed flow design employs a spiral ramp along which particles flow under the influence of gravity and mechanically induced vibration [56]. Models and tests in this work demonstrated that the particles could reach the temperature of 650°C at the outlet after 30 min of radiant power of 5 kW at the aperture. The measured thermal efficiency was about 60%. This design, however, requires beam-down optics, and a significant amount of particle flow may be challenging with this design.

A final obstructed flow design that also employs beam-down optics lifts the particles upward with a screw elevator toward an aperture. The particles are irradiated by concentrated sunlight before spilling into the hollow screw for subsequent heat exchange and reaction. This particle receiver design was developed as part of a thermochemical reactor to reduce particles that are subsequently oxidized to produce either

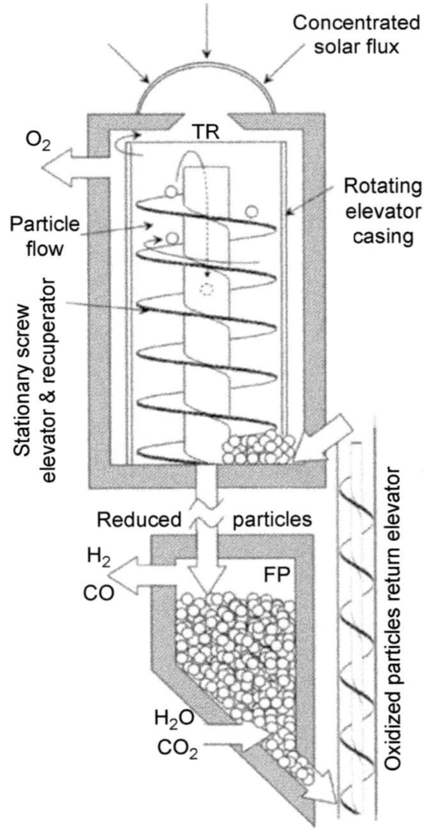

Figure 6.6 Schematic of a moving packed bed particle reactor [57].

hydrogen or carbon monoxide [57]. Fig. 6.6 shows a schematic of the receiver reactor, which also takes advantage of preheating and recuperation since the heated particles that fall through the hollow screw also preheat the oxidized particles being lifted up along the flights of the screw. While analyses have been performed to evaluate the performance and efficiency [58,59], prototype have not yet been tested.

6.2.1.3 Rotating kiln/centrifugal receivers

Rotating kilns were proposed as early as 1980 for use in solar particle heating applications [22]. The general principle is to feed particles into a rotating kiln/receiver with an aperture at one end of the receiver to allow incoming concentrated sunlight. The centrifugal force of the rotating receiver causes the particles to move along the walls of the receiver while they are irradiated by the concentrated sunlight. Early tests by Flamant et al. showed that these systems have a very high absorption factor (0.9–1), but the thermal efficiency was low (10–30%) for heating of $CaCO_3$ at particle mass flow rates of about 1 g/s. During mid-2010, Wu et al. [20,21,24,25] developed a centrifugal particle receiver design and prototype that employs a similar concept (Fig. 6.7). Small bauxite ceramic particles (~1 mm) were introduced into a rotating centrifugal receiver with different inclination angles at mass flow rates of about 3–10 g/s. The particles were irradiated using a 15 kW_{th} solar simulator with an irradiance ranging from about 300 to 700 kW/m^2. For a face-down receiver inclination and incident irradiance of 670 kW/m^2, Wu et al. reported a particle outlet temperature of 900°C and a receiver efficiency of about 75% (±4%) [20]. Challenges include maintaining a constant and sufficient mass flow rate of particles at larger scales, parasitic energy requirements, and reliability associated with a large rotating receiver system.

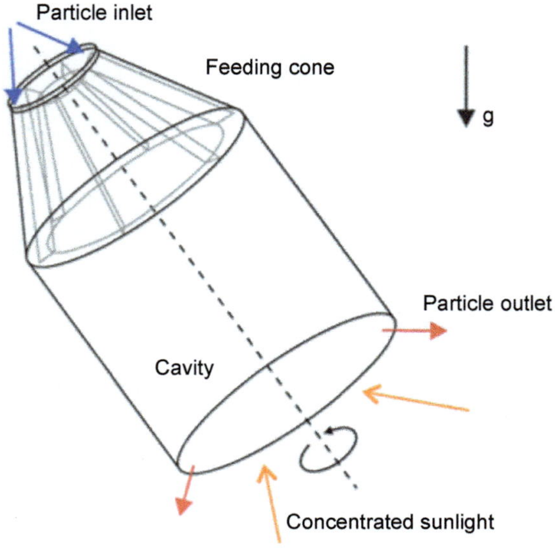

Figure 6.7 Schematic of a rotary kiln/centrifugal receiver [25].

6.2.1.4 Fluidized particle receivers

Fluidization of solid particles in a solar receiver have been proposed for several decades, beginning in the late 1970s and early 1980s by Flamant et al. for thermochemical processing and heating [22,23] and by Sandia for power production [7]. Flamant et al. [22,23] tested a fluidized-bed receiver that consisted of a vertical transparent silica tube (15 cm long × 6.5 cm diameter) that was fluidized with compressed air from the bottom and irradiated at the top. Particles that were tested included zirconia, silica sand, chamotte, and silicon carbide. For a mean flux density of about 500 kW/m^2, the measured equilibrium temperature of the particles ranged from about 1200 K for silica sand to over 1400 K for silicon carbide particles. Thermal efficiencies were reported between 0.2 and 0.4 [23]. The ability to convey the particles and achieve adequate mass flow rates (for power production or continuous processes) may pose a challenge.

During mid-2010s, researchers at the Chinese Academy of Sciences [60–62] performed numerical and experimental studies on the thermal performance of an air receiver with silicon carbide particles in transparent quartz tubes. Air is blown upward through the particles in the quartz tubes while the tubes and particles are irradiated with concentrated sunlight from a 10 kW$_{th}$ furnace (Fig. 6.8). Results of those tests showed that the heated air reached over 600°C with minimum temperature differences between the particles and the air below 10°C, indicating good heat transfer between the air and the particles.

Steinfeld et al. [29] designed and tested a fluidized-bed receiver reactor that employed a vortical flow of air in a conical-shaped receiver. The particle/gas stream was introduced near the aperture, where concentrated sunlight entered the receiver and heated the swirling particles before the particles exited the receiver. The prototype reactor was tested to evaluate the thermal decomposition of calcium carbonate at 1300 K. The mean thermal absorption efficiency was 43% with a peak flux of about 1400 kW/m^2 at the aperture.

Figure 6.8 Images of testing of a quartz-tube particle air receiver [60].

A final type of fluidized particle receiver involves the use of very small carbon particles dispersed in air that flows through the receiver. Concentrated sunlight irradiates and oxidizes the carbon particles, which volumetrically heats pressurized air passing through the receiver for high-temperature Brayton cycles. Abdelrahman et al. [63] and Hunt [64] first introduced this concept in 1979, and Hunt and Brown [65] performed tests on a prototype receiver that heated the air to 1000 K. Miller and Koenigsdorff [66,67] developed theoretical analyses and thermal modeling of the small particle solar receiver. Additional modeling and design optimization of the small particle heating receiver were performed during mid-2010s as well [68–71]. Potential advantages include the following: solar radiation is absorbed throughout the gas volume due to the large cumulative surface area of the particles; higher incident fluxes with no solid absorber that can be damaged; particles are oxidized leaving a particle free outlet stream [66]. Challenges include the development of a suitable window for the pressurized receiver and the development of a solid–gas suspension system that maintains a uniform particle concentration and temperature within the receiver.

6.2.2 Indirect particle heating receivers

6.2.2.1 Gravity-driven particle flow-through enclosures

Ma et al. [15,28,72] proposed an indirectly heated particle receiver with particles flowing downward under the force of gravity around a staggered array of tubes within an enclosure. The tubes were irradiated by concentrated sunlight on the interior surfaces while transferring heat to the particles flowing around the exterior side of the tubes inside of an enclosure (Fig. 6.9). Small-scale tests and models were performed that

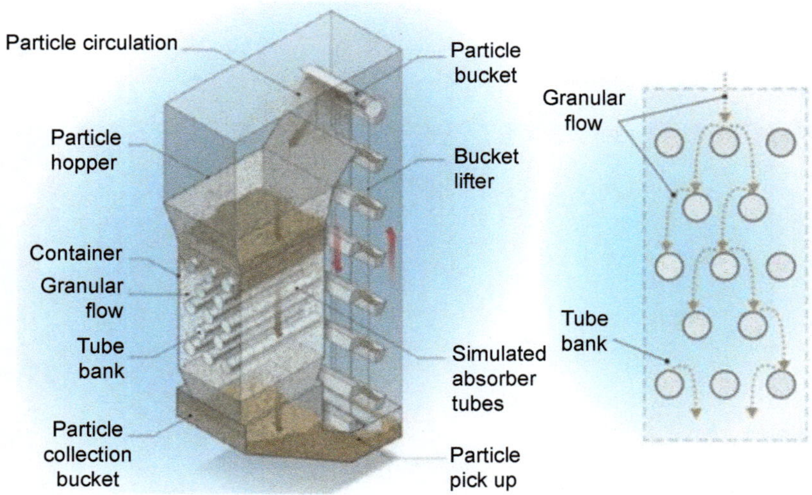

Figure 6.9 Indirect particle receiver with particles flowing inside an enclosure around tubes whose interior surfaces are exposed to concentrated sunlight [28].

showed that the heat transfer to the particles was limited in locations around the tubular structures where the particles lost contact with the heated wall surfaces. Specific data on particle temperatures and thermal efficiencies were not available, and no on-sun tests have been performed. Other limitations included maintaining a sufficient mass flow and obtaining a significant penetration and uniform flux of concentrated sunlight within the tubular cavities. Advantages to this design include no loss of particles through an open aperture and reduced heat losses relative to an open cavity receiver.

6.2.2.2 *Fluidized particle flow-through tubes*

Flamant et al. [26,27,73] have proposed and demonstrated an indirect particle receiver in which the particles are forced upward through irradiated tubes by airflow, which fluidizes the particles and increases heat transfer from the tube walls to the flowing particles. Particle temperature increases of greater than 200°C were recorded in a 50 cm long stainless steel AISI 304L tube with irradiances ranging from about 200 to 400 W/m^2. Suspension temperatures at the outlet of the irradiated tubes were up to 750°C, and the wall-to-suspension heat transfer coefficient was determined to be 420–1100 W/m^2-K for solid mass fluxes of 10–45 kg/m^2-s, respectively. Thermal efficiencies were not reported. Challenges in this system include parasitic energy requirements to fluidize the particles through the receiver tubes with sufficient mass flow to meet desired power requirements. The potential for hot spots and significant tube surface temperatures that radiate energy to the environment also exist.

6.2.3 *Summary of particle receiver technologies*

Table 6.1 summarizes the different types of direct and indirect particle receiver designs. The achievable outlet temperature and thermal efficiency is reported if data were available. The benefits and challenges of each design is also presented, along with relevant references. Overall, each of the particle receiver designs have promising advantages, along with challenges that need to be addressed. Directly heated particle receivers have a significant advantage of direct particle heating, but particle loss may be a problem in open cavities with significant wind effects. Indirect particle receivers have the advantage of particle containment and no particle losses, but additional heat transfer resistance between the irradiated surface and the particles is a challenge. Fluidizing the particles within tubes has been shown to enhance the heat transfer. For large-scale electricity production, which will require significant particle mass flow rates, gravity-driven flow (free-falling or with obstructions) appears to be the most promising.

6.3 Other high-performance receiver designs

Additional novel receiver designs have been proposed to reduce heat losses and achieve higher efficiencies. These include designs to increase light trapping and solar absorption, and air curtains to reduce convective heat losses.

Table 6.1 **Summary of particle receiver designs**

Receiver design	Outlet temperature/ thermal efficiency	Benefits	Challenges/research needs	References
Direct particle receivers				
Free-falling	>700°C/ 50–80%	Capable of achieving high temperatures, direct irradiance of particles reduces flux limitations (on tubular receivers), particles can be stored at high temperatures, particles can be cheaper than molten salt	Need lower radiative and convective heat losses, higher concentration ratios, lower particle attrition, greater solar absorptance, lower thermal emittance, increased particle residence time, more effective particle/fluid heat exchangers	[7,10,14,18,19, 30,32,38,47,49, 51–53,74–77]
Obstructed	>700°C/ 60–90%	Capable of achieving high temperatures, obstructions slow particle flow and increase residence time, flow is more stable than free fall, less particle loss	Hot spots and continuous flow over obstructions may cause deterioration or failure if mass flow and cooling is not maintained; additional cost of fabricating obstructions	[48,49,54,78]
Rotating kiln/ centrifugal	900°C/75%	High particle temperatures, control of residence time via rotational speed of receiver	Maintaining a constant and sufficient mass flow rate of particles at larger scales, parasitic energy requirements, and reliability associated with a large rotating receiver system	[20,21,24,25]

Fluidized bed	>1000 °C/ 20–40%	Excellent heat transfer to fluidized particles with increased residence time	Parasitic energy requirements to fluidize particles, maintaining sufficient mass flow for desired power requirements	[22,23,29,60–62]
Indirect particle receivers				
Gravity-driven flow in enclosures	No data available	High particle temperatures theoretically achievable; no particle loss due to containment	Additional heat transfer resistance from irradiated walls to particles; hot spots on enclosures may cause deterioration or failure if mass flow and cooling is not maintained	[15,28,72]
Fluidized flow in tubes	750°C/ thermal efficiency not reported	Enhanced heat transfer from walls to particles due to fluidization; no particle loss due to containment	Parasitic energy requirements to fluidize particles; maintaining sufficient mass flow for desired power requirements; hot spots on enclosures may cause deterioration or failure if mass flow and cooling is not maintained	[26,27,73]

6.3.1 Light-trapping receiver designs

6.3.1.1 Surface features

Garbrecht et al. [79] proposed an external receiver with an array of numerous pyramidal structures on the exterior of the receiver that intercepted incident radiation and could create a radiation trap (Fig. 6.10). Molten salt flowed from the interior of each pyramid toward the tip, and then along the sides of the pyramid. Simulations showed that for an irradiance of 1 MW/m^2, this receiver design could achieve a thermal efficiency above 90%. The reflective radiative losses could be reduced to about 1%, and thermal emittance was about 3% of incident radiation. Challenges with this proposed design include achieving sufficient heat transfer at the tip of the pyramid, where the greatest flux occurs with the potential for stagnant internal flow. In addition, costs associated with the complexity of the numerous flow-through pyramidal structures may be high.

6.3.1.2 Spiky receiver

Lubkoll et al. [80] describe a spiky central receiver air preheater (SCRAP) design that employs a large number of spikes or tubes that are irradiated from concentrated sunlight (Fig. 6.11). Each spike consists of an inner and outer tube through which air flows. Similar to the pyramidal designs, cold air flows through the inner tube toward the end and then flow back along the outer tube. The temperature of the outer surface of each tube is anticipated to rise from the spike tip to the interior root of each spike, where radiation is minimized. The highest cooling effect from the air occurs at each spike tip, where the irradiance is greatest. CFD simulations showed that radiative losses were only few percent of the total incident power, but convective losses were significantly higher ($\sim$16%) due to the large surface area. The thermal efficiency of a spike irradiated at 1 MW/m^2 with an air outlet temperature of about 800°C was calculated at about 80%. Challenges include reducing the pressure drop within each spike and reducing convective losses.

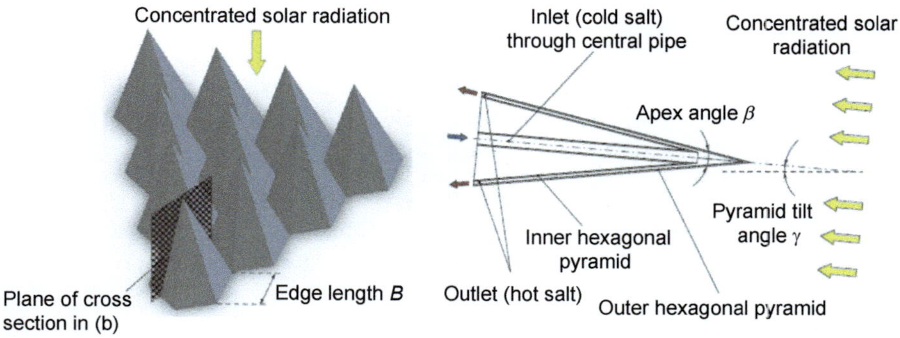

Figure 6.10 Receiver design with array of hexagonal pyramids [79]. (a) Exemplary arrangement of some receiver pyramids, (b) cross section through a receiver pyramid.

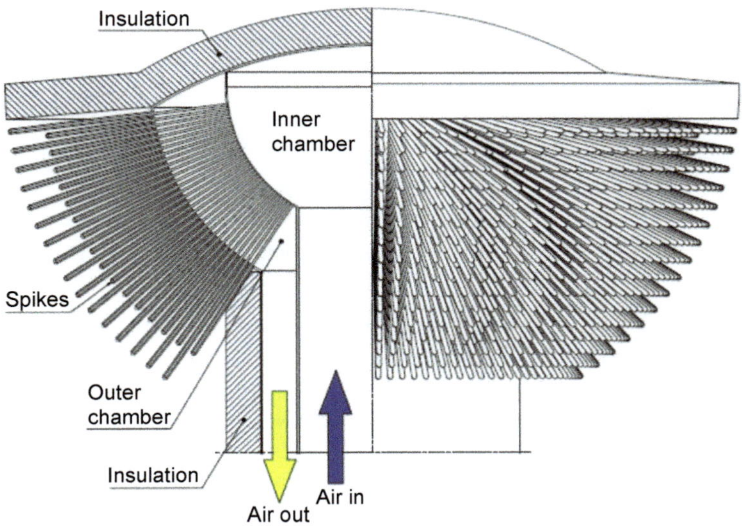

Figure 6.11 Spiky central receiver air preheater (SCRAP) [80].

6.3.1.3 Bladed geometries

Another approach to increase light trapping involves the use of alternative configurations of the tube panels for external receivers. Conventionally, the tube panels are arranged in either cylindrical or cubical fashion. Any incident radiation that is reflected is lost to the environment. By arranging the panels in a bladed configuration, reflected light can be intercepted by the surrounding panels (Fig. 6.12). In addition, by introducing the cold heat-transfer fluid to the outer regions of the panels first, the hottest portions of the panels will be near the interior where radiative heat losses are reduced [81]. CFD simulations showed that the horizontal bladed design yielded the least amount of radiative and convective losses and highest efficiency [82]. For an average irradiance of 500 kW/m^2 (2 MW incident on 4 m^2 of exposed surface area), the thermal efficiency was about 95% with radiative losses of less than 4% and convective losses of less than 1%. Wind effects were not considered. Challenges included structural considerations to prevent dynamic loading and fatigue from wind effects and the proper accommodation of headers and static loads in these novel bladed receiver configurations.

6.3.1.4 Fractal-like geometries

The concept of introducing light-trapping features at multiple length scales was introduced by Ho et al. [83,84]. In this concept, light-trapping features and processes can occur at the macro scale, meso scale, and micro scale. At the macro scale (meters to tens of meters), bladed or spiky receiver geometries can be employed as described in the previous section. At the meso scale (millimeters to tens of centimeters), alternative shapes and arrangements for the tubes carrying the heat-transfer fluid can be

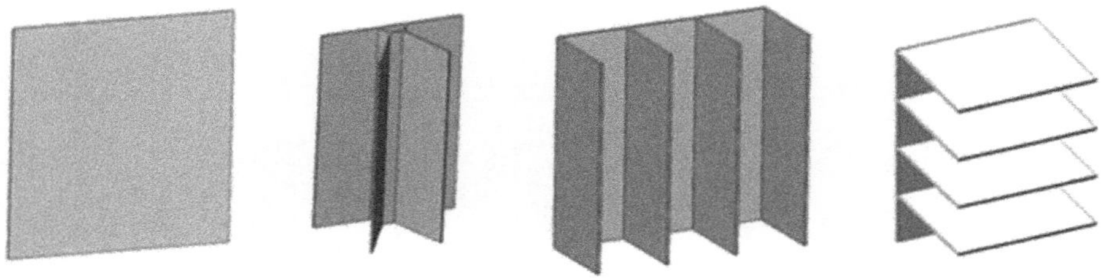

Figure 6.12 Conventional (left) and bladed receiver geometries.

designed and have been shown to increase the effective solar absorptance due to additional light trapping [83,85,86]. At the micro scale (tenths of a millimeter or less), surface features and texturing can be used to increase light trapping and reduce thermal emittance. Combined, these features and geometries and multiple length scales may increase the thermal efficiency significantly. Challenges include reliability and costs associated with novel features, especially at the smaller scales.

6.3.2 Air curtains

Air curtains have been proposed as a way to mitigate convective heat losses from cavity receivers. Air curtains were proposed for cavity-based solid particle receivers as a means to reduce both convective heat losses and particles losses through the aperture [40,42,50–53]. Simulation results showed that the use of an air curtain had the potential to reduce convective losses by several percentage points, but testing showed that the air flow could cause particle instability [42]. The use of air curtains in dish-based cavity receivers was also investigated by Zhang et al. [87]. Different orientations of the air flow across the aperture of the cavity were investigated numerically, and optimal configurations were identified. While air curtains may reduce convective heat losses, a complete demonstration at high temperatures with economic analysis have not been performed.

6.4 Summary and conclusions

Novel receiver designs that can achieve higher temperatures, reduced heat losses, and increased thermal efficiency and performance have been reviewed and presented. Particle receivers enable temperatures significantly higher than conventional receivers employing molten nitrate salts, which are limited to less than about 600°C. Direct particle receivers include free-falling, obstructed-flow, centrifugal, and fluidized designs that irradiate the particles directly. Advantages include the potential for high efficiencies due to direct irradiance of the heat-transfer media. Challenges include maintaining and controlling sufficient mass flow and reducing particle loss. Indirect particle receivers include gravity-driven flow in enclosures and fluidized flow in tubes. Advantages include complete containment of the particles, while challenges include increasing heat transfer between the irradiated surfaces and particles. Other novel receiver designs include features and geometries that trap radiation and minimize thermal emittance by reducing localized view factors in the hottest regions and creating a volumetric heating effect. Finally, the use of air curtains and active airflow have the potential to reduce convective heat losses from cavity receivers.

Acknowledgments

Sandia National Laboratories is a multi-program laboratory managed and operated by Sandia Corporation, a wholly owned subsidiary of Lockheed Martin Corporation, for the US Department of Energy's National Nuclear Security Administration under contract DE-AC04-94AL85000.

References

[1] Kolb GJ, Ho CK, Mancini TR, Gary JA. Power tower technology roadmap and cost reduction plan. Albuquerque (NM): Sandia National Laboratories; 2011. SAND2011-2419.
[2] Bradshaw RW, Carling RW. A review of the chemical and physical-properties of molten alkali nitrate salts and their effect on materials for solar central receivers. Journal of the Electrochemical Society 1987;134(8B):C510−1.
[3] Freeman ES. The kinetics of the thermal decomposition of sodium nitrate and of the reaction between sodium nitrite and oxygen. Journal of Physical Chemistry 1956;60(11): 1487−93.
[4] Ho CK, Iverson BD. Review of high-temperature central receiver designs for concentrating solar power. Renewable & Sustainable Energy Reviews 2014;29:835−46.
[5] Falcone PK. A handbook for solar central receiver design. SAND86-8009. Livermore (CA): Sandia National Laboratories; 1986.
[6] Christian JM, Ho CK. Alternative designs of a high efficiency, north-facing, solid particle receiver. In: SolarPACES 2013, Las Vegas, NV, September 17−20, 2013; 2013.
[7] Falcone PK, Noring JE, Hruby JM. Assessment of a solid particle receiver for a high temperature solar central receiver system. SAND85-8208. Livermore (CA): Sandia National Laboratories; 1985.
[8] Gobereit B, Amsbeck L, Buck R, Pitz-Paal R, Müller-Steinhagen H. Assessment of a falling solid particle receiver with numerical simulation. In: SolarPACES 2012, Marrakech, Morrocco, September 11−14, 2012; 2012.
[9] Ho C, Christian J, Gill D, Moya A, Jeter S, Abdel-Khalik S, Sadowski D, Siegel N, Al-Ansary H, Amsbeck L, Gobereit B, Buck R. Technology advancements for next generation falling particle receivers. In: Proceedings of the SolarPACES 2013 international conference, vol. 49; 2014. p. 398−407. Energy Procedia.
[10] Hruby JM, Steele BR. A solid particle central receiver for solar-energy. Chemical Engineering Progress 1986;82(2):44−7.
[11] Khalsa SSS, Christian JM, Kolb GJ, Röger M, Amsbeck L, Ho CK, Siegel NP, Moya AC. CFD simulation and performance analaysis of alternative designs for high-temperature solid particle receivers. In: ASME international conference on energy sustainability, Washington, DC, USA; 2011.

[12] Khalsa SSS, Ho CK. Radiation boundary conditions for computational fluid dynamics models of high-temperature cavity receivers. Journal of Solar Energy Engineering-Transactions of the ASME 2011;133(3).

[13] Kim K, Moujaes SF, Kolb GJ. Experimental and simulation study on wind affecting particle flow in a solar receiver. Solar Energy 2010;84(2):263–70.

[14] Kolb GJ, Diver RB, Siegel N. Central-station solar hydrogen power plant. Journal of Solar Energy Engineering-Transactions of the ASME 2007;129(2):179–83.

[15] Ma ZW, Glatzmaier G, Mehos M. Fluidized bed technology for concentrating solar power with thermal energy storage. Journal of Solar Energy Engineering-Transactions of the ASME 2014;136(3).

[16] Rightley MJ, Matthews LK, Mulholland GP. Experimental characterization of the heat-transfer in a free-falling-particle receiver. Solar Energy 1992;48(6):363–74.

[17] Röger M, Amsbeck L, Gobereit B, Buck R. Face-down solid particle receiver using recirculation. Journal of Solar Energy Engineering 2011.

[18] Siegel NP, Ho CK, Khalsa SS, Kolb GJ. Development and evaluation of a prototype solid particle receiver: on-sun testing and model validation. Journal of Solar Energy Engineering-Transactions of the ASME 2010;132(2).

[19] Tan TD, Chen YT. Review of study on solid particle solar receivers. Renewable & Sustainable Energy Reviews 2010;14(1):265–76.

[20] Wu W, Trebing D, Amsbeck L, Buck R, Pitz-Paal R. Prototype testing of a centrifugal particle receiver for high-temperature concentrating solar applications. Journal of Solar Energy Engineering-Transactions of the ASME 2015;137(4).

[21] Wu W, Uhlig R, Buck R, Pitz-Paal R. Numerical simulation of a centrifugal particle receiver for high-temperature concentrating solar applications. Numerical Heat Transfer Part A-Applications 2015;68(2):133–49.

[22] Flamant G. Theoretical and experimental-study of radiant-heat transfer in a solar fluidized-bed receiver. AIChE Journal 1982;28(4):529–35.

[23] Flamant G, Hernandez D, Bonet C, Traverse JP. Experimental aspects of the thermo-chemical conversion of solar-energy - decarbonation of $CaCO_3$. Solar Energy 1980;24(4): 385–95.

[24] Wu W, Amsbeck L, Buck R, Uhlig R, Ritz-Paal R. Proof of concept test of a centrifugal particle receiver. In: Proceedings of the SolarPACES 2013 international conference, vol. 49; 2014. p. 560–8.

[25] Wu W, Amsbeck L, Buck R, Waibel N, Langner P, Pitz-Paal R. On the influence of rotation on thermal convection in a rotating cavity for solar receiver applications. Applied Thermal Engineering 2014;70(1):694–704.

[26] Flamant G, Gauthier D, Benoit H, Sans JL, Boissiere B, Ansart R, Hemati M. A new heat transfer fluid for concentrating solar systems: particle flow in tubes. In: Proceedings of the SolarPACES 2013 international conference, vol. 49; 2014. p. 617–26.

[27] Flamant G, Gauthier D, Benoit H, Sans JL, Garcia R, Boissiere B, Ansart R, Hemati M. Dense suspension of solid particles as a new heat transfer fluid for concentrated solar thermal plants: on-sun proof of concept. Chemical Engineering Science 2013;102:567–76.

[28] Martinek J, Ma Z. Granular flow and heat-transfer study in a near-blackbody enclosed particle receiver. Journal of Solar Energy Engineering 2015;137(5):051008.

[29] Steinfeld A, Imhof A, Mischler D. Experimental investigation of an atmospheric-open cyclone solar reactor for solid-gas thermochemical reactions. Journal of Solar Energy Engineering-Transactions of the ASME 1992;114(3):171–4.

[30] Röger M, Amsbeck L, Gobereit B, Buck R. Face-down solid particle receiver using recirculation. Journal of Solar Energy Engineering-Transactions of the ASME August 2011;133(3).
[31] Hruby JM, Steele BR, Burolla VP. Solid particle receiver experiments: radiant heat test. Albuquerque (NM): Sandia National Laboratories; 1984. SAND84-8251.
[32] Hruby JM. A technical feasibility study of a solid particle solar central receiver for high temperature applications. Livermore (CA): Sandia National Laboratories; 1986. SAND86-8211.
[33] Evans G, Houf W, Greif R, Crowe C. Gas-particle flow within a high-temperature solar cavity receiver including radiation heat-transfer. Journal of Solar Energy Engineering-Transactions of the ASME 1987;109(2):134–42.
[34] Meier A. A predictive CFD model for a falling particle receiver reactor exposed to concentrated sunlight. Chemical Engineering Science 1999;54(13–14):2899–905.
[35] Chen H, Chen Y, Hsieh HT, Siegel N. CFD modeling of gas particle flow within a solid particle solar receiver. In: Proceedings of the ASME international solar energy conference; 2007. p. 37–48.
[36] Klein HH, Karni J, Ben-Zvi R, Bertocchi R. Heat transfer in a directly irradiated solar receiver/reactor for solid-gas reactions. Solar Energy 2007;81(10):1227–39.
[37] Martin J, John Vitko J. ASCUAS: a solar central receiver utilizing a solid thermal carrier. Livermore (CA): Sandia National Laboratories; 1982. SAND82-8203.
[38] Khalsa SS, Christian JM, Kolb GJ, Roger M, Amsbeck L, Ho CK, Siegel NP, Moya AC. CFD simulation and performance analysis of alternative designs for high-temperature solid particle receivers. In: Proceedings of the ASME 2011 energy sustainability and fuel cell conference, ESFuelCell2011-54430, Washington DC; August 7–10, 2011.
[39] Gobereit B, Amsbeck L, Buck R. Operation strategies for falling particle receivers. In: Proceedings of ASME 2013 7th international conference on energy sustainability, ES-FuelCell2013-18354, Minneapolis, MN; July 14–19, 2013.
[40] Ho CK, Christian JM. Evaluation of air recirculation for falling particle receivers. In: Proceedings of ASME 2013 7th international conference on energy sustainability, ES-FuelCell2013-18236, Minneapolis, MN; July 14–19, 2013.
[41] Christian JM, Ho CK. System design of a 1 MW north-facing, solid particle receiver. In: Proceedings of the SolarPACES 2014 international conference, (Energy procedia); 2014.
[42] Ho CK, Christian JM, Moya AC, Taylor J, Ray D, Kelton J. Experimental and numerical studies of air curtains for falling particle receivers. In: Proceedings of ASME 2014 8th international conference on energy sustainability, ES-FuelCell2014-6632, Minneapolis, MN; June 29–July 2, 2014.
[43] Knott R, Sadowski DL, Jeter SM, Abdel-Khalik SI, Al-Ansary HA, El-Leathy A. High temperature durability of solid particles for use in particle heating concentrator solar power systems. In: Proceedings of the ASME 2014 8th international conference on energy sustainability, ES-FuelCell2014-6586, Boston, MA; June 29–July 2, 2014.
[44] Siegel N, Gross M, Ho C, Phan T, Yuan J. Physical properties of solid particle thermal energy storage media for concentrating solar power applications. In: Proceedings of the SolarPACES 2013 international conference, vol. 49; 2014. p. 1015–23. Energy procedia.
[45] Ho CK, Christian J, Romano D, Yellowhair J, Siegel N. Characterization of particle flow in a free-falling solar particle receiver. In: Proceedings of the ASME 2015 power and energy conversion conference, San Diego, CA; June 28–July 2, 2015.
[46] Siegel NP, Gross MD, Coury R. The development of direct absorption and storage media for falling particle solar central receivers. ASME Journal of Solar Energy Engineering 2015;137(4):041003–7.

[47] Ho CK, Khalsa SS, Siegel NP. Modeling on-sun tests of a prototype solid particle receiver for concentrating solar power processes and storage. In: ES2009: proceedings of the ASME 3rd international conference on energy sustainability, vol. 2; 2009 [San Francisco, CA].

[48] Ho CK, Christian JM, Yellowhair J, Siegel N, Jeter S, Golob M, Abdel-Khalik SI, Nguyen C, Al-Ansary H. On sun testing of an advanced falling particle receiver system. In: SolarPACES 2015, Cape Town, South Africa; October 13–16, 2015.

[49] Ho CK, Christian JM, Yellowhair J, Armijo K, Jeter S. Performance evaluation of a high-temperature falling particle receiver. In: ASME power & energy conference, Charlotte, NC; June 26–30, 2016.

[50] Kolb GJ. Suction-recirculation device for stabilizing particle flows within a solar powered solid particle receiver. United States Patent 8109265, 12/368,327. Sandia National Laboratories; February 7, 2012.

[51] Chen ZQ, Chen YT, Tan TD. Numerical analysis on the performance of the solid solar particle receiver with the influence of aerowindow. In: Proceedings of the ASME fluids engineering division summer conference -2008, vol. 1. Jacksonville (FL): Pt a and B; 2009.

[52] Tan TD, Chen YT. Protection of an aerowindow, one scheme to enhance the cavity efficiency of a solid particle solar receiver. In: HT2009: proceedings of the ASME summer heat transfer conference 2009, vol. 2; 2009 [San Francisco, CA].

[53] Tan TD, Chen YT, Chen ZQ, Siegel N, Kolb GJ. Wind effect on the performance of solid particle solar receivers with and without the protection of an aerowindow. Solar Energy 2009;83(10):1815–27.

[54] Lee T, Lim S, Shin S, Sadowski DL, Abdel-Khalik SI, Jeter SM, Al-Ansary H. Numerical simulation of particulate flow in interconnected porous media for central particle-heating receiver applications. Solar Energy 2015;113:14–24.

[55] Khayyat AW, Knott RC, Nguyen CL, Golob MC, Abdel-Khalik SI, Jeter SM, Al-Ansary HA. Measurement of particulate flow in discrete structure particle heating receivers. In: Proceedings of the ASME 2015 power and energy conversion conference, San Diego, CA; June 28–July 2, 2015.

[56] Xiao G, Guo KK, Ni MJ, Luo ZY, Cen KF. Optical and thermal performance of a high-temperature spiral solar particle receiver. Solar Energy 2014;109:200–13.

[57] Ermanoski I, Siegel NP, Stechel EB. A new reactor concept for efficient solar-thermochemical fuel production. Journal of Solar Energy Engineering-Transactions of the ASME 2013;135(3).

[58] Ermanoski I. Maximizing efficiency in two-step solar-thermochemical fuel production. International conference on concentrating solar power and chemical energy systems, SolarPACES 2014 Energy Procedia 2015;69:1731–40.

[59] Ermanoski I, Siegel N. Annual average efficiency of a solar-thermochemical reactor. In: Proceedings of the SolarPACES 2013 international conference, vol. 49; 2014. p. 1932–9.

[60] Bai F, Zhang Y, Zhang X, Wang F, Wang Y, Wang Z. Thermal performance of a quartz tube solid particle air receiver. In: Proceedings of the SolarPACES 2013 international conference, vol. 49; 2014. p. 284–94.

[61] Wang F, Bai F, Wang Z, Zhang X. Numerical simulation of quartz tube solid particle air receiver. International conference on concentrating solar power and chemical energy systems, SolarPACES 2014 Energy Procedia 2015;69:573–82.

[62] Zhang YN, Bai FW, Zhang XL, Wang FZ, Wang ZF. Experimental study of a single quartz tube solid particle air receiver. International conference on concentrating solar power and chemical energy systems, SolarPACES 2014 Energy Procedia 2015;69:600–7.

[63] Abdelrahman M, Fumeaux P, Suter P. Study of solid-gas-suspensions used for direct absorption of concentrated solar-radiation. Solar Energy 1979;22(1):45–8.
[64] Hunt AJ. A new solar receiver utilizing a small particle heat exchanger. In: Proceedings of the 14th international society of energy conversion engineering conference, Institute of Electrical and Electronics Engineers, New York; 1979.
[65] Hunt AJ, Brown CT. Solar testing of the small particle heat exchanger (SPHER). Berkeley (CA): Lawrence Berkeley National Laboratory; 1982. Report no. LBL-16497.
[66] Miller F, Koenigsdorff R. Theoretical-analysis of a high-temperature small-particle solar receiver. Solar Energy Materials 1991;24(1–4):210–21.
[67] Miller FJ, Koenigsdorff RW. Thermal modeling of a small-particle solar central receiver. Journal of Solar Energy Engineering-Transactions of the ASME 2000;122(1):23–9.
[68] Crocker A, Miller F. Coupled fluid flow and radiation modeling of a cylindrical small particle solar receiver. In: Proceedings of the ASME 6th international conference on energy sustainability - 2012, Pts a and B; 2012. p. 405–12.
[69] del Campo PF, Miller F, Crocker A. Three-dimensional fluid dynamics and radiative heat transfer modeling of a small particle solar receiver. In: Proceedings of the ASME 7th international conference on energy sustainability; 2014. 2013.
[70] Fernandez P, Miller FJ. Performance analysis and preliminary design optimization of a small particle heat exchange receiver for solar tower power plants. Solar Energy 2015;112: 458–68.
[71] Kitzmiller K, Miller F. Thermodynamic cycles for a small particle heat exchange receiver used in concentrating solar power plants. Journal of Solar Energy Engineering-Transactions of the ASME 2011;133(3).
[72] Ma Z, Zhang R. United States Patent 13/855092, solid particle thermal energy storage design for a fluidized-bed concentrating solar power plant. NREL; April 2, 2013.
[73] Benoit H, Lopez IP, Gauthier D, Sans JL, Flamant G. On-sun demonstration of a 750 degrees C heat transfer fluid for concentrating solar systems: dense particle suspension in tube. Solar Energy 2015;118:622–33.
[74] Chen HJ, Chen YT, Hsieh HT, Kolb G, Siegel N. Numerical investigation on optimal design of solid particle solar receiver. In: Proceedings of the energy sustainability conference; 2007. p. 971–9.
[75] Chen HJ, Chen YT, Hsieh HT, Siegel N. Computational fluid dynamics modeling of gas-particle flow within a solid-particle solar receiver. Journal of Solar Energy Engineering-Transactions of the ASME 2007;129(2):160–70.
[76] Ho CK, Roeger M, Khalsa SS, Amsbeck L, Buck R, Siegel N, Kolb G. Experimental validation of different modeling approaches for solid particle receivers. In: SolarPACES 2009, SAND2009–4140C, Berlin, Germany; September 15–18, 2009.
[77] Siegel N, Kolb G, Kim K, Rangaswamy V, Moujaes S. Solid particle receiver flow characterization studies. In: Proceedings of the energy sustainability conference; 2007. p. 877–83.
[78] Al Ansary H. Solid particle receiver with porous structure for flow regulation and enhancement of heat transfer. United States Patent 2013/0068217 A1. K.S. University; March 21, 2013.
[79] Garbrecht O, Al-Sibai F, Kneer R, Wieghardt K. CFD-simulation of a new receiver design for a molten salt solar power tower. Solar Energy 2013;90:94–106.
[80] Lubkoll M, von Backstrom TW, Harms TM, Kroger DG. Initial analysis on the novel spiky central receiver air pre-heater (SCRAP) pressurized air receiver, international conference on concentrating solar power and chemical energy systems. SolarPACES 2015;2014(69): 461–70.

[81] Ho CK, Christian JM, Pye J. Bladed solar thermal receivers for concentrating solar power. United States Patent Application 14535100. Sandia Corporation; November 6, 2014.
[82] Christian JM, Ortega JD, Ho CK. Novel tubular receiver panel configurations for increased efficiency of high-temperature solar receivers. In: ASME proceedings of the 9th international conference on energy sustainability, San Diego, CA; June 28–July 2, 2015.
[83] Ho CK, Christian JM, Ortega JD, Yellowhair J, Mosquera MJ, Andraka CE. Reduction of radiative heat losses for solar thermal receivers. In: Proceedings of the SPIE optics+photonics solar energy+technology high and low concentrator systems for solar energy applications IX, San Diego; August 17–21, 2014.
[84] Ho CK, et al. Fractal materials and designs with optimized radiative properties. United States Patent Application 62/015052. Sandia Corporation; June 20, 2014.
[85] Ortega JD, Christian JM, Ho CK. Coupled optical-thermal-fluid and structural analyses of novel light-trapping tubular panels for concentrating solar power receivers. In: Proceedings of the SPIE optics+photonics solar energy+technology high and low concentrator systems for solar energy applications X, San Diego; August 9–13, 2015.
[86] Yellowhair JE, Ho CK, Ortega JD, Christian JM, Andraka CE. Testing and optical modeling of novel concentrating solar receiver geometries to increase light trapping and effective solar absorptance. In: Proceedings of the SPIE optics+photonics solar energy+technology high and low concentrator systems for solar energy applications X, San Diego; August 9–13, 2015.
[87] Zhang J, Pye JD, Hughes GO. Active air flow control to reduce cavity receiver heat loss. In: Ninth international conference on energy sustainability, ASME power & energy, san Diego, CA; June 2015.

Next generation of liquid metal and other high-performance receiver designs for concentrating solar thermal (CST) central tower systems

7

M. Romero, J. González-Aguilar
IMDEA Energía Institute, Móstoles, Spain

7.1 Introduction

This chapter focuses on the use of high thermal conductivity liquid metals as excellent heat transfer fluids in central receiver systems. The introduction provides the context of current receiver fluids and the relevance of increasing the solar irradiance onto the receiver aperture. Then a specific section is dedicated to describe the thermophysical properties of liquid metals and subsequently the experience gained on some central receiver systems such as the Central Receiver System (CRS) of the Small Solar Power Systems (SSPS) project in Spain and the Jemalong Solar Thermal Station in Australia. Finally, some innovative power conversion cycles involving liquid metals are presented.

In a solar power tower plant, the receiver is the heat exchanger where the concentrated sunlight is intercepted and transformed into thermal energy useful in thermodynamic cycles [37]. Radiant flux and temperature are substantially higher than in parabolic troughs, and therefore, high technology is involved in the design and high-performance materials should be chosen. The solar receiver should mimic a black body by minimizing radiation losses. To do so, cavities, black-painted tube panels, or porous absorbers to trap incident photons are used. In most designs, the solar receiver is a single unit that centralizes all the energy collected by the large mirror field, and therefore high availability and durability are a must. Just as cost reduction is the priority for further development in the collector field, in solar receivers the priorities are thermal efficiency and durability. Typical receiver absorber operating temperatures are between 500°C and 1200°C and incident flux covers a wide range of flux density, from 300 to over 1000 kW/m^2.

Thermal and optical losses are the key parameters for quantifying the efficiency of a solar receiver:

$$\eta_{REC} = \alpha - \varepsilon \frac{\sigma\left(T^4 - T_a^4\right)}{C\varphi} - U\frac{T - T_a}{C\varphi} \tag{7.1}$$

Advances in Concentrating Solar Thermal Research and Technology
http://dx.doi.org/10.1016/B978-0-08-100516-3.00007-1

This equation is the result of the energy balance of gains and losses in the receiver, with an absorber surface at temperature (T), higher than the ambient (T_a), and for a given direct normal irradiance (φ) that is concentrated C times. Then, the concentrated solar irradiance ($C\varphi$) is absorbed with an efficiency of α (*absorptivity*). The loss terms are of two different types. The most important in a central receiver represents energy thermally radiated by the absorber through the receiver aperture. These radiation losses depend on the emissivity of the absorber (ε). The absorber convective and conductive losses to the atmosphere are determined by the heat transfer coefficient (U), which depends on the temperature and the forced convection due to wind. In good central receiver designs, U can be sufficiently reduced by thermal insulation and decreased aperture area, therefore U is basically expressed as a convective loss heat transfer coefficient. Generally, this coefficient is obtained from the dimensionless Nusselt number (Nu) and subsequently as a function of numbers such as Reynolds (Re), Prandtl (Pr), and Grashof (Gr). Forced convection is determined by combinations of the Re and Pr numbers, while natural convection is characterized by Pr and Gr numbers.

In a solar tower system, the convective heat loss is calculated differently depending on whether the receiver is a cavity or a cylindrical external receiver. A typical simple mixed convection coefficient for an external receiver can be calculated [41] as

$$U_{mix} = \left(h_{forced}^{3.2} + h_{nat}^{3.2}\right)^{\frac{1}{3.2}} \tag{7.2}$$

where $h_{nat} = 9.09$ (W/m^2 K) for an average absorber temperature of 480°C and h_{forced} is separated into three cases depending on the receiver diameter [25]. In all cases, the $\mathrm{Re} = (1.751 \times 10^5)\ D$:

$$h_{forced} = \begin{cases} \dfrac{0.04199}{D}\left\{0.3 + 0.488\mathrm{Re}^{0.5}\left[1.0 + \left(\dfrac{\mathrm{Re}}{282000}\right)^{0.625}\right]^{0.8}\right\} & \text{if } D \leq 4.0\text{ m (Case 1);} \\ 14.0 & \text{if } 4.0\text{ m} < D \leq 125.0\text{ m (Case 2);} \\ 33.75D^{-0.19} & \text{if } D > 125.0\text{ m (Case 3).} \end{cases} \tag{7.3}$$

For a cavity receiver, the convective heat loss can be directly calculated as [25]:

$$Q_{conv} = Q_{forced} + Q_{nat} \tag{7.4}$$

$$Q_{forced} = 7631\frac{A}{W_{ap}^2} \tag{7.5}$$

$$Q_{nat} = 5077A_{cav} \tag{7.6}$$

where A is the aperture area (m^2), W_{ap} is the aperture width (m), and A_{cav} is the approximation to total area inside cavity (m^2).

For more detailed correlations applicable to convection losses in different kinds of solar receivers [3], is recommended.

Fig. 7.1 clearly evidences the dominant effect of radiation losses because of the exponent of the temperature. For typical operation temperatures in the absorber surface between 500°C and 800°C, it becomes clear that average $C\varphi$ values above 500 kW/m^2 are required, with peak flux above 1000 kW/m^2, to approach receiver efficiencies near 90%.

There are different solar receiver classifications criteria depending on the construction solution, the use of intermediate absorber materials, the kind of thermal fluid used, or heat transfer mechanisms. According to the geometrical configuration, there are basically two design options, external and cavity-type receivers. In a cavity receiver, the radiation reflected from the heliostats passes through an aperture into a box-like structure before impinging on the heat transfer surface. Cavities are constrained angularly and subsequently used in polar field (north or south) layouts. External receivers can be designed with a flat plate tubular panel or a cylindrically shaped unit. Cylindrical external receivers are the typical solution adopted for surround heliostat fields. Receivers can be directly or indirectly irradiated depending on the absorber materials used to transfer the energy to the working fluid [38]. Directly irradiated receivers make use of fluids or particle streams that are able to efficiently absorb the concentrated flux. Particle receiver designs make use of falling curtains or fluidized beds. Darkened liquid fluids can use falling films. In many applications, and to avoid leaks

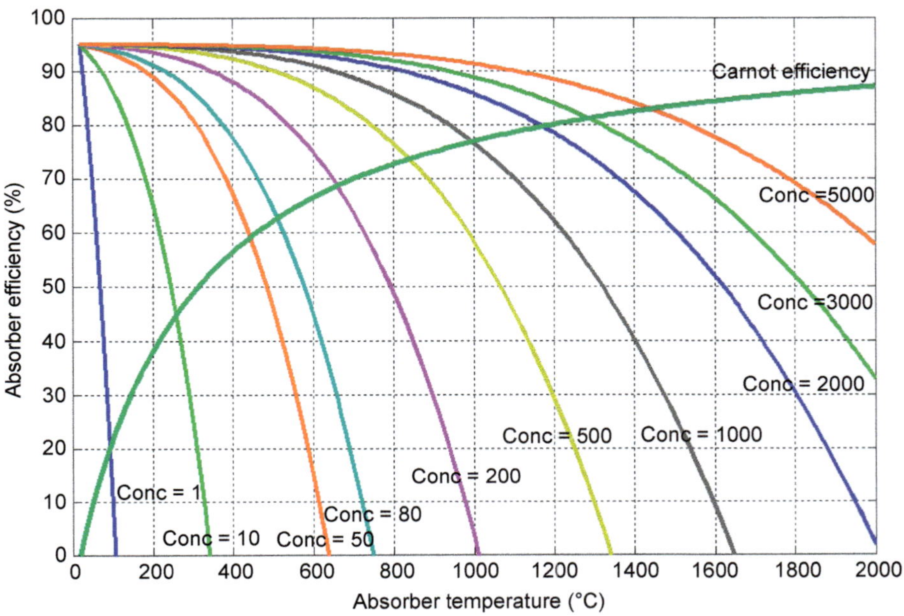

Figure 7.1 Efficiency of the solar receiver versus absorber temperature, and versus solar concentration ratio, assuming $T_{amb} = 20°C$, $\varphi = 770$ W/m^2, and $\alpha = \varepsilon = 0.95$.

to the atmosphere, direct receivers should have a transparent window. Windowed receivers are excellent solutions for chemical applications as well, but they are strongly limited by the size of a single window, and therefore clusters of receivers are necessary.

The key design element in indirectly heated receivers is the radiative/convective heat exchange mechanism. Basically, two heat transfer options are used—tubular panels and volumetric surfaces. In tubular panels, the cooling thermal fluid flows inside the tube and removes the heat collected by the external black panel surface by convection. It therefore operates as a recuperative heat exchanger. Depending on the heat transfer fluid properties and incident solar flux, the tube might undergo thermo-mechanical stresses. Since heat transfer is through the tube surface, it is difficult to operate at an incident peak flux above 600 kW/m^2 when water is used as heat transfer fluid. Table 7.1 shows how only with high thermal conductivity liquids such as sodium it is possible to reach in tubular panels operating fluxes above 1 MW/m^2. Air-cooled receivers have great potential though they have difficulties working with tubular receivers because of the lower heat transfer coefficients. To improve the contact surface, a different approach, based on wires, foams, or appropriately shaped materials within a volume, is used. In volumetric receivers, highly porous structures operating as convective heat exchangers absorb the concentrated solar radiation [19]. The solar radiation is not absorbed on an outer surface, but inside the structure "volume." The heat transfer medium (mostly air) is forced through the porous structure and is heated by convective heat transfer. Volumetric absorbers are usually made of thin heat-resistant wires (in knitted or layered grids) or either metal or ceramic (reticulated foams, monoliths, and so on) open-cell matrix structures. Good volumetric absorbers are very porous, allowing the radiation to penetrate deeply into the structure. Thin substructures (wires, walls, or struts) ensure good convective heat transfer. A good volumetric absorber produces the so-called "volumetric effect," which means that the maximum temperature is achieved inside the absorber. Under specific operating conditions, volumetric absorbers tend to have an unstable mass-flow distribution. Receiver arrangements with mass-flow adaptation elements (e.g., perforated plates) located behind the absorber, as well as appropriate selection of the operating conditions and the absorber material, can reduce this tendency.

Table 7.1 Operating temperature and flux ranges of solar tower receivers

Fluid	Water/ steam	Liquid sodium	Molten salt (nitrates)	Volumetric air
Flux (MW/m^2)				
• Average	0.1–0.3	0.4–0.5	0.4–0.5	0.5–0.6
• Peak	0.4–0.6	1.4–2.5	0.7–0.8	0.8–1.0
Fluid outlet temperature (°C)	490–525	540	540–565	700–1000

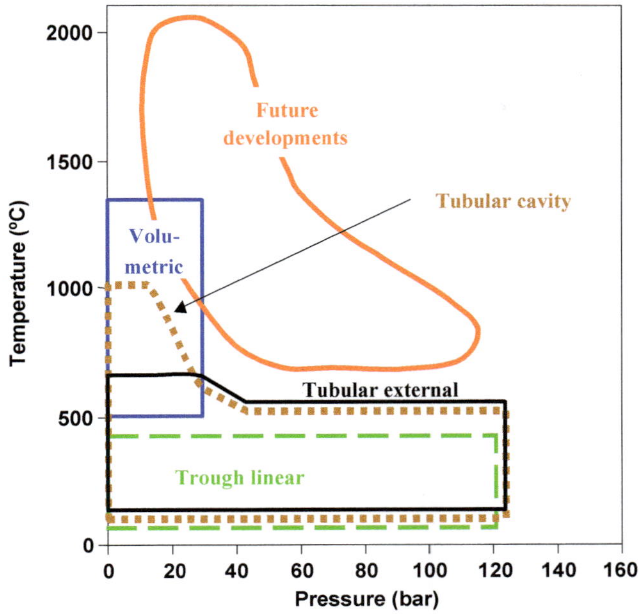

Figure 7.2 Range of operation of tubular, volumetric, and new advanced receivers.

Selection of a particular receiver technology is a complex task, since operating temperature, heat storage system, and thermodynamic cycle influence the design. In general, tubular technologies allow either high temperatures (up to 1000°C) or high pressures (up to 120 bar), but not both [28]. Directly irradiated particle receivers or volumetric absorbers allow even higher temperatures but limited pressures (Fig. 7.2).

7.2 Thermophysical properties of liquid metals

Central receiver systems have a consolidated scheme when subcritical Rankine cycle is used and storage determines the dispatching economic feasibility of the project. In this case molten nitrate salts are becoming the reference design material since nitrates are cheap and provide high storage capacity. As of 2016, solar receivers cooled with nitrate molten salts are referring to thermal efficiencies of 88%. However, there are two essential factors in the short to medium term that make it necessary to look for alternatives, such as the upper temperature limit that as of 2016 is about 580°C and the high melting point that requires trace heating and therefore complicates the operation and maintenance.

As can be observed in Fig. 7.3 and Table 7.2, some mixtures of liquid metals such as Na–K or Pb–Bi eutectic (LBE) may lead to a very wide range of operation

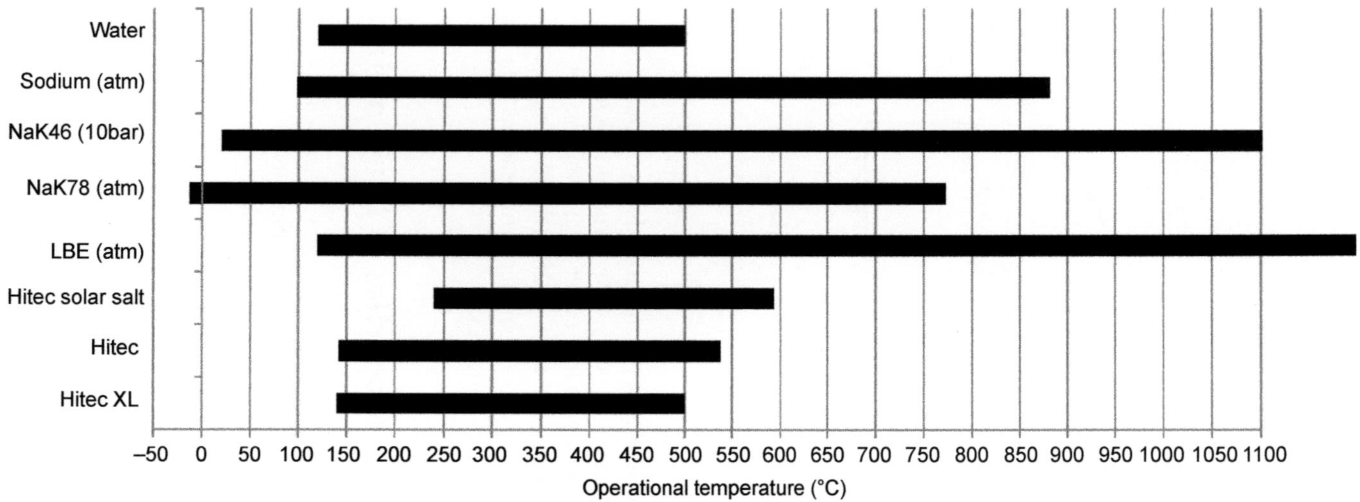

Figure 7.3 Usable temperature range for water, liquid metals, and molten salts.
Adapted from Kotzé JP, Erens PJ, Von Backström TW. NaK as a primary heat transfer fluid in thermal solar power installations. In: Proceeding of SolarPACES 2012, 11–14 September, Marrakech, Morocco; 2012.

Table 7.2 Comparison of physical properties of several liquid metals versus molten nitrate solar salt (60% $NaNO_3$ and 40% KNO_3)

	Sodium	Na–K	LBE	Solar salt
Melting point (°C)	98	−11	125	220
Boiling point (°C)	890	785	1533	565
Thermal conductivity (W/m K)	119.3	26.2	13.7	0.53
Density (kg/m^3)	820	749	10,139	1804
Specific heat capacity (kJ/kg K)	1.256	0.937	0.143	1.52
Dynamic viscosity (Pa s)	0.000149	0.000176	0.00144	0.00169
Prandtl number	0.0016	0.0063	0.015	4.85

temperatures in liquid phase [34]. In some cases like the eutectic composition of Na–K with 78% potassium, the melting temperature is as low as −11°C [17,27], facilitating its use as heat transfer fluid in solar thermal power plants without making use of heat tracing systems that usually increment costs of maintenance and are parasitic.

Sodium gathers extensive experience since it is proposed in nuclear industry for liquid metal fast breeder reactors, existing practical know-how in sensitive aspects such as safety and control, and hardware validation of key components such as pumps, valves, lines, and steam generators. Its boiling temperature is substantially higher than solar salts and presents outstanding thermal conductivity. The vapor pressure at 595°C is only slightly above atmospheric pressure. Its main shortcoming is that it reacts with water and air, and therefore high maintenance costs are associated. The highly reactive nature of sodium and water is an important consideration in the design of sodium components, principally the sodium steam generator, and potentially increases the cost of these components. Melting point is relatively high (98°C), because of which heat tracing is required in the heat transfer fluid loop like in the case of molten salts.

In short, molten salt is cheaper than sodium by a factor of two and has a three-to-one advantage in its volumetric heat capacity, factors that are particularly important in the thermal storage subsystem. On the other hand, sodium has a five times higher heat transfer rate, which means that sodium receivers (such as water/steam receivers) can be single pass; that is, the entire temperature rise of the fluid from about 260°C to 540°C takes place in a single pass through the solar flux [15].

LBE is recently being proposed as a good alternative. Even though thermal conductivity is significantly lower than sodium, it provides a wide range of temperatures for operation. However its high density diminishes the heat capacity in storage systems. In addition, promising fluids such as LBE and Sn present large corrosion rates [35]. LBE presents a high solubility limit for both nickel and copper, up to few weight percent at 600°C [22]. As a consequence, inhibitors or protective layers are required to use nickel-based alloys and steels with high nickel content because of the limited compatibility. Similar problem is found with molten tin above 600°C, since there is lack of information regarding suitable metallic materials for operation, and only some potential options proposed are graphite, molybdenum, tungsten, or rhenium [44] or perhaps the use of some ceramics and refractory materials [35].

Although sodium has much better heat transfer characteristics than LBE, the stringent safety measures associated have motivated formulation of new approaches among the concentrating solar community by adopting LBE as a potential intermediate stage in the process of understanding solarization. A small concentrating solar power system in the 10 kW thermal range is under construction in Germany within a joint initiative launched by the Karlsruhe Institute of Technology (KIT) and the Solar Institute of the German Aerospace Centre (DLR). The system consists of a heliostat parabolic dish solar furnace and liquid metal test loop with a pump, a cooler, an electric heater, and the thermal receiver [16].

Apart from usable range of temperatures of operation, the main asset of the liquid metals is their higher thermal conductivity that makes it possible to reduce the size of the heat exchangers and the solar receiver. Liquid metals present low Pr number leading to turbulent flow with high conductivity. Boerema et al. [8] developed a simple receiver model to determine the influences of the fluids' characteristics on receiver design and efficiency. They found that Hitec has a high Pr number ($\approx$45−1), depending on temperature, while liquid sodium has a very low Pr number ($\approx$0.01−0.004), with only relatively small variation over the temperature range. The study also found for liquid sodium a high heat transfer coefficient (an order of magnitude greater than Hitec) and a low heat capacity (30−50% lower than Hitec salt).

The low value of Pr number and the high conductivity for liquid metals make it difficult to use typical Nusselt correlations, usually expressed as a function of Re and Pr number for the calculation of heat transfer coefficient. Liquid metals do not fall in the validity domain since the dominant parameter is the Peclet number (Pe) that is defined as the ratio of the rate of heat advection over the rate of heat diffusion, which leads to its formula being the product of Re and Pr ($\mathrm{Pe} = \mathrm{Re}\cdot\mathrm{Pr}$). In a 2016 review, Benoit et al. [7] assessed in detail the most usual correlations proposed for liquid metals by Lyon−Martinelli [30], Sleicher−Rouse [42], and Cheng−Tak [12]. They concluded that Lyon−Martinelli correlation tends to overestimate the experimental data for $\mathrm{Pr} = 1$ and recommend the Cheng−Tak correlation in order to obtain the Nusselt number:

$$Nu = A + 0.018Pe^{0.8} \tag{7.7}$$

with

$$A = \begin{cases} 4.5 & \text{when } Pe \leq 1000 \\ 5.4 - 9 \times 10^{-4}Pe & \text{when } 1000 < Pe \leq 2000 \\ 3.6 & \text{when } Pe \geq 2000 \end{cases} \tag{7.8}$$

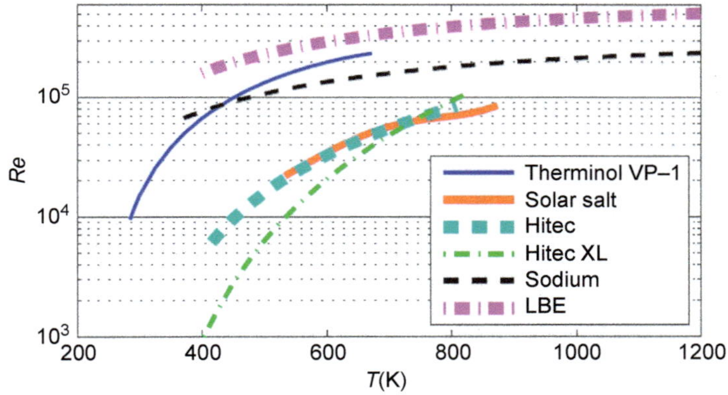

Figure 7.4 Reynolds numbers of liquid heat transfer fluids as a function of temperature for a 0.025 m tube diameter and a 2 m/s velocity [5].

Fig. 7.4 shows a comparison of Re number for different heat transfer fluids at the same velocity. It was assumed a value of 2 m/s as representative of operation of liquid fluids in solar thermal power plants. Sodium and LBE have substantially higher turbulence than molten salts.

The thermal conductivity of liquid metals leads to high heat transfer coefficients. For flow velocities of 1 m/s for a single tube with $D = 12$ mm at 400°C, molten salt reaches about 4600 W/m^2 K, whereas over 50,000 W/m K can be achieved with liquid sodium [18]. Fig. 7.5 shows the evolution of heat transfer coefficient for different liquid metals and solar salts. Coefficients for LBE are twice as much as those of solar salts and five times in the case of Sn. Sodium shows one order of magnitude higher

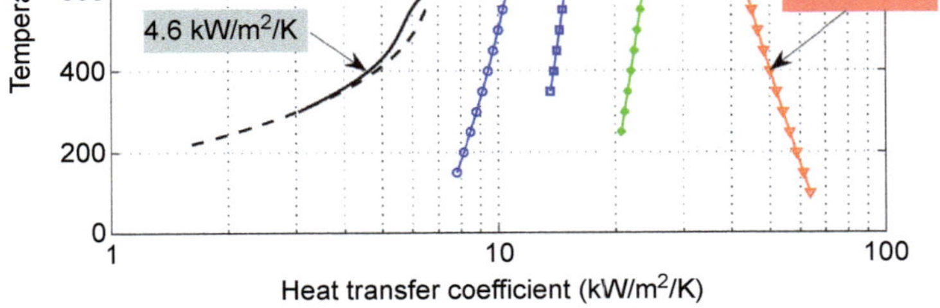

Figure 7.5 Heat transfer coefficient for tubes ($D = 12$ mm; $u = 1$ m/s; and thickness = 0.1 mm) [18].

values. It should be noticed that for the case of sodium the heat transfer coefficient decreases when the temperature increases because thermal conductivity decreases with temperature.

Using a fixed receiver length (4 m) and altering the mass flow rate such that the desired fluid exit temperature is achieved, the efficiency and maximum surface temperature (temperature at pipe exit) were calculated for various pipe diameters in a simplified model [8]. Fig. 7.6 shows that the efficiency and the pipe surface temperature at the receiver exit using liquid sodium are quite stable for increasing pipe diameters, whereas for Hitec both magnitudes are highly dependent on the pipe diameter. This is important as the use of larger diameter pipes reduces the number of pipes needed for the receiver and thus the manufacturing costs.

At the same flux density than molten salts, sodium decreases the thermal gradient of convection on the inner tube wall surface. This also reduces tube wall temperature and the risk of temperature hot spots and thus pipe stresses as well. A goal in the design of solar receivers is to achieve 30-year lifetime. One cycle per day for 30 years would add up to about 11,000 cycles over the lifetime. An issue is the part of lifetime reserved to cover for weather and other transients which also cause receiver thermal cycles. Analysis of 1984 weather data at Barstow combined with thermal hydraulic analysis of the fluid in the tubes and structural analysis of the tubes has led to a recommended peak allowable incident flux of 0.85 MW/m^2 for molten salt and 1.75 MW/m^2 for sodium, in fabricated receivers using 316 stainless steel [26]. However, variations in flow rates or in the location of the peak flux on the receiver will lead to a different flux limit within the ranges as shown on Fig. 7.7.

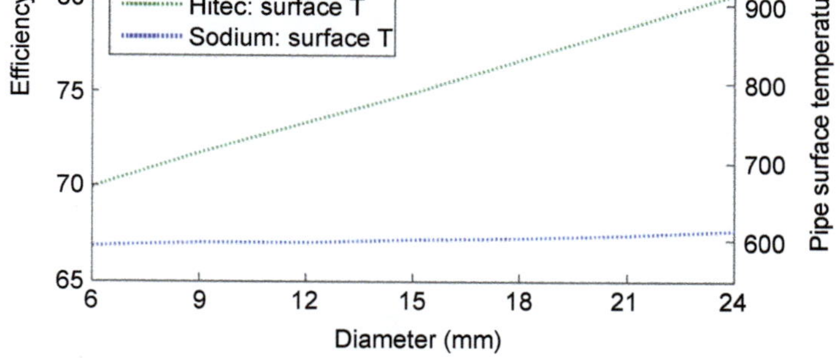

Figure 7.6 Hitec and sodium receiver efficiencies and maximum surface temperatures for various pipe diameters (emissivity = 0.85; concentration ratio = 1300; and pipe length = 4 m) [8].

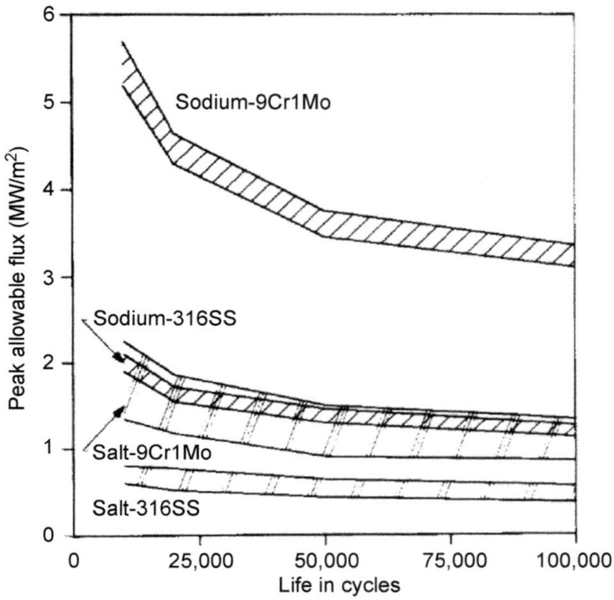

Figure 7.7 Peak receiver tube allowable flux levels as a function of tube life for sodium and molten salt receiver fluids and 9CrlMo and 316SS [15].

Higher flux densities lead to smaller receivers for the same power output. The reduction of absorber surface decreases material and manufacturing costs. According to preliminary receiver performance studies, liquid sodium leads to an absorber area reduction of up to 57%, compared to solar salt and an absolute efficiency increase of 1.1% by utilizing higher concentration ratios [8]. In addition, due to the absorber area reduction, radiation and convection losses might also decrease. Hence, both options imply higher receiver efficiencies and performance.

7.3 Liquid metals in central receiver systems

7.3.1 Experience in central receiver systems

Early experimental solar tower facilities started operation in the 1980s. In general terms, as observed in Table 7.3, they were characterized as being small demonstration systems between 0.5 and 10 MW [15,21,37]. The thermal fluids used in the receiver were at that time liquid sodium, saturated or superheated steam, and nitrate-based molten salts. Important to remark that from the very beginning liquid sodium was considered as a technical option together with steam as heat transfer fluid. The CRS-SSPS facility in Almería (Spain) was promoted by 10 countries within the SSPS Implement Agreement of the International Energy Agency and started operation

Table 7.3 **Early experimental solar tower facilities in the world**

Project	Country	Power (MW_e)	Heat transfer fluid	Storage media	Beginning operation
CRS-SSPS	Spain	0.5	Liquid sodium	Sodium	1981
EURELIOS	Italy	1	Steam	Nitrate salt/water	1981
SUNSHINE	Japan	1	Steam	Nitrate salt/water	1981
Solar One	USA	10	Steam	Oil/rock	1982
CESA-1	Spain	1	Steam	Nitrate salt	1982
MSEE/Cat B	USA	1	Nitrate salt	Nitrate salt	1983
THEMIS	France	2.5	Hitec salt	Hitech salt	1984
SPP-5	Russia	5	Steam	Water/Steam	1986

in 1981. The different projects that are included in Table 7.3 had as main objective to demonstrate the technical feasibility of solar tower power plants.

One of the main conclusions after the comparative assessment carried out, once experimental campaigns were completed, was that solar receivers with proper designs could work at relatively high efficiencies (above 80%) for temperatures high enough in the absorber material to heat up thermal fluids able to produce superheated steam above 500°C. A comparison of solar receivers tested at different power loads in early experimental facilities was done in 1988 [4] and is depicted in Fig. 7.8. At nominal power load, the liquid sodium receiver achieves thermal efficiencies near 90%. As mentioned in previous section, thermal efficiencies at nominal load can be slightly higher than molten salts (about 1%).

The use of liquid metals such as sodium, Na–K, or LBE mixtures in solar receivers may lead to very compact designs of the aperture since the high conductivity improves the heat transfer through the absorber wall [13]. The absorber can work at high fluxes, above 2 MW/m^2, since thermal stress is reduced and temperature gradient between the tube wall and the bulk of the fluid is minimized.

All of them can easily be represented by flow charts, where the main variables are determined by working fluids, with the interface between power block and the solar share. Fig. 7.9 illustrates the flow diagram of a central receiver system with liquid metal as heat transfer fluid. It is very similar to the scheme commonly used for molten salts, and typically includes two circuits the primary solar loop with the HTF and two storage tanks and the secondary loop with water/steam. The heat transfer between loops takes place in the steam generator.

7.3.2 The CRS-SSPS project of the International Energy Agency

The most extensive operational experience of liquid metal solar receivers took place in Almería, Spain, within the framework of the International Energy Agency (IEA) SSPS

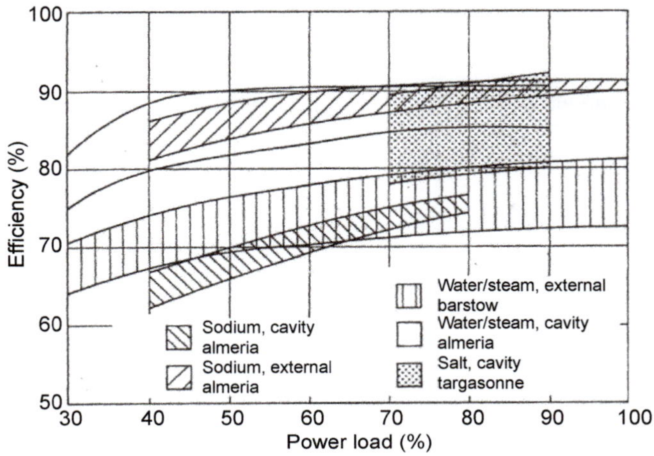

Figure 7.8 Receiver efficiencies in early experimental facilities for water/steam, liquid sodium, and molten salt [4].

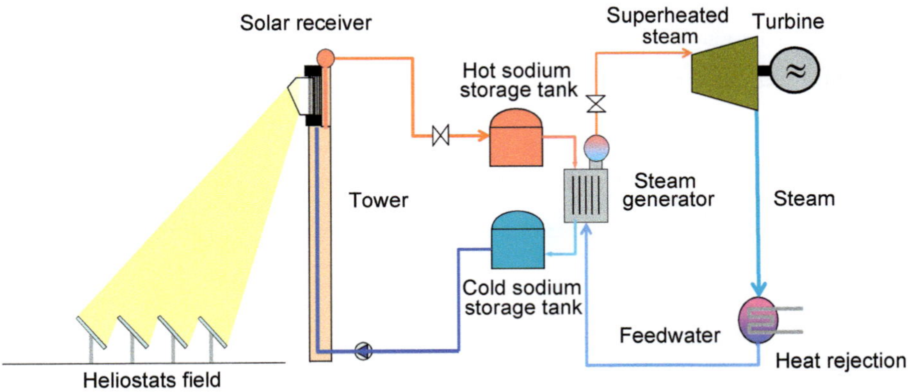

Figure 7.9 Flow diagram of central receiver system using sodium as heat transfer and storage fluid.

project. The project was developed during 1982–1986 and two receivers were analyzed. The first receiver was a north-facing cavity type, having a vertical octagonal-shaped aperture (Table 7.4). The receiver was designed by Interatom (Germany) and manufactured by the Sulzer Company (Switzerland). The absorber panel was a cylinder segment of 120 degrees with 4.5 m diameter. Sodium was flowing through six horizontal parallel tubes (38 mm OD and 35 mm ID) which was winding in a serpentine from the inlet header on the bottom of the cavity to the outlet header at the top (Fig. 7.10). The tubes were not welded along their length but individually supported. Sodium entered the inlet header at 270°C and exited through the outlet header

Table 7.4 **Technical specifications of cavity and external sodium receivers tested in project CRS-SSSP, Almería, Spain**

Receiver specifications	Cavity	External
Number tubes	6	5 × 39
Tube diameter (mm)	38	14
Tube wall (mm)	1.5	1
Tube material	AISI 304 H	AISI 316 L
Aperture area (m^2)	9.7	7.9
Single flow path length (m)	87	23.5
Active surface (m^2)	17	
Total surface (m^2)	62	
Total tube weight (kg)	710	300
Coating: Pyromark	2500	2500
Peak heat flux (MW/m^2)	0.63	1.38
Average heat flux (MW/m^2)	0.16	0.35
Inlet/outlet temperature (°C)	270/530	270/530
Mass flow (design) (kg/s)	7.3	7.3
Pressure (bar)	2.6	6
Pressure drop (bar)		1.5
Input/output power (MW)	2.8/2.4	2.8/2.5

at 530°C. The location of the absorber panel inside the cavity was such that the peak heat flux density reached 0.63 MW/m^2 with an average value of 0.16 MW/m^2 [20]. The cavity receiver accumulated 1005 h of operation during 1981–1983. The receiver performed reliably but with lower thermal efficiency than expected. The calculated instantaneous peak receiver efficiency from steady-state data was 87 ± 5% [2], and daily average efficiency as 66.7%.

The second receiver, so-called advanced sodium receiver or ASR, was external and formed by five panels arranged to form a rectangular absorber (Table 7.4). ASR was designed and manufactured by Franco-Tosi Industriale (Italy) and AGIP SpA (Italy). Each panel consisted of a tube bundle with 39 tubes (14 mm OD and 12 mm ID). The top header could move vertically to accommodate the thermal growth of the panel. The irradiated tubes were assembled in groups of three (Fig. 7.11). These triplets were able to grow axially with respect to the frame and even rotate.

Some concentrated solar flux was passing along the tube gaps and heated up the back-wall structure. Therefore, a double shielding of high refractory alumina-based

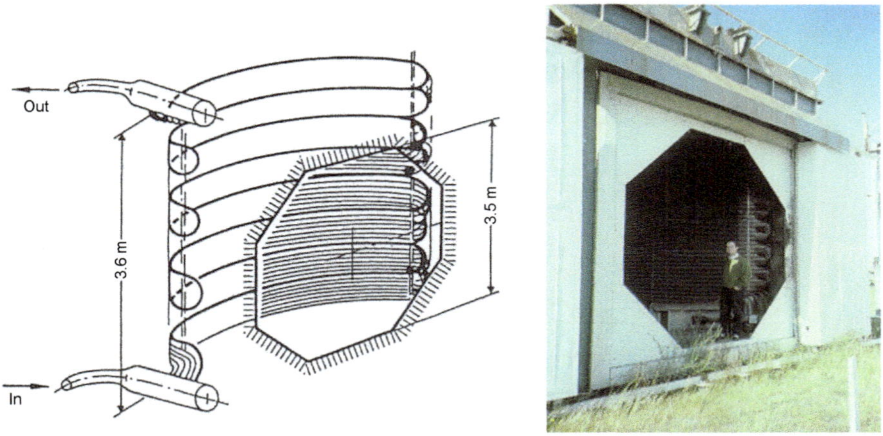

Figure 7.10 Cavity sodium receiver of SSPS project.

Figure 7.11 Billboard external sodium receiver ASR of SSPS project.

material was located behind the tube bundle system to protect the back structure from the incident radiation and to reflect and radiate this energy back to the tubes. The panel was designed for a peak heat flux density of 1.38 MW/m^2, with 0.35 MW/m^2 as average value.

Both receivers were able to produce 2.4–2.5 MW thermal power [3]. In addition, the second receiver ASR could be tested up to its very limits by exerting a peak flux density of 2.5 MW/m^2 on its absorbing surface which is still the highest flux sustained by a solar receiver (see Fig. 7.12).

In the case of the external receiver, there were five panels 2.85 m height and connected in series with each panel consisting of 39 vertical tubes that were connected in

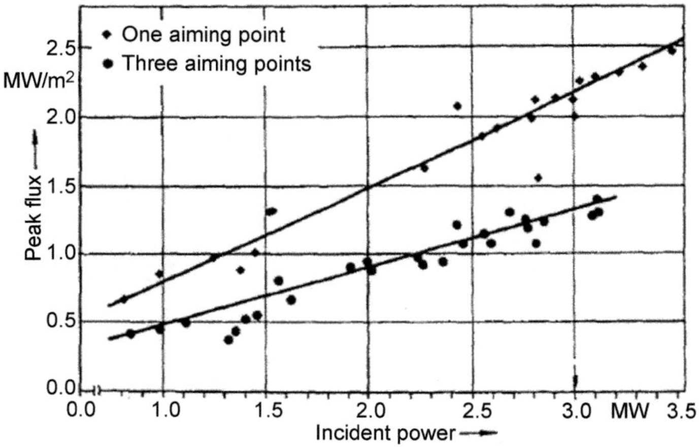

Figure 7.12 Peak flux obtained at the ASR sodium external receiver for different aiming point strategies and incident power [39].

parallel. This means that for the design liquid sodium flow of 7.3 kg/s the corresponding mass flow rate per tube was 0.187 kg/s with a series tube length of about 14 m. The tube internal diameter was 12 mm resulting in a design flow velocity of about 2 m/s, an Re number of 8−104, and a pressure drop of about 3300 Pa/m [40]. Inlet and outlet temperatures tested were 200/480, 270/530, and 280/560°C. At the most irradiated tube sections in the central panel, the maximum tube temperature measured was 690°C (design 590°C) and a maximum temperature difference between the tube surface and bulk sodium of 180°C (design 95°C). Maximum temperature could be reduced by using in the future tailored tube layouts [9]. Regarding temperature difference across the panel width, it was measured 41°C for the three-aiming point strategy and 84°C for one-aiming point strategy. Metallurgical analysis was conducted to detect eventual prompt failures and confirmed that creep damage was not found. This fact contributed to the decision to include high flux tests up to 2.5 MW/m^2. The receiver presented a 5% improvement in average daily efficiency with the one-aiming point strategy, facilitating early startup at 200 W/m^2 DNI instead of 300 W/m^2 at design point (Fig. 7.13).

Unfortunately, a sodium fire occurred at the IEA/SSPS central receiver plant in 1986. It resulted from nonconventional repair procedures undertaken to replace a valve in a sodium line [23]. The consequence was a sodium spray which resulted in 14 tons of sodium being released over 30 min (Fig. 7.14). Due to the nature of sodium fires (short flames, strong aerosol production) fire propagates by natural convection so its spread was limited; however, the accident implied the decommissioning of the facility and the interruption of the R&D activities. Since then much development has been done to model the behavior of sodium combustion and fire extinguishing [33]. In addition the cost of the fluid remains as another important factor against liquid metals since they are relatively expensive [35]. The cost of sodium (US \$2/kg) is four times more expensive than solar salt typically used in current solar towers (60−40%wt $NaNO_3$−KNO_3) and LBE is about 26 times more expensive (US \$13/kg).

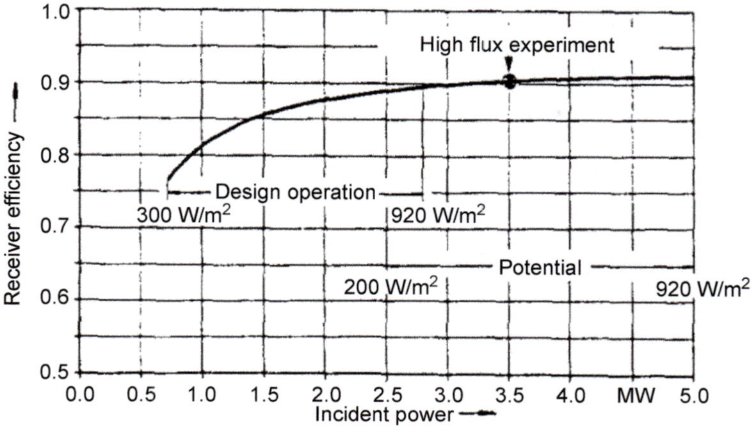

Figure 7.13 Potential performance of improved high flux sodium receivers based upon one-aiming point ASR tests at SSPS project in Almería, Spain [40].

Figure 7.14 Sodium fire at the CRS-SSPS facility in 1986, Almería, Spain.

7.3.3 Other projects with liquid metals in solar receivers

CRS-SSPS has been the only project offering experimental data after an extended test campaign. Few other initiatives can be acknowledged from a literature survey. A pioneering project took place during the early 1980s by Rockwell International and the US Department of Energy whose results were instrumental for the design of the SSPS receivers. The development involved the construction and testing of a 3.6 m^2 sodium-cooled test receiver for evaluation at the Central Receiver Test Facility (CRTF) in Albuquerque, New Mexico [36]. The receiver aperture was 3.0 m high by 1.2 m wide, and consisted of three 21-tube panels operating in parallel (AISI

316, 5.3 m long, 19 mm OD, with wall thickness 1.2 mm). The plant operated for 75 h in 1981/1982 at power levels up to 2.85 MW_{th}, 288°C/593°C inlet/outlet temperatures, solar flux up to 1.53 MW/m^2, and demonstrated satisfactory receiver control with no major receiver subsystem problems. Thermal efficiency reported was higher than 90%, although with large uncertainty in the flux measurement system and the flow meter [13].

As of 2016 the only known project of a solar thermal power plant based upon the use of liquid sodium as heat transfer fluid is Jemalong Solar Thermal Station[1] developed in Forbes, New South Wales, Australia by the company Vast Solar. Jemalong is a small plant of 1.1 MWe (6 MW_{th}) that aims at operating connected to the local grid, with 3 h of thermal storage in two tanks of liquid sodium. It presents the classical flow diagram as described in Fig. 7.9, with receiver operating at 270°C inlet temperature and 560°C outlet temperature.

Vast Solar tested in 2012 employs a sodium loop with a tubular receiver of 1.2 MW_{th} that is able to reach peak flux of 1.5 MW/m^2. The system consists of 700 small heliostats, a thermal receiver mounted on a 25-m tower, a simple heat exchanger (boiler), and energy storage tank (Fig. 7.15).

The 6-MW_{th} project started erection in 2014 and involves a multitower layout composed of five towers with 1.2-MW_{th} receiver each. The pilot plant is planned as the necessary step to scale up to a commercial plant of 30 MW.

Unfortunately, in June 2015 a fire started from a tank leaking metallic sodium, prompting the evacuation of workers and neighbors (Fig. 7.16).

The fire, however, was appropriately handled and the project continued its normal development, with the commissioning of the pilot plant started in 2016.

Figure 7.15 1.2-MW_{th} sodium receiver at Vast Solar's Jemalong Solar Thermal Station, Forbes, Australia.

[1] http://www.nrel.gov/csp/solarpaces/project_detail.cfm/projectID=4284.

Figure 7.16 Sodium fire at Jemalong Solar Station in June 2015.

7.4 Innovative power conversion cycles with liquid metals as heat transfer fluid

After the safety issues with the liquid metal storage system during operation of IEA-SSPS and Jemalong projects, it becomes clear that the use of a sodium/salt binary scheme is needed. It seems necessary to minimize the volume of sodium by restricting its use to the solar receiver loop and then use an intermediate loop with a different heat transfer fluid (e.g., solar salt) for the storage and heat transfer to the steam generator. With this strategy the system will take advantage of the main benefits of sodium and other liquid metals regarding efficiency, size, and aperture area of the solar receiver, where the heat transfer rates are important, and will avoid the low heat capacity and operational risks associated with storage and exchanger involving steam [15]. However, the binary system has an additional level of complexity resulting from the extra heat transfer loop and introduces a new source of risk in case of a leakage in the sodium/salt heat exchanger because the reaction would be strongly exothermic. Another critical issue is that the potential improvement of efficiency in the receiver compared to an all-salt plant would be slightly more than 1%. Because of that, there is little chance that liquid metals replace the well-understood solar salt central receiver systems for typical Rankine subcritical cycles.

As an alternative, future technologies may address receivers based on heat pipe concepts such as the ones already implemented in dish-Stirling systems for distributed generation [31]. Typically, in dishes with solarized Stirling engines, the receiver absorbs the light and transfers the energy as heat to the working gas, usually helium or hydrogen. Thermal fluid working temperatures are between 650°C and 750°C. This temperature strongly influences the efficiency of the engine. Because of the high operating temperatures, radiation losses strongly penalize the efficiency of the receiver; therefore, a cavity design is the optimum solution for this kind of system.

Two different heat transfer methods are commonly used in parabolic dish receivers to be used with Stirling engines or at solarized reactors for thermochemical applications [14]. In directly illuminated receivers, the same gas fluid used inside the engine is externally heated in the receiver through a pipe bundle. Although this is the most conventional method, a good high-pressure, high-velocity, heat-transfer gas such as helium or hydrogen must be used. In indirect receivers, an intermediate fluid is used to decouple solar flux and working temperature from the engine fluid or the chemical reactor. One such method is heat pipes, which employ a metal capillary wick impregnated with a liquid metal heated up through the receiver plate and vaporized. The wick structure distributes sodium across a solar-heated dome, and thermal energy is removed as sodium evaporates typically at a temperature range between 700°C and 850°C. The vapor then moves across the receiver and condenses in a cooler section, transferring the heat to the engine (Fig. 7.17). Evaporation/condensation processes guarantee good temperature control, providing uniform heating of the Stirling engine [32]. Sandia National Labs in USA demonstrated heat pipes at up to 115 kW_{th} throughput and 1 MW/m^2 peak flux. The experience with heat pipe sodium receivers has been successful also in achieving hybrid co-fired designs with extended testing campaigns at the Plataforma Solar de Almería in Spain [29].

A review published in 2015 provides insight into different design alternatives, still at conceptual level, that result from adapting solar dish technology to central receivers in solar towers with liquid–vapor phase change sodium [13]. However, it is advanced that a high risk might be associated with unstable boiling within a tube, producing hot spots and subsequent flash and steep temperature drop. Therefore, experimental testing and research is necessary to assess controllability of temperature gradients on material surface and potential failures prior to scaling up the technology.

If the technical challenges regarding materials and heat transfer in liquid–vapor sodium phase change receivers are eventually solved, the technology maybe used up to 850°C to power high-temperature thermodynamic cycles such as supercritical steam,

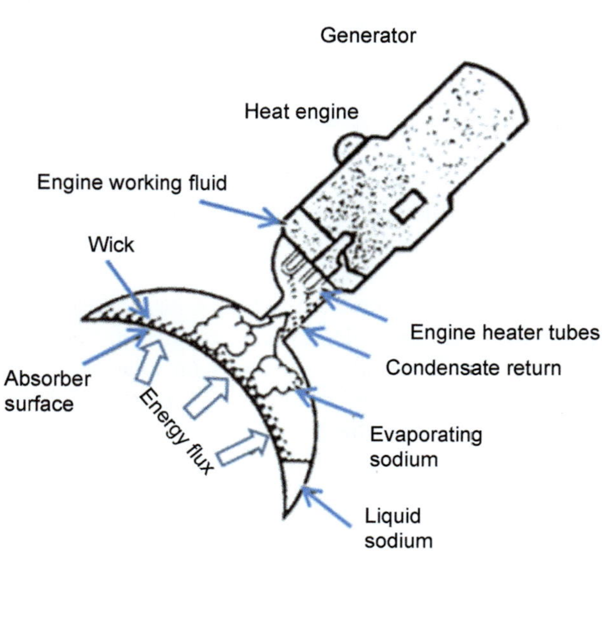

Figure 7.17 Sodium reflux solar receiver used as a buffer to control temperature and thermal transients in dish-Stirling systems [14]. © State of New South Wales (NSW Rural Fire Service). Retrieved from From http://www.abc.net.au/local/stories/2015/06/15/4254876.htm.

Table 7.5 Performance of central receiver systems making use of different thermodynamic cycles (reference case 57 MW receiver thermal outlet, heliostat field with solar multiple 2 and 6 h of thermal storage)

Nominal rate operation	Units	Brayton 1000°C	Combined cycle	Brayton sCO_2	Molten salts
Heliostats efficiency	[%]	67.8	72.1	72.1	72.1
Receiver efficiency	[%]	72.2	83.1	79.7	87.5
Thermal power to storage/ power block	[MW]	20.6	23.7	22.7	25.7
HTX efficiency	[%]	95.0	95.0	95.0	99.0
Net electrical power	[MW]	9.4	21.5	10.4	10.0
Net power cycle efficiency	[%]	47.9	42.6	48.2	40.0
Total efficiency	[%]	22.3	24.2	26.4	24.9

air Brayton, and combined cycles [45]. Higher temperatures, beyond 1000°C, might be achievable with other liquid metals such as LBE in future developments. However, it should be noted that the impact on plant efficiency is only incremental, given the convolution of efficiencies in the different subsystems, such as heliostat field, receiver, heat transfer fluid loop, heat exchangers, storage, and thermodynamic cycles. In Table 7.5 it can be observed the differences in theoretical performance for a generic CRS plant making use of molten salts and a subcritical Rankine cycle and three other advanced solutions like an intercooled-regenerated air Brayton cycle at 1000°C, a combined cycle with solar receiver working at 650°C plus additional fossil backup up to 1000°C, and a supercritical carbon dioxide Brayton cycle operating at 650°C. All plants are dispatching 57 MW receiver thermal outlet in nominal conditions and have a solar multiple of 2 and a 6-h equivalent thermal storage. As it can be observed, less than 2% maximum improvement in nominal efficiency can be achieved. It is therefore not trivial to assess if there is a real niche for technical solutions integrating liquid metals where still costly hardware, technical challenges and risks associated need substantial R&D effort. At present there are strong competitors in other heat transfer fluids also looking for higher operating temperatures like, for example, air streams highly charged in particles [6] or new formulations of molten salts such as carbonates or chloride-based salts. Molten salts are considered nowadays the commercial mature option of CRS plants, though still a major issue is their relatively high corrosive nature to metal alloys. Many new molten salts are being proposed, but their corrosion characteristics are not available in the literature [43].

An important medium- to long-term niche for the application of sodium, and other liquid metals, is the integration into direct conversion systems where sodium might be the working fluid itself in the cycle, and therefore provide disruptive schemes with high efficiency and the removal of the turbomachinery and the entire power block.

In particular there are two technologies that are retained for CSP integration studies at system level, such as AMTEC (Alkali Metal Thermal to Electric Converter) cells and LMMHD (Liquid Metal Magneto-Hydro-Dynamic) power conversion systems. Studies in 2015 focused on conceptual insight and thermodynamic analysis [18].

AMTEC can benefit from previous experience in heat pipes with liquid–vapor phase change in solar receivers. The AMTEC is an electrochemical device that uses a recirculating alkali metal (sodium or potassium) working fluid passing through a solid electrolyte in a closed circuit to produce an electron flow in an external load. Typical solids used are ceramic electrolytes such as β'' or P'' alumina, which are very good conductors of ions but poor conductors of electrons, due to their crystal structure. In addition they withstand high temperature differences between anode and cathode surfaces. The liquid metal is driven around a closed thermodynamic cycle between a heat source and a heat sink held at different temperatures and, during the vapor phase of the cycle, the available work from the isothermal expansion of the working fluid as it passes through the electrolyte is converted directly into electric power. The AMTEC device is characterized by high potential efficiencies and no moving parts except the liquid metal itself. It accepts a heat input in a range from about 600°C to 1000°C and produces direct current with predicted device efficiencies of 10–30%. It can be used as a topping cycle with a bottoming Rankine cycle [18]. Most of solarized liquid–vapor sodium heat pipes in the past made use of embedded systems into or attached to the receiver to improve the isothermal characteristics of the receiver wall or to enhance heat transfer with the heated fluid (thermal carrier); however, 2015 experimental research at laboratory-scale has demonstrated the attractiveness of using loop-pipe configurations. Loop-type heat pipe has the potential to enhance thermal transport capabilities by separating the liquid and vapor lines and thus reducing the fluid dynamic resistance at the liquid–vapor interface that results from liquid and vapor flowing in the opposite direction to one another. Considering that a solar receiver is usually tilted, a thermosiphon heat pipe can be utilized, as long as the condenser section of the heat pipe is located in a higher position in the gravity field [10].

In an LMMHD generator, a highly electro-conductive two-phase mixture composed of a liquid metal and a gas (or vapor) moves across a magnetic field and thus generates electrical power. The two-phase flow is propelled by the expanding gas bubbles and the gas goes through the thermodynamic cycle. Research and development on LMMHD energy conversion systems was started in the beginning of the 1960s and some solar-assisted designs have been proposed since 1980s [24]. In many cases, solar designs have been conceived for low temperatures from 80°C to 300°C [11]. At low temperature operation of the solar assisted–LMMHD power generator, the MHD duct becomes free from many problems, namely electrode and duct life, erosion-corrosion, preheating, cooling, and so on, and the overall system is free from the emission of any type of environmental-unfriendly chemical species.

In order to accelerate the liquid metal flow, there are several possibilities. In a one-component flow, the liquid metal becomes partially gaseous in the receiver and then the changes of density accelerate the flow. In a two-component operation, the thermodynamic working fluid (gas/steam) and electrodynamic fluid (liquid metal) are mixed. Possible combinations are alkali metals with helium or argon, but also lead or lead alloys with water. This mixer is similar to a direct contact heat exchanger and acts in fact

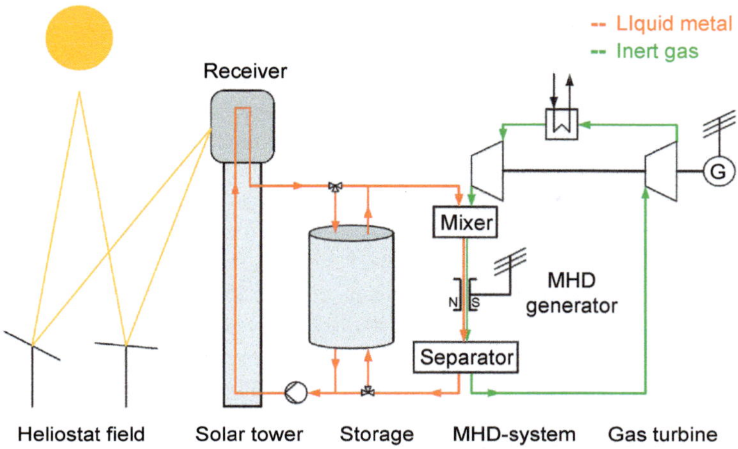

Figure 7.18 Conceptual scheme for the integration of a liquid metal magneto-hydro-dynamic converter into a solar central receiver system [18].

like a pump. A tentative flow diagram integrating a CRS plant with an LMMHD generator is depicted in Fig. 7.18. The LMMHD generator can be used as a toping cycle with a bottoming Brayton or Rankine cycle [18]. The theoretical efficiencies of the device can exceed 60% [1].

7.5 Conclusions and outlook

Sodium has been considered as a good candidate for heat transfer fluid in solar thermal power plants already at early experimental projects such as the IEA/SSPS in Spain. Sodium gathers extensive experience, since it is proposed in nuclear industry for liquid metal fast breeder reactors, existing practical know-how in sensitive aspects as safety and control and hardware validation of key components like pumps, valves, lines, and steam generators. Its boiling temperature (890°C) is substantially higher than solar salts and presents outstanding thermal conductivity. The vapor pressure at 595°C is only slightly above atmospheric pressure. Its main shortcoming is that it reacts with water and air, and therefore high maintenance costs are associated. The highly reactive nature of sodium and water is an important consideration in the design of sodium components, principally the sodium steam generator, and potentially increases the cost of these components. Melting point is relatively high (98°C) and because of that heat tracing is required in the heat transfer fluid loop like in the case of molten salts.

Molten salt is cheaper than sodium by a factor of two and has a three-to-one advantage in its volumetric heat capacity, factors which are particularly important in the thermal storage subsystem. On the other hand, sodium has a five times higher heat transfer rate, which means that sodium receivers (like water/steam receivers) can be single pass. The thermal conductivity of liquid metals leads to high heat transfer coefficients. For flow velocities of 1 m/s for a single tube with $D = 12$ mm at 400°C, molten salt reaches about 4600 W/m^2 K, whereas over 50,000 W/m^2 K can be achieved with liquid sodium. As a consequence, liquid sodium receivers may operate with high peak flux of solar radiation (1.4 MW/m^2), much higher than solar salts (0.8 MW/m^2), and therefore an absorber area reduction of up to 57% can be achieved.

Two 2.5-MW sodium receivers were tested at the IEA/SSPS project between 1983 and 1986, one of them in a cavity and second one external. Extended test campaigns could demonstrate high efficiency of about 90% for inlet/outlet temperature of 280/560°C. High flux tests could achieve peaks of 2.5 MW/m^2. In spite of the outstanding thermal performance of the receiver, a fire accident in 1986 supposed the decommissioning of the facility and the interruption of the R&D activities. Only a significant follow-up project can be reported since then, corresponding to the Jemalong Solar Thermal Station in Forbes, Australia, a multitower system of 6 MW$_{th}$. Once again the risks associated with maintenance and the existence of a fire at the storage tanks are raising uncertainty for the future use of this technology.

Alternative approaches are taking place to move forward with the use of liquid metals efficiently and safely. Other liquid metals like Na−K mixtures or LBE are being explored at laboratory-scale to assessing performance and understanding operational issues. Regarding sodium, after the problems encountered with the operation and maintenance of storage tanks, it seems reasonable to limit its use confined to the receiver that is where the heat transfer properties of liquid metals are advantageous. In that sense, the adoption of evaporation/condensation exchangers as heat pipe solar receivers is proposed for future projects. Liquid−vapor phase change sodium integrated receivers may reach temperatures above 800°C providing CRS plants access to the use of Brayton cycles or advanced direct conversion systems. In particular, there are at present two direct conversion technologies that are retained for CSP integration studies at system level, such as AMTEC cells and LMMHD power conversion systems.

References

[1] Baker R, Tessier M. Handbook of electromagnetic pump technology. New York: Elsevier Science Publishing Co. Inc.; 1987.

[2] Baker AF. International energy agency (IEA) small solar power systems (SSPS) sodium cavity and external receiver performance comparison. Albuquerque (New Mexico): Sandia National Laboratories; 1987. Report number SAND87−8021.

[3] Becker M, Vant-Hull LL. Thermal receivers. In: Winter CJ, Sizmann RL, Vant-Hull LL, editors. Solar power plants. Berlin: Springer-Verlag; 1991, ISBN 3-540-18897-5. p. 163−97.

[4] Becker M, Boehmer M. Achievements of high and low flux receiver development. In: Bloss WH, Pfisterer F, editors. Proceedings ISES solar world congress, Hamburg, 1987. Oxford (UK): Pergamon Press; 1988. p. 1744−83.

[5] Benoit H. Récepteur solaire tubulaire à suspension dense de particules en écoulement ascendant. 2015. Thesis available at: http://www.theses.fr/2015PERP0045.

[6] Benoit H, Pérez López I, Gauthier D, Sans JL, Flamant G. On-sun demonstration of a 750°C heat transfer fluid for concentrating solar systems: dense particle suspension in tube. Solar Energy 2015;118:622−33.

[7] Benoit H, Spreafico L, Gauthier D, Flamant G. Review of heat transfer fluids in tube-receivers used in concentrating solar thermal systems: properties and heat transfer coefficients. Renewable and Sustainable Energy Reviews 2016;55:298−315.

[8] Boerema N, Morrison G, Taylor R, Rosengarten G. Liquid sodium versus Hitec as a heat transfer fluid in solar thermal central receiver systems. Solar Energy 2012;86:2293–305.
[9] Boerema N, Morrison G, Taylor R, Rosengarten G. High temperature solar thermal central-receiver billboard design. Solar Energy 2013;97:356–68.
[10] Boo JH, Kim SM, Kang YH. An experimental study on a sodium loop-type heat pipe for thermal transport from a high-temperature solar receiver. Energy Procedia 2015;69: 608–17.
[11] Branover H, El-Boher A, Yakhot A. Testing of a low-temperature liquid-metal MHD power system. Energy Converservation Management 1982;22:163–9.
[12] Cheng X, Tak N. Investigation on turbulent heat transfer to lead-bismuth eutectic flows in circular tubers for nuclear applications. Nuclear Engineering and Design 2006;236(4): 385–93.
[13] Coventry J, Andraka C, Pye J, Blanco M, Fisher J. A review of sodium receiver technologies for central receiver solar power plants. Solar Energy 2015;122:749–62.
[14] Diver RB. Receiver/reactor concepts for thermochemical transport of solar energy. Journal of Solar Energy Engineering 1987;109(3):199–204.
[15] Falcone PK. A handbook for solar central receiver design. Livermore (USA): Sandia National Laboratories; 1986. SAND86-8009.
[16] Flesch J, Fritsch A, Cammi G, Marocco L, Fellmoser F, Pacio J, Wetzel T. Construction of a test facility for demonstration of a liquid lead-bismuth-cooled 10 kW thermal receiver in a solar furnace arrangement – SOMMER. Energy Procedia 2015;69:1259–68.
[17] Foust OJ. Sodium-NaK engineering handbook. New York: Gordon and Breach, Science Publishers Inc.; 1972.
[18] Fritsch A, Flesch J, Geza V, Singer C, Uhlig R, Hoffschmidt B. Conceptual study of central receiver systems with liquid metals as efficient heat transfer fluids. Energy Procedia 2015; 69:644–53.
[19] Gomez-Garcia F, González-Aguilar J, Olalde G, Romero M. Thermal and hydrodynamic behavior of ceramic volumetric absorbers for central receiver solar power plants: a review. Renewable and Sustainable Energy Reviews 2016;57:648–58.
[20] The IEA/SSPS Solar Thermal Power Plants — Facts and Figures — Final Report of the International Test and Evaluation Team (ITET). Volume 1: Central Receiver System (CRS). Kesselring, Paul, Selvage, Clifford S. (Eds.). Springer Verlag Heidelberg GmbH.
[21] Grasse W, Hertlein HP, Winter CJ. Thermal solar power plants experience. In: Winter CJ, Sizmann RL, Vant-Hull LL, editors. Solar power plants. Berlin: Springer-Verlag; 1991, ISBN 3-540-18897-5. p. 215–82.
[22] Heinzel A, Weisenburger A, Müller G. Corrosion behavior of austenitic steels in liquid lead bismuth containing 10^{-6} wt% and 10^{-8} wt% oxygen at 400–500°C. Journal of Nuclear Materials 2014;448(1–3):163–71.
[23] IEA-SSPS. SSPS-CRS sodium fire incident. International Energy Agency; August 1986 [Small Solar Power Systems Project Report].
[24] Kaushik SC, Verma SS, Chandra A. Solar-assisted liquid metal MHD power generation: a state of the art study. Heat Recovery Systems & CHP 1995;15(7):675–89.
[25] Kistler BL. A user's manual for DELSOL3: a computer code for calculating the optical performance and optimal system design for solar thermal central receiver plants. USA: Sandia National Laboratories; 1986. Sandia Report, SAND-86–8018.
[26] Kistler BL. Fatigue analysis of a solar central receiver design using measured weather data. Livermore: Sandia National Laboratories; 1987. SAND86–8017.
[27] Kotzé JP, Erens PJ, Von Backström TW. NaK as a primary heat transfer fluid in thermal solar power installations. In: Proceeding of SolarPACES 2012, 11–14 September, Marrakech, Morocco; 2012.

[28] Kribus A. Future directions in solar thermal electricity generation. In: Solar thermal electricity generation. Colección documentos CIEMAT. Madrid, Spain: CIEMAT; 1999, ISBN 84-7834-353-9. p. 251–85.

[29] Laing D, Palsson M. Hybrid dish/stirling systems: combustor and heat pipe receiver development. Journal of Solar Energy Engineering 2002;124.

[30] Lyon RN. Liquid metal heat transfer coefficients. Chemical Engineering Progress 1951;47: 75–9.

[31] Mancini T, Heller P, Butler B, Osborn B, Schiel W, Goldberg V, Buck R, Diver R, Andraka C, Moreno J. Dish-stirling systems: an overview of development and status. Journal of Solar Energy Engineering 2003;125:135–51.

[32] Moreno JB, Modesto-Beato M, Rawlinson KS, Andraka CE, Showalter SK, Moss TA, Mehos M, Baturkin V. Recent progress in heat-pipe solar receivers. SAND2001-1079. In: 36th intersociety energy conversion engineering Conference, Savannah, GA; 2001. p. 565–72.

[33] Olivier TJ, Radel RF, Nowlen TK, Blanchat TK, Hewson JC. Metal fire implications for advanced reactors part 1: literature review. Sandia National Laboratories; 2007. SAND2007–6332.

[34] Pacio J, Singer C, Wetzel T, Uhlig R. Thermodynamic evaluation of liquid metals as heat transfer fluids in concentrated solar power plants. Applied Thermal Engineering 2013;60: 295–302.

[35] Pacio J, Wetzel T. Assessment of liquid metal technology status and research paths for their use as efficient heat transfer fluids in solar central receiver systems. Solar Energy 2013;93: 11–22.

[36] Rockwell International. Sodium solar receiver experiment – final report. 1983. Report number SAND82–8192.

[37] Romero M, González-Aguilar J. Solar thermal CSP technology. Wires Energy Environment 2014;2014(3):42–59.

[38] Romero M, Gonzalez-Aguilar J, Zarza E. Concentrating solar thermal power. In: Goswami Y, Kreith F, editors. Energy efficiency and renewable energy handbook. 2nd ed. Boca Raton, Florida: CRC Press Taylor & Francis Group; 2015.

[39] Schiel WJC, Geyer M, Carmona R. The IEA/SSPS high flux experiment. Springer-Verlag; 1987.

[40] Schiel WJC, Geyer MA. Testing an external sodium receiver up to heat fluxes of 2.5 MW/m^2: results and conclusions from the IEA-SSPS high flux experiment conducted at the central receiver system of the Plataforma Solar de Almeria (Spain). Solar Energy 1988;41(3):255–65.

[41] Siebers DL, Kraabel JS. Estimating convective energy losses from solar central receivers. Livermore: Sandia National Laboratories; April 1984. SAND84–8717.

[42] Sleicher CA, Rouse MW. A convenient correlation for heat transfer to constant and variable property fluids in turbulent pipe flow. International Journal of Heat and Mass Transfer 1975;18(5):677–83.

[43] Vignarooban K, Xinhai X, Arvay A, Hsu K, Kannan AM. Heat transfer fluids for concentrating solar power systems – a review. Applied Energy 2015;146:383–96.

[44] Weeks J. Lead, bismuth, tin and their alloys as nuclear coolants. Nuclear Engineering and Design 1971;15:363–72.

[45] Pacioa J, Fritsch A, Singer C, Uhlig R. Liquid metals as efficient coolants for high-intensity point-focus receivers: implications to the design and performance of next generation CSP systems. Energy Procedia 2014;49:647–55.

Part Four

Advances in the power block and thermal storage systems

Supercritical CO_2 and other advanced power cycles for concentrating solar thermal (CST) systems

8

S.M. Besarati, D.Y. Goswami
University of South Florida, Tampa, FL, United States

8.1 Introduction

Over the past few years there has been a growing interest in generating electricity using solar thermal energy due to the potential of low-cost thermal-energy storage. Among the solar thermal power technologies, the power tower has shown the best promise because of its higher operating temperatures, which leads to better thermal-to-electric efficiency and lower overall cost. If molten salt is used as the heat transfer fluid (HTF), the temperature of the fluid leaving the receiver is about 550°C. Higher temperatures are achievable by using other types of HTFs such as air. As of 2016, traditional subcritical steam Rankine cycle (SRC) is being employed in these plants, which limits the efficiency. Therefore, it is clear that the potential of power-tower plants will not be fully utilized until new types of power cycles are put into service [11]. According to goals set in the SunShot program by the US Department of Energy (DOE), next generation of power blocks should offer a thermal efficiency close to 50% in comparison with today's SRC efficiency of 42% or less. Such high thermal efficiency needs to be achieved under dry cooling conditions because solar power plants are typically located in arid areas where water is most scarce [14]. In this chapter, supercritical SRCs, air and helium Brayton cycles, and supercritical carbon dioxide power cycles are described and discussed as the near-term potential candidates, which can fulfill such requirements.

Power conversion efficiency can be further increased by using combined power cycles [9]. In this case, a bottoming cycle utilizes the waste heat from the top cycle resulting in higher efficiency and less heat rejection to the environment. Depending on the operating temperatures, different power cycles can be employed as bottoming cycles including the subcritical SRC. In this chapter, organic Rankine cycle (ORC), supercritical organic Rankine cycle (SORC), and two absorption power cycles, namely Kalina cycle and Goswami cycle, are discussed as potential candidates.

Advances in Concentrating Solar Thermal Research and Technology
http://dx.doi.org/10.1016/B978-0-08-100516-3.00008-3

8.2 Stand-alone cycles

In this section, stand-alone cycles that can be used for power generation in concentrating solar thermal (CST) plants are described. The section is divided into two main categories of SRCs and gas Brayton cycles.

8.2.1 Steam Rankine cycles

The SRC has been used in all existing CST plants. A schematic of a typical SRC cycle including feed water heaters and a reheater is shown in Fig. 8.1. The operating conditions are mainly dependent on the solar technology. In solar energy generating systems (SEGS) plants which use parabolic trough concentrators as the solar collectors and synthetic oil as the HTF, superheated steam is generated at 371°C and 100 bar. Using molten salt as the HTF or direct steam generation in solar power-tower plants increases the steam temperature to about 540°C, resulting in higher thermal-to-electric efficiency in the range of 42% with a wet-cooled condenser [14].

Beside increasing the temperature, elevating the steam pressure to more than its critical pressure (222 bar) leads to supercritical steam Rankine cycles (SSRCs), which give higher thermal efficiency [30]. Although SSRCs have not been utilized in the current CST plants, they have been successfully deployed in fossil fuel power plants [2].

Depending on the cycle configuration and operating temperatures, there is an optimum pressure above which the thermal efficiency does not improve much. For example, in a reheat cycle for the main/reheat temperatures of 600°C/620°C, increasing the main steam pressure above 300 bar does not offer any further economic advantage [8]. As of 2016, the high temperature of SSRCs is limited to about 627°C because of the limitation of ferritic steels. Other materials are under evaluation to operate at higher temperatures while special attention is being given to high-nickel alloys [32].

It is noteworthy that improvement in the efficiency by using supercritical steam, instead of superheated steam, comes at higher material costs. Cheang et al. [2] investigated the associated costs of the two cycles in CST plants. They concluded that using supercritical steam leads to an efficiency improvement of 4.6% while the power block capital cost is increased by 31.9%. They found the cost per kilowatt for the two plants to be about the same, which does not justify using the high-pressure working fluid. This conclusion may not hold for fossil fuel power plants, where saving fuel over the lifetime of the plant, as well as taking advantage of economy of scale (more than 400 MW), results in a lower operating cost of the SSRCs. Therefore, comprehensive lifetime cost analysis is required before deploying SSRC in the CST plants.

8.2.2 Gas Brayton cycles

Although all the operational CST plants are based on using SRCs, there is a growing interest in generating power at higher temperatures, and consequently, higher efficiencies. This requires the use of high-temperature solar receivers. Gas Brayton

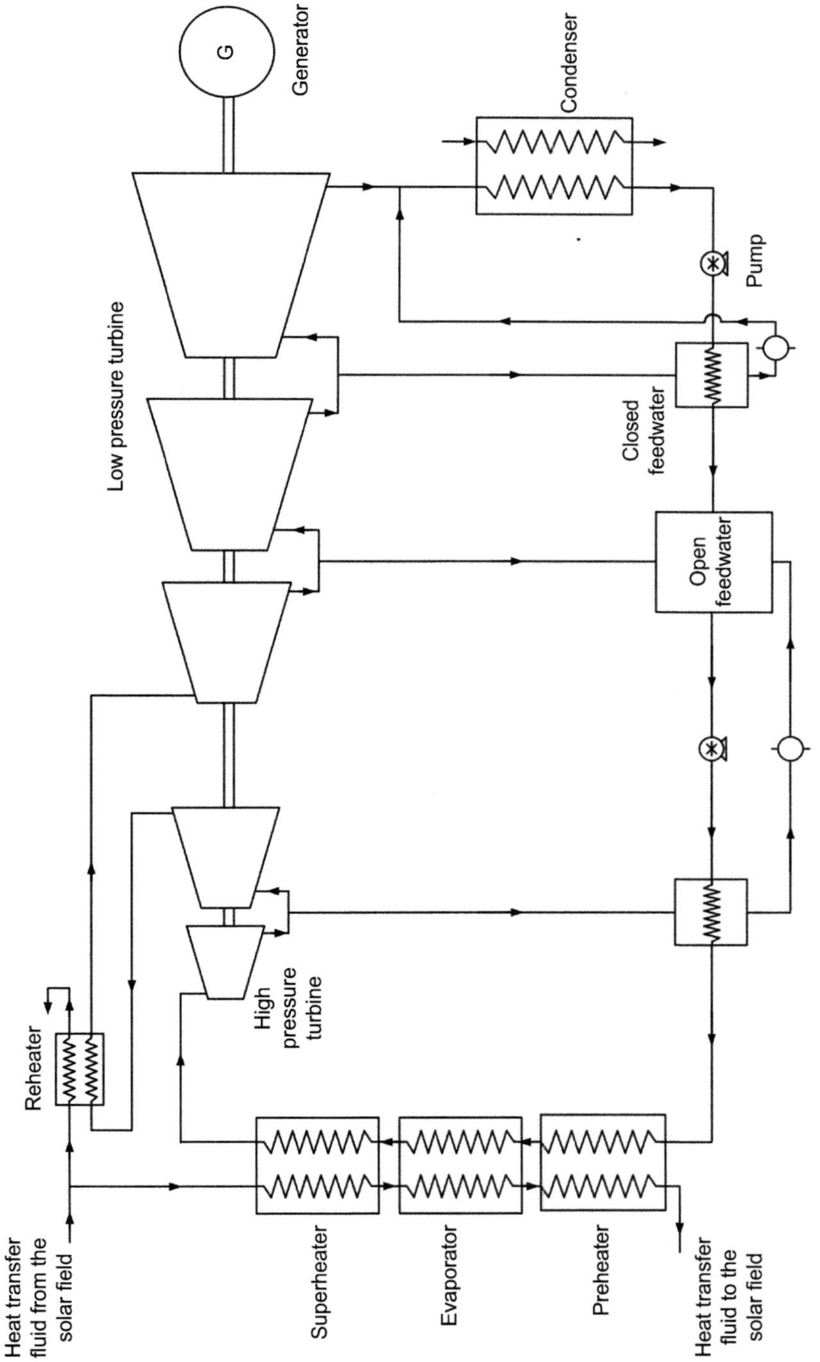

Figure 8.1 Regenerative Rankine cycle.

cycles, which have already been employed for a long time in fossil fuel power plants, are potential candidates for use in the future CST plants. Air, helium, and carbon dioxide can all be used as the working fluids.

8.2.2.1 Air Brayton cycle

Integration of the solar power-tower technology with existing gas turbines is a promising concept, which reduces the capital, and operational and maintenance costs. In this kind of technology, CST is used to preheat the pressurized air before entering the combustion chamber of the gas turbine (Fig. 8.2). Solar air preheating increases the compressed air temperature which leads to less fuel consumption. Moreover, in spite of the fluctuations in solar power input, the turbine inlet temperature remains constant and the output power is fully dispatchable. This reduces the losses due to frequent start-ups and shutdowns and part-load operation in the conventional CST plants, which results in higher system efficiency [28]. As shown in Fig. 8.2, an SRC can also be used as a bottoming cycle to further improve the thermal efficiency.

The temperature of the air entering the combustion chamber is dependent on the type of the solar receiver. Both molten salt [26] and pressurized volumetric receivers [15] can be employed. In the case of volumetric receivers, the operating pressure is in the range of 15–20 bar.

8.2.2.2 Helium Brayton cycle

Closed-loop Brayton cycles are recognized as highly efficient power cycles at relatively moderate top cycle temperatures. The high-temperature fluid leaving the turbine

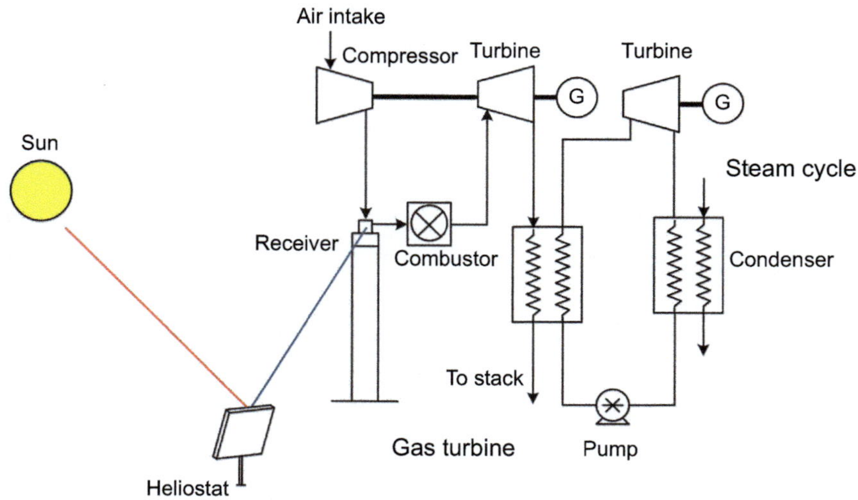

Figure 8.2 Air Brayton cycle with solar preheating.
Adapted from Schwarzbözl P, Buck R, Sugarmen C, Ring A, Crespo MJM, Altwegg P, Enrile J. Solar gas turbine systems: design, cost and perspectives. Solar Energy 2006;80(10):1231–40.

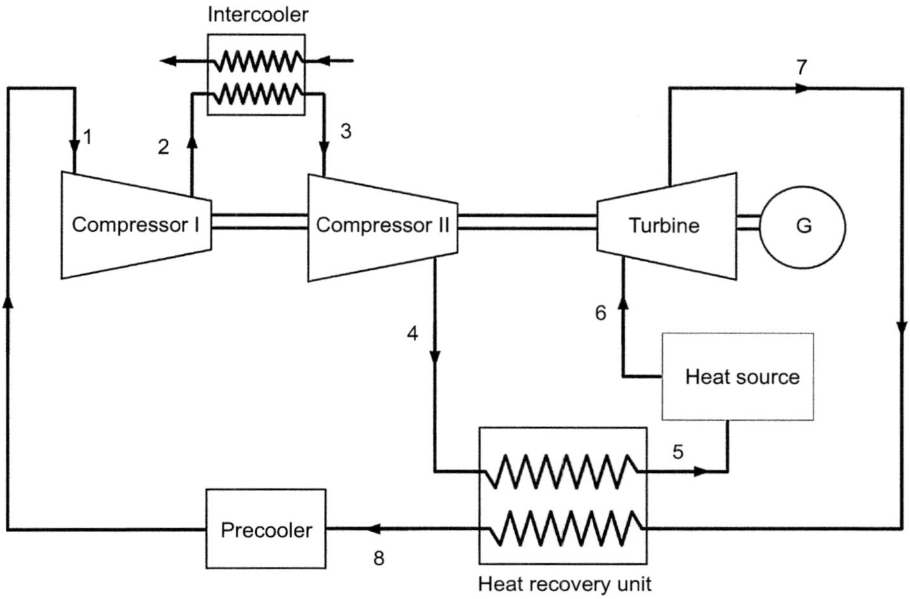

Figure 8.3 Closed-loop Brayton cycle.

goes through a heat recovery unit where its thermal energy is transferred to the stream leaving the last compressor (Fig. 8.3). Using intercooling between the compressors and reheat between the turbines can further improve the efficiency. However, the efficiency benefit of reheating and intercooling decreases with every additional stage [8].

A closed-loop Brayton cycle with helium as the working fluid has been considered for nuclear applications. It has been shown that a net cycle efficiency up to 48% can be achieved when the turbine inlet temperature is around 850°C to 900°C. However, for a sodium cooled fast reactor, the maximum working fluid temperature is limited to 500–550°C [39], which is achievable in the power-tower plants.

8.2.2.3 Supercritical carbon dioxide Brayton cycles

Carbon dioxide is a nontoxic, inexpensive, nonflammable, and highly stable compound with low critical properties, that is, 7.38 MPa and 30.98°C (314.13 K). Around the critical point, CO_2 is not an ideal gas, and its behavior is very sensitive to pressure and temperature. In other words, the fluid properties vary significantly around the critical point. Fig. 8.4 shows the density variations of CO_2 at different operating conditions. As can be seen, the density is very high around the critical point and comparable to liquid density. Therefore, the compression work is considerably reduced if carbon dioxide enters the compressor close to the critical condition, which is the main advantage of supercritical carbon dioxide (s-CO_2) over the other working fluids. Wright et al. [33] compared the density of s-CO_2 in a closed-loop Brayton cycle with that of water. At the specified condition, the density of the CO_2 at the inlet of the

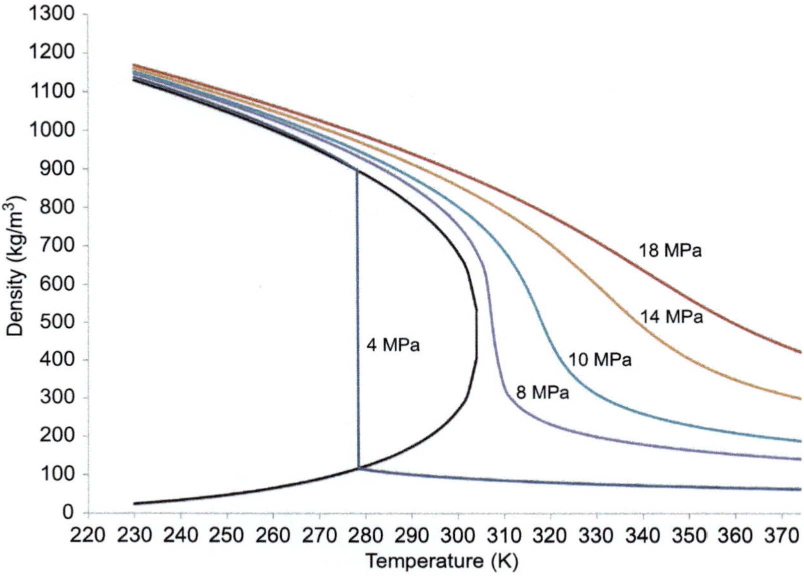

Figure 8.4 Variations of CO_2 density at different temperatures and pressures.

compressor is 60% of the density of water, which results in low compression power requirement.

In addition to the density, other properties of CO_2 also change drastically around the critical point. As can be seen from Fig. 8.5, the thermal conductivity of CO_2 maximizes close to the critical point reaching 148.95 $\frac{\text{mW}}{\text{m-K}}$ at 305 K. From Refprop [16], the

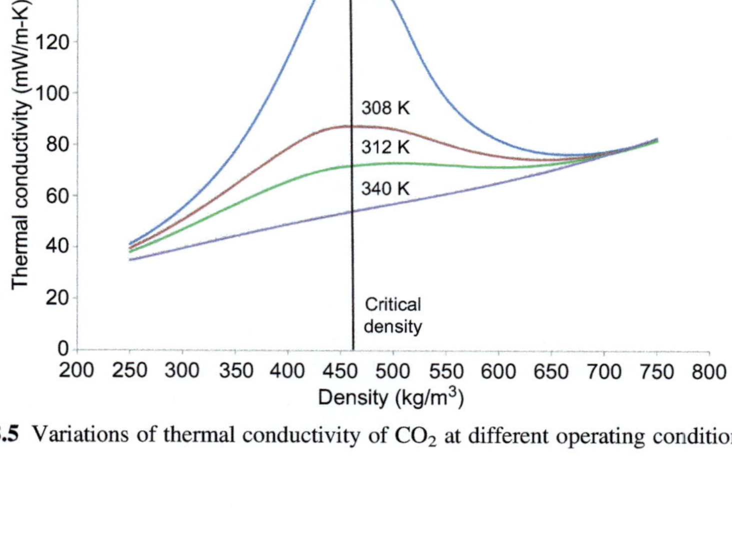

Figure 8.5 Variations of thermal conductivity of CO_2 at different operating conditions.

thermal conductivity of water at 305 K is 618.41 $\frac{\text{mW}}{\text{m-K}}$, while the thermal conductivity of air is 26.355 $\frac{\text{mW}}{\text{m-K}}$ at the atmospheric pressure and 305 K temperature.

Fig. 8.6 shows how the specific heat of CO_2 changes close to the critical point. Large variations in specific heat affect the recuperator design in the power cycle. It is known that for a certain operating condition, a pinch-point exists in the recuperator. The pinch-point is the location where the temperature difference between the hot and cold streams is the lowest. As the specific heat varies radically with the changes in pressure and temperature, the temperature difference between the fluids varies widely within the recuperator. Consequently, the pinch-point location may be found somewhere inside the recuperator. Therefore, more detailed analysis of the temperature profiles is necessary to evaluate the performance of the recuperator [6]. Moreover, the recuperator size and efficiency are directly affected by the operating pressure. Therefore, unlike the recuperators for ideal gases, such as helium, where the temperature difference is almost constant and dependent only on the pressure ratio and temperatures, the operating pressure is also important and has to be optimally determined. In addition, the high specific heat of CO_2 close to the critical point requires high mass flow rate of cooling water in the precooler which increases the parasitic losses [6].

There are multiple configurations of s-CO_2 Brayton cycle in the literature. The three configurations that are introduced in this chapter are, simple, recompression, and partial cooling cycles.

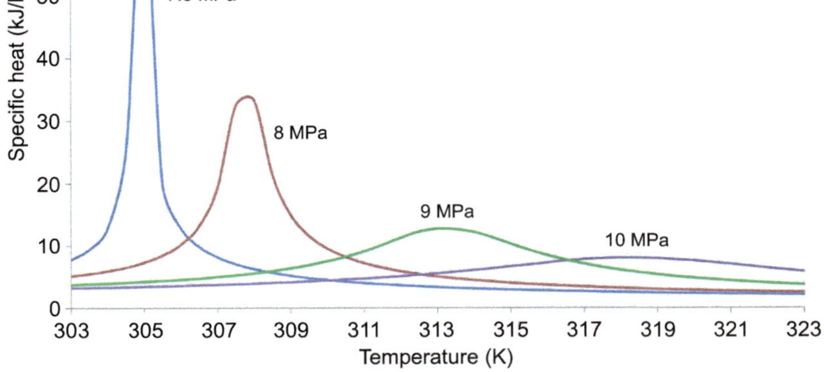

Figure 8.6 Variations of the specific heat close to the critical temperature at different pressures.

Simple cycle

The simple cycle configuration and the corresponding temperature—entropy (*T-s*) diagram are shown in Fig. 8.7. Similar to the helium Brayton cycle, a heat recovery unit is used to transfer thermal energy from the high-temperature fluid leaving the turbine to the low-temperature stream. Intercooling and reheat can further increase the thermal efficiency.

Recompression cycle

In this configuration (Fig. 8.8), the fluid leaving the low-temperature recovery (LTR) unit is divided into two streams. One stream goes through the precooler, main

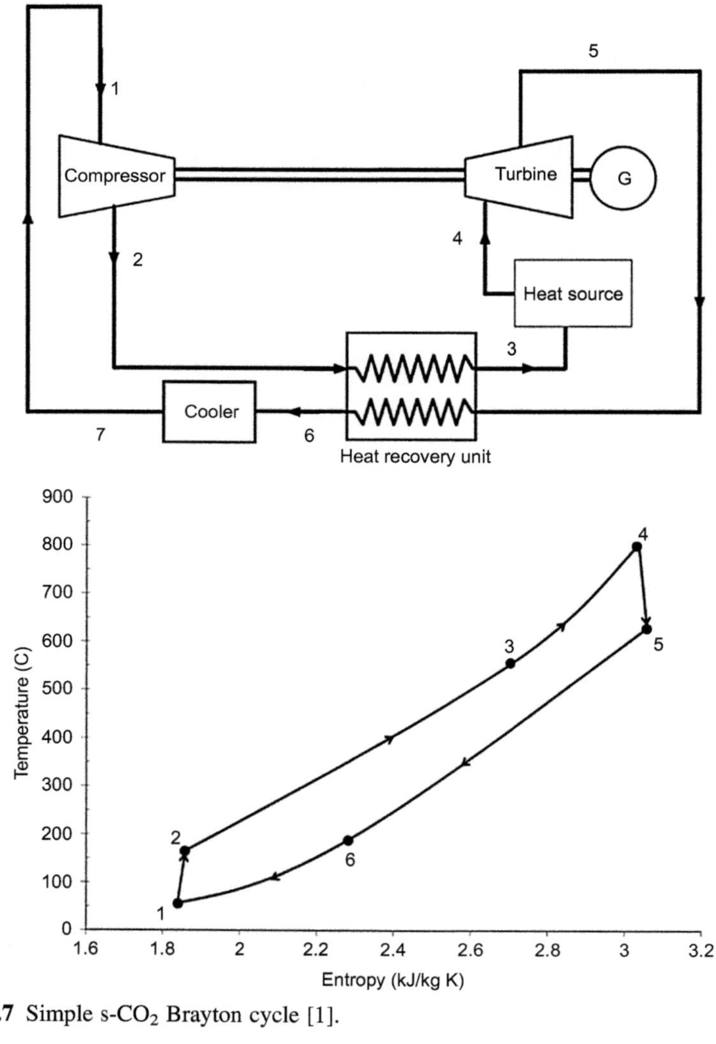

Figure 8.7 Simple s-CO_2 Brayton cycle [1].

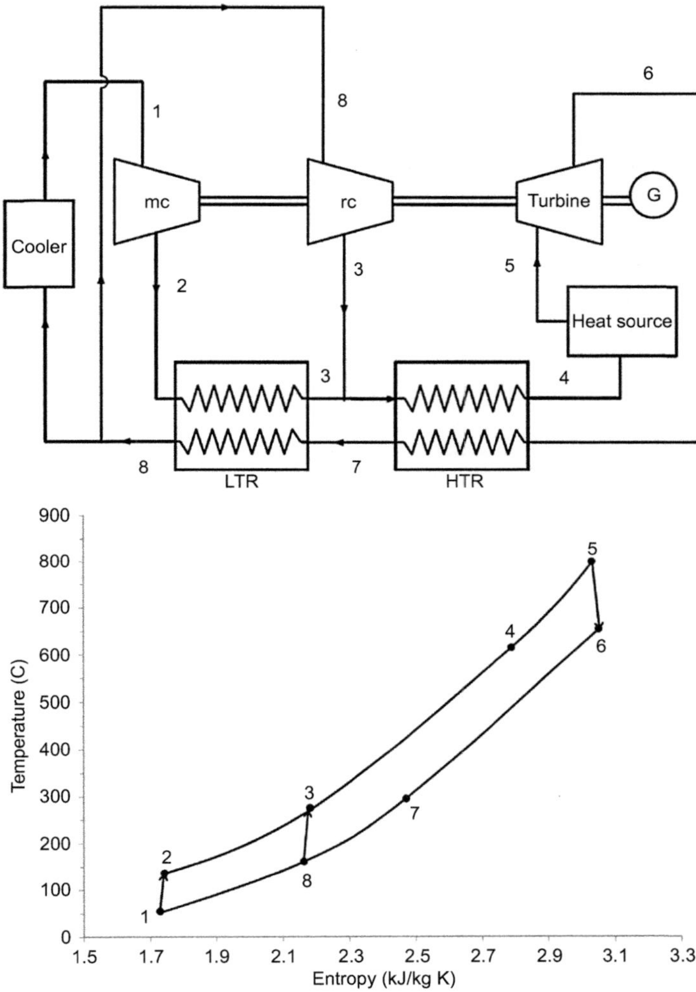

Figure 8.8 Recompression s-CO_2 Brayton cycle [1].

compressor and LTR while the other stream is directly pressurized by a recompression compressor. The two streams are mixed before the high-temperature recovery (HTR). Then, thermal energy is added to achieve the required turbine inlet temperature. After expanding in the turbine, the flow is directed into HTR and LTR to preheat the high-pressure, low-temperature stream. The main advantage of this configuration over the simple cycle is more efficient heat recovery. By splitting the flow, the heat capacity of the high-pressure stream in the LTR decreases, which helps to avoid common pinch-point problems. The fraction of the flow going through the recompression compressor is an important design parameter, which directly affects the cycle thermal efficiency.

Partial cooling cycle

This configuration (Fig. 8.9) has one more compressor and cooler in comparison with the recompression cycle. Before entering the precompressor, the low-pressure stream leaving the LTR is cooled. Then, the flow pressure is increased to an intermediate level. Next, the flow is divided into two streams. One stream enters the main compressor after rejecting heat and the other goes through the recompression compressor. The two streams are mixed before entering the HTR. Two-stage compression and the additional cooler lead to lower compression work in comparison with the previous configurations. Among these configurations the simple cycle is the least efficient. However, its simplicity and fewer numbers of components increase the potential of early commercial market entry. No definite conclusion can be reached yet, as to which of the other two configurations is more advantageous for future CST plants.

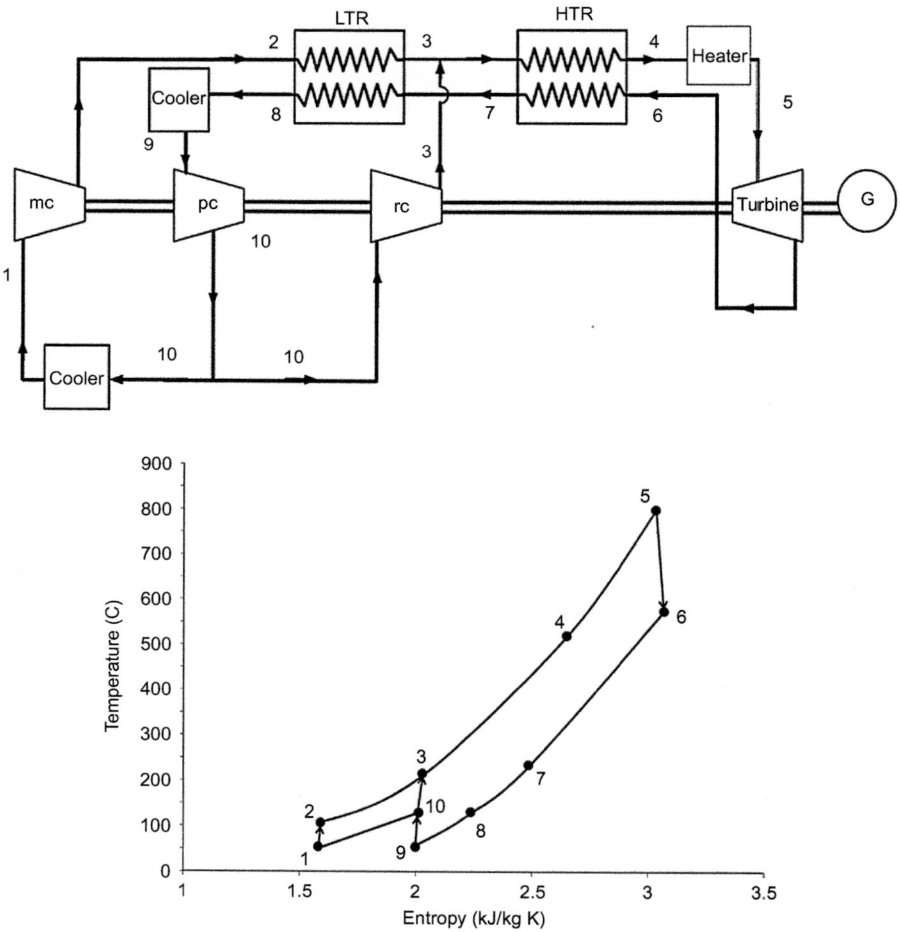

Figure 8.9 Partial cooling s-CO_2 Brayton Cycle [1].

Modeling the two cycles with the same effectiveness values for the heat exchangers leads to quite similar thermal efficiencies. However, more detailed analysis shows that the partial cooling cycle reaches higher efficiency than the recompression configuration when the two are compared, based on equal recuperator conductance values (UA). The difference between the thermal efficiencies can be as high as 4% for low UA values, and becomes almost zero when the UA is higher than 15 MW/K [21]. Therefore, assuming the cost of the cycle is largely dependent on the recuperator conductance, the partial cooling configuration is superior for low conductance values.

In addition, numerical analysis shows that the temperature difference between the inlet and outlet of the primary heater (states 4 and 5) is 23–35% larger in the partial cooling cycles in comparison to the recompression configuration. Higher temperature difference leads to the lower required fluid volume when sensible heat storage is employed. Therefore, the partial cooling cycle results in more cost-effective sensible heat storage. Higher temperature gradients in the receiver reduce the heat loss to the environment by lowering the average operating temperature. Moreover, partial cooling cycles operate at comparatively lower pressures, which is advantageous in terms of component designs [21].

On the other hand, a detailed analysis by [1] showed that the recompression cycle has a higher potential to be used as the top cycle in a combined s-CO_2-ORC configuration. The efficiency benefit is about 2% in comparison with the combined partial cooling s-CO_2-ORC cycle. Moreover, Padilla et al. [22] carried out a comprehensive exergy analysis on different s-CO_2 cycle configurations, and concluded that the recompression cycle with an additional intercooler for the main compressor has the best exergetic performance.

It is noteworthy that this information is based on specific operating conditions and modeling approaches, and cannot be considered as conclusive. More studies are required in this area to further investigate the advantages and disadvantages of each cycle.

8.2.3 Comparison of the presented cycles

Dostal et al. [7] compared the thermal efficiency of a recompression s-CO_2 cycle with a superheated steam Rankine, supercritical steam Rankine, and helium Brayton cycles. More detailed information about the configuration of each cycle can be found in Ref. [7]. As can be clearly seen from Fig. 8.10, s-CO_2 cycle always outperforms the helium Brayton cycle for the entire temperature range. In addition, the s-CO_2 cycle is more efficient than the SRC when the turbine inlet temperature is higher than 550°C. Therefore, the optimal operating temperature range for s-CO_2 cycle is above 550°C. This is the temperature range that can be easily achieved in solar power towers.The expected efficiency for s-CO_2 cycle in the CST plants is in the range of 43–54% under wet-cooling conditions. However, CST plants are usually located in the areas where water resources are limited; therefore, dry cooling would be preferred over wet cooling. Even under Dry-cooling conditions, close to 50% efficiency is achievable, which is consistent with the goals of the DOE SunShot program [25].

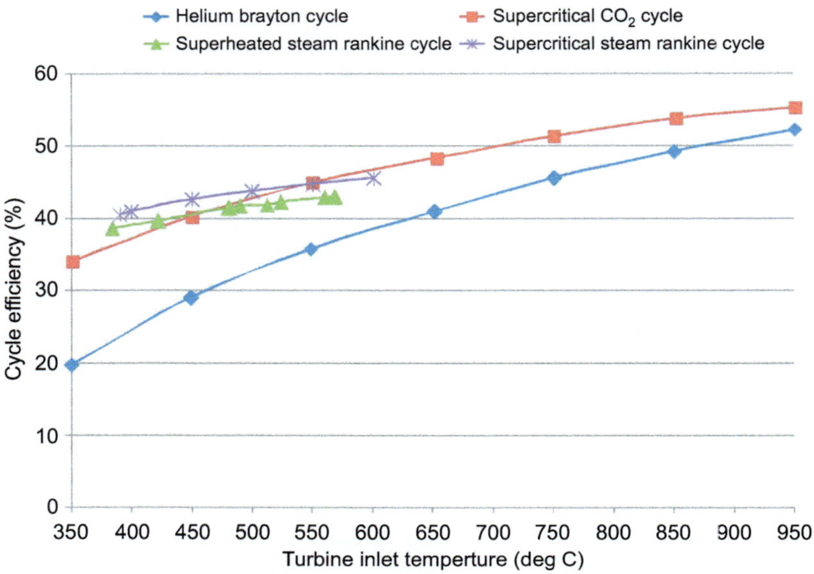

Figure 8.10 Thermal efficiency of different power cycles versus heat source temperature. Adapted from Dostal V, Hejzlar O, Driscoll MJ. The supercritical carbon dioxide power cycle: comparison to other advanced cycles. Nuclear Technology 2006;154(3):283–301.

As stated earlier, the s-CO_2 cycle has high efficiency due to low compression work, as the density of CO_2 increases substantially around the critical point. Higher density of the working fluid also means smaller power conversion components which is another advantage [18]. Table 8.1 shows the turbine size, shaft speed, and CO_2 mass flow rate for power ratings of 0.3, 3, and 300 MW. As can be seen, for 3 MW power, the turbine wheel diameter is only 15 cm with a speed of 50,000 rpm. Therefore, it is possible to place the power block inside the tower. In other words, power can be generated inside the tower right after the receiver and the fluid does not need to flow through long pipes. Hence, the system can be compact, and the pressure drop and heat loss are reduced, which consequently leads to higher efficiency and lower cost.

Table 8.1 CO_2 turbine size at different power rates [18]

Power rate (MW)	Turbine wheel diameter (m)	Desired shaft speed (rpm)	CO_2 flow (kg/s)
0.3	0.04	125,000	3.5
3	0.15	50,000	35
300	1.5	3,600	3,500

8.3 Combined cycles

Combined cycles can improve the overall efficiency of power conversion for CST. In this section, we introduce potential bottoming cycles that can be used in a combined cycle configuration. Any of the stand-alone cycles introduced in the previous section can serve as the topping cycle.

8.3.1 Organic Rankine cycle

The principle of operation for ORC is similar to SRC; however, an organic fluid is used instead of steam as the working fluid. Organic fluids generally have low boiling points, which makes recovering heat from low-grade heat sources possible [31]. The performance of an ORC is substantially affected by the selection of the working fluid.

Organic fluids can be categorized as dry (positive slope), wet (negative slope), and isentropic (vertical slope) fluids, depending on the slope of their saturation curves in *T-s* diagrams (Fig. 8.11).

The wet fluids such as water need to be superheated, while dry and isentropic fluids do not need superheating. The isentropic and dry fluids are for ORCs to protect the turbine blades from liquid droplets during expansion. However, if the fluid is "too dry," the vapor will leave the turbine with substantial "superheat," which adds to the cooling load of the condenser. This energy can be used to preheat the liquid before entering the evaporator to increase the cycle efficiency [4].

There is no best fluid for heat sources with different temperatures and one needs to consider multiple criteria and concerns before selecting a working fluid. These criteria are extensively discussed in [4,27,29] and can be summarized as:

1. *Types of the working fluids*: Dry, wet, or isentropic.
2. *Latent heat, density, and specific heat*: High latent heat, high density, and low liquid specific heat are favorable.

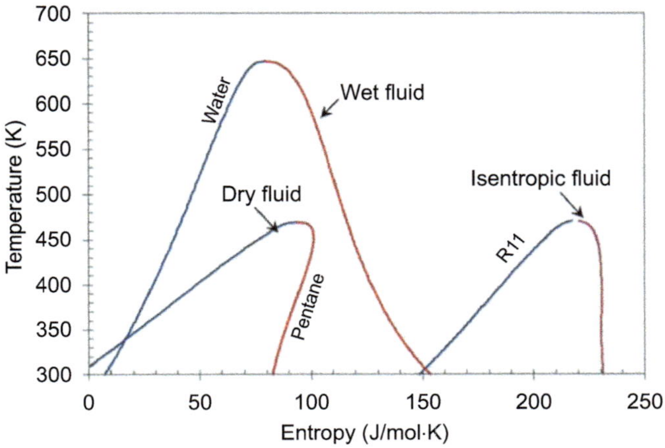

Figure 8.11 Types of organic fluids [4].

3. *Critical properties*: Critical properties of a fluid determine the operating conditions of temperatures and pressures in the cycle. Fluids with low critical temperatures (e.g., below 300 K) have difficulties in condensation and are usually not considered.
4. *Stability of the fluid, compatibility with materials in contact, ozone depletion potential (ODP), global warming potential (GWP), the atmospheric lifetime (ALT), availability, and cost*: All these are of great importance.
5. *Thermal efficiency*: Thermal efficiency of the cycle using the selected working fluid.
6. *Turbine outlet/inlet volume flow ratio*: Lower ratio allows the use of simpler and cheaper turbines.

Chen et al. [4] tabulated the thermodynamic and physical properties of 35 organic fluids which comply with environmental and safety regulations. Using the listed properties, one needs to select the potential working fluids for the heat source and sink temperatures. Then, analyzing the thermodynamic cycles with the selected working fluids can determine the best candidate for the specified conditions.

Besarati and Goswami [1] studied integration of an ORC bottoming cycle with each of the s-CO_2 cycle configurations introduced in the previous sections. The best organic fluids for each configuration were shortlisted based on thermal efficiency and expansion ratio of the ORC turbine. The largest efficiency increase, close to 7%, was achieved by using a simple s-CO_2 as the top cycle. However, as discussed previously, this cycle is less efficient than the recompression and partial cooling cycles. The maximum combined cycle efficiency (i.e., 54%) was obtained by the recompression s-CO_2-ORC cycle using R245fa as the working fluid. The efficiency of the standalone cycle under the same operating conditions was around 49%.

8.3.2 Supercritical organic Rankine cycle

The principles of operation are similar to ORC, however, the working fluid with low critical properties is directly compressed to its supercritical state, bypassing the two-phase region [17]. Fig. 8.12 compares the thermal match between the heat source and two different working fluids, one in ORC and another in SORC, for the same maximum temperature and pinch-point limitation [27]. As can be seen clearly, the mean temperature difference between the heat source and the working fluid is less in the SORC, resulting in less entropy generation (lower irreversibility). On the other hand, SORCs normally operate at higher pressures than ORCs, which may lead to higher equipment costs and safety concerns.

Carbon dioxide has been frequently studied as the working fluid in SORCs [35,37,38]. The main challenge for using CO_2 is its low critical temperature (i.e., about 31°C), which makes the condensation process difficult. Isobutene, propane, propylene, difluoromethane, and R-245fa are some of the other working fluids that have been studied for SORCs [4].

Another approach to reduce irreversibility and improve the system efficiency is using zeotropic mixtures of working fluids [3]. The main advantage is that the condensation process is not isothermal, and there is a better thermal match between the

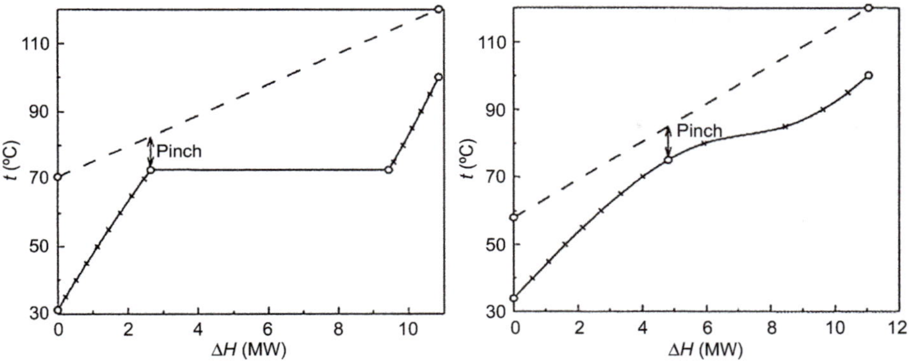

Figure 8.12 (Left) $T-\Delta H$ diagram in an ORC with R152a as the working fluid at 20 bar from 31.16°C to 100°C. (Right) $T-\Delta H$ diagram in an SORC with R143a as the working fluid at 40 bar from 33.93°C to 100°C [27].

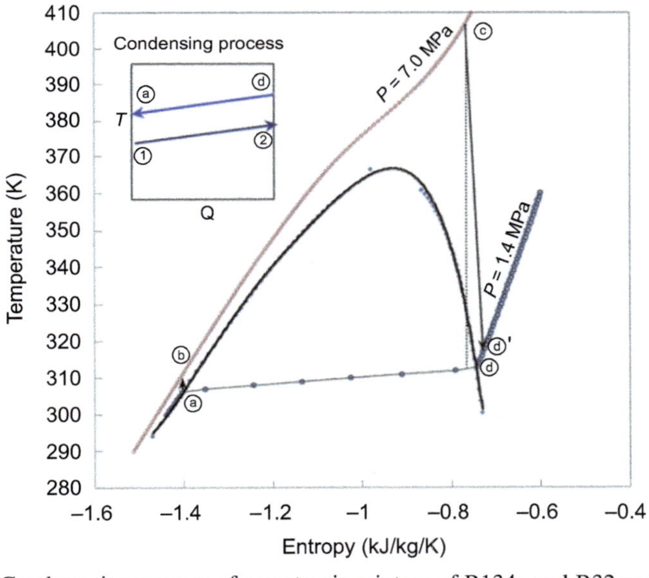

Figure 8.13 Condensation process of a zeotropic mixture of R134a and R32 and its thermal match with the coolant [3].

working fluid and the coolant. Fig. 8.13 depicts the condensation process of a zeotropic mixture of R134a and R32. As can be clearly seen, there is a thermal glide when the mixture is condensed at constant pressure. Therefore, the condensation process can be designed in a way that the temperature profile of the cooling water parallels that of SORC working fluid, resulting in less irreversibility.

8.3.3 Absorption power cycles

In these type of power cycles, an ammonia–water mixture is generally used as the working fluid, where the concentration of ammonia varies at different points of the cycle. As ammonia is a more volatile component than water, it tends to vaporize at a lower temperature than pure water. Therefore, during the heating process, concentration of ammonia in the liquid working fluid decreases, resulting in a better thermal match with the heat source and lower irreversibility. Kalina and Goswami cycles are two absorption power cycles that can be used as the bottoming cycle in CST plants.

8.3.3.1 Kalina cycle

Kalina proposed a novel cycle in 1984 to be used as a bottoming cycle utilizing waste heat from the exhaust of gas turbines [13]. The proposed configuration was called Kalina cycle system 1 (KCS 1). This cycle uses ammonia–water mixture as the working fluid where the concentration of ammonia varies along the cycle. Since then, the cycle configuration has been modified for different applications and each is identified by a unique system number. For example, KCS 2 is intended for low-temperature geothermal applications and KCS 5 is applicable to direct fuel–fired plants. Different Kalina cycles along with their applications are well documented by Ref. [36]. The layout of a Kalina cycle is shown in Fig. 8.14.

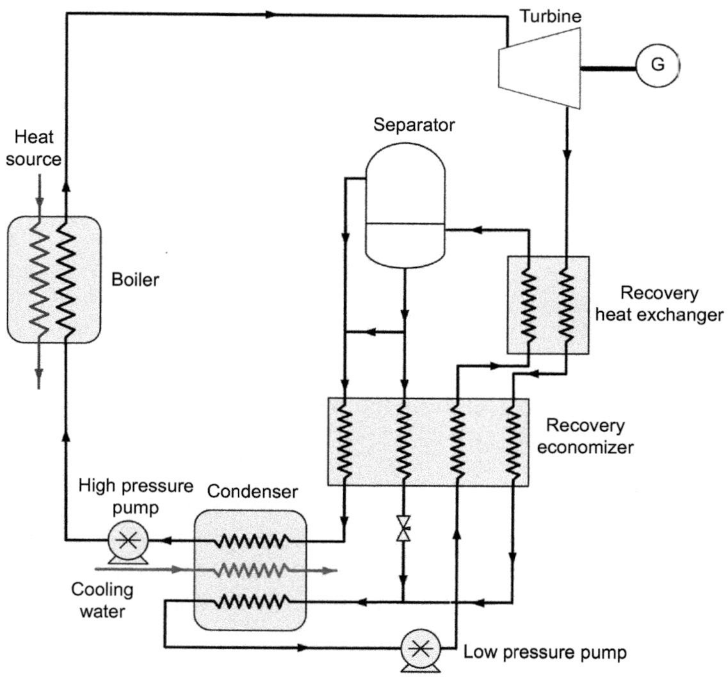

Figure 8.14 A Kalina cycle.

In comparison with the Rankine cycle, the Kalina cycle is more complex and requires more components. On the other hand, the variable concentration of ammonia in the binary mixture provides the possibility to alter the properties of the working fluid at different locations in the cycle in order to maximize the thermal performance.

El-Sayed and Tribus [10] compared the thermal performance of steam Rankine and Kalina cycles when both are used as bottoming cycle with the same thermal boundary conditions. The results show that using the Kalina cycle leads to 10–20% improvement in the thermal efficiency. Other research papers have also indicated the advantages of Kalina over Rankine cycle (in a combined cycle configuration) from both first and second laws points of views [12,19].

Neises and Turchi [20] analyzed the performance of a combined steam Rankine–Kalina cycle in a parabolic trough solar thermal plant. It was concluded that such combined power cycle can overcome some of the drawbacks of the stand-alone SRC. These drawbacks can be summarized as:

1. Limitations in generating power during the low solar insolation periods when the steam temperature and flow rate are not satisfactory.
2. Large condensing unit due to very high specific volume of steam at the turbine outlet.
3. Vacuum pressure at the condenser.
4. Lower performance of the condensing turbines when they are used at small scales.

In their proposed cycle, condensing steam turbines are replaced by back-pressure turbines. The cycles operate in two different modes during high and low insolation modes. The Rankine cycle is bypassed during the low insolation periods (typically 300–400 W/m^2). Under these conditions, the temperature of the HTF can still exceed 300°C, which is high enough for operating the Kalina cycle. When the insolation level is high, the Kalina cycle operates as the bottoming cycle utilizing the waste heat from the SRC. Operating the power block in two different modes improves the plant availability. Moreover, the condensing pressure of the combined cycle is above atmospheric, which is due to the high concentration of ammonia in the condensing unit. Moreover, back-pressure turbines operate efficiently at 5–20 MW size, which makes implementation of the technology possible for small- and medium-size power plants. A techno-economic analysis for a 50 MW parabolic trough power plant indicated that a 4–11% electricity cost saving could be achieved by replacing a stand-alone Rankine cycle with the combined Rankine–Kalina cycle.

In another study Peng et al. [24] investigated using Kalina cycle to utilize waste heat from a gas turbine. A solar power tower was used to preheat the air. A simulation showed that using Kalina instead of SRC as the bottoming cycle reduces the exergy destruction and leads to a peak solar-to-electric efficiency of 27.5% for the turbine inlet temperature of 1000°C.

8.3.3.2 Goswami cycle

Goswami proposed a thermodynamic cycle in 1995 which can be used either as a stand-alone cycle for low-temperature heat sources or a bottoming cycle [11,34]. An ammonia–water mixture is used as the working fluid and the cycle can produce both power and refrigeration, although other fluid pairs could also be used.

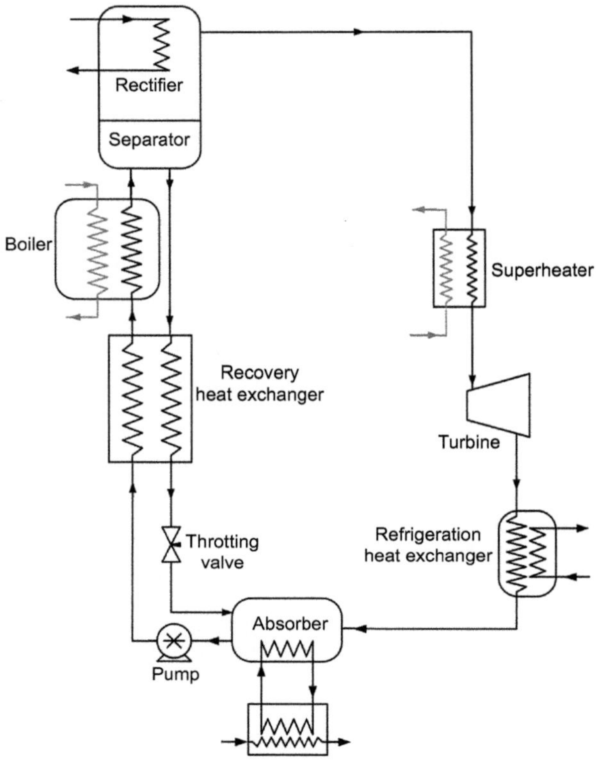

Figure 8.15 Schematic of the Goswami cycle.

Fig. 8.15 shows a schematic of the cycle. The binary working fluid leaves the absorber with a relatively high ammonia concentration. It is then pressurized and receives heat from the returning weak ammonia liquid solution in the recovery heat exchanger. Next, the basic solution is partially boiled to produce a two-phase mixture. The vapor with high ammonia concentration is separated from the liquid and enters the rectifier. The rectifier cools the saturated ammonia vapor to condense out any remaining water. The liquid mixture with low ammonia concentration goes through the heat recovery unit and is throttled to the cycle's low pressure. Additional heat can be transferred to the ammonia vapor in the superheater before expanding in the turbine.

The temperature of the vapor leaving the turbine is significantly below the ambient temperature, which provides cooling in the refrigeration heat exchanger.

Similar to the Kalina cycle, the irreversibilities associated with heat transfer from a sensible heat source are reduced by boiling the ammonia–water mixture. Almost pure ammonia is expanded in the turbine. It is noteworthy that the cycle operating conditions can be optimized depending on the demand profile to generate maximum power, maximum cooling, or a combination of the two [5].

Padilla et al. [23] analyzed a 50-MW parabolic trough solar power plant with Goswami cycle serving as the condenser of a SRC. His analysis indicated that replacing the traditional condenser with the Goswami cycle reduces energy losses, and eliminates the high specific volume and poor vapor quality presented in the last stages of the turbine.

8.4 Summary and conclusions

In this chapter, power cycles that can be utilized in future CST power plants are reviewed. The Supercritical steam cycle is a highly promising candidate as the SRC is already in service at all CST plants. However, more development is required in terms of materials to enable the operation of the cycle at higher temperatures. Moreover, life cycle analysis is necessary to make sure the additional cost of using supercritical cycle does not outweigh the efficiency benefit. The air Brayton cycle is a promising candidate when a CST plant is added to an operational gas turbine to reduce fuel consumption. Direct and indirect receivers are under development with HTF exit temperatures of around 1000°C. Although helium Brayton cycle is being considered in nuclear power plants, its lower efficiency in comparison with s-CO_2 Brayton cycles is a major obstacle for future development. Supercritical CO_2 power cycles are superior to other stand-alone cycles at turbine inlet temperatures greater than 550°C. Both partial cooling and recompression configurations are promising and more detailed comparison is required before deploying in a power plant. Despite all the positive aspects mentioned for these types of power cycles, there are some uncertainties about utilization of s-CO_2 in CST plants. The main concerns are the high pressure of the fluid and lack of experience in operating closed-loop Brayton cycles.

This chapter also describes some lower-temperature power cycles that can be used as potential bottoming cycles. Using a bottoming cycle can further improve the power cycle efficiency by minimizing the rejected heat to the environment. ORC and Kalina cycles are more developed alternatives, which have already been used to generate power from low-temperature heat sources at medium scales. The SORC and Goswami cycle can also be considered as options due to their simplicity and smooth heat addition. The efficiencies shown for each cycle in this study are dependent on the operating conditions and may differ depending on the system requirements.

Although developing new types of power cycles has been an active area of research for a number of years, additional effort is required to model the performance of power cycles, in particular at off-design conditions, which are frequently expected at the solar power plants. In addition, new types of low-cost turbo-machines need to be developed to operate at high efficiencies over a broad range of part-load conditions. The turbomachines must respond quickly to frequent startup/shutdown operations. Moreover, employing high-temperature power cycles demands high-temperature solar receivers. Although new thermal receiver concepts such as, particle and micro-channel receivers are under development, extensive research is required before utilizing them in a solar power plant. In addition, increasing the maximum temperature of the power cycle has to be accompanied by a cost-effective thermal-storage approach. Thermochemical storage can be considered as a viable option for high temperature systems. Furthermore, low-cost recuperators need to be developed to address a challenge with s-CO_2 cycles, which is the required substantial heat transfer area. Moreover, testing different types of materials such as nickel alloys is needed to find their expected thermo-mechanical properties under severe pressure and temperature conditions.

References

[1] Besarati SM, Yogi Goswami D. Analysis of advanced supercritical carbon dioxide power cycles with a bottoming cycle for concentrating solar power applications. Journal of Solar Energy Engineering 2014;136(1):011020.

[2] Cheang VT, Hedderwick RA, McGregor C. Benchmarking supercritical carbon dioxide cycles against steam Rankine cycles for concentrated solar power. Solar Energy 2015;113: 199−211.

[3] Chen H, Goswami DY, Rahman MM, Stefanakos EK. A supercritical Rankine cycle using zeotropic mixture working fluids for the conversion of low-grade heat into power. Energy 2011;88(8):549−55.

[4] Chen H, Goswami DY, Stefanakos EK. A review of thermodynamic cycles and working fluids for the conversion of low-grade heat. Renewable and Sustainable Energy Reviews 2010;14(9):3059−67.

[5] Demirkaya G, Besarati SM, Padilla RV, Archibold AR, Goswami DY, Rahman MM, Stefanakos EK. Multi-objective optimization of a combined power and cooling cycle for low-grade and mid-grade heat sources. Journal of Energy Resources Technology 2012; 134(3):032002.

[6] Dostal V. A supercritical carbon dioxide cycle for next generation nuclear reactors. Massachusetts Institute of Technology (MIT); 2004.

[7] Dostal V, Hejzlar O, Driscoll MJ. The supercritical carbon dioxide power cycle: comparison to other advanced cycles. Nuclear Technology 2006;154(3):283−301.

[8] DTI. Advanced power plant using high efficiency boiler/turbine. 2006. Available at: http://cartechcapstone.wikispaces.com/file/view/BPB010(1).pdf.

[9] Dunham MT, Iverson BD. High-efficiency thermodynamic power cycles for concentrated solar power systems. Renewable and Sustainable Energy Reviews 2014;30:758−70.

[10] El-Sayed YM, Tribus M. A theoretical comparison of the Rankine and Kalina cycles. In: Analysis of energy systems, design and operation, annual meeting of the American Society of Mechanical Engineers. Miami, Florida; 1985.

[11] Goswami DY. Solar thermal power technology: present status and ideas for the future. Energy Sources 1998;20(2):137−45.

[12] Jonsson M, Yan J. Ammonia−water bottoming cycles: a comparison between gas engines and gas diesel engines as prime movers. Energy 2001;26(1):31−44.

[13] Kalina A. Combined-cycle system with novel bottoming cycle. Journal of Engineering for Gas Turbines and Power 1984;106(4):737−42.

[14] Kolb GJ, Ho CK, Mancini TR, Gary JA. Power tower technology roadmap and cost reduction plan. 2011 [Albuquerque, NM].

[15] Kribus A, Zaibel R, Carey D, Segal A, Karni J. A solar-driven combined cycle power plant. Solar Energy 1998;62(2):121−9.

[16] Lemmon EW, Mclinden MO, Huber ML. NIST reference fluid thermodynamic and transport properties — REFPROP. NIST; 2013. Standard reference database 23.
[17] Li C, Besarati S, Goswami Y, Stefanakos E, Chen H. Reverse osmosis desalination driven by low temperature supercritical organic rankine cycle. Applied Energy 2013;102: 1071–80.
[18] Ma Z, Turchi CS. Advanced supercritical carbon dioxide power cycle configurations for use in concentrating solar power systems preprint. In: Supercritical CO_2 for Application in Concentrating Solar Power Systems, Proceedings of SCCO2 Power Cycle Symposium 2009; 2011.
[19] Marston CH, Hyre M. Gas turbine bottoming cycles: triple-pressure steam versus Kalina. Journal of Engineering for Gas Turbines and Power 1995;117(1):10–5.
[20] Mittelman G, Epstein M. A novel power block for CSP systems. Solar Energy 2010; 84(10):1761–71.
[21] Neises T, Turchi C. A comparison of supercritical carbon dioxide power cycle configurations with an emphasis on CSP applications. Energy Procedia 2013;49:1187–96.
[22] Padilla RV, Too YCS, Benito R, Stein W. Exergetic analysis of supercritical CO_2 Brayton cycles integrated with solar central receivers. Applied Energy 2015;148:348–65.
[23] Padilla RV, Archibold AR, Demirkaya G, Besarati S, Goswami DY, Rahman MM, Stefanakos EK. Performance analysis of a rankine cycle integrated with the Goswami combined power and cooling cycle. Journal of Energy Resources Technology 2012; 134(3):032001.
[24] Peng S, Hong H, Jin H, Wang Z. An integrated solar thermal power system using intercooled gas turbine and Kalina cycle. Energy 2012;44(1):732–40.
[25] Price H. Assessment of parabolic trough and power tower solar technology cost and performance forecasts. 2003 [Golden, CO].
[26] Price HW, Whitney DD, Beebe HI. SMUD Kokhala power tower study. In: American Society of Mechanical Engineers - international solar energy conference, San Antonio, TX; 1996.
[27] Saleh B, Koglbauer G, Wendland M, Fischer J. Working fluids for low-temperature organic Rankine cycles. Energy 2007;32:1210–21.
[28] Schwarzbözl P, Buck R, Sugarmen C, Ring A, Crespo MJM, Altwegg P, Enrile J. Solar gas turbine systems: design, cost and perspectives. Solar Energy 2006;80(10):1231–40.
[29] Tchanche BF, Papadakis G, Lambrinos G, Frangoudakis A. Fluid selection for a low-temperature solar organic Rankine cycle. Applied Thermal Engineering 2009; 29(11–12):2468–76.
[30] Tsiklauri G, Talbert R, Schmitt B, Filippov G, Bogoyavlensky R, Grishanin E. Supercritical steam cycle for nuclear power plant. Nuclear Engineering and Design 2005; 235(15):1651–64.
[31] Vélez F, Segovia JJ, Martín MC, Antolín G, Chejne F, Quijano A. Comparative study of working fluids for a Rankine cycle operating at low temperature. Fuel Processing Technology 2012;103:71–7.
[32] Wheeldon JM. Engineering and economic evaluation of 1300°F series ultra-supercritical pulverized coal power plants. Palo Alto, CA: Phase 1; 2008.
[33] Wright SA, Conboy TM, Rochau GE. Overview of supercritical CO_2 power cycle development at Sandia National Laboratories. In: Supercritical CO_2 power cycle symposium; 2011.
[34] Xu F, Goswami DY, Bhagwat SS. A combined power/cooling cycle. Energy 2000;25(3): 233–46.

[35] Yamaguchi H, Zhang XR, Fujima K, Enomoto M, Sawada N. Solar energy powered Rankine cycle using supercritical CO_2. Applied Thermal Engineering 2006;26(17−18): 2345−54.
[36] Zhang X, He M, Zhang Y. A review of research on the Kalina cycle. Renewable and Sustainable Energy Reviews 2012;16(7):5309−18.
[37] Zhang XR, Yamaguchi H, Uneno D, Fujima K, Enomoto M, Sawada N. Analysis of a novel solar energy-powered Rankine cycle for combined power and heat generation using supercritical carbon dioxide. Renewable Energy 2006;31(12):1839−54.
[38] Zhang XR, Yamaguchi H. An experimental study on evacuated tube solar collector using supercritical CO_2. Applied Thermal Engineering 2008;28(10):1225−33.
[39] Zhao H, Peterson PF. Multiple reheat helium Brayton cycles for sodium cooled fast reactors. Nuclear Engineering and Design 2008;238(7):1535−46.

Advances in dry cooling for concentrating solar thermal (CST) power plants

9

K. Hooman, Z. Guan, H. Gurgenci
The University of Queensland, Brisbane, Queensland, Australia

9.1 Introduction

All thermal power plants (including concentrating solar thermal, CST) need a cooling system to cool the turbine exhaust. It is well known that the Carnot cycle efficiency $\left(\eta_{\text{thermal}} = 1 - \frac{T_L}{T_H}\right)$ is maximized with the highest possible heat source temperature T_H and the lowest possible heat sink temperature T_L. According to this correlation, a cooling tower must be designed and built to provide the lowest possible heat sink temperature and is an integral part of a thermal power plant.

Fig. 9.1 shows a simplified CST plant with parabolic trough for electricity generation. As the figure shows, the parabolic trough consists of a linear parabolic reflector that concentrates light onto a receiver (tube) positioned along the reflector's focal line. The reflected sunlight heats up the fluid flowing through the tubes. The hot fluid is used to boil water in a conventional steam-turbine generator to produce electricity. Once the steam has passed through a turbine for electricity generation, it has to be condensed into water by the cold water from the cooling tower. The function of the cooling tower is to continuously provide cold water to condense the turbine exhaust steam. Colder water cools the exhaust steam more effectively and more net power can be generated from the plant.

In general, water and air are the two mostly used cooling media in a cooling tower. According to the primary cooling method used, cooling towers can be classified as

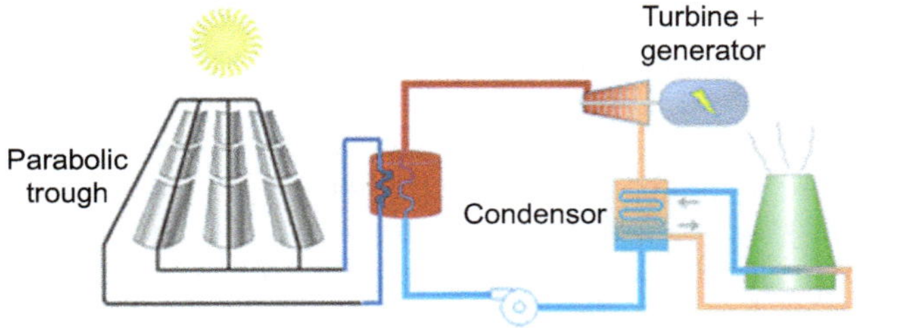

Figure 9.1 Concentrating solar thermal (CST) power plant with parabolic trough.

Advances in Concentrating Solar Thermal Research and Technology
http://dx.doi.org/10.1016/B978-0-08-100516-3.00009-5

either wet or dry cooling. Wet cooling towers reject heat through the direct contact of the water with the flowing air and the cooling is achieved primarily by evaporation. Dry cooling towers transfer heat through air-cooled heat exchangers that separate the working fluid (steam or water) from the cooling air.

Wet cooling is more efficient than the dry counterpart at the expense of water loss to the ambient. For instance, it has been reported that the steam-cycle cooling accounts for over 90% of water consumption in a typical wet-cooled CST plant [1]. The consumption of large quantity of water is to be limited by policy or cost in arid remote area, where CST plants are most likely to be located in future. Therefore, dry cooling will be the cost-effective option for CST plants.

While dry cooling has advantages for water conservation and environmental protection, it suffers from lower efficiency when ambient air temperature is high [2]. According to the experiences gained from coal-fired power plants, it is expected that the CST plants using dry cooling will suffer from the low efficiency during hot periods, which are often the periods of peak system demand and higher electricity sale price. Thus, new technologies offering an increase in electricity generation from CST power plants during high ambient temperature are required.

9.2 Current cooling technologies for concentrating solar thermal power plants

The cooling technologies available for CST plants can be classified into wet or dry cooling with respect to the heat transfer mechanism employed. In either case, either mechanical or natural draft cooling tower can be used to draw air through the tower.

9.2.1 Wet cooling towers

In a wet cooling tower, the water and air are in direct contact, which results in cooling primarily by evaporation.

As shown in Fig. 9.2, hot turbine exhaust steam is condensed in a condenser by the cold water pumped from the bottom of the cooling tower. After the heat exchange between the hot exhaust steam (from the turbine) and the cold water (from the cooling tower), the cold water gains heat and becomes hot while the exhaust steam is condensed. The hot water in the tower loop is then piped to the cooling tower and is distributed onto the fill material. Air is introduced across the fill material by either mechanical draft (fans as shown in Fig. 9.2) or natural draft (tall tower) as shown in Fig. 9.3 [3]. In the fill material, the moving air makes direct contact with the hot water and carries the heat away resulting in cooling the hot water. The cooled water is then collected in a cold water basin below the fill from which it is pumped back to the condenser. The fill material increases both the water/air contact area and the contact time for better cooling performance.

Wet cooling towers transfer the heat from the water stream to the airstream by evaporation, which raises the air temperature and its relative humidity to 100%. The saturated air is then discharged into the atmosphere. Because the performance of wet cooling is

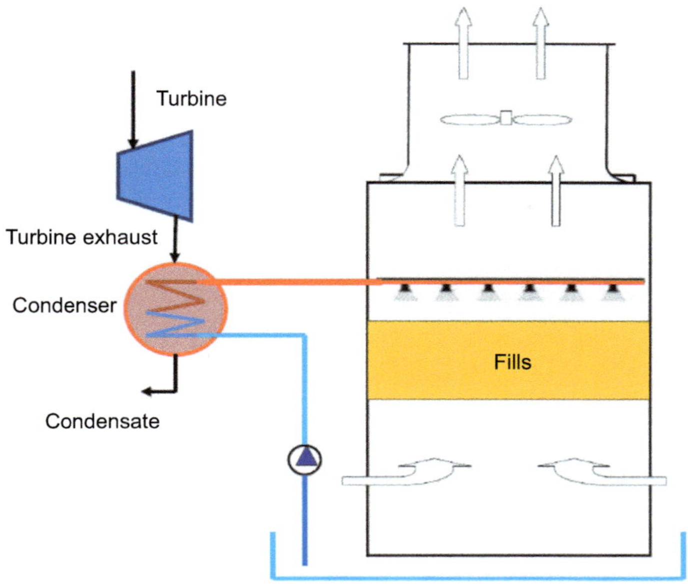

Figure 9.2 Mechanical draft wet cooling tower.

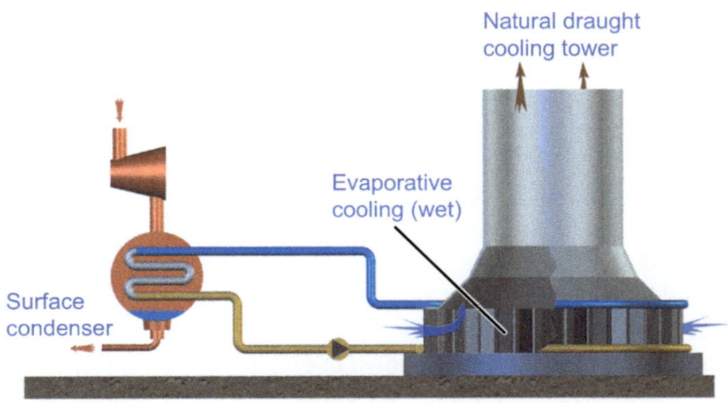

Figure 9.3 Natural draft wet cooling tower (Kelvion Thermal Solutions).

dependent on the ambient air wet bulb temperature, it has higher cooling efficiency than dry cooling which relies on the air dry bulb temperature.

The heat transfer in wet cooling tower is mainly by latent heat through water evaporation and only partially (marginally in most cases) by sensible heat transfer. As such, large quantities of water evaporate into the moving airstream to be discharged into the atmosphere on leaving the tower. The water lost to the evaporation must be continuously replaced. In addition, because evaporation concentrates impurities in the water, some of the circulating water is deliberately drawn off to prevent extensive scale formation. This is called "blowdown." The moving air also carries away some water droplets, which are called drift. The evaporation loss, blown-down water, and drift water correspond to the total water loss in a wet cooling system. Turchi et al. reported that wet cooling takes more than 90% of water consumption in parabolic trough CST power plants [1], and the average water consumption per megawatt power generation for the wet-cooled parabolic trough CST plants is 3.5 m^3/MWh. Williams and Rasul [4] reported the water evaporation rate for a coal-fired power plant of 350 MW capacity in Queensland, Australia, was around 1.8 m^3/MWh power generation. This is about 630 cubic meter water per hour or 5.5 million cubic meters per year for a 350 MW coal-fired power plant. NETL [5] reported that a total makeup water of 5188 gpm was consumed for a 520 MW power plant in USA, among which, the evaporation water was 3891 gpm and the blown-down water was 1297 gpm. The water consumption is about 2.27 m^3/MWh power generation.

Except for the large amount of water consumption, a wet cooling tower requires frequent water treatment to minimize the scaling and fouling; to prevent growth of bacteria, fungi, and algae; and to eliminate the growth of *Legionella* that causes Legionnaires' disease. Wet cooling towers also cause other environment impact such as fog producing.

9.2.2 Dry cooling towers

Dry cooling towers conduct heat transfer through air-cooled heat exchangers that separate the working fluid from the cooling air. Because there is no direct contact between the working fluid and the ambient air, there is no water loss in this system. Dry cooling towers have two basic types: direct and indirect systems.

In a direct dry cooling tower, turbine exhaust steam is condensed directly through an air-cooled condenser (ACC). As shown in Fig. 9.4, the turbine exhaust flows directly into the tubes of ACC. A large-diameter piping is used at the turbine outlet to accommodate the relatively low steam densities (compared with water) and reduce the pressure drop through the route. The exhaust steam flows inside the tubes of ACC and is condensed by transferring the heat to the flowing air outside the surface of the heat exchanger.

With indirect dry cooling tower, as shown in Fig. 9.5 [3], the turbine exhaust steam is condensed in a surface heat exchanger called condenser where the heat of the exhaust steam is transferred to the cold water pumped from the air-cooled heat exchanger located inside the cooling tower. The cooling water gains the heat in the condenser and is pumped back to the heat exchanger inside the cooling tower. Air is introduced to flow across the exterior surface of the heat exchanger to cool the hot water inside the heat exchanger tubes. The cooled water inside the heat exchanger tubes is then pumped back to the condenser. The turbine exhaust is condensed by the circulating water rather by the ambient air directly; hence, it gets the name of *indirect* cooling.

In a dry cooling tower, air can be introduced either by mechanical draft (fans as shown in Fig. 9.4) or by natural draft (tall tower as in Fig. 9.5) to move the air across the ACC/air-cooled heat exchangers. Increasing volumetric flow rate of air through

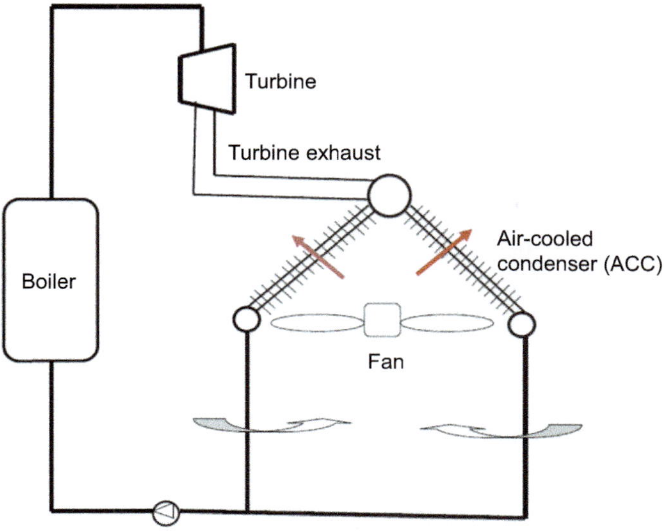

Figure 9.4 Direct dry cooling tower (commonly known as air-cooled condenser, ACC).

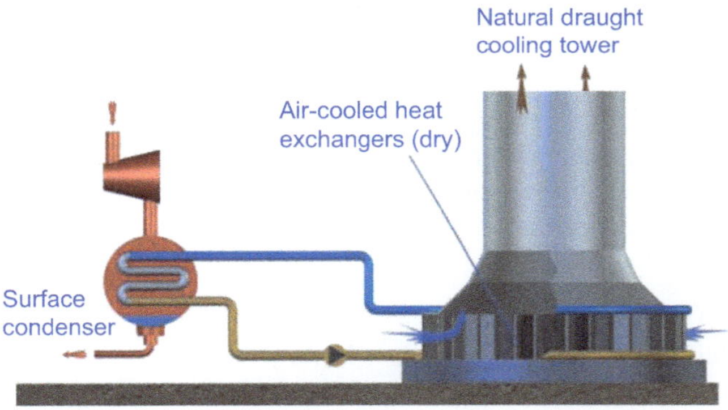

Figure 9.5 Schematic of indirect dry cooling tower (Kelvion Thermal Solutions).

heat exchangers improves the performance of the cooling system which ultimately depends on the air dry bulb temperature.

For a CST power plant, the choice between wet and dry cooling systems involves a number of trade-offs including the availability and cost of water, environmental aspects, and the cost of electric power. Because CST power plants, in general, are located in arid areas, water conservation is a major factor favoring for dry cooling towers.

Although dry cooling has advantages for water conservation and environmental protection, it suffers from lower efficiency when ambient air temperature is high. Thus, new technologies in increasing CST power plant production during high ambient temperatures are required.

9.3 Air-cooled heat exchanger and cooling tower sizing

Air-cooled heat exchanger is a key element of a dry cooling system. Thermal characteristics and pressure drop are the two most important parameters for evaluating the performance of a dry cooling system.

9.3.1 Thermohydraulics of air-cooled heat exchanger

In a dry cooling system, the hot fluid flows inside the heat exchanger tubes while the ambient air flows outside the tubes. Because the air-side heat transfer coefficient is much lower than that of the hot fluid inside the tubes, the surface area of the heat exchanger has to be increased by adding fins onto the bare tubes to achieve the required heat transfer rate. This type of heat exchanger is called finned tube heat exchanger as shown in Fig. 9.6.

The selection of finned tube heat exchanger parameters and the material is based on the fluid temperature, fouling and cleanability, environment (corrosion), and cost.

Models predicting thermohydraulics of air-cooled heat exchangers have been developed by various researchers for sizing (and also rating) the heat exchanger and cooling tower [2,6]. These parameters are also provided by heat exchanger manufacturers, usually in the form of factory test data.

The heat exchange rate of an air-cooled heat exchanger can be expressed in the form of [2]:

$$Q = \mathrm{UAF_T}\Delta T_{\mathrm{lm}} \tag{9.1}$$

where U is the overall heat transfer coefficient referred to the area A, W/(m^2 °C); A is any convenient transfer area pertinent to U, m^2; F_T is the correction factor.

For counterflow arrangement, ΔT_{lm}—logarithmic mean temperature difference, °C—is calculated as:

$$\Delta T_{\mathrm{lm}} = \frac{(T_{\mathrm{h2}} - T_{\mathrm{c2}}) - (T_{\mathrm{h1}} - T_{\mathrm{c1}})}{\ln[(T_{\mathrm{h2}} - T_{\mathrm{c2}})/(T_{\mathrm{h1}} - T_{\mathrm{c1}})]} = \frac{\Delta T_2 - \Delta T_1}{\ln(\Delta T_2/\Delta T_1)} \tag{9.2}$$

Figure 9.6 Finned tube heat exchanger.
Gatton Tower, Gatton campus of the University of Queensland.

where T_{h2} is hot fluid outlet temperature; T_{h1} is hot fluid inlet temperature; T_{c1} is cold fluid (air) outlet temperature; T_{c2} is cold fluid inlet temperature.

The product of the overall heat transfer coefficient and the transfer area (UA) should always be used. When different areas are used, the result of U will be different, but the product of U and A will always be the same.

With finned tube heat exchangers, the total air-side area A_o includes both the fin area and the exposed tube root area. When calculating this total area, fin efficiency must be taken into account. In this case, the outside area A_o is calculated as:

$$A_o = A_r + \eta_f A_f \tag{9.3}$$

where A_r is the exposed root area of the tube, A_f is the fin area, and η_f is the fin efficiency. The fin efficiency is an indication of the effectiveness of the fin.

Based on air-side area A_o of a finned tube heat exchanger, the overall heat transfer coefficient U can be calculated using the following equation:

$$U = \left(\frac{A_o}{h_i A_i} + \frac{A_o \ln(r_o/r_i)}{2\pi L k} + R_{fo} + \frac{1}{h_o}\right)^{-1} \tag{9.4}$$

where A_i is the tube-inside area; h_i and h_o are the convection heat transfer coefficient of the fluid inside and outside; r_o, r_i, and L are the outside radius, inside radius, and length of the tube, respectively; k is the thermal conductivity of the tube material; and R_{fo} is the resistance caused by fouling, corrosion, etc.

U can be either obtained from the heat exchanger manufacturers directly or derived by Eq. (9.4). Because convection heat transfer coefficients h_i (inside fluid) and h_o (outside air) are velocity dependent, various empirical models have been developed to predict these two parameters (h_i and h_o) when Eq. (9.4) is used. These two convection heat transfer coefficients are expressed in terms of dimensionless Nusselt number.

Different equations have been derived for laminar/transition/turbulent flow of fluid inside the tube. Hausen [7] proposed the following equation for turbulent flow in tubes which can be used for calculation of the convection heat transfer coefficient h_i:

$$N_u = \frac{h_i d}{k} = 0.0235\left(R_e^{0.8} - 230\right)\left(1.8P_r^{0.3} - 0.8\right)\left[1 + \left(\frac{d}{L}\right)^{0.667}\right] \tag{9.5}$$

where N_u is Nusselt number, d is the diameter of the tube, R_e is the Reynolds number, P_r is Prandtl number, and L is tube length.

For the convection heat transfer coefficient of air-side h_o, Briggs and Yang [8] proposed a generic equation for a six-row, staggered (set on equilateral triangle) finned tube heat exchangers:

$$N_u = \frac{h_o d_r}{k} = 0.134P_r^{0.33}R_e^{0.681}\left[\frac{2(P_f - t_f)}{d_f - d_r}\right]^{0.2}\left(\frac{P_f - t_f}{t_f}\right)^{0.1134} \tag{9.6}$$

where d_r is fin root diameter, d_f is fin diameter, P_f is fin pitch, and t_f is fin mean thickness.

Eq. (9.6) is valid within the following limits:

$$1000 < R_e < 18000$$

$$11.13\ \text{mm} < d_f < 40.89\ \text{mm}$$

$$1.42\ \text{mm} < \frac{d_f - d_r}{2} < 16.57\ \text{mm}$$

$$0.33\ \text{mm} < t_f < 2.02\ \text{mm}$$

$$1.30\ \text{mm} < P_f < 4.06\ \text{mm}$$

When quoting the value of overall heat transfer coefficient U given by the manufacturers rather than using the previous equation, it is important to identify the reference area associated with the value.

When using Eq. (9.1) to calculate the heat transfer of an air-cooled heat exchanger, one needs to know all four terminal temperatures to obtain the logarithmic mean temperature [Eq. (9.2)]. In practical design of CST power plants, outlet temperatures of the hot fluid, T_{h2}, and the outgoing temperature of the cooling air, T_{c1}, are not known and need to be found by an iterative procedure based on the following heat balance equations.

Heat rejected from the hot fluid inside heat exchanger:

$$Q_w = m_w c_{pw}(T_{h1} - T_{h2}) \tag{9.7}$$

Heat transferred to the air outside heat exchanger:

$$Q_a = m_a c_{pa}(T_{c1} - T_{c2}) \tag{9.8}$$

Heat transfer through heat exchanger can be calculated by Eq. (9.1). The heat transfer rate by heat exchanger, by hot fluid inside and by air-side must be balanced.

With the heat balance, $Q = Q_w = Q_a$

The inlet temperatures of the hot fluid (T_{h1}) and the ambient air (T_{c2}) are known from the power plant design data, which leaves three unknowns: the two exit temperatures (T_{h2}, T_{c1}) and the heat transfer rate (Q). These unknowns can be determined from the three Eqs. (9.1), (9.7), and (9.8) through an iterative procedure.

The effectiveness-NTU method can be used for thermohydraulic performance prediction without applying the previously mentioned iterative procedure. The effectiveness of a heat exchanger, e, is defined as the ratio of the actual rate of heat transfer to the maximum possible rate of heat transfer ($e = Q/Q_{max}$). The maximum possible rate of heat transfer can be expressed as Eqs. (9.9) and (9.10) depending on which of the heat capacity rates (product of the mass flow rate and specific heat—$m_w c_{pw}$ and $m_a c_{pa}$) are smaller.

$$Q_{max} = C_{min}(T_{h1} - T_{c2}) \tag{9.9}$$

where T_{h1} is inlet hot fluid temperature; T_{c2} is ambient air temperature; and C_{min} is the smaller heat capacity rate between the hot fluid $m_w c_{pw}$ and the cooling air $m_a c_{pa}$.

The actual heat transfer is then given by

$$Q = eQ_{max} \tag{9.10}$$

Calculation of the effectiveness, e, can be found in various literatures [2,9]. The equations for calculation of a counterflow heat exchanger are given below:

For counterflow, indirect dry cooling system,

$$e = \frac{1 - \exp[-N(1 - C)]}{1 - C\exp[-N(1 - C)]} \tag{9.11}$$

For ACC (direct cooling),

$$e = 1 - \exp(-N) \tag{9.12}$$

where $N = \text{NTU} = \text{UA}/C_{min}$; $C = C_{min}/C_{max}$; C_{min} is the minimum of $m_a c_{pa}$ and $m_w c_{pw}$; and C_{max} is the larger one.

When using the effectiveness-NTU method for thermal characteristics prediction, the first step is to calculate the effectiveness, e, by using Eq. (9.11) or (9.12). Once the effectiveness has been obtained, the actual heat transfer, Q, can be derived by Eq. (9.10). Once Q is known, the two outlet temperatures of the fluid and air (T_{h2}, T_{c1}) can be calculated using Eqs. (9.7) and (9.8), obviously, without any iterative procedure.

Similar to the heat transfer, the air-side pressure drop of a finned tube heat exchanger is velocity dependent. The friction and contraction cause pressure drop when air flows through the heat exchange. Air-side fouling has a significant impact on air-side pressure drop. The pressure drop data can be either obtained from heat exchanger manufacturers or derived from empirical models reported in the open literature.

In 1966, Robinson and Briggs [10] developed a correlation that was frequently used for pressure drop prediction for staggered or equilateral tube setting pattern:

$$E_u = \frac{\rho_a \Delta P}{G_c^2} = 18.93 n_r \text{Re}^{-0.316} \left(\frac{P_t}{\text{d}_r}\right)^{-0.927} \left(\frac{P_t}{P_d}\right)^{0.515} \tag{9.13}$$

where E_u is Euler number, ΔP is the pressure drop, P_t is transversal tube pitch, n_r is number of tube row, and G_c is mass velocity. P_d is diagonal pitch as given in the following equation:

$$P_d = \left[(P_t/2)^2 + P_l^2\right]^{0.5}$$

Eq. (9.13) is valid within the following limits:

$$2000 < R_e < 50,000$$

$$39.68\ \text{mm} < d_\text{f} < 69.85\ \text{mm}$$

$$18.64\ \text{mm} < d_\text{r} < 40.89\ \text{mm}$$

$$10.52\ \text{mm} < \frac{d_\text{f} - d_\text{r}}{2} < 14.48\ \text{mm}$$

$$42.85\ \text{mm} < P_\text{t} < 114.3\ \text{mm}$$

$$2.31\ \text{mm} < P_\text{f} < 2.82\ \text{mm}$$

$$42.85\ \text{mm} < P_\text{t} < \ 114.3\ \text{mm}$$

$$1.8 < \frac{P_\text{t}}{d_\text{r}} < 4.6$$

The pressure drop of a heat exchanger can be expressed in the form of loss coefficient as in Eq. (9.14):

$$\Delta P = \frac{1}{2} K_\text{he} \rho v^2 \tag{9.14}$$

where K_he is the loss coefficient of heat exchanger.

9.3.2 Mechanical draft cooling tower

When mechanical draft is used in an indirect dry cooling tower, the turbine exhaust steam is directed to and condensed in a condenser, where the cooling water is also directed to the same condenser to take the heat away from the exhaust steam. This heat carried by the water is rejected in the cooling tower through air-cooled heat exchangers. In this case, the hot fluid flowing inside the heat exchanger is the cooling water flowing through the condenser. The airflow around the exterior surface of the heat exchanger is caused by fans.

When mechanical draft is used in a direct dry cooling tower, turbine exhaust steam is condensed directly through ACC as shown in Fig. 9.4.

Design of mechanical draft dry cooling tower (direct or indirect) requires the heat transfer and pressure drop calculations. The former is needed for sizing the heat exchangers while the latter is used for the fan selection.

The heat transfer of the ACC shown in Fig. 9.4 can be expressed as

$$Q = m_\text{a} c_\text{pa}(T_\text{ao} - T_\text{ai}) = m_\text{c} i_\text{fg} = e\, m_\text{a} c_\text{pa}(T_\text{c} - T_\text{ai}) \tag{9.15}$$

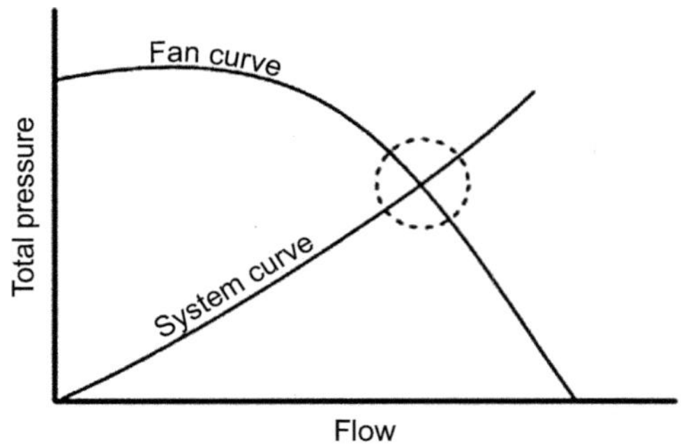

Figure 9.7 A schematic fan selection curve.

$$e = 1 - \exp[\mathrm{UA}/m_{\mathrm{a}}c_{\mathrm{pa}}] \tag{9.16}$$

where m_a is the air mass flow rate; T_{ao} is the air temperature exiting the ACC; T_{ai} is the air inlet temperature; m_c, i_{fg} and T_c are the mass flow rate, latent heat and the temperature of condensate, respectively.

For fan selection, the pressure drop of the cooling system must match the fan performance curve. The fan must deliver airflow rate m_a efficiently to meet the requirement in Eq. (9.15).

The pressure drop of the ACC system consists of pressure drops caused by heat exchanger support ΔP_{ts}, plenum resistance ΔP_{pl}, screen ΔP_{sc}, and heat exchanger ΔP_{he}, etc. The total pressure drop of the ACC system can be expressed as follows:

$$\Delta P_{\mathrm{t}} = \sum \frac{1}{2} K_i \rho_i v_i^2 \tag{9.17}$$

where K_i is the loss coefficient due to the heat exchanger support (K_{ts}), fan plenum (K_{pl}), screen (K_{sc}), heat exchanger (K_{he}), etc., respectively; ρ_i is the air density at these locations; and v_i is the air velocity at these locations.

The pressure drop of the ACC cooling system (system curve) obtained from Eq. (9.17) is drawn on the related fan performance curve as shown in Fig. 9.7 to check if the airflow rate meets the design requirement.

The fans can be selected based on the air volume flow rate m_a/ρ and pressure drop to achieve the maximum fan efficiency.

9.3.3 Natural draft cooling tower

The performance of a natural draft dry cooling tower (NDDCT) is influenced by the characteristics of heat exchanger, the tower geometry, and ambient conditions. An NDDCT must meet the heat balance and draft equation at the specified ambient conditions.

For an indirect dry cooling tower, the heat transfer by air and water and the overall heat transfer are expressed in the following equations:

$$Q = m_a c_{pa}(T_{ao} - T_{ai}) = m_w c_{pw}(T_{wi} - T_{wo}) = \mathrm{UAF_T}\Delta T_{lm} \tag{9.18}$$

$$\Delta T_{lm} = \frac{(T_{wo} - T_{ai}) - (T_{wi} - T_{ao})}{\ln[(T_{wo} - T_{ai})/(T_{wi} - T_{ao})]} \tag{9.19}$$

where T_{wi} and T_{wo} are the inlet and outlet temperatures of the water.

The draft equation for a natural draft cooling tower takes the following form:

$$\sum \text{flow resistance (tower + heatexchanger)} = \text{natural draft force of the tower} \cong (\rho_{ao} - \rho_{ai})gH \tag{9.20}$$

The flow resistance of the tower and heat exchanger are shown in Fig. 9.8 as proposed by Kröger [2] and can be expressed as

$$\Delta P = \sum \frac{1}{2} K_i \rho_i v_i^2 \tag{9.21}$$

K_i represents the loss coefficient at tower support K_{ts}, losses due to separation and redirection of flow at the lower edge of the tower shell K_{ct}, losses at heat exchanger supports K_{hes}, contraction losses K_{ctc}, frictional losses at heat exchanger K_{he}, expansion losses at heat exchanger K_{cte}, and losses in kinetic energy at the outlet of the tower K_{to}, respectively, as shown in Fig. 9.8. The details of calculating these loss coefficients can be found in Ref. [2].

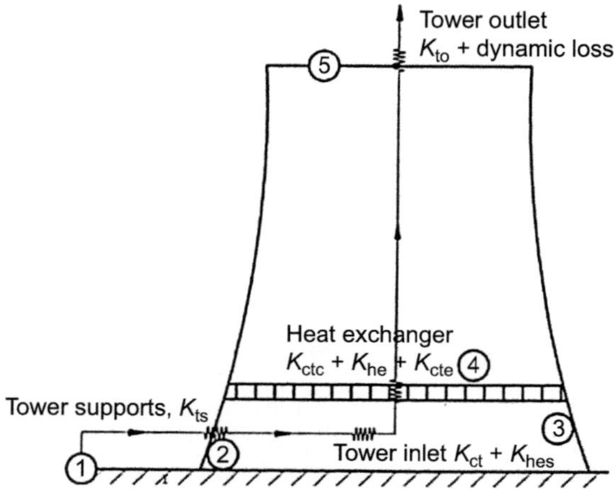

Figure 9.8 Flow resistance of the tower and heat exchanger [2].

On top of the three unknowns of T_{ao}, T_{wo}, and Q, which need to be solved by the equations, there is an extra unknown (m_a), which is also needed to be solved. Therefore, iterative processing has to be applied in NDDCT design.

9.4 Advances in dry cooling technologies for concentrating solar thermal power plants

In a dry cooling system, because the heat transfer coefficient of the air-side is much lower, an improvement on air-side heat exchange rate will increase the net power output of a power plant. According to dry cooling tower design, heat rejection by air-side of the heat exchanger is expressed in Eq. (9.22). To maximize the air-side heat rejection rate, Q_a, by dry cooling system, one can either increase the air mass flow rate (m_a) or decrease the air inlet temperature (T_{ai}).

$$Q_a = m_a c_{pa}(T_{ao} - T_{ai}) \tag{9.22}$$

The given two approaches (increasing m_a or decrease T_{ai}) have been implemented in the advanced hybrid cooling technologies developed by the Queensland Geothermal Energy Center of Excellence (QGECE), the University of Queensland. These advanced hybrid cooling technologies include solar hybrid natural draft cooling, water hybrid cooling, and windbreak wall enhancement hybrid cooling.

9.4.1 Solar hybrid natural draft dry cooling tower

NDDCTs function according to the simple principle of the stack effect. Because warm air inside the tower is less dense than the ambient air outside the tower, warm air rises due to buoyancy and a natural circulation occurs. The driving force for the natural circulation of air is proportional to the difference in air density between the inside and outside of tower and the tower height. Solar energy is used in this technology to heat up the air inside an NDDCT further to increase the density difference.

The solar hybrid NDDCT exploits the solar energy during the hottest periods at which the conventional dry cooling tower would suffer the lowest performance. The concept of the proposed system is shown in Fig. 9.9. The solar-enhanced hybrid cooling system includes a natural draft steel tower, sunroof, and heat exchangers. The sunroofs are arranged radially at the base of the tower, and the heat exchangers are placed vertically at the outside edge of the sunroof. As the Sun heats the sunroof, the air under the sunroof is heated up. Warm air naturally rises through the tower and fresh air is sucked in, thus providing a cooling airflow through the heat exchanger bundles. By this arrangement, it would enhance the performance of an NDDCT by increased m_a due to the added solar energy to the airstream after the heat exchanger. The extra heat from the solar collectors increases the buoyancy of the air inside the tower and helps to drive more air through the heat exchangers. Higher air mass flow rate, $\mathrm{m_a}$ as indicated in Eq. (9.22), has been achieved by this technology.

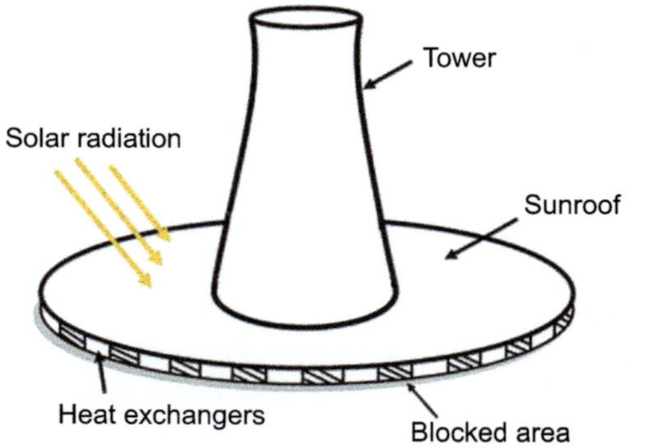

Figure 9.9 Concepts of solar hybrid natural draft dry cooling tower.

Zou et al. [11,12] used a 50-MWe geothermal power plant proposed for Cooper Basin, Australia, as a case study to demonstrate the improved performance of the proposed solar hybrid cooling tower. The global and direct hourly solar irradiance and average air hourly temperature records from Refs. [13,14] were used to represent a typical summer day in Cooper basin as reproduced in Fig. 9.10.

Other parameters selected in this case study include 182 six-row heat exchanger bundles, 4000 kg/s water flow rate inside the heat exchanger, 120 m tower height, 15% efficiency of the geothermal power plant, 27°C ambient temperature at the design point, and 450 m solar collector in diameter.

If the same tower height and heat exchanger area are used for both the solar hybrid natural draft cooling tower and the conventional natural draft cooling tower, solar hybrid cooling system is able to generate more power as shown in Fig. 9.11.

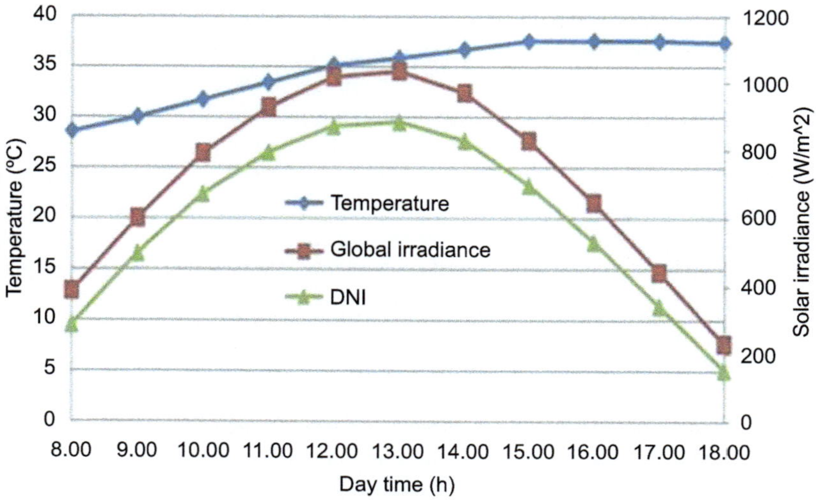

Figure 9.10 Solar irradiance and temperature data near Cooper Basin, Australia.

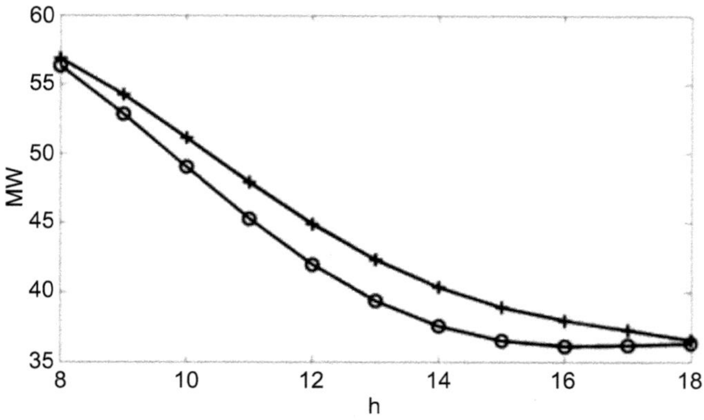

Figure 9.11 Net power generation comparison between the solar hybrid system (+) and the conventional natural draft dry cooling tower (o) at a day time.

From Fig. 9.11, it is shown that the net power output decreases with increasing ambient temperatures no matter which cooling system is used. However, the net power reduction is less with solar hybrid cooling tower than without it (NDDCT) during the hottest part of the day; when the ambient temperature increases from 28.6°C at 8:00 a.m. to 36.7°C at 14:00 p.m., the net power suffers about 50% reduction with a conventional natural draft cooling tower. It is also noted that a relative high solar irradiance, from 12:00 noon to 14:00 p.m. and, when this solar energy is used in solar hybrid system, about 3 MW net power loss can be avoid at 14:00 p.m.

In the design of a cooling system for a particular power plant, the height of the tower has a reverse relationship with the area of heat exchangers, i.e., a higher tower requires smaller area of heat exchangers and vice versa. The trade-off between the heat exchange area and tower height can be optimized based on the minimum cost.

Once the height of the tower and the area of the heat exchangers are determined, selecting the right diameters of the sunroof and the arrangement around the sunroof becomes an important issue in the design of solar hybrid cooling tower.

Theoretically, increasing the solar sunroof size will increase the heat rejection rate as shown in Fig. 9.12. However, bigger solar roof size does not only increase the investment, but also may affect the airflow through the heat exchangers and affect the cooling performance. Large sunroof diameters provide a larger space for placing the heat exchangers, and the selected frontal area of the heat exchangers may not be large enough to cover the entire area provided by the outside perimeter of the sunroof. In this case, parts of the sunroof perimeter will need to be blocked, and the 3D numerical model has confirmed that this blockage will have a significant effect on the airflow and the cooling performance. To maximize the utilization of the sunroof, one design goal is to make the blocked area as small as possible (or the heat exchange coverage area as large as possible).

It seems that partial blockage in solar hybrid cooling system is unavoidable with increased sunroof diameter. The larger the sunroof diameter, the smaller is the heat

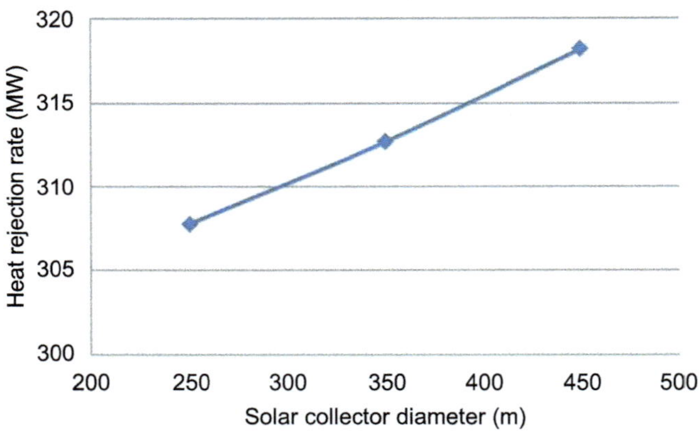

Figure 9.12 Relationship of solar collector size and heat rejection rate.

exchanger coverage ratio, i.e., the ratio of the total heat exchanger frontal area to the flow area available at the perimeter of the sunroof.

3D simulations have been carried out [12] for various coverage ratios in a solar hybrid cooling tower. Six coverage ratios were modeled in this study: 100%, 60%, 55%, 50%, 45%, and 40%. For the heat exchanger type selected (keep the heat exchange area constant), these coverage ratios relate to sunroof diameters of 195 m (a), 325 m (b), 354.4 m (c), 390 m (d), 433.33 m (e), and 487.5 m (f), respectively, as shown in Fig. 9.13.

The heat rejection rate of the proposed solar hybrid cooling tower obtained by 3D numerical model is plotted in Fig. 9.14 when different sizes of solar roof are selected

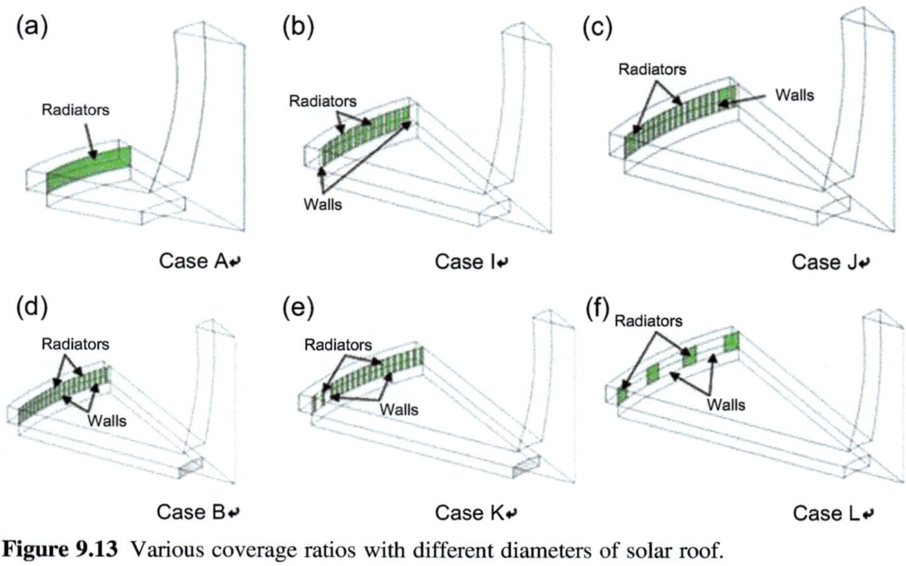

Figure 9.13 Various coverage ratios with different diameters of solar roof.

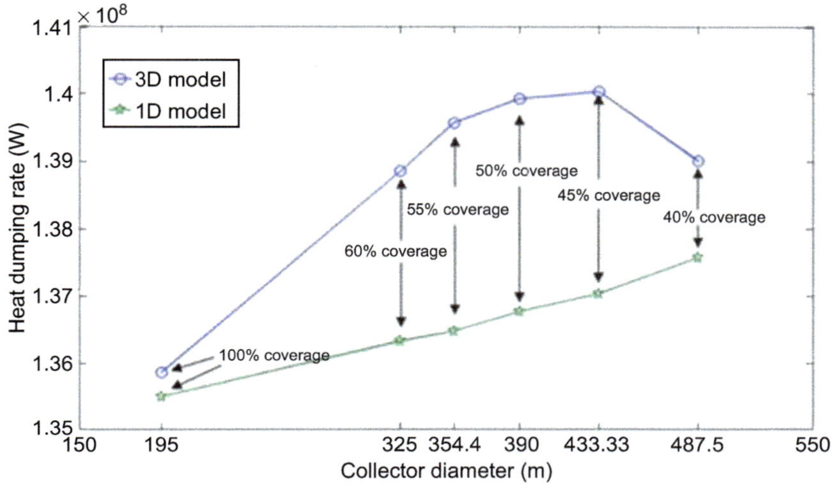

Figure 9.14 Heat rejection and mass flow rates versus size of solar collector.

resulting in different heat exchanger coverage ratios. However, when the diameter of the solar roof increases to a limit in which the selected heat exchanger can only cover 40% of the area provided, there is a drop of the heat rejection rate. The plot of the airflow indicates that vortices are generated at the back of heat exchangers below this coverage ratio owing to the introduction of large partial blockage required to cover the rest of the perimeter. These vortices reduce the effective heat transfer area. These vortices cannot be detected by 1D model as indicated in Fig. 9.14. Therefore, the maximum size of the solar roof can be used in a solar hybrid cooling tower to guarantee a heat exchanger coverage ratio of about 50%. Beyond that ratio, any increase of the sunroof diameter will generate no further gains.

9.4.2 Water hybrid cooling

In this technology, water is introduced into the inlet airstream of the NDDCT. The introduced water evaporates and reduces the entering dry air temperature T_{ai}. This cooler air then cools the dry system more efficiently as indicated in Eq. (9.22). The introduction of water into the entering air can be achieved by either wet media or nozzle spray.

Inlet air cooling by spray has been effectively used in gas turbine power plants to increase power output. The use of spray cooling for power plants has been studied by the Electric Power Research Institute (EPRI) [15].

The inlet air cooling can provide cost-effective alternatives for thermal power plants. Based on the previous test results obtained from both the laboratory and field (field test was on single-cell ACC), the EPRI [15] concluded that the allocation of cooling water in modest quantities (less than 10–15% of full wet cooling consumption) increased the power output by 50% or more during the hottest period. Ashwood

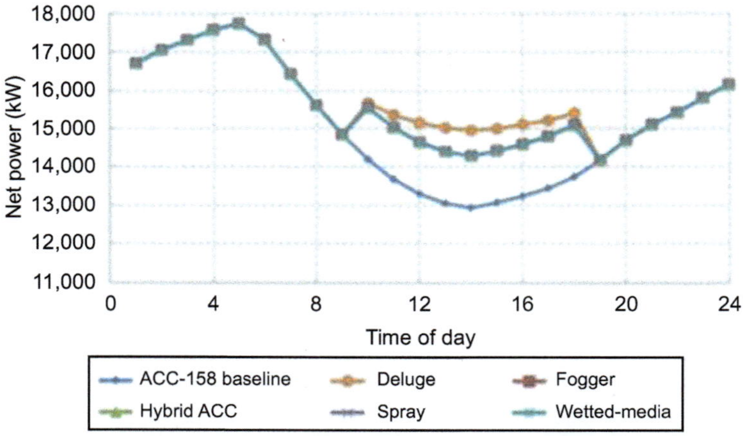

Figure 9.15 Net power for the (158°C resource temperature) air-cooled condenser (ACC) systems.

and Bharathan [16] conducted modeling analysis using ASPEN Plus to evaluate various inlet air cooling methods for low-temperature geothermal power production. A sample of their modeling results is shown in Fig. 9.15. The figure shows that the net power generated during the hottest hours of the day with water spray can be significantly higher than that without water assistance. They also identified that the payback period with spray system could be less than 1 year for the geothermal power plant studied.

Although inlet air cooling have been successfully applied in the process industry, gas turbines, and small-scale mechanical draft cooling power plants, there is little experience for CST power plants, especially in the power plants with natural draft dry cooling systems. Frequent use of water spray in an NDDCT can cause corrosion, scaling, and fouling on the heat exchanger bundles if water droplets are carried by the airstream to the heat exchanger bundles. To avoid this, the system is required to evaporate all water in the airstream to prevent water droplet contact with heat exchanger surface. Special wet media or spray nozzles may be required to meet the requirement. High-pressure nozzles provide small water droplets but at a higher cost. Water quality affects the performance of the nozzles and its maintenance cost.

Fig. 9.16 shows the concept of the inlet air precooling hybrid NDDCT. Either wet media or nozzles are arranged under the tower to introduce the water to the inlet air.

9.4.2.1 Inlet air precooling with wet media

As shown in Fig. 9.16, the heat exchanger bundles are arranged horizontally at a height slightly above that of the tower inlet. The wet media are placed around the tower base to cover the whole inlet area. Water is distributed over the top of the media and flows down by gravity to wet the whole media surfaces uniformly. When ambient air passes through the wet media surfaces, it causes the water to evaporate and the air is then

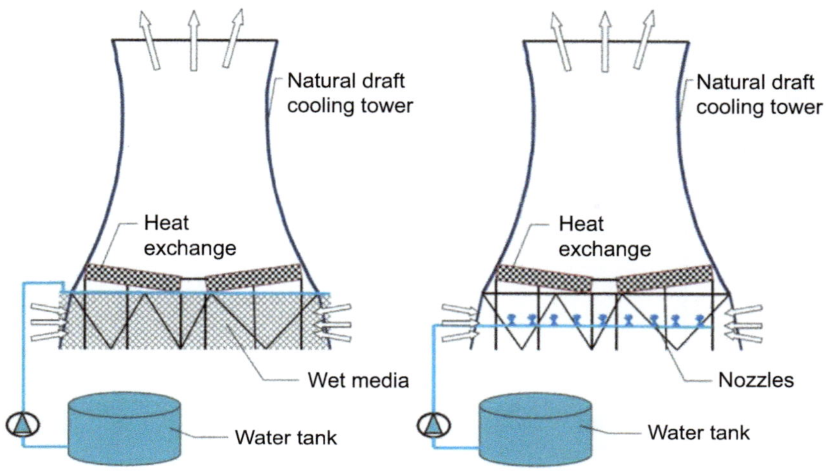

Figure 9.16 Proposed Inlet air precooling hybrid cooling system: wet media (left) and spray nozzles (right).

cooled. The cooler air flows through the heat exchanger and the performance of the cooling system is improved.

One of the main concerns of using wet media for inlet air precooling of an NDDCT is the pressure drop introduced by the wetted media. The additional pressure drop will have a strong effect on the air mass flow rate passing through the heat exchangers, which will then impair the heat rejection. Therefore, there is a trade-off between the cooling performance and the pressure drop when using wetted media for inlet air pre-cooling. Experimental study was carried out to identify the best wet media for this application by QGECE of the University of Queensland.

Numerical model has been developed [17] to predict the performance of an NDDCT with introducing wet media to precooling the inlet air on hot and dry days. Significant benefits from the wet media hybrid cooling have been achieved.

Fig. 9.17 shows the simulation results with four different media applied in a natural draft dry cooling system. The results show that even the air mass flow rate is lower with wet media as shown in Fig. 9.17(a) (extra pressure drop introduced by the wet media); the heat rejection rate is still much higher with the wet media precooling (as shown in Fig. 9.17(b)). Some wet media perform better than other ones.

With the inlet air precooling application for NDDCTs, the high cooling efficiency of wet medium tends to improve the tower performance by reducing the tower inlet air temperature, while the extra pressure drop impairs the tower performance through its contribution to the flow resistances. Because the NDDCT is operated by the balance of both energy and draft equations, the cooling efficiency and extra pressure drop are all crucial to the tower performance. Wet media with high cooling efficiency provides more cooling at the expense of extra pressure drop. The media with low pressure drop have the advantage of small resistance but produce less cooling. There is a trade-off between selecting the cooling efficiency and pressure drop of a wet medium.

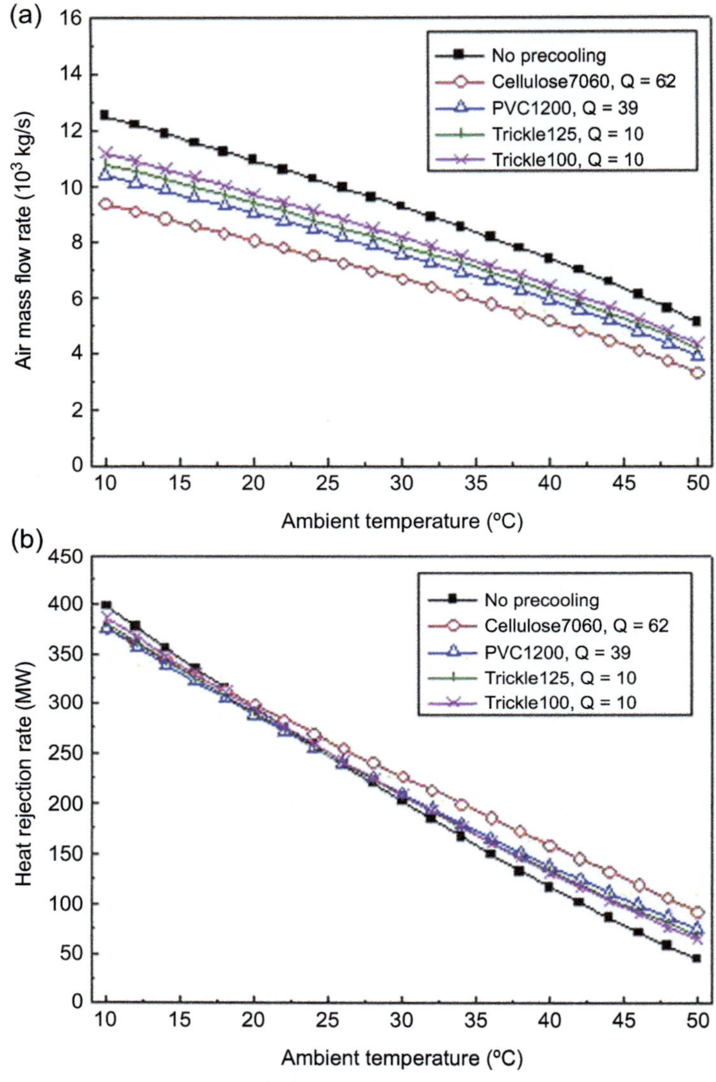

Figure 9.17 Performance of the natural draft dry cooling tower (NDDCT): (a) air mass flow rate and (b) heat rejection rate of the NDDCTs (relative humidity = 20%, l = 300 mm, Q is water flow rate in l/min/m^2).

9.4.2.2 Inlet air precooling with nozzle spray

In this system, nozzles are used to spray water on the entering air only on extremely hot days. The nozzles can be distributed either horizontally under the heat exchanger bundles (as shown in Fig. 9.16) or vertically around the periphery of the tower. During high ambient temperature hours, water is sprayed into the inlet airstream to cool the incoming air by evaporation.

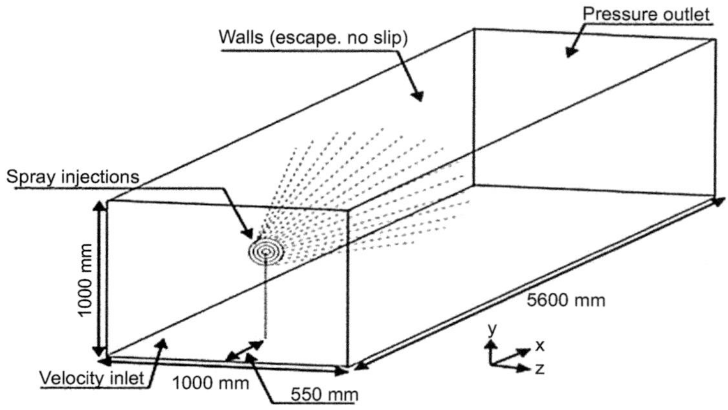

Figure 9.18 3D model of spray cooling.

The major concern for the inlet air precooling with nozzle spray is the water droplet size. To avoid the problem of heat exchanger corrosion, scaling, and fouling caused by water spray, the water is required to be fully evaporated before reaching the heat exchangers to prevent water droplet contact with heat exchanger surface.

Experimental and modeling studies have been carried out to achieve the full evaporation by optimizing the nozzle selection.

A 3D computational fluid dynamics (CFD) model with dimensions 5.6 m long and 1 m × 1 m cross section has been developed by Alkhedhair et al. [18] as shown in Fig. 9.18 to simulate the nozzle spray. The 3D model dimension is the same as the dimensions of the testing facility (wind tunnel). Spray injection is placed 0.6 m above the tunnel floor and 0.55 m from the inlet and directed horizontally in a cocurrent direction with airflow.

The modeling results of droplet trajectories and equivalent air temperature profile along the wind tunnel duct at different cross sections are illustrated in Fig. 9.19, colored by their diameter and local air temperature. It can be seen from this figure that droplets are either airborne with the airstream to the outlet plane or evaporate completely before falling on the duct floor.

Unlike the inlet air precooling by wet media, no extra pressure drop is involved with the inlet air precooling by nozzle spray, and the only negative effect in this system is the reduction of the air density inside the cooling tower as indicated in Eq. (9.20).

To compare the cooling performance of inlet air precooling by nozzle spray, two new dimensionless parameters are introduced: the relative performance η_Q and the benefit efficiency β_Q. η_Q is defined as the ratio of the heat rejection at the off-design to the heat rejection at the design point. β_Q is defined as the ratio of the heat rejection with the inlet air precooling using the nozzle spray to the heat rejection without inlet air precooling. The simulation results on the relative performance η_Q and the benefit efficiency β_Q are plotted in Fig. 9.20. The results were based on the cooling tower height of 38 m with the heat rejection capacity of 25 MW.

Fig. 9.20 shows that the relative performance η_Q (*solid lines*) and the benefit efficiency β_Q (*dashed lines*) at different relative humidity achieved after inlet air

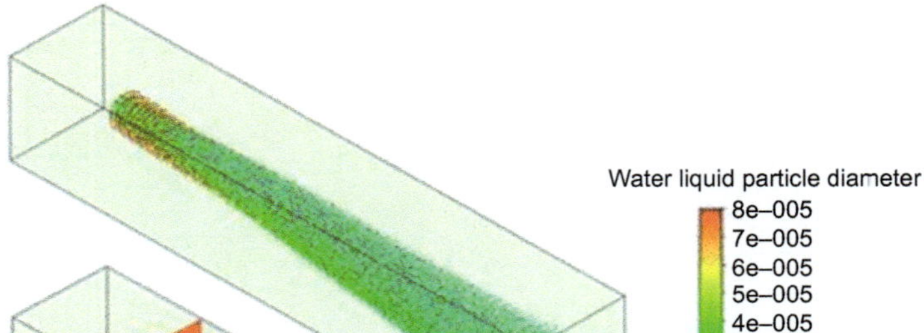

Figure 9.19 Registered droplet trajectories and air temperature profile across the wind tunnel duct.

precooling. The figure demonstrates that the overall performance of the NDDCT with inlet air precooling by nozzle spray has always been improved. More benefits can be achieved at higher ambient temperature.

9.4.3 Windbreak wall hybrid natural draft dry cooling tower

Traditional NDDCTs are normally built over 100 m in height. No small-sized natural draft cooling towers were built to suit small-scale power plants. With the increased desire to build small-scale CST power plants for remote areas in Australia, it is important to develop and demonstrate small, high-performance NDDCTs to improve the efficiency and power output of small CST plants.

A major design issue for small natural draft cooling towers is the negative effect of the crosswind on the cooling performance, which reduces overall plant efficiency. The performance degradation caused by crosswind is much more significant for small towers than for tall ones. QGECE has developed a windbreak wall hybrid solution to address the crosswind problem for small NDDCTs. By introducing a tri-blade-like windbreak wall underneath the heat exchanger bundles, the negative effect of

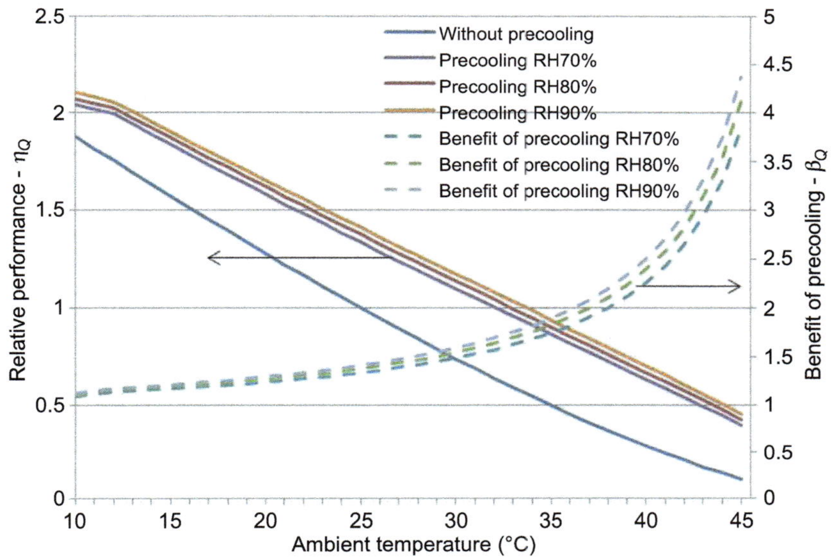

Figure 9.20 Modeling results: the relative performance η_Q and the benefit efficiency β_Q.

the crosswind in a wide range of velocity ratios (up to 40) could be effectively arrested and even converted into a significant performance boost [19].

9.4.3.1 CFD modeling

In small NDDCTs with tower heights less than 30 m, the crosswind effect could be substantial and the CFD modeling aims to identify the effect of crosswind on a small NDDCT of height 15 m.

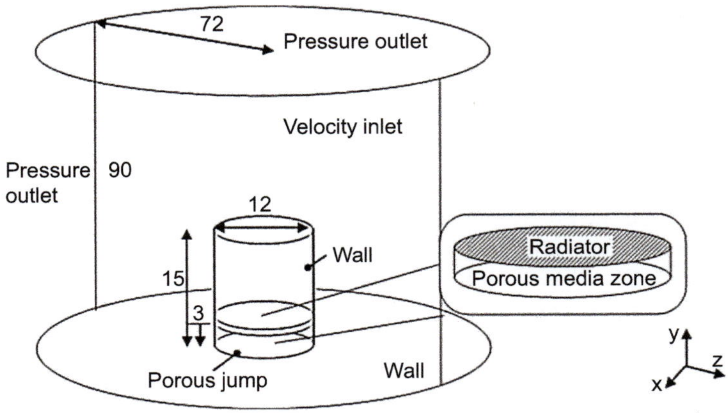

Figure 9.21 Geometry of the CFD models without windbreak walls.

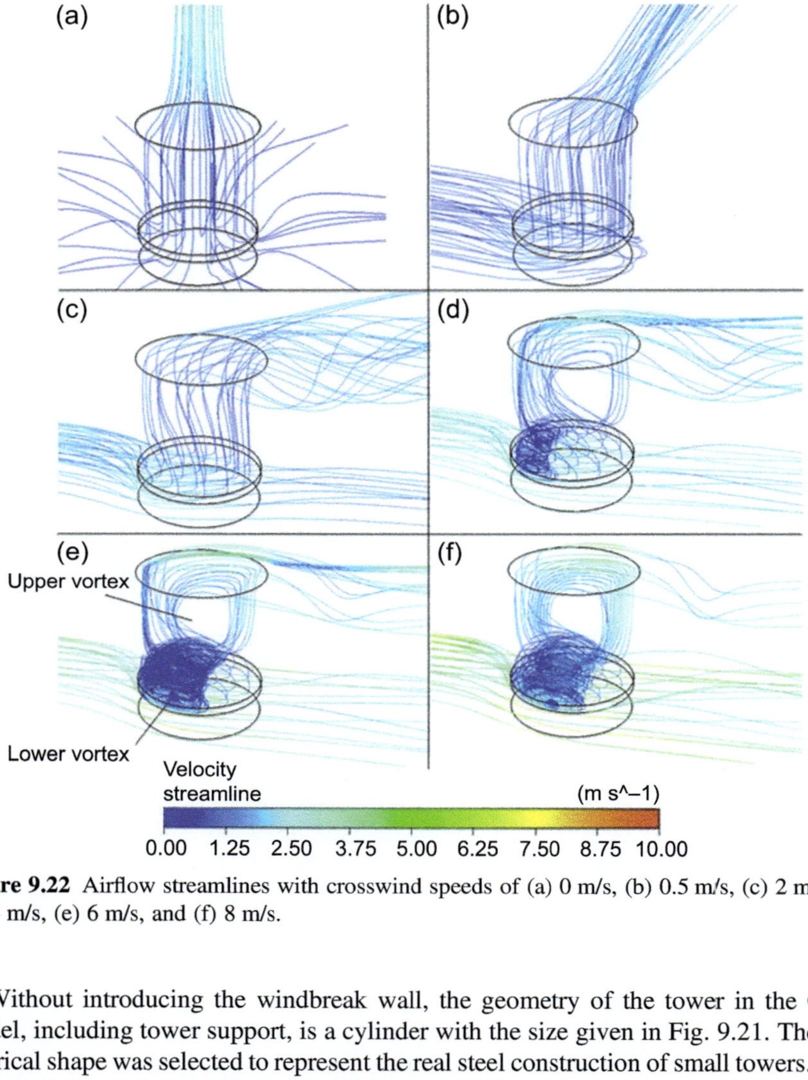

Figure 9.22 Airflow streamlines with crosswind speeds of (a) 0 m/s, (b) 0.5 m/s, (c) 2 m/s, (d) 4 m/s, (e) 6 m/s, and (f) 8 m/s.

Without introducing the windbreak wall, the geometry of the tower in the CFD model, including tower support, is a cylinder with the size given in Fig. 9.21. The cylindrical shape was selected to represent the real steel construction of small towers. The details of the CFD modeling can be found in Ref. [19].

When the crosswind is introduced horizontally, the airflow pattern and the air temperature inside the cooling tower are affected by it. Fig. 9.22 shows the airflow streamlines inside the tower. Fig. 9.23 shows the air temperature contours at the central vertical cross section of the tower at various crosswind speeds.

At low crosswind speeds, such as 0.5 m/s, the airflow pattern inside the tower does not change much from no-crosswind condition. As the speed of crosswind increases, large volumes of vortexes are generated inside the tower which affects the airflow and therefore the uneven temperature distribution. The effect of the crosswind is to reduce the performance of the cooling tower.

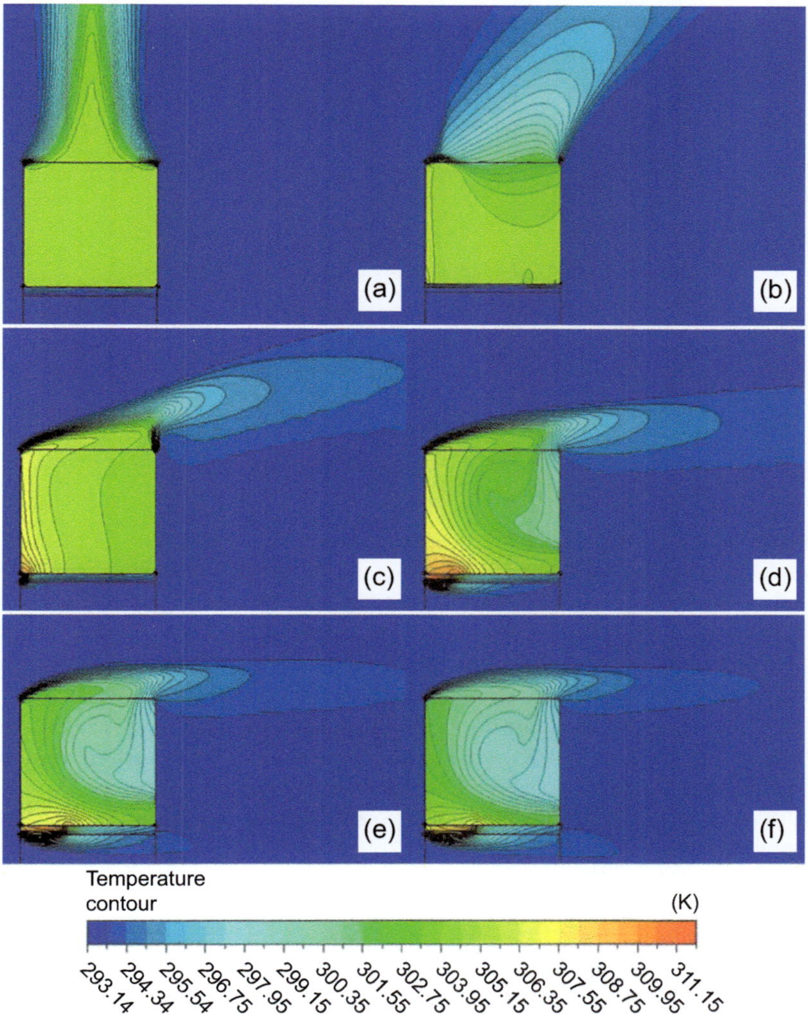

Figure 9.23 The temperature contour at the central vertical plane with crosswind speeds of (a) 0 m/s, (b) 0.5 m/s, (c) 2 m/s, (d) 4 m/s, (e) 6 m/s, and (f) 8 m/s.

To assess the effect of the crosswind on the NDDCT cooling performance quantitatively, the air mass flow rate m_a and total heat transferred Q are computed. Fig. 9.24 plots the m_a and Q against the crosswind speed v_{cw}. The mass flow rate m_a decreases first along with rising crosswind speed and remains nearly constant beyond 10 m/s. The variation of the heat transfer rate with crosswind is more interesting. For this small tower, the crosswind does not always exert a negative effect on the cooling tower performance in terms of total heat transfer rate. Q reaches its lowest value at a crosswind speed around 5 m/s and then increases with increasing crosswind speed. When the crosswind speed is larger than 15 m/s, the positive effect is shown.

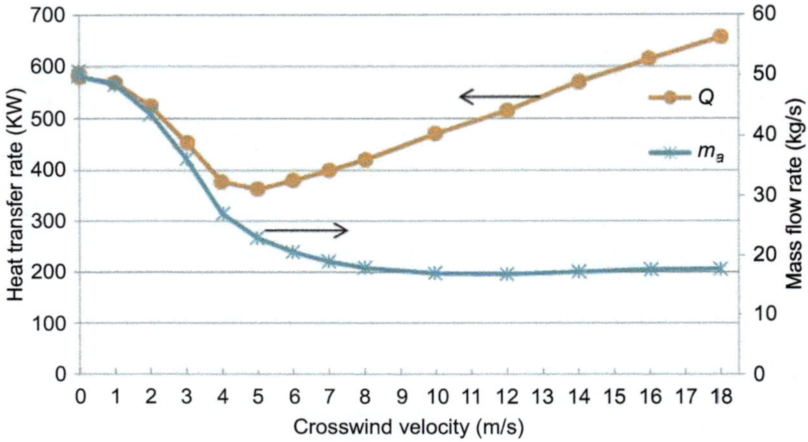

Figure 9.24 The change of m_a and Q against various crosswind speeds.

The change of Q in this small tower is better understood by examining the two cooling components separately: the heat carried away by the air flowing through the tower vertically and exiting at the top, and the heat carried away by the air flowing through the bottom part of the tower horizontally. The total heat rejected from the heat exchangers, Q, is equal to the sum of these two components. When there is no crosswind, the second component is zero and all heat is dissipated through the air exit on the top of tower.

The change in Q as shown in Fig. 9.24 indicates that, when crosswind speed is larger than 5 m/s, the heat transfer rate through the bottom of the tower becomes influential in the overall heat transfer rate. This phenomenon does not exist in large NDDCTs because in a tall tower the heat rejected by the air inside the tower is dominant due to the large buoyancy force.

It is also noted, from Fig. 9.24, that with the existence of crosswind at speed about 5 m/s, the total heat rejection, Q, could decrease by 37% compared with no-crosswind condition, which leads to a significant drop in net power generation at this crosswind speeds. At some other high-speed levels, the total heat rejected, Q, was increased from its lowest level. It is very hard to design a small cooling tower under unpredictable crosswind without some controllable means to be implemented.

A tri-blade-like windbreak wall is introduced and placed underneath the heat exchanger bundles for the crosswind effect mitigation. Fig. 9.25 shows the CFD model with the proposed windbreak wall. The effect of the angle of attack of the crosswind was also examined. The variations of the heat transfer rates at different velocity ratios are examined [20].

Two dimensionless quantities were introduced by dividing the air mass flow rate m_a and total heat rejection rate Q_r obtained under crosswind to their corresponding values under no-crosswind m_{aN} and Q_{rN}, respectively. These dimensionless quantities are plotted against the velocity ratio δ for different angles of attack using solid lines, as shown in Figs. 9.26 and 9.27, respectively. For comparison purposes, the results without the windbreak walls are plotted in dashed lines in the figures.

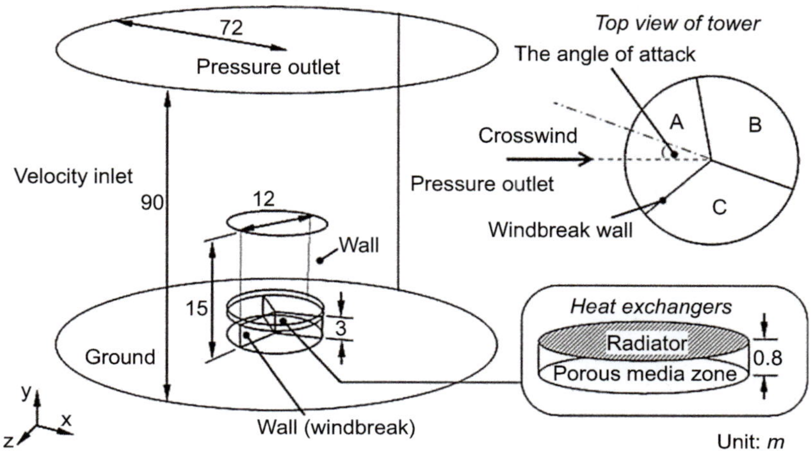

Figure 9.25 The CFD model with windbreak wall.

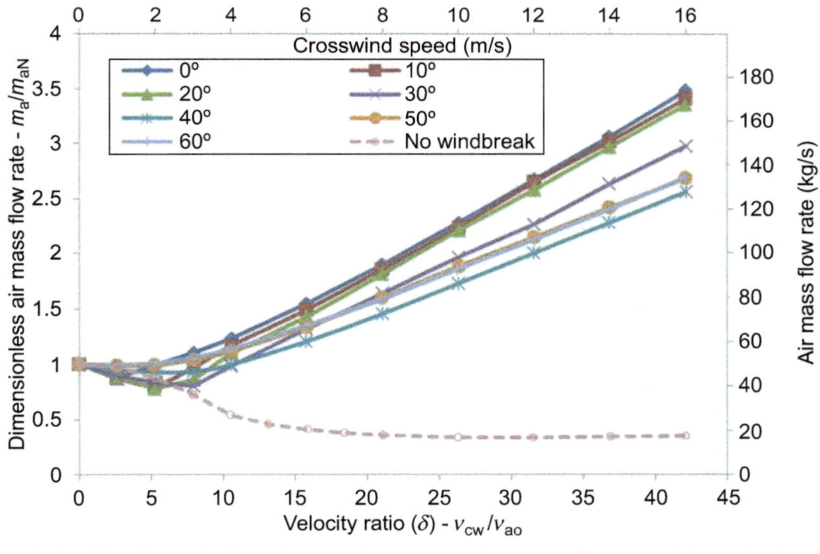

Figure 9.26 The dimensionless air mass flow rate m_a/m_{aN} as a function of the velocity ratio δ (speed of crosswind/speed of vertical). The secondary x-axis and y-axis show the corresponding wind speed and air mass flow rate, respectively.

Different attack angles show different declinations. Above the critical value, the trend is reversed, which indicates the benefits of the windbreak walls. The troughs of these curves, depending on the wind attack angle, occur in the velocity ratio range of 2.5–10. In this 15-m-high NDDCT, an air velocity ratio of 10 corresponds to a crosswind speed of ~4 m/s. A comparison between the *solid lines* and the *dashed lines* shows the significant effectiveness of the windbreak walls at high velocity ratio ($\delta > 10$).

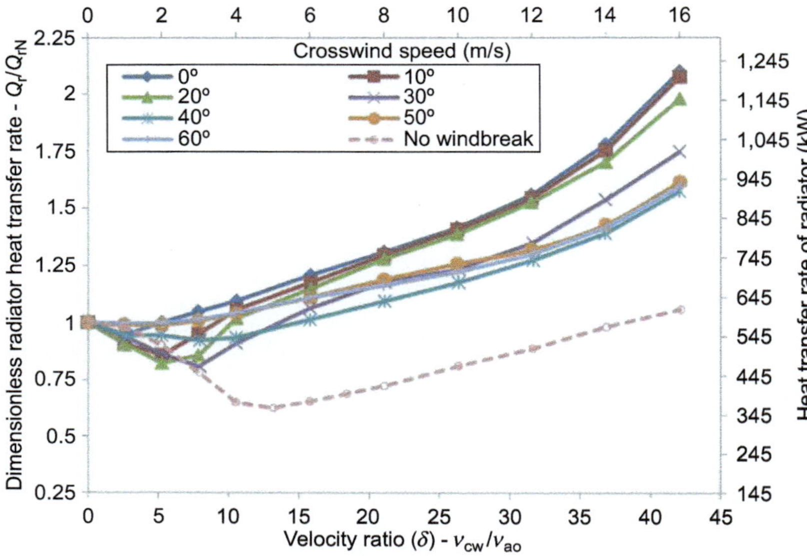

Figure 9.27 The dimensionless heat transfer rate Q_r/Q_{rN} as a function of the velocity ratio δ. The secondary x-axis and y-axis show the corresponding wind speed and the heat transfer rate, respectively.

It is noted that at the attack angles of 50 degrees and 60 degrees, the cooling performance is almost unaffected at low-velocity-ratio ($\delta < 10$) regions, which implies a great advantage compared with other wind attack angles. Once the wind velocity ratio exceeds $\delta > 10$ (v_{cw} >4 m/s), a significant advantage is observed for the attack angles of 0 degree, 10 degrees, and 20 degrees, where the wind direction is closer to one of the walls.

The effectiveness of the windbreak wall is further investigated by the experimental study.

9.4.3.2 Experimental study

A scale model (based on applying similarity laws) of the mentioned tower was built and tested in UQ's Gatton wind tunnel [20].

The scaled tower model is shown in Fig. 9.28. A circular finned tube electric heater was used to represent the air-cooled heat exchanger.

In the experimental study, temperature, pressure, and speed sensors were employed to measure the temperature, pressure, and the air speed at different levels as shown in Fig. 9.29.

Similarity between the scaled tower model and the 15-m-high tower cannot be satisfied completely as certain dimensionless parameters are not the same for both. Therefore, a CFD model has been built for the scaled tower placed in the wind tunnel also. The purpose of building up this extra CFD model is to compare the CFD results with

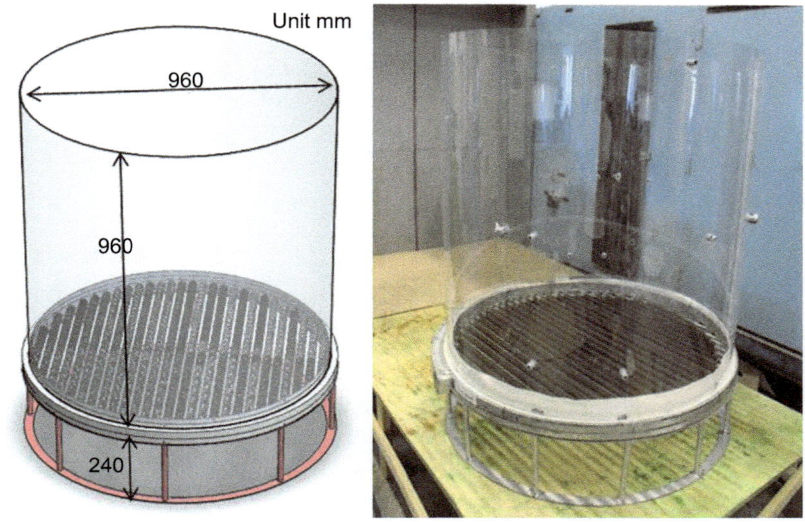

Figure 9.28 The scaled cooling tower model used for experimental study.

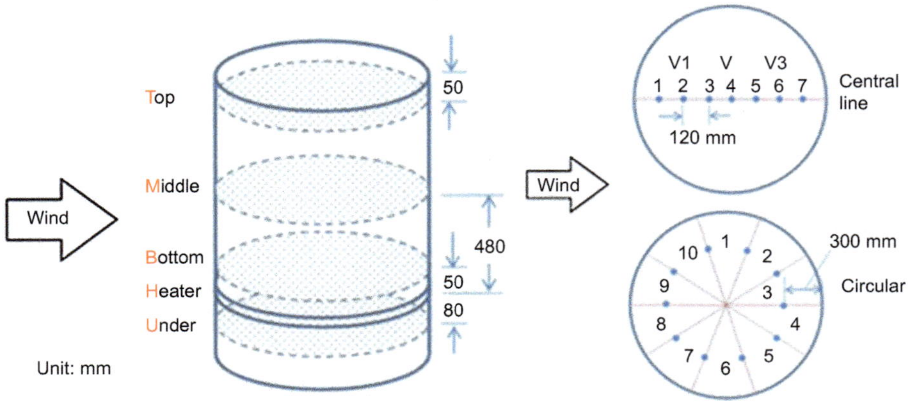

Figure 9.29 Five levels for sensor position.

the experimental ones directly, which can validate the correctness of the CFD modeling methodology used in the approach.

Fig. 9.30 shows the air velocity ratio v_a/v_{aN} (v_a is the natural draft air velocity inside the tower under crosswind conditions and v_{aN} is the natural draft air velocity without crosswind) varies with the ratio of v_{cw}/v_{aN} for the experimental result and also the results predicted by the two CFD models without windbreak walls.

It can be seen that the result of CFD model for the 15-m-high tower is different from the other two. This is because the similarity between the scaled model and the 15-m tower is not completely satisfied; for instance, the Reynolds numbers do not satisfy the similarity.

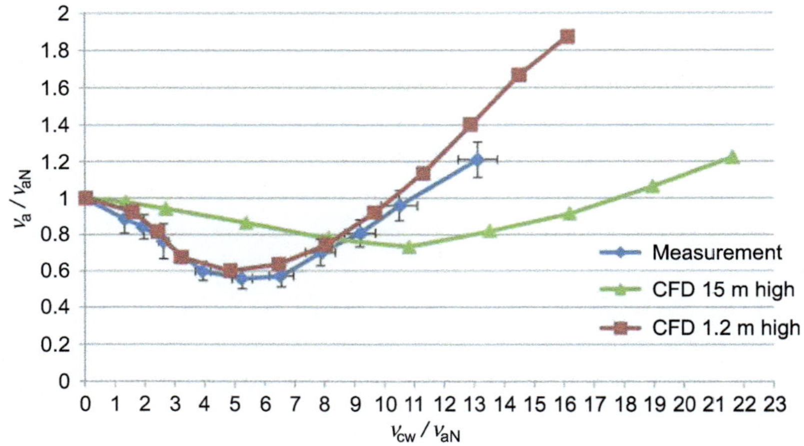

Figure 9.30 Air velocity v_a/v_{aN} changes with the crosswind at constant heat dumping rates, both experimental and numerical results. The error bars are shown on experimental results.

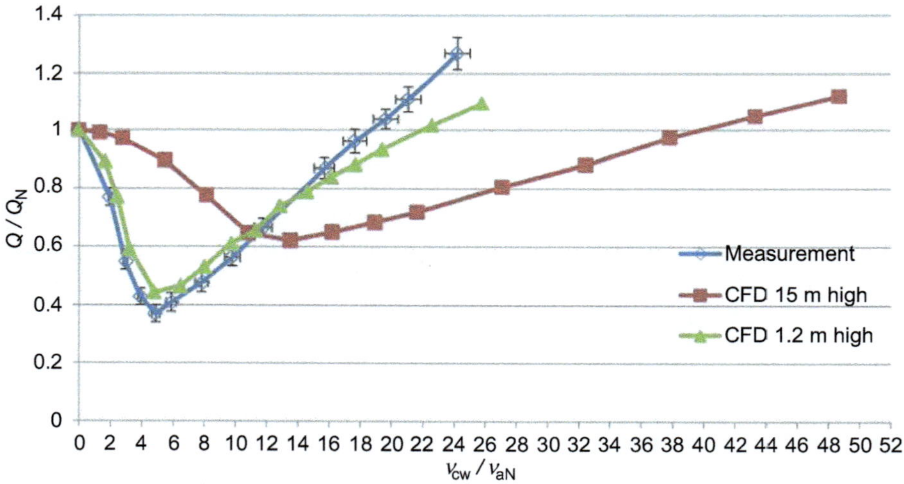

Figure 9.31 The total heat rejected by the cooling tower at different crosswind speeds when the heater/heat exchanger surface temperatures are constants.

Fig. 9.31 shows the relationship between the dimensionless parameter Q/Q_N (Q is the heat rejection with crosswind and Q_N is the heat rejection without crosswind) and the speed ratio v_{cw}/v_{aN} without the windbreak wall.

With the introduction of a tri-blade-like windbreak wall underneath the heater in the tower, Fig. 9.32 presents the results in the same way as in Fig. 9.30. Three dashed lines are the CFD results with the 15-m-high tower.

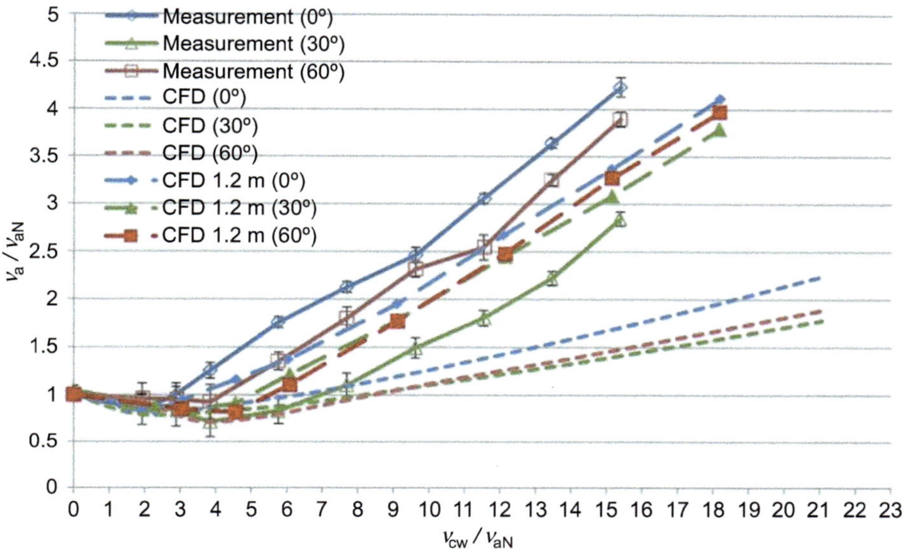

Figure 9.32 Air velocity as a function of crosswind speed ratio for different attack angles at constant heat dumping rates.

It is found that the windbreak wall increases the air velocity flowing inside the cooling tower as the crosswind speed increases.

Fig. 9.33 shows the dimensionless heat transfer rate Q/Q_N as a function of crosswind speed ratio v_{cw}/v_{aN}, obtained from the experiment and CFD results.

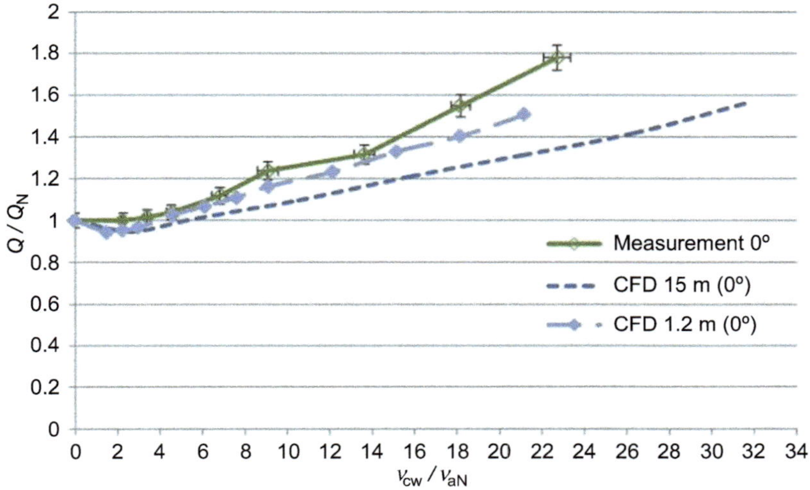

Figure 9.33 Heat transfer rate Q/Q_N as a function of crosswind speed ratio.

The figures show that the heat rejection rates are greatly increased by the introduction of the windbreak wall. The benefit of windbreak walls under the heat exchangers has been verified by the experimental results. The cooling performance of the tower even exceeds the performance of the no-crosswind case when the walls are arranged at attack angles of 0 degree and even 60 degrees. The windbreak wall hybrid small natural draft cooling tower is an effective new technology for the CST power plants.

9.4.4 Advances in tower structure

In a traditional NDDCT, the tower structure is made of reinforced concrete (RC). The erection of conventional RC tower requires advanced construction methods and skilled work force. Building such a tower is expensive and time-consuming. To overcome the typical expenses associated with RC natural draft cooling towers, and improve economics for small-scale towers, the QGECE has pioneered a modular construction from a steel frame and PVC polymer membrane. These materials dramatically reduce costs relative to concrete while still providing the same structural integrity and resilience to wind loading. The modular design also substantially reduces construction time (by an order of magnitude or more) and thereby reduces construction costs. The modular design enables rapid deployment to remote sites.

One of the structure frames built by QGECE is shown in Fig. 9.34 located at the Gatton campus of the University of Queensland. The frame is made of tubular members and each individual member is joined together by welding. The cost of the tower structure is greatly reduced.

Figure 9.34 Configuration of Gatton cooling tower.

9.5 Conclusions

CST power plants using dry cooling system have advantages of water conservation, flexible selection of site locations, and environmental protection. However, dry-cooled power plants suffer lower efficiencies when the ambient air temperature is high [2]. Up to 20% net power reduction was recorded during the hottest hours compared with the operations at average metrological conditions for coal-fired power plant. It is expected that CST plants using dry cooling will suffer the same way. High ambient temperatures usually coincide with peak system demand and higher electricity sale prices.

Because water conservation ranks top priority in future research on power plant cooling, new dry cooling technologies to address high ambient temperature are essential to increase the electricity production for CST plants. Although there are several step-changing cooling technologies, such as adsorption chillers, having potential for power plants' water conservation application, they are either still in its early stage or have high risk financially. Future research on dry cooling should be focused on dry cooling technology innovation, improving thermal efficiency of air-cooled heat exchanger and cooling tower and reducing the costs.

The concept of solar hybrid NDDCTs has been considered a highly innovated technology by world cooling experts in an international cooling tower conference. Further study on the design optimization will be required to demonstrate the cost-effectiveness on the levelized cost of electricity (LCOE).

In spray cooling system, results showed that inlet air can be precooled by up to 80% of the difference between the wet bulb and dry bulb temperatures. This difference is 15–20°C in most future CST plant locations in Australia. Therefore, the technology has the potential to have a large impact on plant performance on hot days of the year. It is expected that multiple nozzles will be involved and optimization of the nozzles arrangement will be needed for a tower design to minimize the LCOE. The key for the optimization is the interaction study between multinozzles. This will also involve turbulence enhancement on the nozzle spray. To further save freshwater, saline water could be used, which requires further study for spray cooling.

The accurate cost study will be required based on the optimized design of the advanced dry cooling systems. The study should be based on the detailed cost study/ comparison with different cooling technologies.

References

[1] Turchi CS, Wagner MJ, Kutscher CF. Water use in parabolic trough power plants: summary results from WorleyParsons' analyses. Technical Report NREL/TP-5500–49468, December. 2010.

[2] Kröger DG. Air-cooled heat exchangers and cooling towers, vol. 2. Tulsa, Okla: PennWell. Courtesy PennWell Publishing; 2004.

[3] Cooling technology in power generation. Kelvion Thermal Solutions (Pty) Ltd. www.kelvion.com.

[4] Williams CR, Rasul MG. Feasibility of a hybrid cooling system in a thermal power plant. In: Proc third Iasme/Wseas Int Conf Energy & Environ; 2008. p. 124–9.

[5] NETL. Estimating freshwater needs to meet future thermoelectric generation requirements. 2009 [Update].

[6] Kakac S, Liu H. Heat exchangers selecting, rating, and thermal design. CRC Press; 2002.

[7] Hausen H. Extended equation for heat transfer in tubes at turbulent flow. Waerme- Stoffuebertragung 1974;7(4):222–5.

[8] Briggs DE, Young EH. Convective heat transfer and pressure drop of air flowing across triangular pitch banks of finned tubes. Chemical Engineering Progress Symposium Series 1963;59(41):1–10.

[9] Kays WM, London AL. Compact heat exchangers. New York: McGraw Hill Book Co; 1984.

[10] Robinson KK, Briggs DE. Pressure drop of air flowing across triangular pitch banks of finned tubes. Chemical Engineering Progress Symposium Series 1966;62(64):177–84.

[11] Zou Z, Guan Z, Gurgenci H, Lu Y. Solar enhanced natural draft dry cooling tower for geothermal power applications. Solar Energy 2012;86(9):2686–94.

[12] Zou Z, Guan Z, Gurgenci H. Numerical simulation of solar enhanced natural draft dry cooling tower. Solar Energy 2014;101:8–18.

[13] Lee TR, Oppenheim D, Williamson TJ. Australian solar radiation data handbook. Canberra (Australia): Energy Research and Development Corporation; 1995.

[14] Australian Bureau of Meteorology. Australian hourly temperature, humidity and pressure Data. Canberra (Australia): Australian Bureau of Meteorology; 2009.

[15] Electric Power Research Institute (EPRI). Comparison of alternate cooling technologies for U. S. power plants: economic, environmental, and other trade-offs. Palo Alto (CA): EPRI; 2004. 1005358.

[16] Ashwood A, Bharathan D. Hybrid cooling systems for low-temperature geothermal power production, NREL technical report; 2011.

[17] He S, Guan Z, Gurgenci H, Jahn I, Lu Y, Alkhedhair AM. Influence of ambient conditions and water flow on the performance of pre-cooled natural draft dry cooling towers. Applied Thermal Engineering 2014;66:621–31.

[18] Alkhedhair A, Gurgenci H, Jahn I, Guan Z, He S. Numerical simulation of water spray for pre-cooling of inlet air in natural draft dry cooling towers. Applied Thermal Engineering 2013;61(2):416–24.

[19] Lu Y, Guan Z, Gurgenci H, Zou Z. Windbreak walls reverse the negative effect of crosswind in short natural draft dry cooling towers into a performance enhancement. International Journal of Heat and Mass Transfer 2013;63:162–70.

[20] Lu Y, Guan Z, Gurgenci H, Hooman K, He S, Bharathan D. Experimental study of crosswind effects on the performance of small cylindrical natural draft dry cooling towers. Energy Conversion and Management 2015;91:238–48.

High-temperature latent heat storage for concentrating solar thermal (CST) systems

10

K. Nithyanandam [1], *J. Stekli* [2], *R. Pitchumani* [3]
[1]University of California, Los Angeles (UCLA), Los Angeles, California, United States; [2]United States Department of Energy, Washington, D.C., United States; [3]Virginia Tech, Blacksburg, Virginia, United States

10.1 General introduction

Concentrating solar thermal (CST) technology is one of the promising renewable energy solutions for sustainable power production that converts sunlight into thermal energy prior to generation of electricity in a power block. A distinct advantage of CST technology, in comparison to other intermittent forms of renewable energy such as wind or photovoltaics, is the ability to integrate cost-effective thermal energy storage (TES) for increased dispatchability and load shifting. Furthermore, the addition of TES to a CST power plant has been shown to reduce its levelized cost of electricity (LCOE) [1] because of the increase in capacity factor as it allows electricity to be generated even when solar radiation is not available. Hence, CST-based power plants integrated with TES can be a key component of the renewable energy portfolio that can function as a peaking or a load-following plant.

TES systems can be broadly categorized into three types: sensible heat storage (SHS), thermochemical heat storage, and latent heat storage (LHS). SHS involves storing heat by raising the temperature of storage material that is in a solid, liquid, gaseous, or supercritical state [2]. In thermochemical storage, thermal energy is typically stored in the form of bond energy of a reversible chemical reaction involving one or more chemical compounds as the storage material. In LHS, energy is stored via phase transition of the storage material, commonly referred to as phase change material (PCM).

The U.S. Department of Energy (DOE) SunShot Initiative, launched in 2011, endeavors to enable large-scale deployment of CST power plants by driving innovations in various components of CST technology so as to achieve subsidy-free LCOE of less than 6 ¢/kWh, to be on par with the cost of electricity from fossil-fueled power plants [3]. Achieving the SunShot target calls for TES solutions with an exergetic efficiency greater than 95% and storage cost less than \$15/kWh$_t$. Similar cost reduction initiatives and targets are launched globally. As various TES technologies are developed toward the goal of cost-competitive CST, this chapter focuses on the principles of LHS technology with particular emphasis on elucidating the advantages of LHS; the technical challenges associated with LHS and a balanced compilation of approaches taken to

Advances in Concentrating Solar Thermal Research and Technology
http://dx.doi.org/10.1016/B978-0-08-100516-3.00010-1

address them; cost and performance analysis of different configurations of LHS that are applicable to CST systems; and research priorities necessary for the widespread application of LHS systems in CST applications.

10.2 Introduction to latent heat storage

Latent heat storage is accompanied by the storage of energy in a material via a phase transition from one state of matter to another. The phase change transition may vary between solid–solid, liquid–gas, and solid–liquid states. For CST applications, the phase transition most studied for LHS is between the solid and liquid states due to its low volumetric expansion compared to liquid–gas phase transition and high volumetric energy density compared to solid–solid phase transition. The material undergoing the phase change in an LHS system is commonly referred to PCM.

The energy (Q) stored in a PCM of mass m is given by

$$Q = m\left[c_{p,s}(T_m - T_s) + h_{sl} + c_{p,l}(T_l - T_m)\right] \tag{10.1}$$

where $c_{p,l}$ and $c_{p,s}$ are the average specific heats in the liquid and solid phases, respectively, h_{sl} is the enthalpy of phase change due to solid–liquid transition, or latent heat of fusion, which is defined as the heat required to melt a solid substance of unit mass, T_m is the melting temperature of PCM, T_s is the temperature of PCM in solid phase, and T_l is the temperature of PCM in liquid phase. The first and third terms in Eq. (10.1) represent the sensible energy storage in solid and liquid phases, respectively.

The primary advantages of an LHS system are:

1. *High volumetric energy density*: For a given temperature difference (ΔT), the heat stored in an LHS system is greater than that in an SHS system due to the inclusion of latent heat of fusion (Fig. 10.1(a)). For instance, the energy required to melt 1 kg of KNO_3 (latent heat) is 95 times higher compared with the energy required to raise the temperature of 1 kg of KNO_3 by 1 K (sensible heat). Fig. 10.1(b) compares the energy densities of some PCM molten salt candidates [4–7] with those of solar salt and oil [8,9] that work only in the sensible energy regime between operating temperatures of 300 and 600°C. LHS, by virtue of its higher energy density, requires a smaller volume of PCM to store energy and, in turn, offers compact energy storage, less construction material, and ultimately lower cost advantages over SHS counterpart.
2. *Isothermal operation*: Owing to the constant temperature solid–liquid phase transition, high power cycle efficiency could be obtained if the selected PCM melts at a temperature close to the maximum operating temperature of the power block.
3. *Low system pressure*: As mentioned earlier, LHS depends on the absorption and release of heat during solid–liquid phase transformations. Unlike liquid–vapor phase transformation, solid–liquid phase change transformation is accompanied only with a moderate change in density, which does not lead to a large increase in the pressure inside the PCM storage vessel. A low system pressure does not require a thick-walled vessel, which leads to savings in construction material cost.

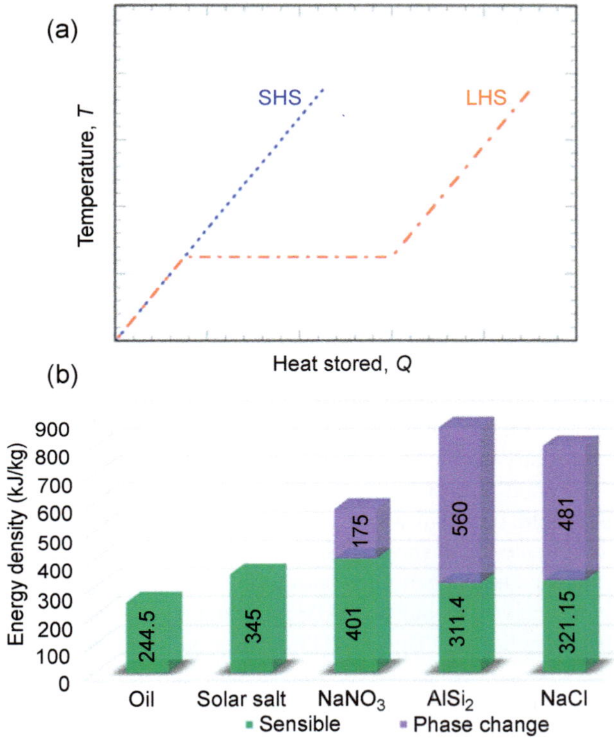

Figure 10.1 (a) Energy stored in a sensible heat storage (SHS) and latent heat storage (LHS) system, (b) Comparison of energy densities of phase change materials ($NaNO_3$, $AlSi_2$, NaCl) and sensible storage materials (oil, solar salt) between operating temperatures of 300 and 600°C.

The main challenges of LHS are:

1. *Low thermal conductivity*: During extraction of heat from the PCM, a solid layer of PCM quickly forms on the wall of the heat exchanger. The solid PCM typically has a low heat transfer coefficient, and therefore heat transfer out of the system once this solid layer has formed is comparatively much slower.
2. *Thermal stability*: Some of the PCM mixtures and eutectics do not exhibit ideal mixing behavior leading to incongruent melting and change in thermophysical properties during subsequent thermal cycles. Lack of reliable data of PCM subjected to thermal cycling will lead to nonoptimal design of the LHS.

The general challenges associated with LHS in CST applications, regardless of the specific configuration of LHS, are discussed in Section 10.3. Specific configurations of LHS for CST applications are then described in Section 10.4.

10.3 General challenges for concentrating solar thermal latent heat storage systems

CST technologies can be primarily categorized into four system configurations: parabolic trough, linear Fresnel, dish engines, and power towers [3]. Linear Fresnel and parabolic trough systems focus sunlight onto a linear receiver while dish and power towers focus sunlight to a point receiver. The majority of CST power plants are deployed as either parabolic trough or power tower configurations. The following discussion is presented in the context of a power tower CST plant configuration.

The working of LHS in a CST power plant fundamentally involves the exchange of heat between a heat transfer fluid (HTF) and the PCM during the charging and discharging processes (Fig. 10.2). A schematic of the integration of LHS in a power tower CST power plant working on Rankine cycle is shown in Fig. 10.2(a) for a typical charging operation and Fig. 10.2(b) shows the discharge operation of a power tower CST system based on advanced supercritical carbon dioxide (s-CO_2) Brayton cycle [3,10]. As shown in Fig. 10.2, a power tower CST plant comprises of four subsystems namely, the heliostat field, central tower/receiver, power block, and TES system. A heliostat field constitutes a series of sun-tracking flat-shaped mirrors, which collect thermal energy from the sun and concentrates onto a central tower/receiver (Fig. 10.2(a)). In the external tower/receiver, molten salt HTF is circulated to absorb the solar thermal energy reflected by the heliostats. The hot HTF exiting the receiver is then fed into the power block, which works on either conventional steam Rankine cycle or s-CO_2-based Brayton cycle to convert the thermal energy into electrical energy. The pump, steam generator, superheater, and steam turbine and electrical generator in Fig. 10.2(a) make up the Rankine cycle—based power block components, while the low- (LTR) and high-temperature recuperators (HTR), compressors, and s-CO_2 turbines and generators make up the components of a Brayton cycle—based power block. The typical operating pressure and temperature of the working fluid of the power cycle at the inlet and outlet of each of the power block components (Fig. 10.2) can be found in Ref. [11] for Rankine cycle and Ref. [10] for s-CO_2 cycle.

LHS in a CST power plant operates in one of the two modes, namely, charging and discharging, depending on the solar thermal energy resource at the particular hour of the day and the energy available in the LHS. Charging occurs whenever the incident solar thermal energy is in excess to that required by the operation of the power block at the rated capacity. Excess hot HTF from the solar receiver enters the LHS from one end and transfers heat to the PCM, thus effecting the melting of PCM at a constant temperature. The cooled HTF exits the tank to return to the solar field, completing a closed loop of the HTF flow (Fig. 10.2(a)). The excess hot HTF is directed to the storage system until the cooled exit temperature of the HTF reaches a certain charge cutoff temperature, T'_C, which determines the extent to which the TES system can be charged. The charge cutoff temperature value is limited by the maximum flow rate achievable by the HTF pumps in the receiver section. The increase in HTF flow rate through the receiver for a higher cutoff temperature increases undesirable parasitic and thermal losses [11].

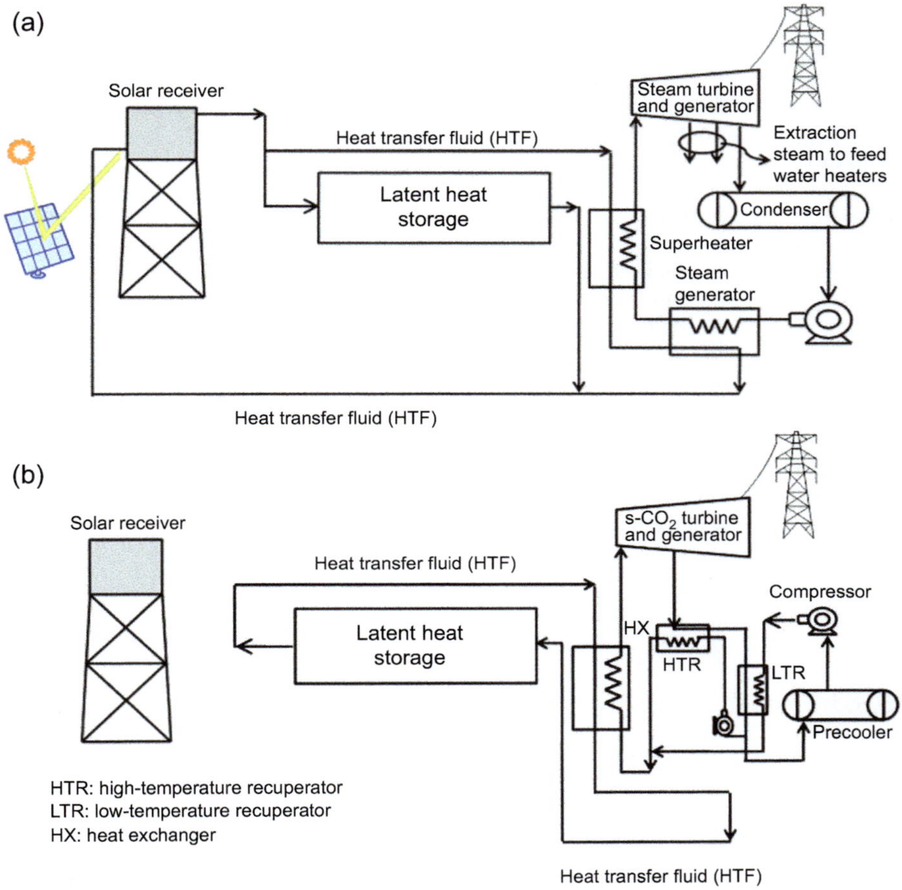

Figure 10.2 Schematic illustration of (a) CST power plant integrated with a LHS system working on steam-powered Rankine cycle, (b) CST power plant integrated with LHS system working on a s-CO_2 cycle. The HTF flow path during (a) charging and (b) discharging are also shown.
Adapted from Nithyanandam K, Pitchumani R. Cost and performance analysis of concentrating solar power systems with integrated latent thermal energy storage. Energy 2014;64:793–810.

As illustrated in Fig. 10.2(a), excess hot HTF fluid from the receiver outlet is stored in the TES, which can be subsequently removed to smooth fluctuations in power output during cloud episodes and to extend power generation after sunset (Fig. 10.2(b)). During discharging, cold HTF is pumped from the other end of LHS, resulting in the solidification of the PCM as heat is drawn from the PCM to the fluid, and the hot fluid exiting the tank is directed to generate steam to drive a turbine and produce electricity (Fig. 10.2(b)). Similar to the charging process, the strong dependence of the power block cycle efficiency on the hot HTF temperature [11] requires the termination of discharge process when the HTF exit temperature from LHS reaches a certain minimum discharge cutoff temperature, T'_D.

The U.S. DOE SunShot Initiative goal calls for the LHS system (as with other TES systems) to meet the following technical targets: cost <\$15/kWh$_t$, exergetic efficiency >95%, and material degradation due to corrosion <15 μm/year [3,12]. In addition, the required PCM candidates for LHS must have the following desirable properties:

1. phase change temperatures within the upper and lower limit of the attainable temperatures in the solar field;
2. high value of specific heat, high latent heat of fusion, high density, and small to moderate change in density during phase transformation;
3. high working temperatures (>500°C);
4. high thermal conductivity for enhanced heat transfer into and out of the PCMs;
5. congruent melting;
6. insignificant supercooling;
7. low vapor pressure;
8. low corrosivity and toxicity;
9. flame and fire safety;
10. thermal stability at the operating temperatures;
11. chemical stability at the operating temperatures;
12. earth-abundant material supply and low material cost.

10.3.1 Phase change material selection

For high-temperature CST applications, inorganic salts and salt eutectics as well as metals and metallic alloys are promising candidates as PCM due to their high density, high heat capacity, low vapor pressure, and low viscosity (low pumping power requirements) [13]. Kenisarin [6] published a comprehensive review of various PCMs including chlorides, fluorides, nitrates, and carbonates, among others with phase transitions in the temperature range between 200 and 1000°C. From the reported PCM compilation, it is observed that the melting point and heat capacity increase in the following order: nitrates, chlorides, carbonates, and fluorides. The latent heat of fusion, PCM material cost, and thermal conductivity for various inorganic salts and metallic PCMs are obtained from literature [4,6,7] and plotted against their melting temperatures in Fig. 10.3(a)−(c), respectively. Fig. 10.3 helps in the identification of suitable PCM candidates for the temperature range of interest. However, the corrosion and thermal stability of these PCMs have not all been studied, which makes choosing the right PCM difficult. In addition, there is a wide discrepancy in the literature for the reported thermophysical properties for a particular chemical composition of molten salt PCM, and moreover, their properties at high temperatures are mostly incomplete and contradictory [14]. For instance, Liu et al. [14] showed that for some candidate PCMs suitable for CST applications (Li_2CO_3−35K_2CO_3−33Na_2CO_3, $MgCl_2$−48 NaCl, Li_2CO_3−71.5K_2CO_3, Li_2CO_3−65K_2CO_3 in wt%) there was a wide discrepancy in the phase change temperatures during multiple melt−freeze cycling tests between the calculated and experimental values. Therefore, verification of the PCM properties

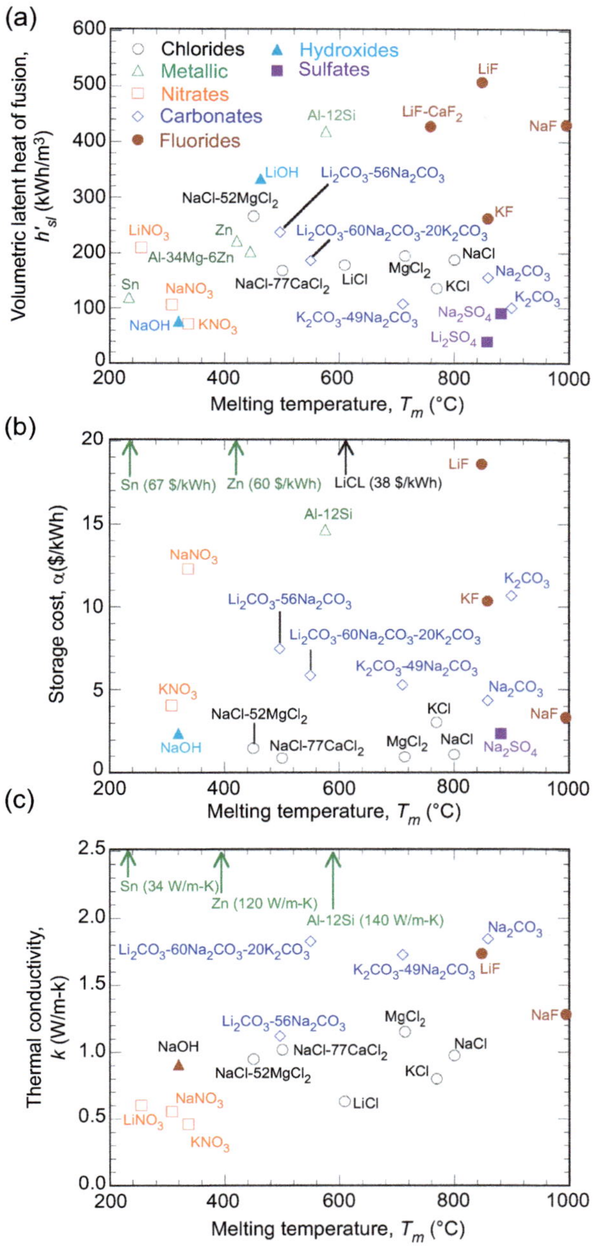

Figure 10.3 (a) Volumetric latent heat storage capacity, (b) storage media cost, and (c) thermal conductivity of various phase change materials.

is essential to design the LHS system accurately. Efforts to create binary or ternary eutectics having desirable properties are also being undertaken [15,16].

10.3.2 Corrosion and containment compatibility

One critical issue with high-temperature applications is corrosion, which decreases the life cycle of the system and also reduces the thermal performance. Both high temperature stability and corrosion resistance at high temperatures are critical in a TES system designed for a service lifetime of 30 years without failure [3,12]. Any degradation of the PCM would alter its thermodynamic properties, such as phase change temperature, which can have detrimental impact on the efficiency of the power block, especially if the phase change temperature drops. In the literature, there is limited information on high-temperature salt and metal corrosion with cyclic charging and discharging.

Compatibility of containment materials with various PCMs is therefore a major technical challenge for CST plants. PCM candidates considered for high-temperature applications are chlorides, fluorides, carbonates, nitrates, their eutectics, or mixtures [4]. Currently, iron-based stainless steels (304SS and 316SS) are used as containment materials for TES in power tower CST plants, with temperatures up to 565°C. Advancing CST technologies to achieve the SunShot cost target of <$0.06/kWh requires developing various components of the system to operate at high temperatures (>650°C) [3]. To achieve a lifetime of 30 years, as needed for CST plants, material degradation rate due to corrosion must be less than 15 μm/year [3,12]. At such high temperatures, iron-based stainless steels no longer have the required corrosion resistance and mechanical strength. For instance, at 650°C, corrosion rate of stainless steel by solar salt (40 wt.% KNO_3, 60 wt.% $NaNO_3$) is substantially higher that it cannot provide the required 30-year lifetime [17,18]. Kruizenga and Gill [18] tested 321SS and 347SS with stabilizing additives, titanium and niobium respectively, for stress corrosion cracking. It was found that at 680°C, the corrosion rate was excessive for it to withstand the 30-year lifetime.

A viable material choice that will withstand the corrosive environment at elevated temperatures is nickel-based alloys; however, their cost is prohibitively high (approximately four times more expensive than stainless steel–based alloys) [12,18] to meet the SunShot TES cost target of <$15/kWh$_t$ [3]. As detailed in Ref. [12], alternative strategies such as shot peening, deposition of protective layers through cladding, and internal insulation of the high-temperature vessels should be considered. Shot peening is a cold working process used to produce a compressive residual stress layer in an alloy and make it more corrosion resistant by impacting the surface with a separate material. Shot peening does not require any additional material to be added to the alloy and is a widely used process in large industries such as the automotive industry [19]. Cladding involves layering a thin corrosion resistive material such as nickel, titanium, etc. onto base metal such as stainless steel through laser application [20], explosive welding [21], roll welding [22], or electroplating [23]. In the internal insulation approach, a low thermal conductivity, chemically inert, and low-cost refractory material acts as a thermal and chemical buffer between the PCM storage and containment material.

The available studies on corrosion kinetics of various PCM salts at high temperatures are reviewed in Ref. [24]. Nitrates, carbonates, sulfates, and hydroxides act in a similar way in that they form protective oxide layers, which inhibit corrosion from the base alloy. Chloride salts are considered to be highly corrosive to steel in the presence of moisture and oxygen. Grabke et al. [25] explained that the corrosion mechanism involves the reaction of chlorine gas in the salt with metal oxides on the metal surface or with moisture in the system. The mobility of Cl gas is high enough to penetrate through the oxide scale and form metal chlorides at the metal/metal oxide interface. Due to the volatile nature of the metal chlorides at high temperatures, it sublimates and diffuses through the scale to the oxidative environment where they oxidize again to liberate chlorine again. The chlorine can diffuse back toward the metal/metal oxide interface and cause further corrosion.

Goods and Bradshaw [17] studied the effect of thermal cycling and the presence of impurities on the corrosion of the two austenitic stainless steels Type 304SS and Type 316SS. The impurities typically present in commodity grades of nitrates are NaCl, $KClO_4$, and Na_2SO_4. Thermal cycling increased the corrosion rates of stainless steels compared to isothermal immersion in the molten salt [17], but the increases were moderate. The total metal losses experienced by these alloys ranged from about 5–16 μm after more than 4000 h and 500 thermal cycles of exposure to the molten salt. These rates fall well within the SunShot target corrosion rates of <15 μm/year [3]. The corrosion kinetics for the stainless steel specimens exposed to molten salt mixtures with low concentrations of dissolved chloride ions under thermal cycling conditions were expressed by a parabolic rate law:

$$\text{metal loss} = k_p(time)^{1/2} \times 10^{-7} \tag{10.2}$$

where k_p is the parabolic rate constant in cm^2/s. However, relatively high concentrations of dissolved chloride ions in the nitrate salt increased corrosion rates and tended to cause corrosion kinetics to follow linear rate equations for stainless steel specimens under thermal cycling conditions. The linear rate law is given by

$$\text{metal loss} = k_l(time) \times 10^{-7} \tag{10.3}$$

where k_l is the linear rate constant in cm/s. The values for rate constant k_p were found to increase from 1.3 to 1.4 for 316SS and from 2.3 to 2.4 for 304SS with increase in chloride concentration. With increase in dissolved chloride ions, the values for linear rate constant k_l were found to increase from 0.49 to 0.64 and from 0.65 to 0.90 for 316SS and 304SS, respectively.

10.3.3 Latent heat storage sizing for a concentrating solar thermal power plant

The sizing of storage system for a power plant of rated capacity, P (in MW), depends on the required discharge duration (t_D) needs and the rated thermal to electric power conversion efficiency. The cycle conversion efficiency (η_t) depends on the power cycle configuration and temperature range of the working fluid. The net thermal capacity (P/η_t) of a power plant decides the mass flow rate of the HTF required as given by Eq. (10.4):

$$\frac{P}{\eta_t} = \dot{m}_f c_f \Delta T \tag{10.4}$$

where c_f is the specific heat of the HTF and ΔT is the difference between the maximum and minimum operating temperatures of the CST power plant. The storage capacity is determined by $Q = P \times t_D$ given in MWh, which also equals $Q = m(c_p \Delta T + h_{sl})$, where m is the mass, c_p is the average specific heat, and h_{sl} is the latent heat of fusion of the storage material. Thus, the storage capacity of the LHS determines the mass of the storage material required, which indirectly decides the volume of the storage material using the following correlation: $m = \rho_p \times V$, where V is the volume of the storage material and ρ_p is the storage density. In nondimensional terms, the storage capacity is given as $Q^* = 1 + 1/Ste$, where the Stefan number $Ste = c_p \Delta T/h_{sl}$ and $Q^* = Q/mc_p$. Fig. 10.4 shows the influence of the Stefan number on a nondimensional system storage capacity. Note that the Stefan number for a SHS is ∞, for which the dimensionless system storage capacity, $Q^* = 1$, as observed in Fig. 10.4. It is seen that a smaller value of the Stefan number is essential to realize the high volumetric energetic density benefits of LHS compared to the sensible thermal storage system. A larger value of the Stefan number, on the other hand, mitigates the high volumetric energy density benefits of LHS over SHS.

10.3.4 Understanding charge/discharge characteristics

Unlike SHS, which involves storing energy in a single phase (mostly solid such as concrete, rocks, etc.), the multiphase nature of LHS leads to different fluid and heat

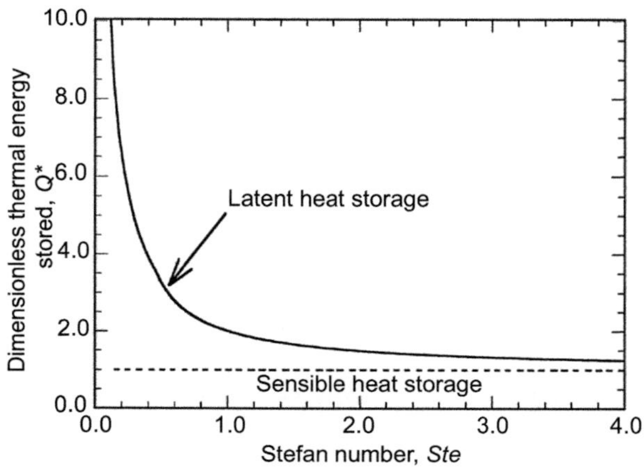

Figure 10.4 Variation of energy stored with the Stefan number for latent and sensible heat storage.

transfer dynamics during charging and discharging. Insights into the charging and discharging characteristics of PCM in an LHS are required to understand the heat transfer rate-limiting regime to optimally design the storage system. The major nondimensional parameters that influence the performance of the system are the Rayleigh number, Ra_D, and the Stefan number, Ste, defined as:

$$Ra_D = \frac{g\beta D_i^3 (T_w - T_o)\rho^2 c_p}{\mu k} \tag{10.5}$$

$$Ste = \frac{c_p (T_w - T_o)}{h_{sl}} \tag{10.6}$$

where g is the gravitational acceleration, D_i is the inner diameter of the tube storing the PCM, T_w is the wall temperature, and T_0 is the initial temperature. The notations ρ, c_p, μ, k, β, and h_{sl}, respectively, represent the density, specific heat, viscosity, thermal conductivity, thermal expansion coefficient, and latent heat of fusion of PCM.

Computational modeling [26] may be used to explain the influence of Rayleigh number and Stefan number on the charge and discharge process. Fig. 10.5 shows the charging dynamics of PCM stored inside the tube of diameter 60 mm. The melt fraction (blue representing solid PCM and red representing liquid molten PCM) and temperature profile inside the PCM at time instant of 50 s are illustrated in Fig. 10.5(a) and (b), respectively. The PCM is assumed to melt at 300°C while the outer wall temperature is maintained at 600°C. The melting of the PCM is aided by the formation of buoyancy-driven convection currents within the melt that occurs as a result of the temperature difference between the heated wall and the relatively cold solid phase. The lighter melt at the bottom of the tube rises along the hot tube wall to the top, being heated up in the process, and upon descending along the cold solid PCM exchanges thermal energy, thus entraining more fluid and forming a clockwise (counterclockwise) vortex in the melt region to the left (right) of the tube. The recirculation ensures relatively faster melting of the PCM at the top of the tube compared to the bottom.

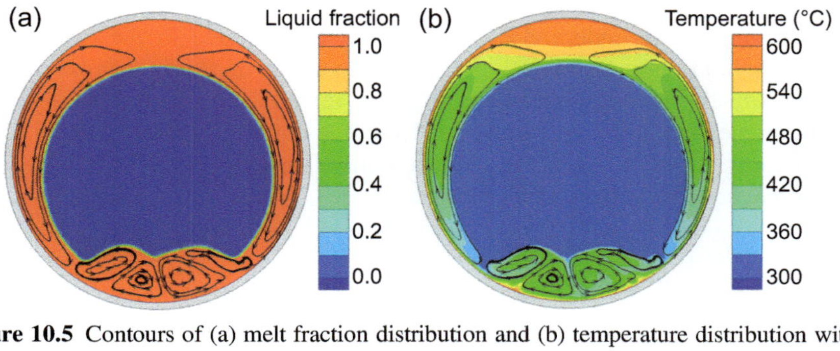

Figure 10.5 Contours of (a) melt fraction distribution and (b) temperature distribution within phase change material during charging at time instant of 50 s (*gray shade* indicates solid wall).

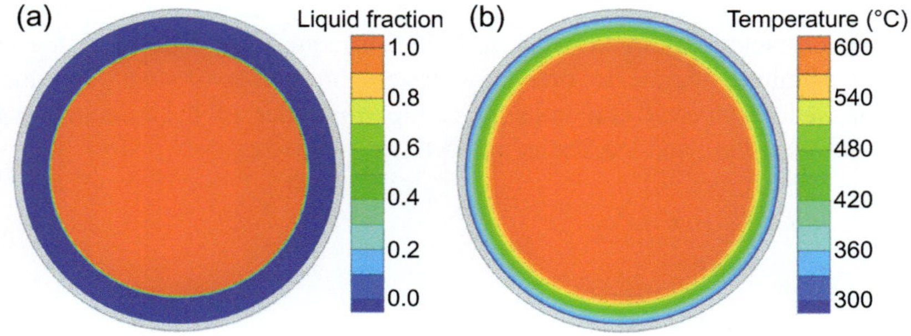

Figure 10.6 Contours of (a) melt fraction distribution and (b) temperature distribution within phase change material during discharging at time instant of 50 s (*gray shade* indicates solid wall).

Fig. 10.6 shows the discharging dynamics of PCM stored inside the tube of diameter 60 mm. The PCM is assumed to melt at 600°C while the outer wall temperature is maintained at 300°C, thus maintaining the same temperature difference $\Delta T = 300°C$ as in charging. Discharging of the energy available in the PCM to the HTF results in the solidification of the PCM. The solidification of the PCM is seen to be uniform throughout the solid fraction adjoining the tube (Fig. 10.6(a)). This is primarily due to the absence of buoyancy-driven convection effects within the solidified PCM and the purely conduction-dominated solidification that accounts for the very slow discharging rate of the PCM compared with the charging rate. The corresponding temperature profile inside the PCM is shown in Fig. 10.6(b).

For the design of LHS, it is important to understand the implications of Rayleigh numbers and Stefan numbers on the heat transfer coefficient during charging and discharging. Fig. 10.7(a) and (b) compares the transient charging and discharging heat transfer performance, represented by the Nusselt number (defined as: $Nu = hD_i/k$, where h is the heat transfer coefficient obtained from the numerical simulations) for various Rayleigh numbers as a function of temperature difference, $\Delta\theta$, between the volumetric average temperature of PCM which varies temporally and the wall temperature. From Fig. 10.7(a), it is observed that the Nusselt number and, in turn, the heat transfer intensifies with increase in Rayleigh number during charging due to stronger buoyancy-driven convection currents (Fig. 10.5). However, in the case of discharging, the Nusselt number is independent of Rayleigh number due to the conduction-dominated solidification process (Fig. 10.6). The effect of Stefan number on the discharge Nusselt number is shown in Fig. 10.7(c) and (d), and it has very little influence on the heat transfer performance. This shows that PCM with a smaller value of the Stefan number can improve the heat storage capacity of the LHS with only marginal reduction in heat transfer performance.

10.3.5 Exergetic efficiency

Exergy is used to determine the reversible energy obtainable in thermal systems based on the second law of thermodynamics. Exergy analysis can help the performance

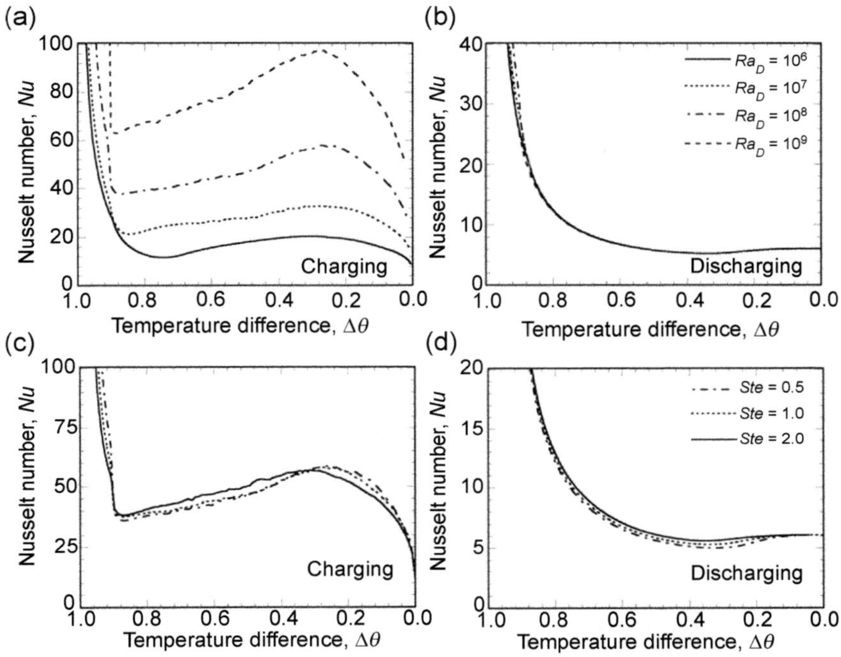

Figure 10.7 Variations in the Nusselt number as a function of temperature difference during charging and discharging (a, b) for various Rayleigh numbers and (c, d) Stefan numbers.

assessment and optimization of the LHS system. In developing TES technologies, round-trip energy efficiency is often cited as a key metric of performance; however, it is, in fact, most critical that the exergetic efficiency be very high to ensure that heat quality is maintained after storage. The SunShot targets for TES require an overall exergetic efficiency of >95% [3]. The round-trip exergetic efficiency of the LHS is defined as the exergy recovered from the HTF during discharging process as a fraction of the total exergy supplied by the HTF during the charging process [1,12]:

$$\xi = \frac{\int_{t_C}^{t_D} \left[T_f(PB,t) - T_D - T_{ref} \cdot \ln\left(\frac{T_f(PB,t)}{T_D} \right) \right] \cdot dt}{\int_0^{t_C} \left[T_C - T_f(SF,t) - T_{ref} \cdot \ln\left(\frac{T_C}{T_f(SF,t)} \right) \right] \cdot dt} \tag{10.7}$$

where T_{ref} is the reference or ambient temperature, $T_f(PB,t)$ in the numerator refers to the exit temperature of the HTF from LHS during discharging, which is fed in to the power block (*PB*), while the term $T_f(SF,t)$ in the denominator refers to the temperature of the HTF at the outlet of LHS, which goes back to the solar field (*SF*) for charging. T_C and T_D are the hot and cold inlet temperatures during charging and discharging, respectively.

Initial studies [27–29] showed that the optimal melting temperature ($T_{m,opt}$) for a complete charging–discharging cycle based on maximum exergy recovery is equal to the geometric average of the ambient (reference) temperature and the hot HTF inlet temperature during charging ($T_{m,opt} = \sqrt{T_C T_{ref}}$). However, the studies assume that the HTF exiting the LHS during charging was delivered to the environment, which limits the practical usefulness of the analyses for CST application where the HTF is rerouted to the solar collector field. Several investigations have concluded that cascading PCMs in the order of decreasing melt temperatures from the hot HTF inlet side provide for the highest exergy and energy efficiency without increasing the tank cost [30,31]. A thorough review of the exergy analyses for LHS systems is presented by Jegadheeswaran et al. [30]. LHS can be optimized for use with a sensible HTF by using a cascade of PCMs with equally spaced melting points. In an effort to optimize the overall exergetic efficiency, the optimal melting temperatures of PCMs were identified by Gong et al. [31], assuming negligible thermal resistances associated with PCM, as an approximate geometrical regression, i.e., $\frac{T_{m1}}{T_{m2}} = \frac{T_{m1}}{T_{m2}} = \frac{T_{m2}}{T_{m3}} = \ldots = \frac{T_{mn-1}}{T_{mn}}$, where n is the number of PCMs and T_m is the phase change temperature of the PCM.

Shabgard et al. [32] concluded that exergetic efficiency is not an appropriate criterion while the exergy recovered during the discharge process is the relevant second-law figure of merit for use in LHS design. In other studies, Nithyanandam and Pitchumani [33,34] evaluated the performance of single PCM and cascaded PCM configuration taking into account the practical constraints encountered in a CST power generation. Based on combined heat transfer analyses, it was found that the maximum exergy recovered increases only slightly beyond a 2-PCM cascaded configuration. The optimal LHS design was, therefore, obtained to be a 2-PCM cascaded arrangement with 75% of the LHS volume from the hot HTF inlet section during charging filled with PCM of high melting temperature (T_{mH}) while the remaining 25% of the volume filled with PCM of low melting temperature (T_{mL}). The optimal high and low PCM melting temperatures for maximum exergy recovery can be approximated as $T_{mH,opt} = \sqrt{T_C T'_D}$ and $T_{mL,opt} = \sqrt{T_D T'_C}$, respectively [33,34], where T'_C and T'_D are the charging and discharging cutoff temperatures, as introduced in Section 10.3. Fig. 10.8 schematically represents the optimal cascaded configuration for a CST power tower plant. The optimal LHS design in the analyses reported in Refs. [33,34] showed an exergetic efficiency greater than 95%, under the assumption of negligible heat loss to the surroundings. In addition, higher exergetic efficiency can also be achieved using a single PCM with higher melting temperature (close to the charging

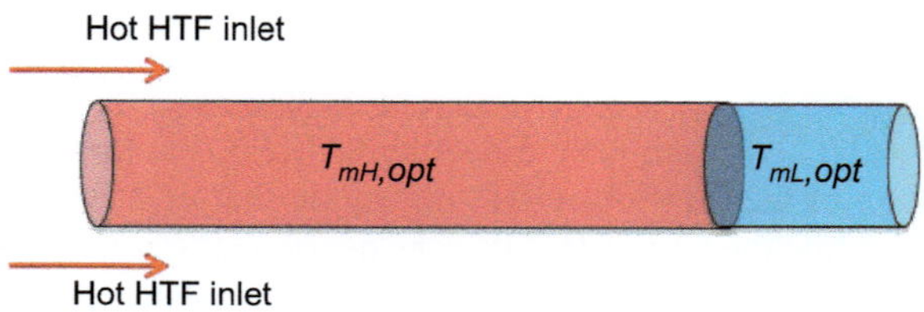

Figure 10.8 Schematic of optimal cascaded phase change material arrangement in latent heat storage system.

hot HTF inlet temperature) as only a very small fraction of exergy is degraded under this condition; however, heat transfer enhancement techniques are necessary to ameliorate the diminished heat transfer rate during charging resulting from the smaller temperature difference between hot HTF inlet and PCM melting temperatures [33,34].

10.4 Latent heat storage configurations for concentrating solar thermal applications

LHS systems can be broadly considered in two configurations: (1) tank phase change material latent heat storage (TPCM-LHS) and (2) encapsulated phase change material latent heat storage (EPCM-LHS). These are discussed in detail in Sections 10.4.1 and 10.4.2.

10.4.1 Tank phase change material latent heat storage

TPCM-LHS systems are simply large tanks filled with a PCM, with an HTF flowing around and exchanging energy with PCM. The most straightforward, market-ready design of TPCM-LHS is the shell and tube heat exchanger, in which two different configurations are possible: (1) HTF flowing in the tubes and PCM stored inside the shell around the HTF pipes (Module 1) or (2) HTF flowing in the shell and PCM stored inside the tubes (Module 2). One such configuration with HTF flowing inside tubes surrounded by PCM is shown in Fig. 10.9.

Most of the TPCM-LHS systems use salts as the material undergoing phase change [4,6] although some researchers have explored the use of metals [5,7]. These systems were limited in their effectiveness due to the poor heat transfer out of the system during discharge. In TPCM-LHS, as heat is removed from the system, the PCM solidifies on the surface of the heat exchanger. The salts being used in these systems have relatively low thermal conductivity values when they are in the solid state, and as a result, heat transfer out of the system is greatly reduced. As the solid salt builds in thickness on the heat exchanger wall, the heat transfer eventually falls to the point that insufficient energy is being pulled from the LHS system to maintain operation of the power block.

To address the poor heat transfer out of TPCM-LHS systems, a number of researchers have evaluated the addition of more piping through the PCM block or the addition of fins/heat pipes (HPs) to the heat exchanger surface to increase the effective heat transfer surface area [35–41]. Others have incorporated the use of a more thermally conductive material into the PCM block, such as graphite or metallic foams [42–45]. Comprehensive assessment of different thermal performance enhancement techniques for high-temperature LHS systems is available in the literature [46,47]. All of these methodologies have succeeded in increasing the thermal conductivity, and therefore heat transfer, of the TPCM-LHS system. However, all the solutions also served to greatly increase the cost of the TPCM-LHS system, as they all require greater use of metals in the system, which are a significant cost addition. Some of the prominent approaches for improving the heat transfer in a tank-based LHS configuration are discussed briefly here.

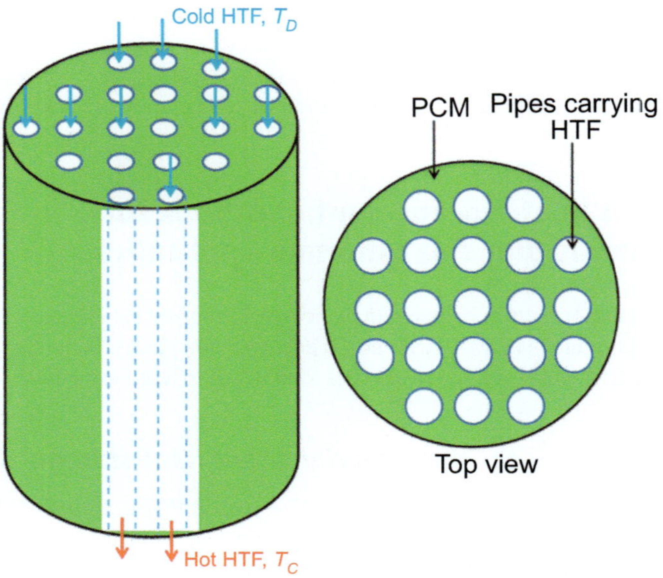

Figure 10.9 Schematic of the tank phase change material latent heat storage system.

10.4.1.1 Finned tubes

The use of fins to improve the thermal conductivity of PCM has been explored widely in the literature [46,47]. A shell and tube heat exchanger configuration with PCM on the shell side and HTF circulating in the tubes is widely studied. The usage of fins has been shown to improve the melting and solidification rate of PCM. Velraj et al. [35] concluded that adding fins reduced the solidification time by a factor of $1/N_{fins}$, where N_{fins} is the number of fins. However, the literature predominantly reports on the usage of aluminum fins, which are only suitable for low-temperature applications due to the susceptibility of Al to aggressive corrosion at high temperatures. Stainless steel is a good option for high-temperature applications; however, its low thermal conductivity ($\sim$20 W/m$^2\cdot$K), being an order of magnitude lower compared to Al ($\sim$200 W/m$^2\cdot$K), makes it less viable for cost-competitive applications.

10.4.1.2 Heat pipe/thermosyphons

The use of HPs, which can efficiently transfer heat along their length by means of evaporation and condensation of an internal working fluid between the HTF and the PCM, has been investigated as a means for increasing the heat transfer rate [36–41]. The operation of an HP involves the evaporation of a working fluid at the high-temperature (evaporator) end creating a high vapor pressure in its vapor core. Due to the pressure difference, the vapor migrates along the length of the central vapor core to the low-temperature (condenser) end where it condenses back to liquid onto a porous wick and flows back to the high-temperature end by means of capillary action.

Because the return mechanism is not gravity assisted, HPs are effective in any orientation with respect to gravity and are capable of transferring large amounts of energy nearly isothermally.

The concept of HP/thermosyphon embedded LHS for TES in CST systems was widely studied both experimentally and computationally by Shabgard et al. [37], Nithyanandam and Pitchumani [38–41], and Robak et al. [35]. A shell and tube LHS of two different configurations namely, Module 1 (Fig. 10.10(a)) in which HTF flows inside the tube with PCM surrounding it and Module 2 (Fig. 10.10(b)) in which HTF flows outside the shell with PCM stored inside the tubes, is studied. Detailed parametric studies were carried out using a thermal resistance network model to assess the influence of the HP and LHS geometric and operational parameters on the performance of the system during charging and discharging. The performance metrics that were evaluated are the effectiveness of the storage system defined as: $\varepsilon(t) = Q(t)/Q_0(t)$, where $Q(t)$ and $Q_0(t)$ are the energy stored/discharged in time (t) for LHS with and without HPs, respectively. The optimum design conditions showed that effectiveness (ε) is significantly higher for Module 1, while highest energy charge/discharge rate is obtained for Module 2 [39]. Furthermore, based on a three-dimensional computational modeling of different HPs/thermosyphons locations and orientations (Fig. 10.11), Nithyanandam and Pitchumani [38,40] systematically analyzed the charging and discharging performance of the LHS system, to elucidate the complex interplay between the governing heat transfer and fluid dynamics phenomena. The results indicated that the highest volumetric energy charge and discharge rate per unit cost was obtained for Module 2 with two vertical HPs/thermosyphons configuration [40].

Building upon the results from the previous studies [37-40], Shabgard et al. [9], and Nithyanandam and Pitchumani [33] considered the design of a large-scale LHS embedded with vertical gravity-assisted HPs as shown in Fig. 10.12 for CST power plant operation. The flow channels carry the HTF where the heat transfer between the HTF and PCM takes place. The schematic of one channel of the LHS system

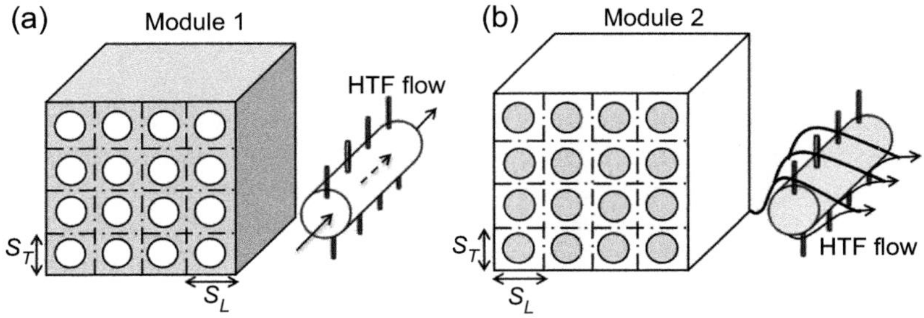

Figure 10.10 Two different configurations studied for latent heat storage with embedded heat pipes.
Adapted from Nithyanandam K, Pitchumani R. Computational studies on a latent thermal energy storage system with integral heat pipes for concentrating solar power. Applied Energy 2013;103:400–15.

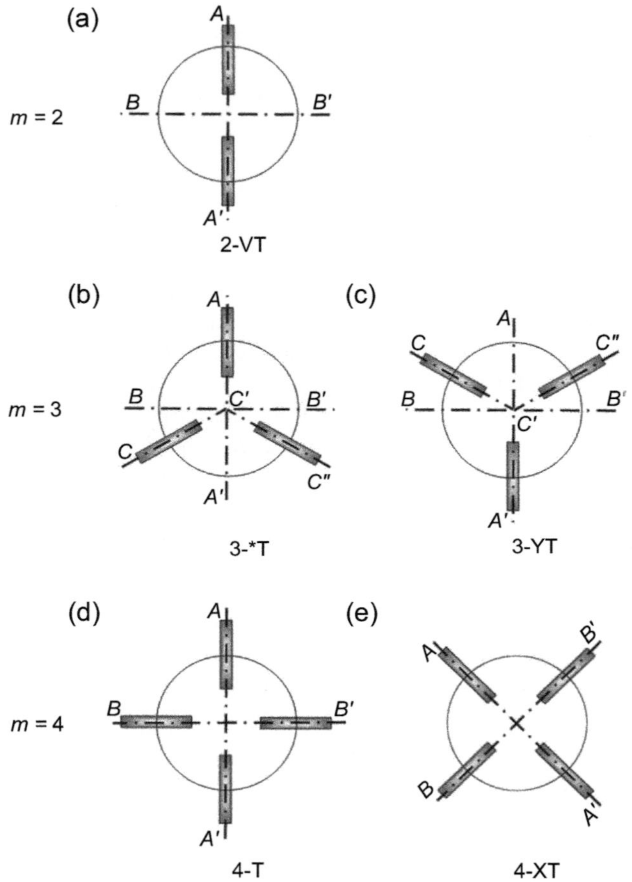

Figure 10.11 Schematic diagram of different arrangements of heat pipes/thermosyphons. *HTF*, heat transfer fluid.
After Nithyanandam K, Pitchumani R. Thermal energy storage with heat transfer augmentation using thermosyphons. International Journal of Heat and Mass Transfer 2013;67:281–94.

with HPs embedded in a staggered fashion and accompanied by a PCM unit is also shown in Fig. 10.12(a). During charging, hot HTF enters the bottom channel from the left (Fig. 10.12(b)) while cold HTF during discharging enters the top channel from the right (Fig. 10.12(b)). The heat transfer pathways during charging and discharging are shown in Fig. 10.12(c) and (d), respectively. The air gap is provided to accommodate the volumetric expansion of the PCM during melting. Fig. 10.12(a) and (b) also shows the length of the various sections of the HP, which include the adiabatic length (L_a), the condenser length (L_c), and evaporator length (L_e). Based on a detailed parametric investigation on the dynamic behavior of the system, optimum design configurations that provide the maximum energy discharge rate and exergetic efficiency were deduced and reported in Ref. [33].

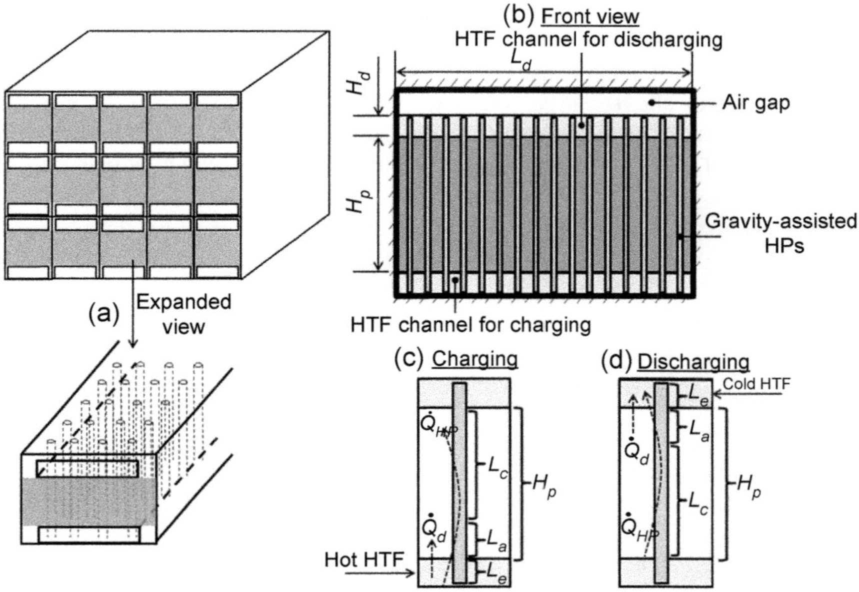

Figure 10.12 Schematic illustration of large-scale gravity-assisted heat pipes (HP) embedded latent heat storage (LHS) and an expanded view of HP-LHS channel, (b) front view of HP-LHS channel and heat transfer pathways during (c) charging and (d) discharging of HP-LHS system. *HTF*, heat transfer fluid.
Adapted from Nithyanandam K, Pitchumani R. Design of a latent thermal energy storage system with embedded heat pipes. Applied Energy 2014;126:266–80.

10.4.1.3 Particles and metal structures

Addition of high conductivity particles such as copper or aluminum is another means to enhance the thermal conductivity of inorganic salt PCMs [46]. A major drawback with the approach is agglomeration of particles and lengthy preparation process to achieve homogenous dispersion. The dispersed particles reduced the volume fraction of PCM and subsequently its heat storage capacity while improving the thermal conductivity. The optimum mass fraction of the particles due to the trade-off between decrease in heat storage capacity and increase in thermal conductivity is widely discussed in the literature. Incorporating metal structures made of steel or stainless steel in various geometries, such as cylindrical and spherical, has been studied in the literature [46,47]. It has been widely reported that a large volume of the metal structures per unit volume of PCM is required to see an appreciable change in the charge/discharge rate, which decreases the volumetric heat storage capacity of the storage system.

10.4.1.4 Metal foams

PCM stored within the framework of open-celled metal foam can substantially enhance the heat transfer rate. Metal foam is a porous lightweight structure with continuous metal matrices of high thermal conductivity and typically available in high porosities (>85%). They are generally manufactured using open-cell polymer foam as templates to create investment-casting molds of desired size and shape into which a variety of metals or alloys can be cast [48]. Very few studies are focused on using metal foam to enhance the effective thermal conductivity of PCM for efficient heat exchange process [42,44]. However, most of the studies reported involve operation at low temperatures <200°C.

Nithyanandam and Pitchumani [45] considered the performance of PCM impregnated with metal foam made of silicon carbide for high-temperature applications, while Kim et al. [43] focused on the application of aligned graphitic foams with very high thermal conductivities for improving the heat transfer rate. By means of computational analysis, Nithyanandam and Pitchumani [45] elucidated the effect of foam porosity and pore density on the performance of system. It was found that the augmentation in heat transfer rate during charging decreased with increase in pore density of metal foam due to restriction in the formation of buoyancy-induced convection currents. During discharge, an enhanced heat transfer rate was observed for LHS with metal foam of high pore density due to larger effective surface area for heat exchange interaction between the PCM and solid metal foam matrix. The performance of the system during discharge was found to decrease with increase in foam porosity due to decrease in the surface area of the metal foam matrix per unit PCM volume with increase in foam porosity. However, during charging, based on the trade-off between increase in the surface area of the metal foam matrix per unit PCM volume and decrease in latent storage capacity, and restriction of buoyancy-driven convection currents with decrease in porosity, an optimum value of foam porosity of 0.9 maximized the charging heat transfer rate.

In addition, Nithyanandam and Pitchumani [45] reported on a combination of HPs and metal foams to improve the thermal performance of high-temperature LHS system. In their design, the HPs were used to reduce the thermal resistance between HTF and PCM, while metal foams contributed to reducing the thermal resistance of the PCM itself.

10.4.2 Encapsulated phase change material latent heat storage

EPCM-LHS represents a promising approach to increase the heat transfer area by incorporating the PCM mixture in small capsules using suitable shell materials [49–51]. Encapsulating PCM material inside small capsules increases the specific surface area, and using HTF in direct contact with the capsules increases the heat transfer coefficient. For example, PCM stored in spherical capsule diameter of 10 mm offer surface area of more than 600 square meters per cubic meter of capsules.

The configuration of a single-tank thermocline storage system packed with spherical capsules containing PCM is schematically depicted in Fig. 10.13. In an EPCM-LHS, a tank of height H_t and radius R_t is packed with spherical capsules filled with PCM. For the sake of clarity in illustration, an ordered arrangement of capsules is depicted in Fig. 10.13; however, in reality, the packing could be disordered with the porosity for most packing of interest falling into the range of 0.36–0.40 [52]. The inner radius of the capsules, represented by R_c, is filled with PCM (shaded), while the

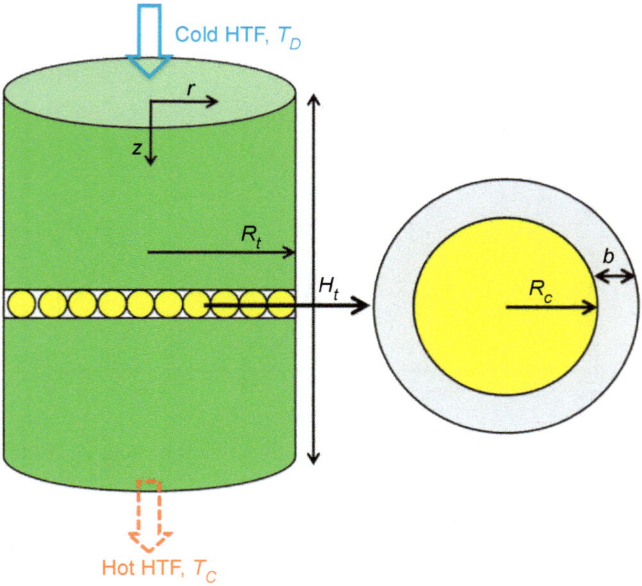

Figure 10.13 Schematic of encapsulated phase change material latent heat storage system. After Nithyanandam K, Pitchumani R, Mathur A. Analysis of a latent thermocline storage system with encapsulated phase change materials for concentrating solar power. Applied Energy 2014;113:1446–60.

thickness of the capsule wall is denoted by b in Fig. 10.13. Generally, hot HTF from the solar receiver enters the LHS from the top while cold fluid is pumped from the bottom of the LHS packed bed during discharge. Buoyancy forces ensure stable thermal stratification of hot and cold fluids within the tank during both charging and discharging processes.

The various temperature regimes that are formed during the transient heat transfer process inside an EPCM storage system is shown in Fig. 10.14 [49]. Fig. 10.14 presents the axial HTF temperature distribution (solid line) and the PCM melt fraction (dashed line) inside the tank at various time instants during the charge process for PCM melt temperature of $\theta_m = 0.5$. The nondimensional temperature (θ) and the nondimensional axial position (z^*) are defined as $\theta = \frac{T-T_D}{T_C-T_D}$ and $z^* = z/H_t$, where H_t is the height of the EPCM-LHS system (Fig. 10.13). The HTF temperature distribution presented in Fig. 10.14 at any time consists of four zones: (1) a constant low-temperature zone ($\theta_f = 0$) near the bottom of the tank ($z^* = 1$), (2) a constant high-temperature zone ($\theta_f = 1$), which prevails near the top of the tank ($z^* = 1$), (3) a constant melt-temperature zone ($\theta_f = \theta_m$), and (4) two intermediate heat exchange zones ($\theta_m < \theta_f < 1$ and $0 < \theta_f < \theta_m$). In the constant low- and high-temperature zones, the molten salt and PCM capsules are in thermal equilibrium; in the heat exchange zones, sensible energy transfer between the PCM capsules and the HTF occurs; and in the constant melt-temperature zone, latent energy transfer between the HTF and PCM takes place.

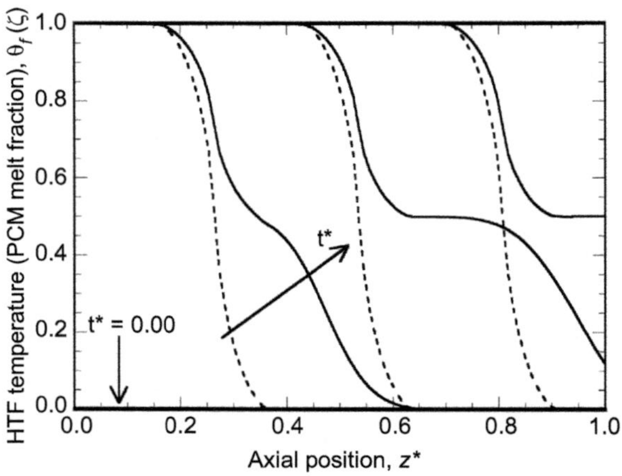

Figure 10.14 Variation in the transient evolution of heat transfer fluid (HTF) temperature and phase change material (PCM) melt front along the height of encapsulated phase change material latent heat storage system during charging.
Adapted from Nithyanandam K, Pitchumani R, Mathur A. Analysis of a latent thermocline storage system with encapsulated phase change materials for concentrating solar power. Applied Energy 2014;113:1446–60.

There exist technoeconomic trade-offs between the cost of the encapsulation method and the reliability of the capsule. Individual assembly of millions of metal canisters loaded with PCM would likely produce the most robust capsules. The manufacturing processes utilized for creation of individual assembly EPCMs include stamping, robotic welding, and other highly reliable methods, which contribute to the robustness of the capsule [12,53]. However, as previously mentioned, due to the amount of labor associated with most of the manufacturing processes likely to be used in the assembly process, these EPCMs would likely not meet the cost targets of the SunShot Initiative. In comparison, although the manufacturing methods used for batch assembly EPCMs are more likely to result in a less costly EPCM, these methods need further development with a high degree of process control [53]. A further consideration with EPCMs is their susceptibility to rupture when loaded in a TES system. To prove the reliability of the batch assembly EPCMs as a viable TES solution, rigorous reliability tests must be undertaken to demonstrate the 30-year lifetime required of CST plants.

A major technical barrier with encapsulating salts is that a void must be created inside the shell when it is produced so as to accommodate the volumetric increase of the PCM during charging. In addition, the shell used to encapsulate the salt must be compatible with a molten salt HTF heated to temperatures above 600°C and must be robust to withstand over 10,000 thermal cycles between 300 and 600°C. The breakage rate, if any, must be less than 0.1% per year. Many proposals for manufacturing batch-assembled EPCMs involve methods such as electroplating, self-assembly, or spray coating (sol gel), allowing them to be created in large batches with comparatively little need for labor and, in turn, cost [12].

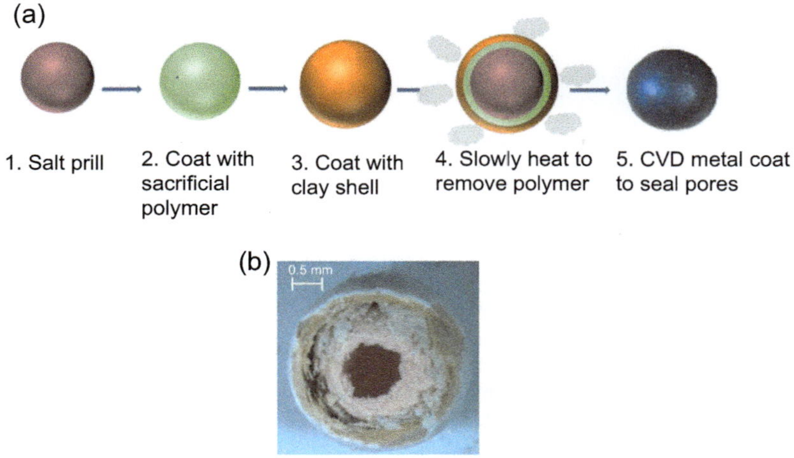

Figure 10.15 (a) Manufacturing process of spherical phase change material capsules for encapsulated phase change material latent heat storage system and (b) dissected section of a spherical capsule.

One such process developed by Mathur et al. [50] is illustrated in Fig. 10.15, which involves five steps to make the capsules. In step 1, the salt prills of KNO_3 were coated to a predetermined thickness with a high molecular weight polymer selected such that it decomposes to gases at temperature lower than the melting point of salt (volume of polymer at least equal to salt volume expansion). Methyl cellulose with Ceramabind binder was used. In step 2, a 300- to 500-μm-thick coating of the shell material was added on top of the polymer coat. Montmorillonite clay mixed with proprietary additives was used. In step 3, the clay-coated capsules were heated very slowly in a furnace to about 280°C. At this temperature, the polymer decomposes to gases and escapes through the pores in the capsules leaving an open space or void. In step 4, the capsules were heated further beyond the melting point of the salt prill. As the salt melts and is heated beyond its melting point, it expands and occupies the free space left by the polymer. In step 5, a thin coating of nickel metal is chemically deposited over the shell to seal the pores. Fig. 10.15(b) shows a dissected capsule examined in an optical microscope showing the void in the middle.

10.4.3 Heat transfer comparison between the tank phase change material latent heat storage–based and encapsulated phase change material–based latent heat storage

To effectively compare the performance of the TPCM- and EPCM-based LHS, a constant solar salt (HTF) mass flow rate ($\dot{m}_f$) of 518.52 kg/s is considered, which corresponds to the required flow rate to power a 100-MWe solar power plant operating at

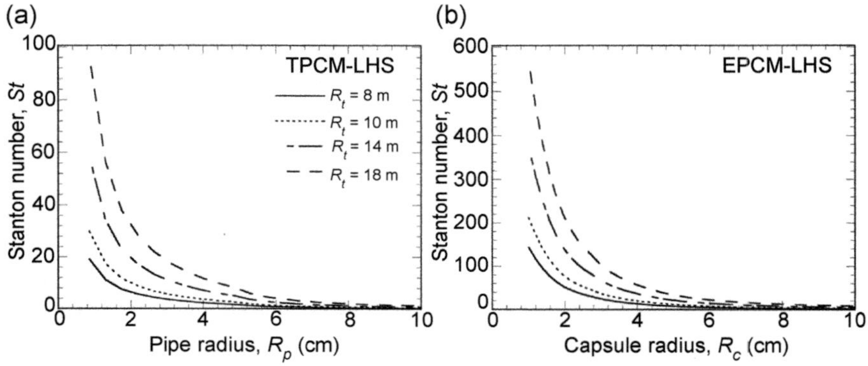

Figure 10.16 Variation of Stanton number with tank radius and (a) pipe radius for tank phase change material latent heat storage (TPCM-LHS) system, (b) capsule radius for encapsulated phase change material latent heat storage (EPCM-LHS) system.

41% thermal efficiency. The height of the tank (H_t) and the porosity of the tank are kept constant at 20 m and 0.37 [54], respectively. The porosity in a TPCM-LHS reflects the volume of the tank occupied by the HTF pipes, while in an EPCM-LHS it denotes the volume of the tank occupied by HTF. This ensures that the volume of PCM is the same in both cases. A plot of the Stanton number (*St*) defined as the ratio of the overall heat transfer to the heat-carrying capacity of the HTF ($St = h_{eff}A_s/\dot{m}_f c_f$) for varying pipe radius in TPCM-LHS and capsule radius in EPCM-LHS is shown in Fig. 10.16. The effective heat transfer coefficient, h_{eff}, includes contributions from convection heat transfer due to HTF flow and conduction heat transfer in the PCM during solidification. A_s denotes the total surface area for interaction of HTF with PCM and c_f is the specific heat of the HTF. It is seen from Fig. 10.16 that for a given PCM volume, the Stanton number is at least six times higher in the case of EPCM-LHS compared to TPCM-LHS. For a given tank radius (R_t), the Stanton number decreases with increase in pipe and capsule radius because of decreases in effective heat transfer coefficient and surface area per unit volume for interaction of HTF with PCM. This figure also reiterates the importance of heat transfer enhancement techniques for TPCM-based LHS systems.

10.4.4 System integration of latent heat storage with concentrating solar thermal power plants

To achieve the SunShot Initiative goals, a critical that the exergetic efficiency be very high to ensure that heat quality is maintained after storage.system-driven approach that examines the cost and efficiency is necessary to understand the benefits of CST with TES. The Solar Advisor Model (SAM) [11] developed by the National Renewable Energy Laboratory (NREL) and the Sandia National Laboratories provides the framework to investigate the impact of geographical, geometrical, and operating parameters on the performance of various CST technologies including parabolic trough, molten salt power tower, and dish Stirling power plants. As per the U.S.

DOE SunShot Initiative requirements, the optimal design of a TES for integration into CST power plants should yield Page 24round-trip exergetic efficiency (ξ) greater than 95% and a storage capital cost (ψ) less than \$15/kWh$_t$ for a minimum discharge period (t_D) of 6 h [3]. Although the round-trip exergetic efficiency of sensible TES can attain close to 100%, the storage capital cost of current 2-tank sensible TES is reported as \$27/kWh$_t$ [12]. Nevertheless, an optimally designed LHS has the potential to reach both storage capital cost (ψ) less than \$15/kWh$_t$ and exergetic efficiency greater than 95%.

The SunShot Initiative's goal of an LCOE of 6 ¢/kWh, without subsidies, is expected to be achieved with molten salt power tower CST plant of nameplate capacity 200-MW$_e$ [3]. To this end, Nithyanandam and Pitchumani [1] evaluated the performance of the CST power plant with LHS systems (net annual energy production, storage capital cost, capacity factor, and LCOE) by integrating the physics-based model of the storage systems with the performance model of the molten salt power tower CST plant. The dynamic model uses actual weather data at time intervals of 1 h and calculates the net electrical energy output at every time step during an entire year, thus simulating the hourly, monthly, and annual energy output of a solar thermal power plant. More detailed descriptions of the system and cost models may be found in Ref. [1]. A systematic analysis of the various design configurations of TPCM-LHS embedded with HPs (Fig. 10.12 in Section 10.4.1) and EPCM-LHS (Fig. 10.13 in Section 10.4.2) on CST power plant operation based on s-CO_2 and Rankine power cycles is conducted, and for the first time, a methodology for deriving design envelopes of the two different types of LHS based on the aforementioned constraints ($\xi > 95\%$, $\psi <$ \$15/kWh$_t$, $t_D > 6$ h) was illustrated.

An optimum design configuration of the LHS system based on minimum LCOE from the design windows is reported in Ref. [1] and compared with the CST power plant operation based on Rankine cycle and current state-of-the-art two-tank direct sensible energy storage system, as presented in Tables 10.1(a)–(c). Important results pertaining to the analysis can be summarized as follows: smaller capsule radius of EPCM-LHS system has lower storage capital cost and higher exergetic efficiency and provides higher capacity factor and least LCOE of CST power plant for a given storage capacity while there exists an optimum longitudinal spacing between the HPs in a TPCM-LHS with HPs for which the LCOE is minimum. Based on a systematic parametric analysis on the various performance metrics, feasible operating regimes and design conditions were identified which meets the SunShot Initiative 2020 requirements: (1) LCOE less than 6 ¢/kWh, (2) exergetic efficiency greater than 95%, and (3) storage cost less than US\$87.95 MM (equivalent to \$15/kWh$_t$). The minimum LCOE obtained from the parametric study and the design window was less than the SunShot 2020 target of 6 ¢/kWh for EPCM-LHS system at 5.37 ¢/kWh. Also, it is found that the maximum cost of HPs embedded in LHS, which satisfies the SunShot requirements, is \$4/unit HP. The results provide direct usable information for optimum operation and design of the latent thermal storage systems for CST power plants. The methodology presented in Ref. [1] may be used to obtain the optimum designs for other LHS and CST power plant configurations.

Table 10.1a **Optimum 2-Tank design configuration for least levelized cost of electricity (LCOE)**

Power plant cycle	Storage type	Tank radius, R_t (m)	Tank height, H_t (m)	Annual net electric output, Q_e (GWh)	Capacity factor, β (%)	Storage capital cost (US$ MM)	LCOE (¢/kWh)
$s\text{-}CO_2$	No TES	–	–	553.5	31.6	–	11.07
	2-Tank	*15.3*	*20*	*1138.7*	*65.0*	*158.1*	*6.49*
Rankine	No TES	–	–	565.5	32.3	–	17.46
	2-Tank	23.9	20	1160.3	66.2	204.6	9.92

TES, thermal energy storage.
Adapted from Nithyanandam K, Pitchumani R. Cost and performance analysis of concentrating solar power systems with integrated latent thermal energy storage. Energy 2014;64:793–810.

Table 10.1b Optimum encapsulated phase change material latent heat storage (EPCM-LHS) (Fig. 10.13) system design configuration for least levelized cost of electricity (LCOE)

Power plant cycle	Encapsulation cost ($/kg-PCM)	Tank radius, R_t (m)	Tank height, H_t (m)	Capsule radius, R_c (cm)	Annual net electric output, Q_e (GWh)	Capacity factor, β (%)	Exergetic efficiency, ξ (%)	LCOE (¢/kWh)	Storage capital cost (US$ MM)
s-CO_2	**(1) *0.75***	***15***	***15***	***0.5***	***1204.9***	***68.77***	***96.04***	***5.37***	***42.8***
	(2) *1.50*	*15*	*15*	*0.5*	*1204.9*	*68.77*	*96.04*	*5.47*	*58.1*
Rankine	(1) 0.75	20	15	0.5	11,517.6	66.07	98.42	9.05	74.6
	(2) 1.50	20	15	0.5	1157.6	66.07	98.42	9.24	101.8

PCM, phase change material.
Adapted from Nithyanandam K, Pitchumani R. Cost and performance analysis of concentrating solar power systems with integrated latent thermal energy storage. Energy 2014;64:793–810.

Table 10.1c Optimum heat pipes embedded latent heat storage (HP-LHS) (Fig. 10.12) system design configuration for least levelized cost of electricity (LCOE)

Power plant cycle	Heat pipe cost ($/unit)	Channel length, L_d (m)	Channel Width, W_d (m)	No. of heat pipes	Annual net electric output, Q_e (GWh)	Capacity factor, β (%)	Exergetic efficiency, ξ (%)	LCOE (¢/kWh)	Storage capital cost (US$ MM)
s-CO_2	(1) *2.0*	*25*	*1.2*	*15,624,164*	*1158.7*	*66.14*	*97.05*	*5.77*	*78.61*
	(2) *4.0*	*25*	*1.2*	*15,624,164*	*1158.7*	*66.14*	*97.05*	*6.05*	*110.5*
Rankine	(1) 2.0	30	1.6	24,898,663	1122.1	64.05	97.61	9.69	124.7
	(2) 4.0	30	1.6	24,898,663	1122.1	64.05	97.61	10.05	175.7

Adapted from Nithyanandam K, Pitchumani R. Cost and performance analysis of concentrating solar power systems with integrated latent thermal energy storage. Energy 2014;64:793–810.

10.4.5 Large-scale demonstrations

An example of a case study of an active pilot-scale demonstration of LHS module was presented in Refs. [55,56]. To date, the world's largest LHS module integrated with water/steam test loop was commissioned by German Aerospace Center (DLR) at Endesa power plant Litoral in Carboneras, Spain, in the year 2010. The LHS module is combined with a concrete sensible storage system for preheating and superheating water/steam while PCM storage is used for water evaporation. Fig. 10.17(a) shows the thermodynamic characteristics of combined sensible/latent storage with direct steam generation in the solar collectors. An overview of the three-part TES system for direct steam generation

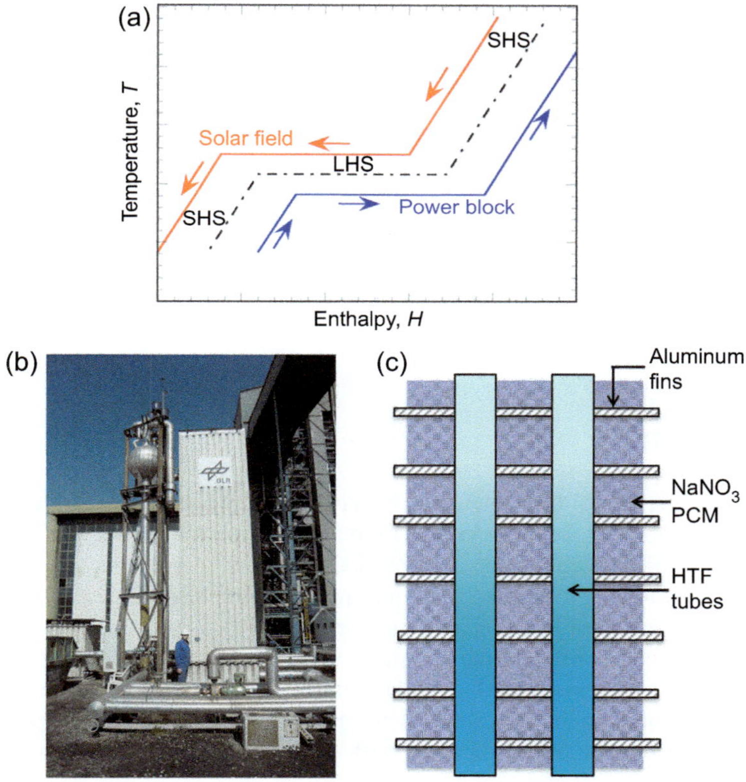

Figure 10.17 (a) Temperature–enthalpy characteristics with combined sensible/latent heat storage for direct steam generation, (b) 700 kWh phase change material (PCM) storage in the water/steam test loop at the Endesa power plant Litoral in Carboneras, Spain, and (c) schematic of latent heat storage (LHS) module with aluminum fins. *HTF*, heat transfer fluid; *LHS*, latent heat storage; *SHS*, sensible heat storage.

(a) Adapted from Laing D, Bahl C, Bauer T, Lehmann D, Steinmann WD. Thermal energy storage for direct steam generation. Solar Energy 2011:85(4):627–33; (b) After Laing D, Bauer T, Breidenbach N, Hachmann B, Johnson M. Development of high temperature phase-change-material storages. Applied Energy 2013;109:497–504.

Table 10.2 System characteristics of German Aerospace Center (DLR) demonstration

Storage capacity	700 kWh
Storage medium weight	14,000 kg
Outer storage unit dimensions, w/o insulation	1.7 m × 1.3 m × 7.4 m
Tube dimensions	152 tubes of length 6 m
Fin type	Radial, aluminum
Steam conditions	128 bar, 400°C
PCM	$NaNO_3$ ($T_m = 306°C$
Insulation type	40-cm-thick mineral wool

PCM, phase change material.
Adapted from Laing D, Bauer T, Breidenbach N, Hachmann B, Johnson M. Development of high temperature phase-change-material storages. Applied Energy 2013;109:497–504.

combining SHS and LHS is shown in Fig. 10.17(b). The LHS module uses $NaNO_3$ as the PCM sandwiched between aluminum fins for enhanced heat transfer and has a total storage capacity of 700 kWh [55,56] as shown in Fig. 10.17(c). The system operates under temperature and pressure conditions similar to that of direct steam generation in parabolic trough power plants and the details are provided in Table 10.2.

10.5 Summary

This chapter presented the potential challenges and opportunities of LHS as a cost-effective approach for improving the efficiency and dispatchability of CST technologies. The LHS systems involve phase change of the storage material, commonly referred to as PCM, between solid and liquid state. Storing thermal energy in the form of latent heat of fusion of PCM, in addition to sensible heat, significantly increases the energy density, in turn, offering the opportunity for reducing the storage size and cost. The general challenges associated with LHS for CST technologies in the areas of PCM requirements, compatibility of containment material, heat transfer performance, and system integration were discussed along with a presentation of the different configurations of LHS for CST applications.

Various PCMs were categorized based on their temperature ranges of application, volumetric energy density, and cost ($/kWh). It was pointed out that more studies are needed to accurately characterize various molten salt PCMs to resolve the discrepancy in the thermophysical properties reported in the literature. Research efforts are also focused on creating binary or ternary eutectic PCM with enhanced thermophysical properties and low cost ($/kWh). High-temperature, long-term thermal and chemical stability (corrosion) of construction material in PCMs is an important requisite for safe and cost-effective LHS design in CST applications. Further research required in the area of containment material compatibility includes measuring the temperature dependence of corrosion rate, fully understanding mechanisms of corrosion, and devising methods for mitigating corrosion, especially at the higher temperatures, so that PCM materials can be viable for high-temperature CST.

A fundamental challenge with LHS using molten salt PCMs for high-temperature CST applications is its intrinsically low thermal conductivity that limits the discharge heat transfer rate during the conduction-dominated solidification of PCM. The poor discharge rate results in a huge penalty in overall exergetic efficiency of CST system. This chapter detailed the research efforts focused on alleviating the poor heat transfer performance of PCMs by categorizing into two classes: TPCM-LHS system with performance enhancement techniques and EPCM-LHS system that promote high heat transfer rate by encapsulating the PCM in small spherical capsules thus increasing the surface area per unit volume for interaction with HTF. Of the various performance enhancement techniques such as extended surfaces (HPs/thermosyphons), dispersing high conductivity particles, impregnating metal or graphitic foam structures that have been explored for TPCM-LHS, PCMs with graphite foams and heat pipes show great potential [1, 43] to meet aggressive cost and performance targets [3]. Opportunities for ECPM-LHS systems lie in improved manufacturing methods with high reliability and low production cost.

There has been significant amount of exploratory research in the area of LHS for CST applications; however, the demonstration of the systems is only limited to bench-scale studies. Future work should focus on pilot-scale demonstration of the systems to derisk the technology, attract investors, and secure project financing for large-scale demonstrations. The chapter presented an example system-level study [1] that analyzed the technoeconomic feasibility of different configurations of LHS integrated with a CST power tower plant, based on key performance metrics of exergetic efficiency, storage cost, and LCOE. Further system-level analysis and optimization is necessary for designing and implementing a full-scale LHS with a CST power plant.

References

[1] Nithyanandam K, Pitchumani R. Cost and performance analysis of concentrating solar power systems with integrated latent thermal energy storage. Energy 2014;64:793–810.

[2] Louis AT, Ganapathi GB, Wirz RE, Lavine AS. Spatial and temporal modeling of sub-and supercritical thermal energy storage. Solar Energy 2014;103:402–10.

[3] US Department of Energy. SunShot vision study. US Department of Energy; 2012. http://www1.eere.energy.gov/solar/pdfs/47927.pdf.

[4] Hoshi A, Mills DR, Bittar A, Saitoh TS. Screening of high melting point phase change materials (PCM) in solar thermal concentrating technology based on CLFR. Solar Energy 2005;79(3):332–9.

[5] Li F, Hu YJ, Zhang RY. The influence of heating-cooling cycles on the thermal storage performances of Al-17 wt.% Si alloy. Advanced Materials Research July 2011;239: 2248–51.

[6] Kenisarin MM. High-temperature phase change materials for thermal energy storage. Renewable and Sustainable Energy Reviews 2010;14(3):955–70.

[7] Kotzé JP, Von Backström TW, Erens PJ. High temperature thermal energy storage utilizing metallic phase change materials and metallic heat transfer fluids. Journal of Solar Energy Engineering 2013;135(3):035001.
[8] Kearney D, Herrmann U, Nava P, Kelly B, Mahoney R, Pacheco J, Cable R, Potrovitza N, Blake D, Price H. Assessment of a molten salt heat transfer fluid in a parabolic trough solar field. Journal of Solar Energy Engineering 2003;125(2):170–6.
[9] Shabgard H, Robak CW, Bergman TL, Faghri A. Heat transfer and exergy analysis of cascaded latent heat storage with gravity-assisted heat pipes for concentrating solar power applications. Solar Energy 2012;86(3):816–30.
[10] Turchi CS, Ma Z, Neises TW, Wagner MJ. Thermodynamic study of advanced supercritical carbon dioxide power cycles for concentrating solar power systems. Journal of Solar Energy Engineering 2013;135(4):041007.
[11] System advisor model Version 2012.5.11 (SAM 2012.5.11), User documentation. Golden, CO: National Renewable Energy Laboratory.
[12] Stekli J, Irwin L, Pitchumani R. Technical challenges and opportunities for concentrating solar power with thermal energy storage. Journal of Thermal Science and Engineering Applications 2013;5(2):021011.
[13] Raade JW, Padowitz D. Development of molten salt heat transfer fluid with low melting point and high thermal stability. Journal of Solar Energy Engineering 2011;133(3): 031013.
[14] Liu M, Gomez JC, Turchi CS, Tay NHS, Saman W, Bruno F. Determination of thermophysical properties and stability testing of high-temperature phase-change materials for CSP applications. Solar Energy Materials and Solar Cells 2015;139:81–7.
[15] Mantha D, Wang T, Reddy RG. Thermodynamic modeling of eutectic point in the $LiNO_3$-$NaNO_3$-KNO_3 ternary system. Journal of Phase Equilibria and Diffusion 2012;33(2): 110–4.
[16] Mantha D, Wang T, Reddy RG. Thermodynamic modeling of eutectic point in the $LiNO_3$–$NaNO_3$–KNO_3–$NaNO_2$ quaternary system. Solar Energy Materials and Solar Cells 2013;118:18–21.
[17] Goods SH, Bradshaw RW. Corrosion of stainless steels and carbon steel by molten mixtures of commercial nitrate salts. Journal of Materials Engineering and Performance 2004; 13(1):78–87.
[18] Kruizenga A, Gill D. Corrosion of iron stainless steels in molten nitrate salt. Energy Procedia 2014;49:878–87.
[19] Flour L. Contact fatigue of automotive gears: evolution and effects of residual stresses introduced by surface treatments. Fatigue & Fracture of Engineering Materials & Structures 2000;23(3):217–28.
[20] Goswami GL, Kumar S, Galun R, Mordike BL. Laser cladding of Ni-Mo alloys for hardfacing applications. Lasers Engineering 2003;13(1):1–12.
[21] Manikandan P, Hokamoto K, Fujita M, Raghukandan K, Tomoshige R. Control of energetic conditions by employing interlayer of different thickness for explosive welding of titanium/304 stainless steel. Journal of Materials Processing Technology 2008;195(1): 232–40.
[22] Yahiro A, Masui T, Yoshida T, Doi D. Development of nonferrous clad plate and sheet by warm rolling with different temperature of materials. ISIJ International 1991;31(6): 647–54.
[23] Olson L, Sridharan K, Anderson M, Allen T. Nickel-plating for active metal dissolution resistance in molten fluoride salts. Journal of Nuclear Materials 2011; 411(1):51–9.

[24] Liu M, Tay NS, Bell S, Belusko M, Jacob R, Will G, et al. Review on concentrating solar power plants and new developments in high temperature thermal energy storage technologies. Renewable and Sustainable Energy Reviews 2016;53:1411–32.

[25] Grabke HJ, Reese E, Spiegel M. The effects of chlorides, hydrogen chloride, and sulfur dioxide in the oxidation of steels below deposits. Corrosion Science 1995;37(7):1023–43.

[26] Brent AD, Voller VR, Reid KTJ. Enthalpy-porosity technique for modeling convection-diffusion phase change: application to the melting of a pure metal. Numerical Heat Transfer, Part A Applications 1988;13(3):297–318.

[27] De Lucia M, Bejan A. Thermodynamics of energy storage by melting due to conduction or natural convection. Journal of Solar Energy Engineering 1990;112(2):110–6.

[28] De Lucia M, Bejan A. Thermodynamics of phase-change energy storage: the effects of liquid superheating during melting, and irreversibility during solidification. Journal of Solar Energy Engineering 1991;113(1):2–10.

[29] Bjurström H, Carlsson B. An exergy analysis of sensible and latent heat storage. Journal of Heat Recovery Systems 1985;5(3):233–50.

[30] Jegadheeswaran S, Pohekar SD, Kousksou T. Exergy based performance evaluation of latent heat thermal storage system: a review. Renewable and Sustainable Energy Reviews 2010;14(9):2580–95.

[31] Gong ZX, Mujumdar AS. Thermodynamic optimization of the thermal process in energy storage using multiple phase change materials. Applied Thermal Engineering 1997;17(11): 1067–83.

[32] Shabgard H, Bergman TL, Faghri A. Exergy analysis of latent heat thermal energy storage for solar power generation accounting for constraints imposed by long-term operation and the solar day. Energy 2013;60:474–84.

[33] Nithyanandam K, Pitchumani R. Design of a latent thermal energy storage system with embedded heat pipes. Applied Energy 2014;126:266–80.

[34] Nithyanandam K, Pitchumani R. Optimization of an encapsulated phase change material thermal energy storage system. Solar Energy 2014;107:770–88.

[35] Velraj R, Seeniraj RV, Hafner B, Faber C, Schwarzer K. Experimental analysis and numerical modelling of inward solidification on a finned vertical tube for a latent heat storage unit. Solar Energy 1997;60(5):281–90.

[36] Robak CW, Bergman TL, Faghri A. Enhancement of latent heat energy storage using embedded heat pipes. International Journal of Heat and Mass Transfer 2011;54(15): 3476–84.

[37] Shabgard H, Bergman TL, Sharifi N, Faghri A. High temperature latent heat thermal energy storage using heat pipes. International Journal of Heat and Mass Transfer 2010; 53(15):2979–88.

[38] Nithyanandam K, Pitchumani R. Computational studies on a latent thermal energy storage system with integral heat pipes for concentrating solar power. Applied Energy 2013;103: 400–15.

[39] Nithyanandam K, Pitchumani R. Analysis and optimization of a latent thermal energy storage system with embedded heat pipes. International Journal of Heat and Mass Transfer 2011;54(21):4596–610.

[40] Nithyanandam K, Pitchumani R. Thermal energy storage with heat transfer augmentation using thermosyphons. International Journal of Heat and Mass Transfer 2013;67: 281–94.

[41] Nithyanandam K, Pitchumani R. Computational modeling of dynamic response of a latent thermal energy storage system with embedded heat pipes. Journal of Solar Energy Engineering 2014;136(1):011010.

[42] Fernandes D, Pitié F, Cáceres G, Baeyens J. Thermal energy storage: "How previous findings determine current research priorities". Energy 2012;39(1):246−57.
[43] Kim T, France DM, Yu W, Zhao W, Singh D. Heat transfer analysis of a latent heat thermal energy storage system using graphite foam for concentrated solar power. Solar Energy 2014;103:438−47.
[44] Zhao CY, Lu W, Tian Y. Heat transfer enhancement for thermal energy storage using metal foams embedded within phase change materials (PCMs). Solar Energy 2010;84(8): 1402−12.
[45] Nithyanandam K, Pitchumani R. Computational studies on metal foam and heat pipe enhanced latent thermal energy storage. Journal of Heat Transfer 2014;136(5):051503.
[46] Liu M, Saman W, Bruno F. Review on storage materials and thermal performance enhancement techniques for high temperature phase change thermal storage systems. Renewable and Sustainable Energy Reviews 2012;16(4):2118−32.
[47] Xu B, Li P, Chan C. Application of phase change materials for thermal energy storage in concentrated solar thermal power plants: a review to recent developments. Applied Energy 2015;160:286−307.
[48] Ashby MF, Evans T, Fleck NA, Hutchinson JW, Wadley HNG, Gibson LJ. Metal foams: a design guide. Elsevier; 2000.
[49] Nithyanandam K, Pitchumani R, Mathur A. Analysis of a latent thermocline storage system with encapsulated phase change materials for concentrating solar power. Applied Energy 2014;113:1446−60.
[50] Mathur A, Kasetty R, Oxley J, Mendez J, Nithyanandam K. Using encapsulated phase change salts for concentrated solar power plant. Energy Procedia 2014;49:908−15.
[51] Goswami Y. Development of low cost industrially scalable PCM capsules for thermal energy storage in CSP plants. 2013. SunShot presentation.
[52] Zhang W, Thompson KE, Reed AH, Beenken L. Relationship between packing structure and porosity in fixed beds of equilateral cylindrical particles. Chemical Engineering Science 2006;61(24):8060−74.
[53] Jacob R, Bruno F. Review on shell materials used in the encapsulation of phase change materials for high temperature thermal energy storage. Renewable and Sustainable Energy Reviews 2015;48:79−87.
[54] Torquato S, Truskett TM, Debenedetti PG. Is random close packing of spheres well defined? Physical Review Letters 2000;84(10):2064.
[55] Laing D, Bahl C, Bauer T, Lehmann D, Steinmann WD. Thermal energy storage for direct steam generation. Solar Energy 2011;85(4):627−33.
[56] Laing D, Bauer T, Breidenbach N, Hachmann B, Johnson M. Development of high temperature phase-change-material storages. Applied Energy 2013;109:497−504.

Thermochemical energy storage for concentrating solar thermal (CST) systems

11

L. Irwin[1], J. Stekli[2], C. Pfefferkorn[3], R. Pitchumani[4]
[1]ManTech International Corporation, Arlington, Virginia, United States; [2]United States Department of Energy, Washington, D.C., United States; [3]United States Nuclear Regulatory Commission, Washington, D.C., United States; [4]Virginia Tech, Blacksburg, Virginia, United States

11.1 Introduction to thermochemical energy storage

This chapter briefly introduces the first law of thermodynamics to frame the role of heat, temperature, and mass transfer in dictating the progression of a chemical reaction. These are important considerations for realization of storage of solar thermal energy at magnitudes that are meaningful for the target application. The second law of thermodynamics is then used to establish that not all heat is created equal—that higher temperature heat is more useful for conducting work, given a fixed temperature for the heat sink. Since heat-to-work conversion processes are not perfect, the various contributing factors to inefficiencies (losses) are described in the context of *exergy*—a term used to quantify a system's ability to do useful work.

We then describe the general advantages and disadvantages of thermochemical energy storage (TCES). This is followed by a discussion of specific challenges that occur in TCES systems. Many of these challenges revolve around the high temperatures of operation and include corrosion, sintering, changes in mechanical properties, and competing thermodynamic processes. Each of these concerns requires explicit treatment as they stand to restrict a TCES system's ability to efficiently and effectively conduct the desired chemical transformation.

If left unaddressed, the aforementioned challenges of high-temperature operation may lead to TCES systems that become inoperable over the extended lifetime of a concentrating solar thermal (CST) plant. Consequently, CST systems with TCES require careful consideration of an integrated solar power plant and a thermal chemical plant operation. Little information exists in this area, which calls for thoughtful integration of the first and second laws of thermodynamics. These, combined with Le Châtelier's principle of chemical equilibria, allows for use of heat, temperature, and mass transfer to satisfy the power demands of the energy generating cycle. Several classes of TCES reactions are described, along with specific examples of chemical transformations that have been previously investigated at research and pilot scale.

Despite all of the known challenges, the incredibly high energy density, dynamic, adjustable character of the heat output, and flexibility of TCES system warrant continued pursuit and development.

Advances in Concentrating Solar Thermal Research and Technology
http://dx.doi.org/10.1016/B978-0-08-100516-3.00011-3

11.1.1 Energy and exergy analysis for thermochemical energy storage systems

TCES is defined as the storage of heat energy through the cycling of matter through two or more thermodynamic states. This transition between states is achieved via the making and/or breaking of chemical bonds. In this chapter endothermic and exothermic chemical processes are the focus of discussion, but it should be realized that the chemical process is but one contributor to the total thermal energy contained in a TCES system. Total energy stored may be defined by Eq. (11.1):

$$Q = mC_p\Delta T + mH_l + m\Delta H_{rxn} \tag{11.1}$$

where m is the mass of the storage media, C_p is the heat capacity of the storage media, ΔT is the change in temperature of the storage media, H_l is the latent heat associated with any phase change of the storage media, and ΔH_{rxn} is the heat of reaction for the chemical transformation. Eq. (11.1) is consistent with the first law of thermodynamics.

The second law of thermodynamics guides how useful the energy contained in a thermal energy storage (TES) system is for producing work. The efficiency with which thermal energy can be converted to work is bounded by the Carnot limit:

$$\eta = 1 - \frac{T_C}{T_H} \tag{11.2}$$

where T_C represents the absolute temperature of the "cold" energy sink and T_H represents the absolute "hot" temperature of the system.

In considering TCES systems as driving forces for Carnot cycles we must keep in mind that TCES may involve processes more complex than just simple changes in temperature. TCES may involve changes in pressure, concentration, and even counts of molecular units. These are all factors that ultimately contribute to the entropy of the system. This makes the second law of thermodynamics especially useful for comparing two or more TCES systems as it explicitly treats entropy. This comparison, often referred to in the literature as an exergy analysis, has been widely applied to evaluate energy conversion processes in TCES systems [1–5]. Exergy can be thought of as the potential work that could be obtained from a TCES system at the system's current state given the larger environment the system exists in. A simplified exergy analysis for a generic, closed, perfectly reversible storage system is provided in Fig. 11.1.

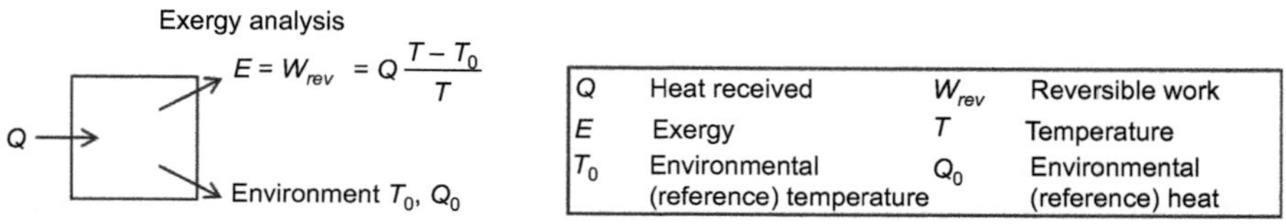

Figure 11.1 A simplified exergy analysis for a closed and reversible storage system.

During a thermochemical cycle multiple exergy transfer processes occur. Expressions describing exergy transfer from heat (E_{heat}), work (E_{work}), and materials (E_{mat}) are provided below:

$$E_{heat} = Q\left(\frac{T - T_0}{T}\right) \tag{11.3}$$

$$E_{work} = W_{rev} \tag{11.4}$$

$$E_{mat} = (H - H_0) - T_0(S - S_0) \tag{11.5}$$

$$e_{chm} = \sum_i n_i e_{chmi} \tag{11.6}$$

$$e_{ph} = m(h_1 - h_2 - T_0(s_1 - s_2)) \tag{11.7}$$

where H and H_0 are the total enthalpy of the substance and S and S_0 are the total entropy of the substance at temperatures T and T_0, respectively. The total exergy of the material(s) used in a TCES system is given by its Gibbs free energy (Eq. 11.5) which itself is comprised of separate chemical exergy (Eq. 11.6) and physical exergy (Eq. 11.7) subcomponents. The chemical exergy, e_{chm}, is the sum of the product of partial molar chemical exergy ($e_{chm\ i}$) scaled by the number of moles of each substance (n_i). The physical subcomponent (e_{ph}) is comprised of terms associated with mixing and changes in temperature and pressure. These equations provide us with an explicit treatment for the system-level concerns previously mentioned as complexities in the comparative analyses of TCES systems. It is worth noting that TCES systems that employ combustion have additional exergy concerns that must be treated explicitly. That treatment is not provided here as additional information regarding second law analysis of combustion processes can be widely found in the literature [6,7].

The Carnot cycle, which produces work from heat, is a theoretical thermodynamic cycle. Practical engineering requires that we evaluate the benefits and tradeoffs of reducing any given TCES system to its real-life embodiment. To do so one must consider deviations from theoretical performance due to irreversible processes. These irreversible processes create entropy and reduce the system's capacity to do useful work. They are typically grouped and labeled as exergy destruction (E_{loss}). Their detrimental contributions may be evaluated when calculating the exergetic efficiency (η_E) of a process (Eq. 11.8):

$$\eta_E = 1 - \frac{E_{loss}}{E_i} \tag{11.8}$$

where i is the relevant component of exergy (heat, work, or substance). Irreversibilities leading to exergy destruction include heat loss to environment, pressure drop across the system, and processes associated with the chemical reaction such as reactant and product mixing. Handling of these matter and energy transfer processes should be

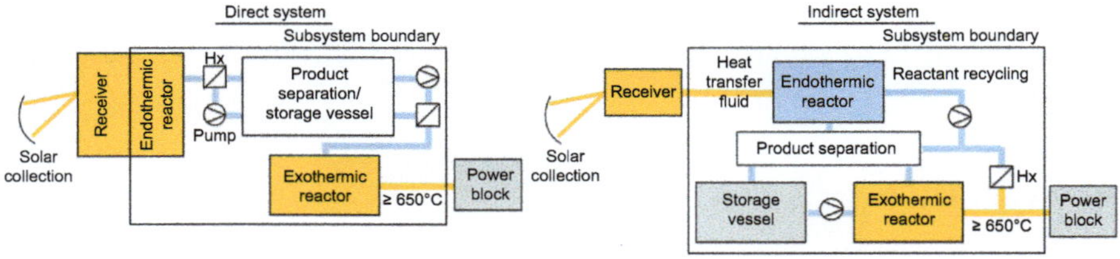

Figure 11.2 Simplified diagrams of two solar-driven TCES systems. Left: Direct system—solar receiver and endothermic chemical reactor are same vessel. Right: Indirect system—solar receiver and endothermic chemical reactor are different vessels necessitating an inert heat transfer fluid. "Hx" denotes a heat exchanger.

considered when making a choice between direct and indirect TCES systems, as each of these has subsequent effects on total exergy. Simplified schematics for possible direct and indirect solar TCES systems, including endothermic and exothermic reactor and storage unit are presented in Fig. 11.2.

For a storage media to be useful in a TCES system the reaction should have minimum exergy losses, both the forward and the reverse chemical reactions should occur at a rate appropriate for the time scales of energy collection and power generation, and the temperature of the chemical reaction should be well matched to the heat engine with which the storage system will be integrated. The extent of the chemical reaction, which is how far the equilibrium may be shifted during the charge and discharge process, impacts the overall cost of the system by limiting its actual energy density. In many instances it is difficult and/or impractical to achieve 100% conversion of reactant to products or vice versa. The cost of driving an equilibrium further to completion should be carefully weighed against the benefit gained from extracting the additional heat.

11.1.2 Advantages and disadvantages of thermochemical energy storage for CST systems

There are a number of potential advantages that TCES has over sensible or latent heat storage for use in CST systems. The first potential advantage is the relatively higher energy density that could possibly be attained with TCES systems. Fig. 11.3 shows the range in energy densities for latent, sensible, and TCES materials, as evaluated for CST applications. As the figure shows, the energy density of TCES systems can be many multiples of the density of alternative TES systems. A second advantage that TCES possesses, and one that is shared with latent heat storage systems, is that the heat released from a TCES system may be isothermal. Most heat engines have optimum efficiency at a specific design temperature and the isothermal nature of TCES can guarantee that heat from the system is consistently well matched with the needs of the heat engine paired with it.

The final major advantage for TCES systems is the ability for greater operational flexibility. Plant operators with TCES systems have more options in deciding how and when heat is released from the system as compared to other types of TES. How heat is released is governed by Le Châtelier's principle of chemical equilibrium. Pressure swings may be used for a gaseous reactant(s) to achieve a greater output temperature from the TCES system as compared to the original temperature the system was

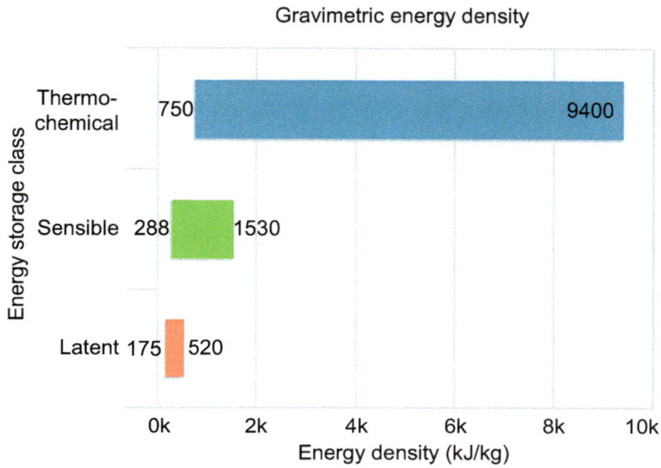

Figure 11.3 Typical ranges for gravimetric energy density in latent, sensible, and thermochemical energy storage materials. The 9400 kJ/kg value for thermochemical is achieved via combustion of sulfur as part of an S/H_2SO_4 cycle.

charged at. This could allow for a lower-quality (lower-temperature) heat source to use higher-efficiency (higher-temperature) heat engines, effectively making the TCES system a heat pump.

TCES systems also offer greater flexibility in the timing of the heat discharge. While traditional TES systems do offer limited temporal flexibility, TCES systems are, in theory, capable of providing "seasonal storage." Seasonal storage, which is defined as the ability to store collected energy for a period of months, is not likely to be viable for sensible TES since the heat is stored at an elevated temperature relative to the surrounding environment. Since all real TES systems obey the first law of thermodynamics, sensible TES will gradually lose energy through transfer to the surrounding environment, even within a well-insulated container. This limits the length of time that heat can be economically stored. TCES systems may also suffer from this type of energy loss; however, if the sensible heat is recuperated (e.g., used to preheat inputs) the heat in a TCES system can be efficiently stored in chemical bonds. These bonds are stable so long as we do not reintroduce the reactant and/or expose the material to a catalyst to reverse the reaction. Therefore, as long as the system is deprived of the conditions required to reverse the reaction, the chemical energy can be stored indefinitely.

High energy densities, operational flexibility, and seasonal storage are all important advantages of TCES systems. However, TCES systems also present certain additional challenges in their operation. For example, TCES systems have not been demonstrated on a commercial scale. The lack of a demonstration means that some of the operational challenges with a TCES system are likely unknown. Getting to know these challenges will take time, especially when we consider the fact that TCES systems require the integration and operation of both a traditional thermal CST plant and the more complex operations of the conjoined chemical plant. Additional experience with mass transport, separations, recombinations, and safety will be required.

We now visit some of these operational challenges in greater detail. We do so by walking through the charging of a TCES system via an endothermic chemical reaction in a direct solar receiver/reactor (see the left side of Fig. 11.2). Immediately upon leaving the receiver/reactor the products must be separated (for uncatalyzed reactions) to prevent the acquired charge from spontaneously reversing. Once separated the sensible energy in the product stream must be recuperated so the materials may be efficiently stored. This storage must take place in separate vessels until such time as the exothermic reaction is required to release the heat. During release, the exothermic reactor requires careful recombination of reactants at ratios, pressures, and mixing rates that result in output appropriate for the thermal cycle. This means that both the amount of energy released and the temperature at which the energy is released has to be carefully managed. To be cost-effective, this energy output needs to occur in a vessel or heat exchanger of manageable dimensions. As we discuss later all of these vessels must have acceptable corrosion resistance and mechanical properties at the temperatures and pressures of operation so as to last for the lifetime of the power generating facility.

Another major challenge for TCES systems is that many of the reaction cycles of interest degrade or otherwise lose capacity over time. This means that less energy can be stored with each subsequent cycle. There are a number of potential causes of this degradation. The first is due to the fact that for nearly all heterogeneous chemical reactions, high surface area is required for high reaction yield. To achieve this, strategies such as placing the solid reactant on a mesh screen substrate or pulverizing the solid reactant into small particles are utilized when the TCES system is first put into service. However, due to the high temperature of operation particle sintering or degradation of the support providing the high surface area occurs. Consequently, the availability of sites for the chemical reaction to take place is reduced. A second cause of degradation, for those systems requiring a catalyst, is catalyst poisoning. Over operational cycles, many catalysts will lose their effectiveness due to introductions of other chemical species that form undesirable complexes with the catalyst. This causes less of the catalyst to be available for the desired reaction. While many catalysts are capable of being "regenerated" for TCES, it may be economically infeasible to do this on anything but an infrequent basis.

11.2 General challenges for CST thermochemical storage systems

The challenges associated with TCES for CST applications may be described through a careful consideration of the mass and energy balances for the solar receiver when the receiver is considered an endothermic chemical reactor (this approach considers a system where the endothermic reaction takes place at the focal point of the sun's rays, which is the design for the majority of TCES systems for CST applications). We describe the challenges of integrating CST with chemical processes by considering generic requirements for effective chemical reactions. Refer back to Fig. 11.2 for an illustration of the various subsystem interfaces.

Initially, we consider a direct solar TCES system, shown on the left in Fig. 11.2. In this system the solar receiver and the endothermic chemical reactor are the same device. An initial necessary condition to realizing a viable TCES system of this kind is that the receiver must be thermally and optically efficient (considerations to realizing such a receiver are presented in earlier chapters). In addition to this condition, direct TCES systems present the additional requirement that the receiver design be co-optimized to also serve as a reaction vessel. An important consideration in performing this additional optimization step is that heat received from the solar field will not be uniform upon the entire reactor surface and the heat flux upon the receiver may change with both time of day and time of year. This movement and associated nonuniformity of the heat flux provided to drive the endothermic reaction is not a common consideration in a typical commercial-scale chemical reaction, and therefore requires careful consideration when designing an efficient direct TCES receiver/reactor.

The primary considerations in the design of a TCES system are similar to those required for any commercial-scale chemical reaction: effective heat transfer, mixing, transport, and, in certain cases, separations. Efficient heat transfer through the receiver wall into the chemically reacting substance can be challenging due to the fact that efficient receiver designs often attempt to minimize surface area in order to reduce convective and radiative losses to the atmosphere. This surface area minimization imposes a corresponding requirement for the chemically reactive media to absorb a high power flux ($J/s \cdot m^2$). Matching this large energy flux passage requirement is a requisite, if exergetic efficiency is to be optimized; however, the heat transfer coefficients for a typical chemically reactive media within a TCES system are low enough to make this a difficult outcome to achieve. Recall from Eq. (11.3) that an exergetic penalty is realized if there is a significant degradation in temperature from the receiver surface (T_0) to the chemical reacting media (T). This temperature difference becomes large if the energy flux attainable by the chemical media does not match well with the incoming solar radiation (which is assumed to be matched well with the amount of power the CST system must produce to be economically viable).

Of course, a temperature gradient must exist for heat to flow, but we find that the physical phase of the chemically reacting substance (solid, liquid, or gas) can greatly impact the magnitude of this gradient. Consequently, thermal gradients can vary considerably based upon selection of TCES material and solar receiver/reactor designs. These thermal gradients must be taken into consideration as the incredibly high solar flux placed upon the outer surface of the receiver (1×10^6 W/m^2) can quickly lead to localized overheating and failure in the vessel's material of construction. In theory the endothermic chemical reaction occurring in the receiver should help reduce the likelihood of runaway overheating, but this can only occur if heat can effectively be transferred into the reacting media.

The fundamental physics of heat transfer using gases, liquids, and solids [8] has been treated extensively elsewhere. It is worth mentioning that specific attention has been paid to heat transfer using solids in packed beds [9], fluidized beds [10], and in diffuse flow [11]. Heat transfer in these systems is of particular relevance to CST TCES due to the relatively high temperatures (>650°C) of CST applications. It becomes evident that both liquids and gases of appropriate thermophysical and chemical character are difficult to come by at these temperatures. Organic compositions are typically excluded from use due to irreversible side reactions and general decomposition. Inorganic compositions may possess chemical stability, but the thermophysical properties tend to be unfavorable. Typically, the melting temperature is too high or the viscosity of the melt is too high. If a liquid TCES material of appropriate reactivity, thermal conductivity, and viscosity could be identified, the system design would stand to benefit from the ease by which liquid may be moved through complex systems.

Gases may benefit from ease of transported through the TCES systems; however, their low thermal conductivities make it difficult to match power fluxes ($J/s \cdot m^2$). Microchannel heat exchanger receiver reactors are typically required. These high surface area-to-volume devices have been successfully employed in methane and ammonia conversions, which is discussed in later sections. It is worth noting that gases suffer from high cost in storage due to what is typically their noncondensable character.

In summary, a great body of knowledge exists for pumping chemically reacting gases and fluids through endothermic reactors; however, there are few known liquids that can be chemically cycled between two thermodynamic states while withstanding extended exposure to temperatures >650°C, as are typically encountered in CST tower applications.

11.2.1 Particle sintering

Within the constraints dictated by the needs of the power block and by what constitutes an efficient receiver design, a TCES system must again be constraint by the needs of the chemical reaction, including adequate mass transport (mixing) and thermal energy transfer. For those TCES systems that utilize particulates as the reacting media sintering can have significant and detrimental impact on these factors. Sintering is the process by which small particles combine into larger particles. The resultant particle typically has a larger mean diameter. This has important consequences on both the molecular, microscopic and macroscopic scales.

On the molecular scale, larger particle sizes correspond to an increased path length over which reactants/products must diffuse to access/egress the particle core. Consequently, there may be a decrease in overall system reaction rates and, since residence times in the receiver may not be indefinite, there is a decrease in thermochemical storage capacity.

On the microscopic scale sintering may result in changes to the average heat transfer coefficient for particulates moving through the solar receiver/reactor. As was discussed earlier, changes in heat transfer can result in localized hot spots and failure of materials used to define the containment boundary of the TCES. On the macroscopic scale sintering can alter the flowability of the particles, thereby hampering bulk transport of material through the TCES system.

Efforts to overcome sintering include choice of materials whose melting temperature is far above the operational temperature. Other strategies to overcome sintering include the coating of the reactive chemical with a thin layer of porous yet structurally rigid shell material. Yet other strategies include the loading of the reactive thermochemical substance onto a high surface area, inert, high-melting substrate. This strategy seeks to minimize diffusion path lengths by having a high surface area to volume ratio for the chemically reactive species; however, it also effectively dilutes the thermochemical payload giving back a significant percentage of the enhanced energy density argued as an advantage for TCES systems in Fig. 11.3.

11.2.2 Catalyst poisoning

Some TCES systems require a catalyst to affect dissociation or recombination of the chemically reactive species. In later sections we use Eq. (11.i) to describe the endothermic dissociation of ammonia into hydrogen and nitrogen as just one example. In these systems the reactive species is typically flowed through a packed bed of catalyst that is loaded onto high surface area particles. Alternatively, the walls of high surface area microchannel heat exchangers may be internally coated with catalyst to achieve a similar result. Regardless of the specific approach for introducing the catalyst into the TCES system, the efficacy of the catalyst invariably changes with time. The nature of this change is itself variable, often most strongly depending upon degree of use, time at temperature, and to a lesser extent number of thermal cycles. The field of heterogeneous catalysis is vast and diverse with many studies detailing the specific causes for change in behaviors. Suffice it to say that high temperatures often contribute to sintering (reduction in exposed catalyst), undesirable thermodynamic sinks (chemical side reactions), and ultimately a decrease in catalyst effectiveness via decay in turnover counts (lower chemical yields).

11.2.3 Side reactions

As was touched on in the previous section there can exists certain chemical reactions that are not the desired outcome of the TCES process. These undesirable reactions, referred to as side reactions, may occur with or without the presence of a catalyst. They are expressed as consequences of chemical impurities, imperfect catalysis, competing chemical equilibria, and as changes in the condition of the TCES system itself (corrosion). We touch on few of these briefly here.

Chemical impurities are likely to exist in practical TCES systems. This is due to the fact that high-purity chemicals are typically cost prohibitive for practical TCES. The increased costs typically originate from chemical purification processes. Each step of purification may consume time, energy, and/or materials—each adding cost to the final product. If impurities are left in the thermochemical substance they may serve as reagents for undesired side reactions. The implications of this vary. For closed-loop systems, in which the same chemical substance is cycled through the TCES system repeatedly, the impurities may be considered limiting reagents in that they are eventually consumed in generating the products of the side reaction. If the products of the side reaction are a benign thermodynamic sink, then the side reaction will halt. These TCES systems may thereafter expect an unchanging content of the TCES payload to exist as unproductive side products. For open TCES systems in which the chemical substance is periodically replenished to achieve balance with chemicals leaving the system, chemical impurities may continue to build up with time of operation. In extreme cases the continued buildup of impurities may carry consequences that ultimately render the TCES system inoperable.

Even pure chemical species may be prone to generating undesirable or unanticipated side products if there exists a competing chemical equilibria unbeknownst to the system designers. These equilibria may be achieved on timescales significantly longer than the average daily cycles of a TCES system. Consequently, the impact of their expression may only be realized several month or years after CST TCES commissioning.

11.3 Power plant and chemical plant

As mentioned earlier, one of the challenges with running a CST TCES system is that it must be treated as both a solar thermal plant and a chemical plant. Taken as individual operations there exists knowledge on how to operate each; however, there is far less information available on how to integrate and successfully operate their combination. Here we discuss how some of the chemical plant aspects may exasperate the challenges already encountered in thermal plant operations. These challenges include increased corrosion, high-temperature containment material stability, matching thermochemical input/output with thermal engine requirements, and unit operation and maintenance. First we touch on corrosion.

11.3.1 Corrosion

Corrosion of structural alloys in current CST plants is typically mitigated by the fact that the act of corrosion forms an adherent, dense metal oxide film. This film then serves as a corrosion inhibition layer by greatly limiting further transport of oxygen to the native alloy. Ultimately this results in a significant decrease in annualized corrosion rates. The complication in CST TCES systems is made apparent by realizing that many of the same chemical processes that form the dense oxide layer are also occurring for the TCES substance. For a TCES system oxidation may constitute the addition of oxygen atoms to a central element while reduction would then constitute the removal of oxygen atoms from that element. These oxygen transport processes are similar and in some ways competitive with those same processes that occur in the formation of the protective metal oxide scale on the structural alloy. One must carefully consider this competition for oxygen via evaluation of both thermodynamic and kinetic scenarios. For those TCES systems whose structural alloys show an enduring, timely preference for formation of the protective metal oxide film on the structural alloy one must further consider the possibility that this film could erode during transport of the TCES media. This risk becomes especially pronounced if the TCES material is comprised of moving solid particles.

The challenges of corrosion extend beyond the simple occurrence and depletion of the oxide film. For those TCES systems that use carbonates or sulfur species there exists the possibility that carbide or sulfides may form inside the structural alloy, respectively. There may be significant mechanical consequences that result from these phenomena, and they are discussed in the high-temperature containment stability section.

11.3.2 High-temperature containment stability

Herein high-temperature containment stability is defined to be a mechanical concern—one that considers a material's ability to withstand exposure to stresses, loads, and high temperatures. The concept of evaluating materials in this context is well established in many professions. Creep, fatigue, and embrittlement are typical terms used to describe containment stability phenomena.

Creep, which is the slow movement of an alloy to a condition of permanent deformation under applied stress, is exasperated when materials are subjected to high temperatures for extended periods. The occurrence of creep, then, is a phenomenon that is not unique to TCES systems. Sensible, latent, and TCES systems may require that their materials of containment be exposed to stress at high temperatures for extended periods. TCES offers an advantage here in that it is possible to cool the thermochemical material prior to placing it in storage. Since creep is strongly dependent upon temperature its occurrence in TCES storage can be greatly mitigated.

Fatigue, which is the weakening of an alloy due to repeated cycling of loads, is again a phenomenon that is strongly dependent upon temperature and one that is not unique to TCES. One of the most pertinent loads in all CST systems is pressure, and pressurization may occur in latent, sensible, and thermochemical storage. Latent systems may experience cyclic loads via the repeated expansion and contraction of the storage material upon melting and cooling. This load maybe mitigated by appropriate sizing of the storage vessel. Sensible energy storage systems may experience cyclic loads due to the repeated filling and draining of the storage tank with liquid storage fluid. These hydrostatic pressures are often low and may be mitigated by carefully considering the storage tank height. TCES systems that store particles are subject to loads similar to those experienced by sensible systems. TCES systems that attempt to store gases, on the other hand, may be subject to some of the highest loads of the three TES classes. Typical gases are CO_2, O_2, H_2, and NH_3. Of these only NH_3 may be consider relatively condensable at low pressures. Consequently, if gases are not cooled prior to storage these TCES containment materials will be subjected to some of the highest fatigue risks of all three storage classes. Precooling gases prior to storage may help to reduce fatigue risks, but this requires large, costly gas heat exchangers.

11.3.3 Difficulties matching optimal rate of reaction with needs of power production

To be of high energetic and exergetic efficiency TCES systems often recuperate heat from various material flows. While this recuperated heat can be used in a flexible way to preheat reagent streams and reactor beds, the largest amount of thermal energy (stored in the chemical substance) must be released according to the predefined specifications of the chosen power cycle. Power cycles use hardware that is optimized for a certain operation. Consequently, it is important that the selection, qualification, and optimization of a TCES consider the constraints of an identified power cycle explicitly. Here we generalize that consideration according to a "goodness of fit" figure of merit, temperature × power, and we track the endurance of that term over time.

Fig. 11.4 illustrates three possible scenarios for matching the thermal output from TCES with a heat engine or other such load. They are "Good" (Fig. 11.4(a)), "Acceptable" (Fig. 11.4(b)), and "Bad" (Fig. 11.4(c)). In each case there is a region deemed well matched to the specifications of the power cycle, indicated as "In Spec." This region is bound above and below by operational outputs that are not well matched to the specifications of the power cycle, deemed "Out of Spec." In Fig. 11.4(a) we see a system with small, random variations, but we observe that these variations remain "in specification" for the life of the plant. This represents an ideal match between the temperature × power term and also an enduring capacity. This TCES is well matched with its load.

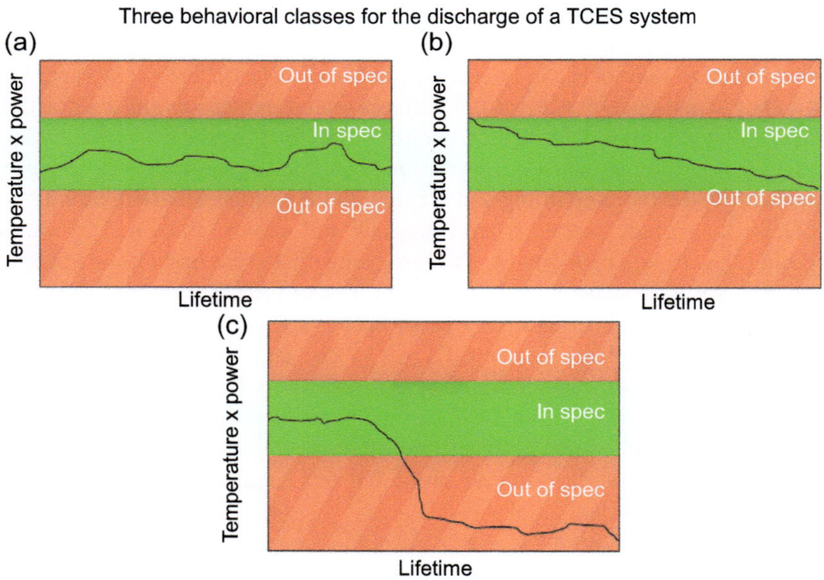

Figure 11.4 Both the power and temperature outputs of a TCES system must be well matched to the load or power cycle: (a) A "Good" temperature–power product is stable (does not trend) and is within the power cycle's specifications. (b) An "Acceptable" temperature–power product may exhibit some trending or degradation, but this may be OK so long as the rate (slope) is low and constant. (c) A "Bad" temperature–power product is one that is unpredictable, and/or exhibits rapidly changing properties.

In Fig. 11.4(b) we see an example of what may be a more realistic performance for TCES systems. In this case the temperature–power product is again exhibiting relatively small, random fluctuations along any increment; but, more importantly, there is a global trend to a lower temperature and/or power over time. This may be interpreted as a degradation in capacity. However, since the output from the TCES never goes out of specification it is conceivable that operational adjustments may be made to this system to adapt to the predictable change in TCES performance. The TCES depicted in Fig. 11.4(b) may be suffering from one or more of the limitations previously described in this chapter. For example, if this TCES system were an open system then the degradation in total power may be a result of impurities in the chemical feedstock diluting the payload over time. There are then, in any given moment in time, less chemical species capable of undergoing the discharge reaction and so a decrease in power output is observed.

The preceding discussion indicates that close attention must be paid to defining the root cause of change in the temperature–power term. This is especially true since one phenomena may contribute to both the temperature and power components. For example, consider the case in which TCES particles are sintering together thereby limiting full reaction via mass transport constraints. This impacts the rate of the chemical reaction which degrades the power output of the TCES. There may well be a second consequence, and this on the heat transfer side. In flowing particle systems larger particles, such as those arrived at by sintering, may be less effective at heat transfer. Consequently, the average temperature acquired as an output of the TCES may be less. This last example presents an instance were both temperature and power degrade simultaneously. Despite all of these practical challenges it is worth noting that TCES systems that experience degradation of the sort portrayed in Fig. 11.4(b) may still be viable if the degradation is predictable and of relatively low rate (or if the system is capable of regeneration).

The third behavioral class of TCES output is shown in Fig. 11.4(c). This plot depicts a material that is not suitable for TCES. Note that the temperature–power term is changing in an unpredictable fashion and that when it does so it occurs at a relatively rapid rate. The value of this system is marginalized by the fact that operators will find it difficult to exploit the system in a way that is optimized with a relatively constraining set of demands of the heat engine.

11.3.4 Lack of history of operational systems

TCES systems combined with CST systems also suffer challenges in commercialization simply by the fact that there is a lack of operational history for these systems. While there has been some successful demonstration of TCES and CST for fuel production, these systems do not utilize the same reaction materials as TCES systems for heat storage, and they also are not constrained by the need to produce heat at a rate sufficient to integrate with a heat engine. Additionally, the limited previous experiments of TCES heat systems with CST have been of a size of less than 1 MW. This leaves uncertainty around the ability of these systems to scale to the size of a typical CST system (50–300 MW). While these challenges present an opportunity for continued development of TCES heat storage systems, it also presents a challenge in raising the necessary capital to demonstrate these TCES systems at a commercially meaningful scale.

11.4 Le Châtelier's principle and thermochemical energy storage

Le Châtelier's principle pertains to the ability of a dynamic chemical equilibrium to spontaneously shift in a rebalancing act. This rebalancing occurs in response to changes in the composition and/or the environs of the chemically reacting system. The changes may be concentration, energy, pressure, heat, and so on. Since TCES utilizes endothermic and exothermic chemical reactions to capture and store heat, we explicitly present Le Châtelier's principle in this context (Eq. 11.i):

$$\mathrm{A} + \mathrm{B} + \mathrm{heat} \Leftrightarrow \mathrm{C} + \mathrm{D} \tag{11.i}$$

If we supply heat to the left side of Eq. (11.i) reactants A and B will chemically transform so as to create more products C and D thereby rebalancing the equilibrium. Conversely, if one extracts heat from the left side of Eq. (11.i) then products C and D will revert to reactants A and B. This dynamic process of using heat energy to cycle matter through two or more thermodynamic states (A + B) vs. (C + D) is a core, enabling concept of TCES.

It is important to note that Le Châtelier's principle is a principle of chemical thermodynamics. It does not speak to the kinetic accessibility of the reaction that must occur to shift the equilibrium. In other words, it is conceivable that a dynamic chemical equilibrium may exist as described in Eq. (11.i) but that the usefulness of that reaction in the TCES application will be limited by how quickly the reaction responds to the addition or extraction of heat. To minimize this limitation, one must be mindful of the terms that influence the kinetics of a chemical reaction. These are concentration, temperature, proximity (mass transport), catalysis, as well as some other concepts covered in basic chemistry texts. These concepts should be reviewed before proceeding to the next section which describes several specific chemical reactions for CST TCES.

11.4.1 Metal oxides

Metal oxides represent a broad class of chemical substances. These materials are typically comprised of one or more transition metal atoms bonded to one or more oxygen atoms. When the metal oxides are comprised of more than one metal atom, these atoms may be the same or different metals. A generic example of a metal oxide TCES reaction is presented in Eq. (11.ii):

$$\mathrm{MO_x} + \mathrm{heat} \Leftrightarrow \mathrm{MO_{(x-2y)}} + \mathrm{YO_2} \tag{11.ii}$$

This chemical reaction is consistent with the example presented for Le Châtelier's principle in Eq. (11.i), but we see that only a single reactant is required. MO_x undergoes a spontaneous thermal reduction when sufficient heat is applied to drive the system to the correct temperature. The thermal reduction of MO_x releases oxygen gas (O_2) and produces some reduced form of the metal, $MO_{(x-2y)}$. This reduction may occur for values of y up to and including $2y = x$. When $2y = x$ the metal oxide is fully reduced to the elemental metal. While theoretically possible, the reduction of metal oxides to elemental forms of the constituent metals is fairly rare in TCES applications.

In TCES applications metal oxides typically exist as particulates. Use of these particulates are subject to all the aforementioned challenges, such as sintering, heat transfer, and erosion of the protective oxide scales that inhibit corrosion of the containment materials. Various approaches have been taken to mitigate the detrimental impact of these challenges. For example, attempts to limit sintering of manganese−cobalt oxides have exploited a rotary kiln reactor such that the particles are constantly in motion [12]. In theory this constant motion should limit the enduring proximity of one particle to any singular neighbor particle. The lack of proximity limits the ability to sinter. A second approach to limit sintering has involved the inclusion of inert diluents such as silica (SiO_2), alumina (Al_2O_3), and titania (TiO_2). Again, these inert materials serve to prevent an enduring and close proximity of two metal oxide particles thereby discouraging sintering. It is worth noting that the addition of inert diluents decreases the energy density of the TCES system—a drawback previously mentioned in this chapter.

Despite these efforts some sintering in the manganese−cobalt oxide systems still seems to occur. The consequence of sintering is manifest as a correlation between the decay in total capacity and TCES cycle count. The sintering appears to lock away some of the TCES payload into a kinetically inaccessible regime. As previously discussed, this may be a consequence of the longer path lengths over which oxygen must diffuse to allow the reaction to proceed. Regrinding the particulates seems to restore some of the lost capacity. Presumably the higher surface area delivers an improvement of reaction rates. As was described in preceding sections the sintering of particulates can dramatically impact the ability of these materials to transfer heat.

11.4.2 Nonmetal oxides

Nonmetal oxides represent to a broad class of chemical compounds. They are comprised of main group elements that are bound to one or more oxygen atoms. Examples include boron (B), phosphorus (P), and sulfur (S) oxides, among others. Eqs. (11.iii−11.v) describe a multistep thermochemical process for the interconversion of sulfur as has been explored for CST TCES applications:

$$S^0 + O_2 \Leftrightarrow SO_2 + \text{heat} \tag{11.iii}$$

$$SO_2 + \frac{1}{2}O_2 \Leftrightarrow SO_3 + \text{heat} \tag{11.iv}$$

$$SO_3 + H_2O \Leftrightarrow H_2SO_4 + \text{heat} \quad (11.v)$$

The first step in the sulfur TCES cycle is the combustion of elemental sulfur to form SO_2 (Eq. 11.iii). This is an extremely exothermic process and the heat is used to boil water and drive a steam turbine. During the combustion process gaseous SO_2 is not typically isolated; rather, it is oxidized again to form gaseous SO_3 (Eq. 11.iv). Sulfur trioxide, SO_3, is then combined with water to make liquid sulfuric acid—one of the most important industrial chemicals on planet Earth (Eq. 11.v). The process of converting S^0 into H_2SO_4 is well understood and occurs in high yield. Reversing this process using CST is much more difficult outcome to efficiently realize and is discussed in detail below.

The first step in converting H_2SO_4 back to S^0 is to evaporate and disassociate H_2SO_4. As can be imagined CST systems work well for providing the heat input required for evaporation. Once vaporized further thermal input causes the water molecule to disassociate from SO_3. Separations and/or unit operations become important concerns since the water must be removed or the SO_3 must be quickly reacted; else, the two spontaneously recombine into H_2SO_4 upon cooling. All the while corrosion can be significant during this step since the system involves acid at high temperatures.

The second step in converting H_2SO_4 back to S^0 is to chemically split off one of the oxygen atoms that were added during the combustion process. This approach often utilizes thermochemical reactors possessing a transition metal oxide catalyst. These catalysts are typically loaded onto high surface area supports so that as the SO_3 gas flows over them the chemical reaction has sufficient opportunity to occur. As mentioned earlier, these catalysts degrade over time. One route of degradation is the generation of sulfate ion $\left(SO_4{}^{2-}\right)$ as an undesired side product. Sulfate is a thermodynamic sink, which upon excessive buildup, renders the system ineffective. The catalyst system may also suffer from sintering of the catalyst particles reducing their effectiveness. The presence of oxygen gas as a product of successfully cracking SO_3 into SO_2 also leads to corrosion and issues with stability of the materials of construction. Exasperating the corrosion problem is the fact that the thermodynamics guides us to higher reaction temperatures in order to achieve higher chemical yields. Cracking SO_3 into SO_2 represents one of the principle challenges that must be solved if sulfur is to be used in CST TCES.

The third and final step in converting H_2SO_4 back into S^0 is to convert SO_2 to S^0. This typically occurs in aqueous solution through a disproportionation reaction involving a halide catalyst, such as iodide (I^-). The products of the disproportionation reaction are S^0 and SO_3. The reaction proceeds at relatively low temperatures ($\sim 100°C$) and moderate pressures of SO_2 but provides only low to moderate chemical yields. The elemental sulfur thus produced occurs in the molten state. Fortunately, the molten sulfur is of higher density than the aqueous solution and settles out to the bottom of the reactor where it may be drained off. Subsequent processing is required to remove the iodide that is entrained. The sulfur may then be dried and stored as a pile on the ground until the combustion process starts the cycle over again.

11.4.3 Carbonation

Carbonation reactions are a special subset of metal oxide chemistries that involve the addition or removal of CO_2 from a metal oxide center. The most typical metal carbonates used in CST TCES is calcium carbonate, $CaCO_3$ (limestone). Developing knowledge on the manipulation of this substance has benefited from the fact that it is used in cement and more recently in efforts to sequester CO_2 from the emissions of fossil fuel power plants. The generic thermochemical reaction for metal carbonates in given in Eq. (11.vi).

$$MO + CO_2 \Leftrightarrow MCO_3 + \text{heat} \qquad (11.vi)$$

Many of the same aforementioned challenges exist for this system, and therefore they are not covered in detail here. Suffice it to say that sintering, achieving effective heat transfer, and storing CO_2 gas (which is a relatively noncompressible gas that requires costly high pressure vessels to contain) all must be overcome.

11.4.4 Synthesis reactions

If the synthesis of sulfuring acid is considered the most important chemical reaction in the industrialized world, then the synthesis of ammonia may be considered the second most important. The production of ammonia (NH_3) from nitrogen (N_2) and hydrogen (H_2) gases occurs according to the Haber–Bosh process (Eq. 11.vii).

$$N_2 + 3H_2 \Leftrightarrow 2NH_3 + \text{heat} \qquad (11.vii)$$

This is a high-pressure, high-temperature chemical reaction that requires fine particulates of modified iron oxide as a catalyst. N_2 and H_2 are passed over the catalyst and NH_3 is subsequently formed. Ammonia is an important feedstock for producing fertilizers, among other things. As with all chemical equilibria the reaction does not proceed to 100% completion. Most of the ammonia thus produced is collected by condensation and the incompressible, unreacted gases are fed back into the reactor along with new feedstock. CST TCES systems that attempt to use of the ammonia synthesis reaction may benefit from nearly 100 years of infrastructure development. Still, there are significant hurdles that must be overcome.

While the reaction of nitrogen and hydrogen to form ammonia (and the reverse process) both benefit from the ease by which gases may be pumped through the TCES reactor, the low density of these materials is a significant drawback. Heat transfer can become challenging and the size of heat exchangers to transfer energy to or from a dense phase to a gas phase can be cost prohibitive. For the ammonia synthesis reaction to proceed at a meaningful rate a catalyst must be used. This catalyst deactivates over time and must be regenerated or replaced. This is complicated by the fact that the catalyst is often loaded onto the inner surfaces of the solar receiver.

Finally, the enthalpy of reaction for the ammonia synthesis reaction is relatively small when compared with other TCES systems. At −92.4 kJ/mol the ammonia synthesis reaction delivers only about half that of some metal hydrides, the materials that are discussed in the following section.

11.4.5 Metal hydrides

Metal hydrides encompass a broad group of chemical substances. They are typically comprised of one or more transition metal and one or more hydrogen atoms or hydrogen molecules. Due to their high enthalpies of reaction and relatively high densities, metal hydrides may possess high volumetric energy densities. A simplified representation of the chemical reaction that occurs for these systems is shown in Eq. (11.iii).

$$M^0 + H_2 \Leftrightarrow MH_2 + \text{heat} \quad (11.\text{viii})$$

In previous sections we discussed how impurities can detrimentally impact the stability of TCES systems by resulting in side reactions that reduce overall TCES capacity over time. Metal hydrides typically suffer from such side reactions when water and/or oxygen enter the system. These impurities may react with the metal and/or with the metal hydride to form metal oxides, metal hydroxides, water, and various adducts and combinations thereof. These materials are thermodynamic sinks. To limit side reactions metal hydrides are typically operated as a closed or looped system. Still, leaks and other (sometimes intentional) compromises of the system boundary are sure to occur.

Even when closed-loop systems are operated with pure hydrogen the metal hydrides often suffer from a degradation in total capacity upon extended cycling. There have been reports in the literature that would seem to suggest stability over a few hundred cycles; however, extended cycling into the several hundreds or even thousands of counts typically reveals a meaningful degradation in capacity. One must keep in mind that seemingly small changes in TCES cycle capacity become significant in the context of CST applications. This becomes evident in considering that current financial models for CST plants assume that the facility will operate for 30 years. Assuming one TCES cycle per day for 365 days a year we arrive at nearly 11,000 cycles over the life of the plant. This means that an unabated degradation in capacity as small as 0.005% per cycle translates to a total capacity degradation of 55% over the life of the plant. Such a loss in capacity can quickly make a metal hydride (or any TCES) seem uneconomical.

Regarding cyclic stability and whether or not the degradation is abated over time, it is worth noting that one must carefully consider the mechanism by which the capacity is degrading. If capacity degradation is caused by oxygen and/or water contamination during the initial charging of the system, then we may have a limiting reagent scenario in which capacity loss ceases once these impurities are consumed. If, on the other hand, we have separation of the functional metal hydride into two or more less-operational substances and/or phases, then this is a much more fundamental failure of the material that would require further materials development research to overcome.

Still, metal hydrides show some of the greatest promise for use as TCES systems. Materials development has benefited from interest in metal hydrides from other industries such as automotive, consumer electronics, and defense. Metal hydrides have favorable thermodynamics and fast kinetics. The major challenges remaining include cyclic stability/capacity, cost of the hydride material, and identification of containment vessels of acceptable cost and stability. Embrittlement is known to occur for structural alloys exposed to hydrogen over extended periods of time. This can cause loss of mechanical properties and adds to the risk of these technologies.

11.5 Conclusions

CST TCES holds significant promise as a means to store thermal energy from the sun. The value proposition for TCES is perhaps the greatest when high energy densities and isothermal energy outputs are required. Advanced power cycles such as the supercritical CO_2 cycle stand to perhaps benefit the most from TCES development and deployment. To date there are no instances of TCES deployed at greater than 1 MW. Significant operational challenges need to be addressed in integrating what amounts to a thermal power plant and a chemical processing plant. This integration requires careful treatment of the chemical thermodynamics, kinetics, and materials challenges so that the output of the TCES may be well matched with the load it is meant to drive.

Both the first and second laws of thermodynamics must be considered in designing CST TCES systems. The first law determines how much energy we may hope to convert to work, while the second law and the concept of exergetic efficiency determines how much useful work we may actually realize. Our ability to do useful work depends not only on the design of our system but also on the chemistries selected, as described in Eqs. (11.3−11.7).

TCES systems are capable of delivering heat isothermally, and we may leverage Le Châtelier's principle to operate the system like a heat pump and deliver heat at a prescribed temperature. TCES systems present the option of seasonal storage. Still, pursuit of these significant benefits requires that we explicitly consider the challenge of interoptimizing a solar thermal plant and a chemical plant. Specifically, we must address heat transfer, corrosion, materials strength, and cyclic stability, as well as safety, permitting, and other regulatory concerns.

This chapter has discussed how sintering, catalyst poisoning, and side reactions may all detract from system performance. We have shown how each of these factors manifest in specific chemistries by discussing metal oxides, nonmetal oxides, carbonation and synthesis reactions, as well as metal hydrides. The results are summarized in Table 11.1. As of 2016 no flawless TCES system has been identified; however, the substantial advantages offered by this technology will ensure that research will continue in the years to come.

Table 11.1 Summary of strengths, weaknesses, and future research for the TCES classes

TCES class	Strengths	Weaknesses	Future research
Metal oxides	High temperatures, moderate cost, simple chemistry	Cyclic stability[1], slow kinetics[1], solid-state heat transfer difficult	Methods to prevent sintering, improve heat transfer
Nonmetal oxides	High temperatures, low cost	Low chemical yields, requires catalyst, chemistry can be complex, corrosion	New and improved catalysts, corrosion resistance
Carbonation	Moderate to high temperatures, low cost	Cyclic stability, slow kinetics, solid-state heat transfer difficult, requires storage of gas	Methods to prevent sintering, improve heat transfer
Synthesis	Moderate temperatures, low cost, chemistry can be well studied	Low enthalpy, requires storage of gases, stability of containment vessels, requires catalyst	New and improved catalysts, methods to prevent sintering
Metal hydrides	Moderate temperatures, high reaction enthalpy, high energy density, fast kinetics	Costly, low cyclic stability, chemistry impacts containment vessel, solid-state reaction makes heat transfer difficult	Methods to prevent sintering, improve heat transfer, new materials to reduce cost

[1]Excludes metal oxide perovskites.

References

[1] Xu C, Wang Z, Li X, Sun F. Energy and exergy analysis of solar power tower plants. Applied Thermal Engineering 2011;31:3904–13.

[2] Abedim AH, Rosen MA. Closed and open thermochemical energy storage: Energy- and exergy-based comparisons. Energy 2012;41:83–92.

[3] Lovegrove K, Luzzi A, McCann M, Freitag O. Exergy analysis of ammonia-based solar thermochemical power systems. Solar Energy 1999;66:103–15.

[4] Kreetz H, Lovegrove K. Exergy analysis of an ammonia synthesis reactor in a solar thermochemical power system. Solar Energy 2002;73:187–94.

[5] Koca A, Oztop HF, Koyun T, Varol Y. Energy and exergy analysis of a latent heat storage system with phase change material for a solar collector. Renewable Energy 2008;33: 567–74.

[6] Som SK, Datta A. Thermodynamic irreversibilities and exergy balance in combustion processes. Progress in Energy and Combustion Science 2008;34:351–76.

[7] Abam FI, Ugot IU, Igbong DI. Effect of operating variables on exergetic efficiency of an active gas turbine power plant. Journal of Emerging Trends in Engineering and Applied Sciences 2012;3:131–6.
[8] Incropera FP, DeWitt DP, Bergman TL, Lavine AS. Introduction to heat transfer. John Wiley and Sons, Inc.; 2007.
[9] Singh H, Saini RP, Saini JS. A review on packed bed solar energy storage systems. Renewable & Sustainable Energy Reviews 2010;14:1059–69.
[10] Bi HT, Ellis N, Abba IA, Grace JR. A state-of-the-art review of gas-solid turbulent fluidization. Chemical Engineering Science 2000;55:4789–825.
[11] Morris AB, Pannala S, Ma Z, Hrenya CM. A conductive heat transfer model for particle flows over immersed surfaces. International Journal of Heat and Mass Transfer 2015;89: 1277–89.
[12] Tescari S, Neises M, de Oliveira L, Roeb M, Sattler C, Neveu P. Thermal model for the optimization of a solar rotary kiln to be used as high temperature thermochemical reactor. Solar Energy 2013;95:279–89.

Thermal energy storage concepts for direct steam generation (DSG) solar plants

12

L. Valenzuela
CIEMAT, Plataforma Solar de Almería, Tabernas (Almería), Spain

Nomenclature

Acronyms	
CRS	Central receiver system
CST	Concentrating solar thermal
DNI	Direct normal irradiance
DSG	Direct steam generation
EHX	External heat exchanger
HP	High pressure
HTF	Heat transfer fluid
IHX	Integrated heat exchanger
LCOE	Levelized cost of electricity
LFR	Linear Fresnel reflector
LTES	Latent heat thermal energy storage
LP	Low pressure
PCM	Phase change material
PTC	Parabolic trough collector
STES	Sensible heat thermal energy storage
SHX	Screw heat exchanger
TES	Thermal energy storage

Advances in Concentrating Solar Thermal Research and Technology
http://dx.doi.org/10.1016/B978-0-08-100516-3.00012-5

12.1 Introduction

Direct steam generation (DSG) solar power plants are currently one of the suitable and best options to foster the commercial development of concentrating solar power technologies, including parabolic troughs, linear Fresnel collectors, and solar tower systems. However, one of the existing handicaps for the better grid integration and dispatchability of power plants using DSG technology is the availability of adequate, reliable, and cost-effective thermal energy storage (TES) systems.

Clear benefits of using TES in concentrating solar thermal (CST) power plants are:

- Delivery of thermal energy can be extended to periods when solar radiation is not available.
- Buffer storage is available for transient weather conditions.
- Annual plant capacity factor increases.

This chapter describes what solutions are commercially available or under investigation for integrating a TES system with a DSG solar field and the power cycle and assures an effective operation of the whole system. Section 12.2 reviews the state of the art of existing commercial CST power plants using DSG technology. Section 12.3 describes some basic thermodynamic considerations to have in mind when speaking of TES for DSG solar fields, relevant storage materials, and some other technical aspects to take into account for TES systems. Finally, Section 12.4 focuses on aspects related to the integration of sensible or latent heat storage concepts in DSG solar plants.

12.2 Overview on direct steam generation solar plants

Commercial DSG solar plants feature examples of systems using line-focusing systems, i.e., parabolic trough collectors [14] or linear Fresnel systems [24], and point-focusing systems, in particular central receiver systems (CRSs) [26,34]. Investigation on DSG using parabolic dishes has been conducted in recent research projects [11,30], but there are still no reliable and cost-effective solutions using this concentrating solar technology for large-scale DSG systems.

Table 12.1 summarizes some relevant commercial projects in operation around the world. Steam conditions in the solar collector field or solar receiver are listed, and also the existence or not of TES in the solar power plant. Examining these data shows that a standardized storage solution for DSG solar plants cannot be adopted because, in general, operating conditions in the steam power cycle vary from one DSG solar plant to another. Even for similar technology, e.g., solar tower system (CRS), there are large differences in steam cycle conditions from one project to another.

In any case, existing commercial DSG solar plants do not have high-capacity TES systems. In general, the TES solution implemented is based on steam accumulators where surplus steam produced in the solar collector field or central solar receiver is fed into a pressurized liquid water volume (see Section 12.4.2).

Table 12.1 Main characteristics of direct steam generation (DSG) solar plants in operation [10,25]

CSP plant	Technology type	Turbine capacity (net)	Solar field/receiver inlet/outlet conditions	Thermal storage
Planta Solar 10, PS-10 (Abengoa Solar)	CRS	11 MW_e	257°C/45 bar (saturated steam)	Yes (steam accumulator; 1 h partial load)
Planta Solar 20, PS-20 (Abengoa Solar)	CRS	20 MW_e	257°C/45 bar (saturated steam)	Yes (steam accumulator; 1 h partial load)
Puerto Errado 1, PE1 (Novatec Solar)	LFR	1.4 MW_e	55 bar/270°C (saturated steam)	Yes (steam accumulator; 1 h partial load)
Puerto Errado 2, PE2 (Novatec Solar)	LFR	30 MW_e	55 bar/270°C (saturated steam)	Yes (steam accumulator; 0.5 h)
Sierra SunTower (eSolar)	CRS	5 MW_e	218°C/440°C (superheated steam; pressure n.a.)	No
Thai Solar Energy 1, TSE-1 (Solarlite GmbH)	PTC	5 MW_e	201°C/340°C (30 bar)	No
Ivanpah Solar Electric Generating System, ISEGS (BrightSource Energy)	CRS	377 MW_e	249°C/565°C (165 bar); reheat at 485°C/41 bar	No
Khi Solar One (Abengoa Solar)	CRS	250 MW_e	235 t/h 530°C/120 bar +105 t/h saturated steam/130 bar for storage	Yes (molten salts—2 h of storage capacity)

CSP, concentrating solar power; *CRS*, central receiver system; *LFR*, linear Fresnel reflector; *n.a.*, not available; *PTC*, parabolic trough collector.

12.3 Basic considerations on thermal energy storage

TES allows storage of heat to be used later. To be able to retrieve the heat after some time, the method of storage needs to be reversible. TES can be divided into physical or chemical processes. To solve the storage problem in DSG solar plants, both at commercial and research levels, the focus is mainly on physical processes that can be divided into sensible heat and latent heat.

12.3.1 Thermodynamics considerations

When heat is transferred at constant pressure to/from a pure substance, either the temperature of the substance changes with phase remaining the same or its phase changes while temperature remains constant. In the former case, when the heat transferred causes change in temperature, it is referred to as sensible heat. In the latter case, when the heat transferred causes change in phase, it is referred to as latent heat. Sensible heat transfer is apparent as it is noticed in the form of a change in temperature, whereas latent heat transfer remains latent and is noticed only by a change of phase of the substance.

For quantitative evaluation of these heat energies, we have specific heats and latent heats relating to sensible heat transfer and latent heat transfer processes, respectively.

The specific heat or heat capacity of the substance is the amount of heat required to change the temperature of a unit mass of the substance by one degree. And the sensible heat, Q, gained or lost by the substance in changing temperature from T_a to T_b is therefore

$$Q = m \cdot \int_{T_a}^{T_b} c_p \cdot dT = V \cdot \int_{T_a}^{T_b} \rho \cdot c_p \cdot dT \tag{12.1}$$

where m is the mass, c_p is its specific heat, V is the volume of substance, and ρ is its density. From Eq. (12.1) it is seen that higher the c_p and ρ of the material, more energy would be stored in a certain volume of material.

Concerning the latent heat transfer to a substance, the process of fusion is driven by the difference of free energy:

$$\Delta G = \Delta H - T_m \cdot \Delta S \tag{12.2}$$

where G is the free energy, H is enthalpy of the substance, T_m is the transition temperature, and S is the entropy. If there is equilibrium, $\Delta G = 0$, and

$$\Delta H = T_m \cdot \Delta S \tag{12.3}$$

So, for a given transition temperature T_m, the higher the entropy change, the higher is the latent heat of the phase change material (PCM). Also, for a pure substance in thermodynamic equilibrium, the properties are related by

$$T \cdot dS = dH - V \cdot dP \tag{12.4}$$

where V and P are the volume and pressure, respectively. If the pressure remains constant during the melting of the PCM,

$$T \cdot dS = dH \tag{12.5}$$

From the practical point of view, in the constant pressure process, the heat, Q, required to completely change the phase of saturated liquid of mass m into saturated vapor at constant temperature and pressure is

$$Q = m \cdot \int_{h_l}^{h_g} dh = m \cdot h_{lg} \tag{12.6}$$

where h_{lg} is the enthalpy or latent heat of vaporization of the substance. In a similar manner, we also have h_{sl} as the enthalpy or latent heat of fusion and h_{sg} as the latent heat of sublimation. At the critical point of the substance, the latent heats of vaporization, fusion, and sublimation are zero.

For sensible heat storage, apart from specific heat capacity and density, there are other parameters that affect the performance of the TES systems such as thermal conductivity, thermal diffusivity, viscosity, and vapor pressure. The rate of heat penetration into any specific storage media (e.g., into concrete) is dependent on the thermal diffusivity. This is a combination of two factors, the thermal conductivity (how well heat is conducted through the substance) divided by the heat capacity (how much heat it takes to increase the temperature of the substance). For latent heat storage, the importance of most of these parameters is also noticeable [13].

The two ways of TES, sensible and latent heat storage, are currently available or under investigation for DSG power plants. Both options present advantages and disadvantages.

Sensible heat storage presents a lower storage capacity than latent heat storage but is more dynamic, which means the system can be charged and discharged faster and more easily than current latent heat storage systems. *Latent heat storage* systems store higher heat compared with a similar volume of sensible heat storage system, but the process of charging and discharging is slower than sensible heat system. From a thermodynamic point of view, latent heat storage assures maximum heat energy is stored and released at a certain minimum temperature. The energy efficiency is closer to the Carnot parameters (within the same temperature range) compared to sensible heat storage.

The desired characteristics of any type of thermal storage systems can be summarized as follows [1,21]:

- Compact, large storage capacity per unit mass and volume
- High storage efficiency
- Heat storage media with suitable properties in the operating temperature range
- Uniform temperature
- Capability of charge and discharge with the largest heat input/output rates but without large temperature gradients

- Ability to undergo large number of charging/discharging cycles without loss in performance and storage capacity
- Small self-discharging rate, i.e. negligible heat loss to the surroundings
- Long lifetime
- Low cost

12.3.2 Relevant materials with thermal storage capabilities

12.3.2.1 Liquid materials

Different types of media can be used as storage materials for solar thermal applications, e.g., water, organic liquids, molten salts, or liquid metals, which exhibit a wide range of properties from the heat storage point of view. A survey by Kuravi et al. [16] analyzes the current development of liquid materials that have been evaluated for using as storage media in sensible thermal energy storage (STES) systems in terms of cycle life and stability.

Nitrates-based molten salts are currently used in commercial solar power plants. Different mixtures based in the mixture of $NaNO_3$ and KNO_3 (60% $NaNO_3$ + 40%KNO_3—solar salt) or ternary mixtures containing, e.g., sodium nitrite (7% $NaNO_3$ + 53%KNO_3 + 40%$NaNO_2$—Hitec) and calcium or lithium nitrates [e.g., 7%$NaNO_3$ + 45%KNO_3 + 48% $Ca(NO_3)_2$—Hitec XL] can be used as storage materials in STES systems developed for DSG solar plants. Density of these various molten salts varies from around 1600–2000 kg/m^3 at 300°C. And their working temperature ranges from about 140°C (minimum operating temperature above freezing point) to 565°C (maximum operating temperature below the maximum admissible).

An intensive development in new formulations of molten salts is being conducted by different research groups and laboratories around the world to increase the effectiveness of these liquid materials for heat storage in CST applications.

Liquid metals, in particular, alkali metals that present minimum temperature below 100°C and maximum temperatures higher than 750°C [27], could be also an option for STES, especially for heat storage in the high-temperature range of (ultra) supercritical steam cycles, which work with steam temperatures over 565°C, where the use of existing molten salts used in commercial solar plants is not feasible.

State of the art of STES in solar power plants using liquid materials are two-tank molten salts technology, using sensible heat for temperature up to 565°C. The size of the storage tank is around 28 m^3/MWh$_{th}$ for a standard 50 MW$_e$ CST plant with parabolic troughs and thermal oil as HTF located in Spain [25].

12.3.2.2 Solid materials

Solid materials have also been used in CST applications and offer some benefits compared with liquid materials. Solid materials can be used in a wide temperature range. Density of available materials ranges between 1000 kg/m^3 and 2500 kg/m^3. Natural solid materials, such as sand or rocks, are abundant and cheap, and depending on the temperature use, it is important to focus on their thermomechanical stability. Materials such as granite or basalt can exhibit suitable properties for high-temperature energy storage [9].

Also the use of manufactured solid materials has been investigated or is under investigation. For high-temperature applications, the use of refractory bricks based on oxides (silica, magnesia, alumina, and feolite), carbonates, and their mixtures are commercially available.

For example, the use of concrete for CST applications was proven during WESPE and ITES projects [18,19,20]. Long-term stability of concrete, with a density of 2750 kg/m^3 and specific heat capacity at 350°C of around 916 J/kg K, was proven at lab scale up to 500°C and validated at pilot scale in a 20-m^3 test module of operation between 200°C and 400°C.

The use of solid materials in STES systems for DSG solar plants has been proposed for feedwater preheating and steam superheating [18,19].

12.3.2.3 Phase change materials

The use of PCMs for DSG solar plants requires state transitions at temperatures close to steam working conditions in the solar field, and the most suitable ones for TES are the solid–liquid and solid–solid transitions. The problem of solid–gas transitions is that they involve very large volume changes. Most of the research done up to now has been directed to the use of inorganic nitrate salts with solid to liquid transitions in the range from 200 to 340°C and enthalpies around 100–300 kJ/kg [39].

Thermotropic liquid crystals (TLCs) are alternative PCMs. TLCs are organic-based molecular compounds that, instead of exhibiting a single transition from solid to liquid, present a cascade of transitions involving intermediate fluid phases (called meso-phases) that keep a certain crystalline structure [7].

Although significant advances have been made in the development of TES systems using PCMs for DSG solar plants, some major issues still remain in the development of reliable and practical storage systems:

- Difficulties in obtaining an optimal balance between transition region and operating range, because of the small number of materials available in the working temperature range of DSG solar plants.
- Uncertainties concerning the long-term thermal behavior, despite testing over a number of cycles generally much below the number of cycles that can be expected during the lifetime of DSG solar plant.
- Cost-effectiveness compared to other storage materials.

The evaluation and selection of a PCM requires the consideration of several aspects, including freezing or solidification, supercooling, nucleation, thermal cycling, encapsulation, and compatibility [12].

12.3.3 Technical aspects in the design of thermal energy storage systems

Any TES should have at least the following three components: (1) a storage material in the desired operating temperature range; (2) a suitable container to store the heat; and (3) a suitable heat exchanger (HX) for an efficient heat transfer from the heat source to the storage material and from the storage material to the working fluid of the steam cycle.

Liquid materials used for sensible heat storage in existing CST solar power plants are contained in one tank (pressurized liquid water in steam accumulators) or two tanks (molten salts) [31] made of materials completely compatible with the storage media. Recently, a single-tank storage system for molten salt has been tested in a commercial solar power plant. The system uses a single tank to contain both hot and cold molten salts, which are separated by an insulation barrier that can float according to the volume of hot and cold salts inside the tank [29]. Independently of the configuration, the storage material moves from the cold zone (or tank) to the hot zone (or tank) through an external heating section (which contains the HX) during the storage charging process, and vice versa during discharging process.

In STES systems using solid materials investigated for DSG applications, the storage media are commonly stationary and the heat transfer fluid (HTF) charges and discharges the storage flowing through the system in direct contact with the storage material, or it is separated through HX walls that are inside the storage container (STES with integrated HX). Storage configurations with moving storage material, e.g., small particles such as sand, are also an option; in this case the storage material moves from/to a hot tank to/from a cold tank in a similar manner to the two-tank molten salts configuration (STES with external HX).

Due to the high cost of an LTES compared to STES, its application is of interest when the load demand is such that energy is required at a constant temperature or within a small range of temperatures, the storage size required should be limited, and of course, high energy density or high volumetric energy capacity is desired. Therefore the integration of LTES could be recommended to the steam generation and not to be used for preheating, superheating, or reheating of the steam in DSG solar plants. However, the use of single LTES systems is recently analyzed [28] for offering the complete heat storage functionality in DSG solar plants as described in Section 12.4.2. Cascaded LTES was analyzed by Michels and Pitz-Paal [23] as an alternative solution to the two-tank configuration in CST plants using thermal oil as HTF in the receiver of parabolic troughs. This approach could be also potentially analyzed for DSG solar plants.

The development of an LTES involves the consideration of two essential diverse subjects: the heat storage materials and the HX. One of the major issues using PCMs is to extend the heat transfer surface in the storage media during the charging/discharging processes. Different implementations have been investigated, such as the use of finned tube HXs embedded into the PCM or macroencapsulation of the PCM [4,9]. Also an increase of the thermal conductivity of the latent heat storage volume can be achieved by adding a material having high thermal conductivity. The development of containers where the PCM is integrated into matrices made of aluminum or graphite has been tested at pilot scale [6,18,19]. Finally, although existing solutions of storage systems with PCMs for DSG solar plants consider that the storage media are stationary, new ideas explore options where the PCM is transported through a heating section where the charging/discharging processes occur [7,37].

12.4 Integration of thermal energy storage systems in direct steam generation solar plants

In DSG solar technology, the integration of steam from the solar field in the power cycle is direct, not indirect as it occurs in commercial PTC solar power plants with thermal oil technology and CRS using molten salt receivers or volumetric/pressurized air receivers. The steam produced in the solar field can be saturated or superheated steam. State of the art of turbines for CST power plants go from single-state turbines working with dry saturated steam to multistage turbines working with superheated steam and reheating, which exhibits higher thermal-to-electric conversion efficiencies. The solar receiver can be working as economizer, evaporator, and even steam superheater, depending on how the solar field/receiver is integrated in the steam cycle.

The distribution of energy needed for feed water preheating, evaporation, and steam superheating can range between 15–35%, 55–65%, and 10–30%, respectively, depending on the operating conditions, i.e., temperature and pressure in the solar collector field or solar receiver in Refs. [17,36]. This distribution of thermal energy required along with the operating temperature and pressure is the main parameter taken into account when evaluating different options of TES configurations for DSG plants.

12.4.1 Operation of thermal energy storage in direct steam generation solar plants

The operation modes of DSG solar power plants with a TES system can be simplified to three basic operation modes. In the first mode the steam generated in the solar receiver or solar collector field directly drives the turbine. When the thermal energy collected in the solar receiver exceeds the amount of thermal energy accepted by the turbine, the extra amount of thermal energy can be transferred to a thermal storage system to be used later. Therefore in this second operation mode, the solar steam runs the turbine and charges the TES system. In the third operation mode, the solar field is not running (during nights or bad weather conditions) and the TES system is discharged to generate steam and run the turbine. In commercial CST systems using thermal oil or molten salts, there are other operation modes which combine, e.g., simultaneous operation of the turbine using the solar field and the TES system.

This section describes configurations options for TES systems in DSG solar plants considering basically their discharging for the economizer, evaporator, or steam superheater of the steam cycle. If there is steam reheating in the cycle, the integration of TES for any intermediate step could be also an option for increasing the capacity factor of the power plant. The solutions implemented for the superheating section could be also applied for reheating steam in intermediate steps of the turbine. However, the integration of the corresponding HX and how TES for reheating steam is charged/discharged can affect to the design of the whole system, including the solar field.

The integration of steam from the storage in DSG plants is not as easy as in CST plants with indirect steam generation. Different integration options have been explored recently, and in all the cases the complexity of the power plant increases, and also its complete operation and control.

12.4.2 Thermal energy storage systems based on sensible heat storage

12.4.2.1 Steam accumulators

Ruths accumulators [22] use sensible heat storage in pressurized liquid water at the boiling temperature appropriate to the pressure in the vessel. Steam is generated by flashing, i.e., lowering the pressure, of the saturated liquid during discharge [35]. Steam accumulators provide good volumetric storage capacity of the liquid water for sensible heat due to its high specific heat capacity. About 90% of the volume is filled with saturated liquid, while the remaining volume is filled by saturated steam. They are preferably arranged horizontally so as to give the largest possible surface of water for the liberation, as flash, of the stored steam (see Fig. 12.1).

Steam accumulators are commercially available since many years ago and are commonly used in the conventional process industry. They are also installed in some of the existing DSG solar plants in operation (see Table 12.1). However, in DSG solar plants the volume of the tanks is limited and the sliding pressure during TES discharging decreases the power block efficiency. They are installed to facilitate the operation of the power plant during short periods of bad weather conditions and not for extending the DSG plant operation after sunset or during long periods of cloudy conditions; e.g., the PS-10 solar tower plant in Spain uses four separate tanks with a TES capacity of about 20 MWh to operate the turbine at 50% load for 50 min [9].

Steinmann and Eck [35] proposed the use of a cascade TES system combining a steam accumulator and a second sensible heat storage system for steam superheating. For this additional TES unit candidate concepts use solid storage media such as concrete or nonpressurized liquids such as molten salt. In a DSG solar plant the thermal energy needed for superheating steam ranges from 10 to 30%, so the contribution of the additional storage system is smaller than the share of the steam accumulator.

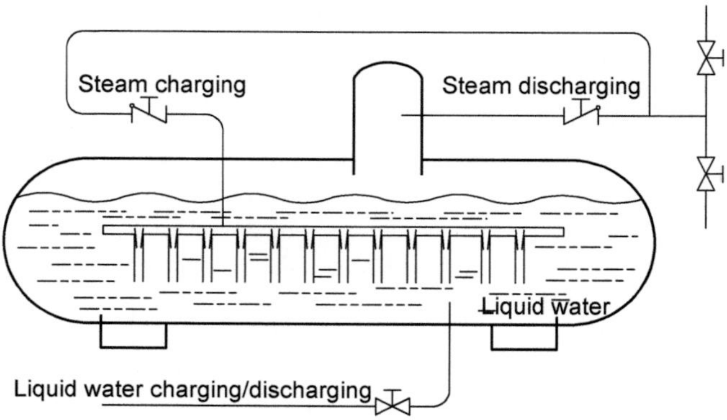

Figure 12.1 Ruths steam accumulator.

12.4.2.2 Two-tank molten salts thermal energy storage system

Koretz et al. [15] developed a system based on two-tank molten salt STES technology, which is considered a proven concept from the experience in the Spanish solar plants, but varying the configuration and charging/discharging operation modes applied in the existing solar power plants using molten salts or thermal oil as HTF. The molten salt mixture considered is the standard solar salt (60% $NaNO_3$ + 40% KNO_3). For a two-tank configuration, where the hot tank operates with an operating temperature of ~560°C and the cold tank with a temperature of ~290°C, with the improved method suggested, when charging the storage, the molten salts are heated only by de-superheating and partially condensing live steam, which means that only the sensible heat above the water-steam saturation temperature, along with a small fraction of latent heat of evaporation, would be transferred to the salts via the HX. In a typical 2-h storage system, the partially condensed steam leaving the HX during charging would contain ~30% wetness. This wet steam is then further condensed and subcooled by the boiler feedwater using a recovery HX. Considering the key system design parameters in the case study proposed by Koretz et al. [15], the 2-h discharge capacity and the 8-h charging rate were selected to achieve the largest temperature difference possible while providing an economically viable storage solution.

The main drawback of using two-tank molten salts for DSG solar plants is that steam pressure in discharge mode is limited by the linear behavior of the cooling molten salt. The live steam pressure during TES discharge operation is considerably lower than during direct production due to the pinch point losses in transferring heat from the water-steam (PCM) to the molten salt (a sensible heat medium) (see Fig. 12.2). Koretz et al. [15] reported that maximum storage capacity for DSG solar plants using two-tank molten salts technology is limited to 2–3 h with existing components technology and power block characteristics and performance.

Aga et al. [2,3] also performed a detailed analysis on the ability of using the existing two-tank molten salts TES technology to be taken as TES solution also for DSG solar

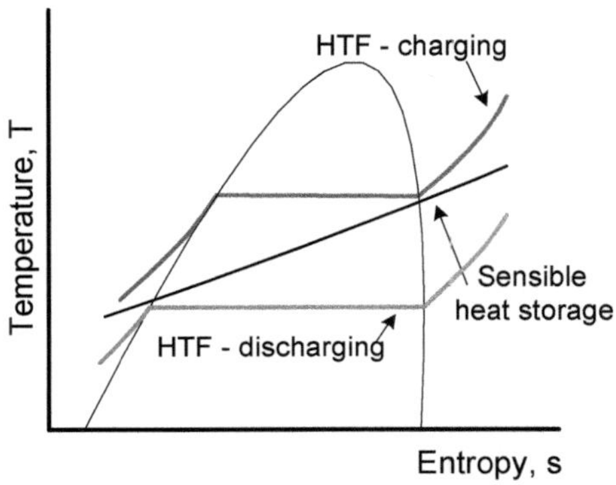

Figure 12.2 Temperature–entropy diagram for the charging and discharging processes with water/steam as heat transfer fluid (HTF) using a two-tank molten salt sensible heat thermal energy storage system.

plants. If integration of the DSG solar field and two-tank molten salts is done with the existing steam turbines installed in solar power plants, the live steam pressure during TES discharge operation is considerably lower than during direct production as it has been previously mentioned (see Fig. 12.2). An alternative solution is the use of a turbine specifically designed to achieve nominal rated power output with the reduced steam parameters during TES discharge. With TES discharging, the steam is injected into the turbine at a section downstream of the high-pressure (HP) stage. During direct production, this section located downstream the HP stage operates at a lower pressure than when TES discharging. Following this operation scheme, the mass flow through the turbine stage located downstream the injection is increased compared to direct production.

12.4.2.3 Three-tank molten salts thermal energy storage system

A solution for the case in which the market adequately values lengthier storage from solar thermal energy is the three-tank STES system [15,33], which incorporates a hot tank, a warm or intermediate tank, and a cold tank. Comparing to the two-tank configuration, this approach with three tanks would allow two different mass flows; a higher mass flow in the lower temperature range and a smaller mass flow in the higher temperature range should be defined. This way the live steam temperature and pressure reduction is not as large as in the two-tank configuration leading to a more moderate reduction of the power block efficiency during discharging (see Fig. 12.3). The three-tank system maximizes exergy utilization in the combined charge/discharge process, increasing the power block efficiency of the storage discharge cycle. Koretz et al. [15] analyzed the performance of a three-tank molten salt STES system for DSG solar tower plants working with subcritical water-steam conditions but also with supercritical conditions. Three-tank STES systems formulated for supercritical plants materially resemble that of STES for subcritical conditions, but differ in its effective charging of the storage liquid material to reach a higher weighted-average temperature.

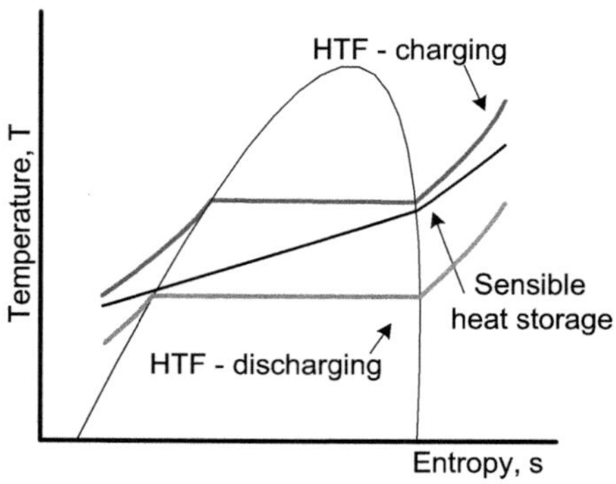

Figure 12.3 Temperature–entropy diagram for the charging and discharging processes with water/steam as heat transfer fluid (HTF) using a three-tank sensible heat thermal energy storage system with different mass flows.

This is because the latent heat associated with the phase change that occurs in the subcritical charging process does not especially charge the molten salt (storage medium), while the energy in the supercritical steam, which is almost all sensible heat, favors heat transfer to the storage liquid material with much greater efficiency.

Two-tank STES systems, with adapted operation of an optimized power block during the charging/discharging storage processes, or three-tank STES concepts, could increase efficiency to specific energy capacity ratio of DSG solar plants; the increase in investment and operational costs would be compensated by energy yield output of the power plant. The sizing of the TES tanks and integration of the various components (DSG solar field, TES system, and power block) require good expertise and knowledge to predict the annual energy output according to the local site conditions, which include direct normal irradiance and incentive schemes.

12.4.3 Thermal energy storage systems based on latent heat storage

This option numerically analyzed by Pirasaci and Goswami [28] consists in using a PCM for preheating, steam generation, and superheating to simplify the structure of the TES system in DSG plants. The effectiveness of the storage is considered as the design criterion during the modeling of the TES, which consists of *M* pieces of serially connected insulated tanks filled with the PCM (Fig. 12.4). The feedwater enters the first module of the TES and is heated, evaporates, and becomes superheated steam as it flows through the *N* number of tubes inserted in the PCM. The PCM selected for the study is NaCl + $MgCl_2$ eutectic mixture which presents a melting temperature of 549°C, latent heat of 316 kJ/kg, thermal conductivity of 0.95 W/mK, specific heat of 1000 kJ/kgK. Initially the PCM is in the liquid phase having 600°C. The operating temperature of the water/steam considered in the study is from 325°C to 600°C. Results of this study show that the length of the storage has a major impact on the effectiveness; the flow rate of the water/steam fluid is important and, depending on the design, an optimum value can be calculated with the model; the tube diameter also

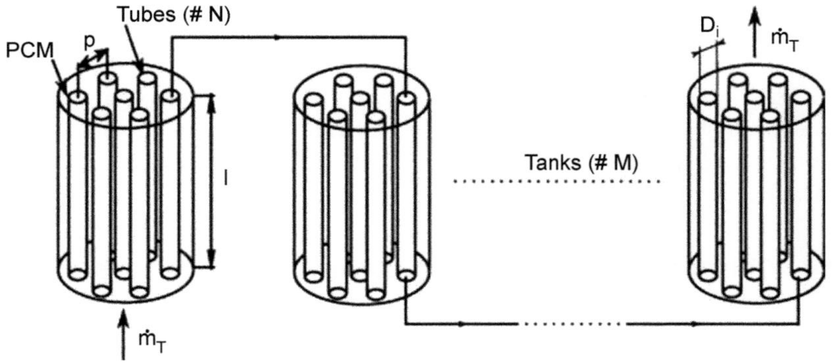

Figure 12.4 Schematic description of a phase change material (PCM) thermal energy storage system for direct steam generation plants [28].

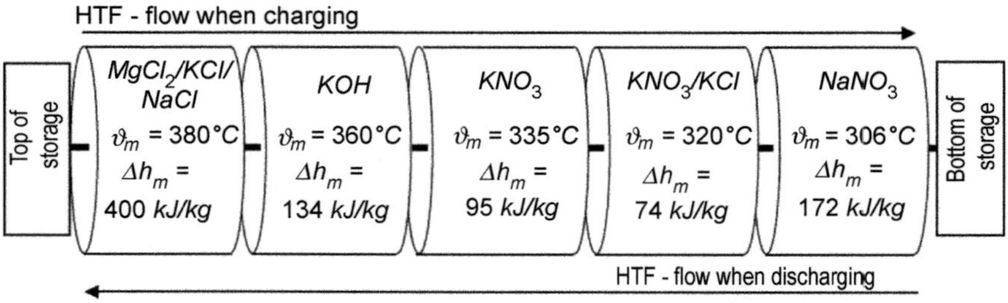

Figure 12.5 Cascaded latent heat thermal energy storage with five phase change materials [23].

has an important role on the effectiveness; and finally, the distance between the tubes does not have a significant effect on the effectiveness but plays an important role on the storage capacity. The reliability of this TES solution has not been tested yet, so further experimental research is needed to prove its suitability to real-scale DSG solar plants.

A similar approach but using cascade LTES was proposed by Michels and Pitz-Paal [23] as alternative solution for storage in solar plants with parabolic troughs and thermal oil as HTF (see Fig. 12.5). This approach also suggests the use of modular units containing the PCM. The difference is that different PCMs are used depending on the temperature range. Although the analysis was performed considering thermal oil as HTF in the storage system (single-phase HTF), an adaptation of materials and design conditions could be potentially applicable to DSG solar plants.

An approach considering the transport of the PCM is the one proposed by Bayón and Rojas [7]. This solution focuses on the research of latent heat storage material itself more than the design of the HX studied in the nextPCM project (see Section 12.4.4). The use of TLCs, which are organic-based molecular compounds that exhibit a cascade of transitions involving intermediate fluid phases, as PCMs are presented by Rojas et al. [32]. The benefits expected of using liquid crystals are to have the energy exchange in an external HX and assure a constant discharging power during discharging process, nor like the inorganic PCMs. A two-tank TES configuration is applicable with a cold tank containing the liquid crystal and a hot tank with the isotropic liquid (see Fig. 12.6). Rojas et al. [32] analyze different liquid crystals in terms of clearing temperature and enthalpy as possible candidates for DSG solar plants with working pressure in the solar receiver of 30 bar ($T_{sat} = 234°C$), 70 bar ($T_{sat} = 286°C$), and 100 bar ($T_{sat} = 311°C$). The suitability of this solution for feedwater preheating and steam superheating is not analyzed in this preliminary study. Further research is required to confirm if this LTES solution is applicable to real-scale DSG solar plants.

12.4.4 Thermal energy storage systems combining sensible and latent heat storage

Solid storage media are considered mainly for cost considerations. Compared to commercially available molten salts used in existing solar power plants, the costs of the equivalent mass of solid materials might be one order of magnitude lower or

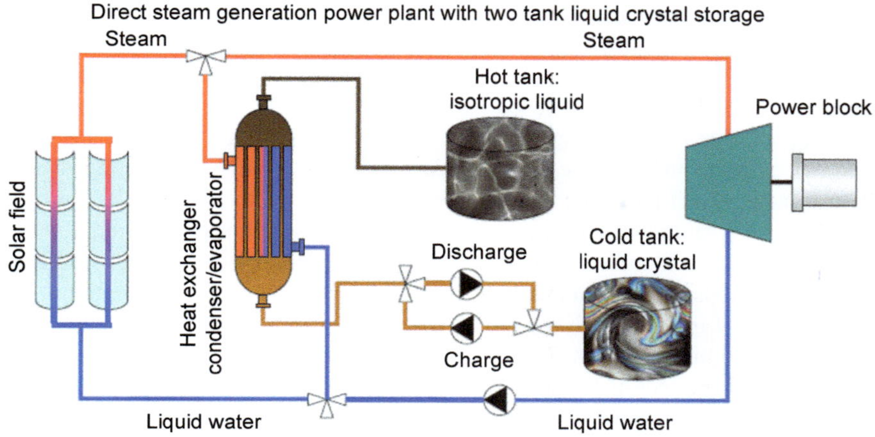

Figure 12.6 Scheme of a direct steam generation solar plant with a liquid crystal–based thermal energy storage [32].

even lower if natural materials are considered [9]. In concepts using the sensible heat of solid media, the storage material contains a tube register HX to transfer the thermal energy to or from the HTF. This solution of STES using solid materials and optimizing the tube register HX to transfer heat to the fluid was considered in the project WESPE [20,36]. A castable ceramic, which was principally composed of a binder including Al_2O_3 and iron oxide aggregates and various other materials, was developed. This solid material with sensible heat storage capabilities presents a density of 3600 kg/m^3 and its thermal conductivity is slightly higher than that of concrete, which was also experimentally studied within the same project. The variation of thermal conductivity, volumetric heat capacity, diameter of flow channels, and distance between flow channels were analyzed for optimizing the STES unit design. As a result of the

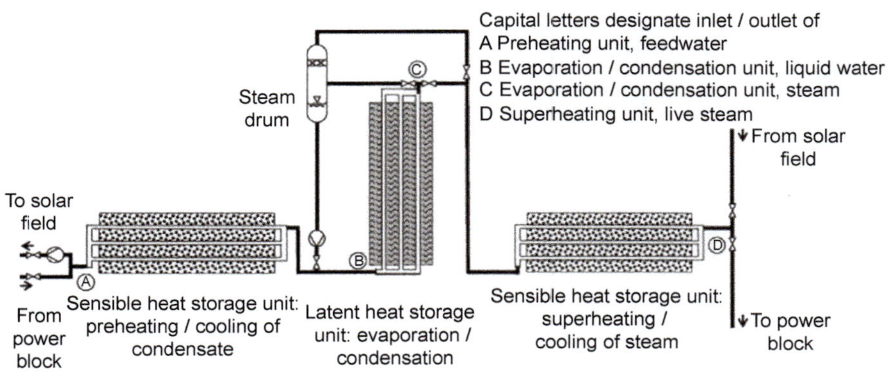

Figure 12.7 Three-part thermal energy storage system for direct steam generation solar plants combining sensible and latent heat storage [18,19].

sensitivity analysis, the distance between flow channels was important, and the choice for the strategy for charging and discharging the STES represented a key element for the optimization of the STES system. In ITES project [18,19], a tree-part TES system where PCM latent heat storage is used for evaporation, while concrete storage is used for storing sensible heat (see Fig. 12.7). A prototype with a total storage capacity of ~1 MWh, with 22 m^3 of concrete and a 8.5 m^3 PCM storage, was coupled to a DSG test facility with parabolic troughs specially connected to a conventional power plant. The testing of the system was performed at a pressure level of around 100 bar. For this pressure level, sodium nitrate ($NaNO_3$) with a melting temperature of 306°C was selected for being most suitable as PCM. A sandwich concept using aluminum fins was implemented in the LTES system to increase thermal conductivity during PCM charging/discharging processes. This solution was also investigated at pilot plant scale in the DISTOR project [6]. For the STES system, a high-temperature concrete specially developed to support up to 500°C can be applied. However, it was tested only up to 400°C [18,19].

In the DSG-Store project, Seitz et al. [33] study a complete three-part TES system that combines the use of PCM storage for evaporation of water and sensible heat storage using molten salts for pre- and superheating (see Fig. 12.8). The STES system consists of a cold tank, a hot tank, and an intermediate or warm tank. If an LTES system is used to store the latent heat of water/steam, the temperature profile in the

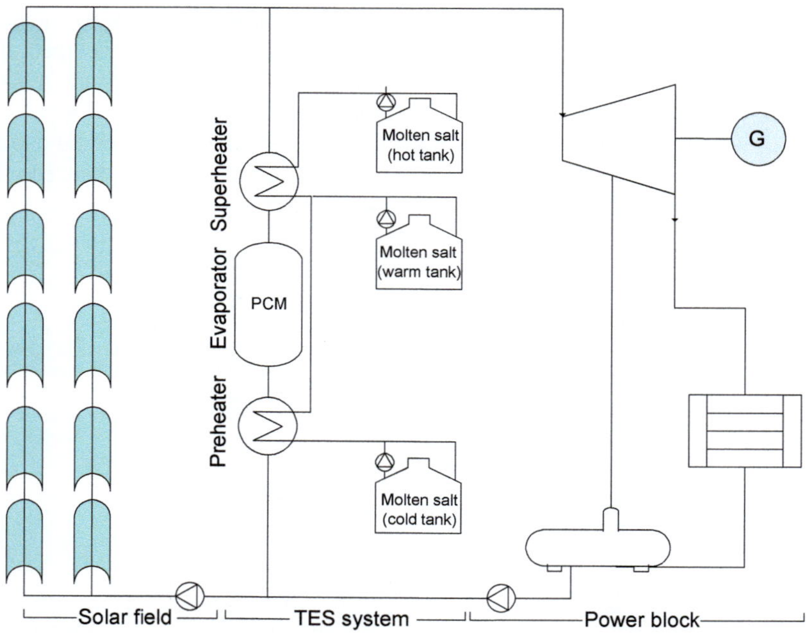

Figure 12.8 Scheme of direct steam generation solar plant layout with three-part storage which combines three-tank STES and LTES. *PCM*, phase change material; *TES*, thermal energy storage.

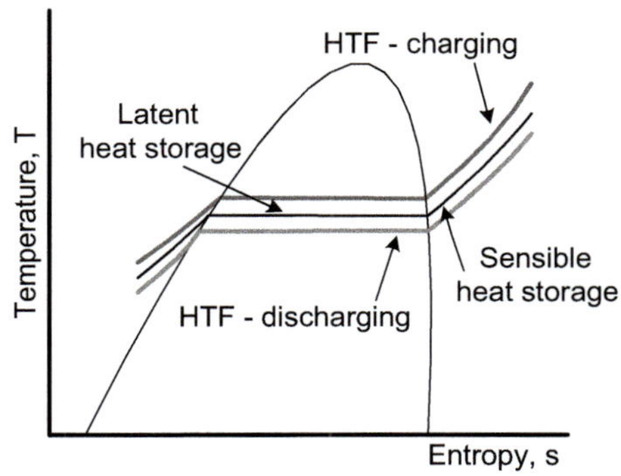

Figure 12.9 Temperature–entropy diagram for the charging and discharging processes with water/steam as heat transfer fluid (HTF) using three-tank sensible heat thermal energy storage for pre- and superheating and latent heat thermal energy storage for evaporation.

complete TES system is matched to the temperature profile of the HTF during charging and discharging (see Fig. 12.9). Compared to the two-tank STES and three-tank STES concepts, this approach leads to the highest live steam parameters and thus better power block efficiency during TES discharging at the expenses of an increased system complexity.

Birnbaum et al. [8] analyzed the interactions between the solar field, a combined STES-LTES system, and the power block in DSG plants with parabolic troughs considering different configurations. The use of LTES for evaporation of water implies that the pressure for the discharge process is lower than the pressure level required for charging. The integration of a TES system in DSG plants increases the complexity of the whole system both from the technical and control point of view if it works with the conventional steam power blocks used in commercial solar plants with PTCs (see Fig. 12.10). Power block optimization was suggested to maintain the power block efficiency in DSG plants [8]. Two configurations for the steam parameters, 110 bar/400°C and 156 bar/500°C, were analyzed based on available steam turbines manufactured by Siemens and installed in many of the CST plants in operation. An

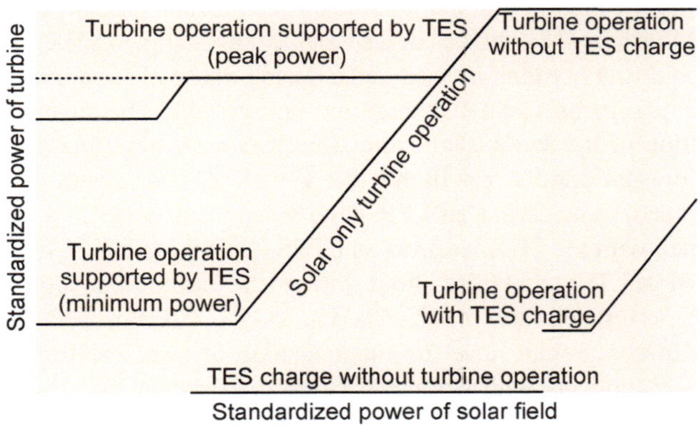

Figure 12.10 Practical operating point diagram for a direct steam generation solar plant with thermal storage and a conventional steam power block likes the one installed in commercial CST plants with parabolic troughs [8]. *TES*, thermal energy storage.

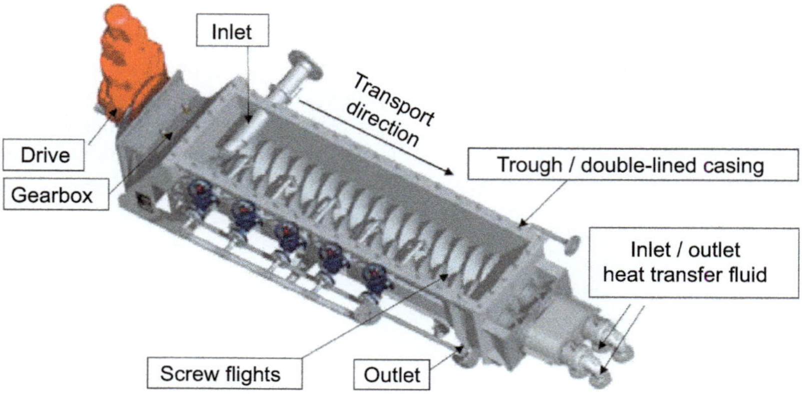

Figure 12.11 A double screw heat exchanger [37].

optimization process conducted to change (increase) the final feedwater temperature to 260°C for both configurations of steam parameters and using of a steam–steam reheater with condensation. The gross power block efficiency was maintained over 40% for the configuration 110 bar/400°C.

Seitz et al. [33] conducted preliminary analyses of different solutions using a three-tank STES system combined with LTES for evaporation comparing them with a two-tank STES system also with LTES. However, as in previous solutions detailed, a complete assessment must also consider investment costs and a detailed prediction of annual electricity production for representative sites to confirm the benefit of using a three-tank configuration.

Bachelier et al. [5] performed a systematic analysis of Fresnel CSP plants using molten salts or water-steam as HTF and considering different options of TES. As result of the study, two concepts, namely direct molten salt concept (molten salts work as HTF and storage media) and the DSG concept combined with PCM storage, offer similar efficiency to specific energy–capacity ratio. Further investigations are required again to determine which option exhibits better technoeconomical potential.

In NextPCM project [38], the use of a screw heat exchanger (SHX) to transport the PCM from a cold to a hot tank or vice versa during phase change is under investigation. This concept separates heat exchange and storage tank. The material can be transported by rotation of the screw shafts and simultaneously heated/cooled by the HTF inside the hollow shaft and screw flights (see Fig. 12.11). As result of the numerical analysis performed, considering an LFR DSG solar plant with (1) a two-tank STES system combined with the SHX working as LTES, (2) a three-tank STES system combined with an SHX LTES, and (3) without storage, the minimum levelized cost of electricity (LCOE) estimated are 0.143 €/kWh, 0.134 €/kWh, and 0.129 €/kWh, respectively. As storage can raise the market value of solar electricity, an increase in LCOE for the concepts with storage is considered acceptable. By decreasing the storage cost by 20%, the LCOE can be decreased by 5%. To achieve this reduction, the SHX concept should be further optimized, e.g., increasing the thermal efficiency of the SHX by optimizing charging/discharging processes parameters.

12.5 Conclusions

The main limitation associated with DSG solar plants is the current unavailability of commercially proven TES systems of enough storage capacity and cost-effectiveness. However, this drawback will likely disappear in the future because several projects have demonstrated the potential of combining sensible and latent heat storage systems to allow for storage with DSG solar plants [2,15,18,19,33].

More than 50% of the energy needed in DSG solar plants is for water evaporation. Therefore, the development of reliable and cost-effective latent heat storage systems could benefit a wider deployment of this type of CST systems. Current research follows two paths, the search of new PCMs on one hand, including adequate design of LTES containers with higher heat transfer capabilities, and on the other hand the design of equipment to transport the latent heat storage material during charging–discharging processes.

Besides, the thermodynamic cycle optimization of steam turbines designed for DSG solar plants with TES is another relevant subject, given that the design of the intermediate stages of the turbine could increase the efficiency of the plant compared with existing solar power plants. Hence, it will be one important issue that should be followed closely.

References

[1] Abhat A. Short term thermal energy storage. Revue de Physical Appliquee 1980;15: 477–501.

[2] Aga V, Ehrsam A, Boschek E, Breschi D, Girard R. Thermal energy storage and turbine solution for direct steam solar power plants with nominal power output during storage discharge. In: Proceedings of the 18th SolarPACES Conference, 11–14 September, 2012, Marrakech, Morocco; 2012.

[3] Aga V, Ehrsam A, Boschek E, Simiano M. Adaptation of a direct steam solar tower plant with molten salt storage for optimum value creation under different incentive schemes. Energy Procedia 2014;49:1097–106.

[4] Alam TE, Dhau JS, Goswami DY, Stefanakos E. Macroencapsulation and characterization of phase change materials for latent heat thermal energy storage systems. Applied Energy 2015;154:92–101.

[5] Bachelier C, Selig M, Mertins M, Stieglitz Zipf V, Neuhäuser A, Steinmann WD. Systematic analysis of Fresnel CSP plants with energy storage. Energy Procedia 2015;69: 1201–10.

[6] Bayon R, Rojas E, Valenzuela L, Zarza E, Leon J. Analysis of the experimental behaviour of a 100 kW latent heat storage system for direct steam generation in solar thermal power plants. Appl Therm Eng 2010;30:2643–51.

[7] Bayón R, Rojas E. Liquid crystals: a new approach for latent heat storage. International Journal of Energy Research 2013;37(14):1737–42.

[8] Birnbaum J, Eck M, Fichtner M, Hirsch T, Lehmann D. A direct steam generation solar power plant with integrated thermal storage. Journal of Solar Energy Engineering 2010; 132:031014. 1–5.

[9] Cabeza LF. Advances in thermal energy storage systems, methods and applications. Woodhead publishing; 2015.
[10] CSP Today, Global Tracker. 2016. http://social.csptoday.com/tracker/projects.
[11] Cumpston J, Pye J. Exergoeconomic optimization of steam networks connecting solar-thermal dish arrays. Solar Energy 2015;119:383–98.
[12] Dincer I, Rosen M. Thermal energy storage, systems and applications. 2nd ed. John Wiley and Sons; 2011.
[13] Garg HP, Mullick SC, Bhargava AK. Solar thermal energy storage. D. Reidel Publishing Company; 1985.
[14] Khenissi A, Krüger D, Hirsch T, Hennecke K. Return of experience on transient behavior at the DSG solar thermal power plant in Kanchanaburi, Thailand. Energy Procedia 2015; 69:1603–12.
[15] Koretz B, Afremov L, Chernin O, Rosin C. Molten salt thermal energy storage for direct steam tower systems. In: Proceedings of 17th Solar PACES Conference, 20-23 September, 2011, Granada, Spain; 2011.
[16] Kuravi S, Trahan J, Goswami DY, Rahman MM, Stefanakos EK. Thermal energy storage technologies and systems for concentrating solar power plants. Progress in Energy and Combustion 2013;39:285–319.
[17] Laing D. Storage development for direct steam generation power plants. Golden CO, USA: Parabolic Trough Technology Workshop; March 9, 2007.
[18] Laing D, Bahl C, Bauer T, Lehmann D, Steinmann WD. Thermal energy storage for direct steam generation. Solar Energy 2011;85:627–33.
[19] Laing D, Eickhoff M, Fiss M, Hempel M, Meyer-Grünefeldt M, Bahl C. Test results of a combined storage system for parabolic trough power plants with direct steam generation. In: Proceedings of the ASME 2011 5th International Conference on energy Sustainability, August 7–10, 2011, Washington DC, USA; 2011.
[20] Laing D, Steinmann WD, Tamme R, Richter C. Solid media thermal storage for parabolic trough power plants. Solar Energy 2006;80:1283–9.
[21] Lorsch HG, Kauffman KW, Denton JC. Thermal energy storage for solar heating and off-peak air conditioning. Energy Conversion 1975;15(1–2):1–8.
[22] Lyle O. Heat storage. chapter 16. In: The efficient use of steam, London; 1947.
[23] Michels H, Pitz-Paal R. Cascaded latent heat storage for parabolic trough solar power plants. 2007. p. 829–37.
[24] Novatec Biosol Information Brochure. World's first Fresnel solar power plant in commercial operation. 2011. http://www.novatecsolar.com/files/mne0911_pe1_broschure_english.pdf.
[25] NREL. Concentrating solar power projects. 2016. http://www.nrel.gov/csp/solarpaces/. (Accessed 10.04.16).
[26] Osuna R, Olavarría R, Morillo R, Sánchez M, Cantero F. PS10 Construction of a 11 MW solar thermal tower plant in Seville, Spain. In: Proceedings of 13th Solar PACES Conference, June 20–23, 2006, Seville, Spain; 2006.
[27] Pacio J, Singer C, Wetzel T, Uhlig R. Thermodynamic evaluation of liquid metals as heat transfer fluids in concentrated solar power plants. Applied Thermal Engineering 2013;60: 295–302.
[28] Pirasaci T, Goswami DY. Influence of design on performance of a latent heat storage system for a direct steam generation power plant. Applied Energy 2016;162:644–52.
[29] Querol P, Olano J, Pereña A, Velasco T, Arevalo JE, Lata J. Single tank thermal storage prototype. In: Proceedings of the 18th Solar PACES Conference, September 11–14, 2012, Marrakech, Morocco; 2012.
[30] Reddy KS, Veershetty G. Viability analysis of solar parabolic dish stand-alone power plant for Indian conditions. Applied Energy 2013;102:908–22.

[31] Relloso S, Gutiérrez Y. Real application of molten salt thermal storage to obtain high capacity factors in parabolic troughs. In: Proceedings of 14th SolarPACES Conference, March 4–7, 2008, Las Vegas, USA; 2008.

[32] Rojas E, Bayón R, Zarza E. Liquid crystals: a different approach for storing latent energy in a DSG plant. Energy Procedia 2015;69:1014–22.

[33] Seitz M, Cetin P, Eck M. Thermal storage concept for solar thermal power plants with direct steam generation. Energy Procedia 2014;49:993–1002.

[34] Silberstein E, Magen Y, Kroyzer G, Hayut R, Huss H. Brightsource solar tower pilot in Israel's Negev operation at 130 bar@530°C superheated steam. In: Proceedings of the 15th SolarPACES Conference, 15-18 September, 2009, Berlin, Germany; 2009.

[35] Steinmann WD, Eck M. Buffer storage for direct steam generation. Solar Energy 2006;80: 1277–82.

[36] Tamme R, Laing D, Steinmann WD. Advanced thermal energy storage technology for parabolic trough. In: Proceedings of ISEC 2003, Hawaii, March 15–18, 2003; 2003.

[37] Zipf V, Neuhäuser A, Willert D, Nitz P, Gschwander S, Platzer W. High temperature latent heat storage with a screw heat exchanger: design of prototype. Applied Energy 2013;109: 462–9.

[38] Zipf V, Neuhäuser A, Bachelier C, Leithner R, Platzer W. Assessment of different PCM storage configurations in a 50 MWel CSP plant with screw heat exchangers in a combined sensible and latent storage—simulation results. Energy Procedia 2015;69:1078–88.

[39] Bauer T, Doerte L, Tamme R. Overview of PCMs for concentrated solar power in the temperature range 200 to 350°C. Advances in Science and Technology 2010;74: 272–7.

Part Five

Advances in the control an operation of CPS plants

Forecasting and nowcasting of DNI for concentrating solar thermal systems

13

L. Ramírez, J.M. Vindel
CIEMAT, Madrid, Spain

13.1 Introduction

The accurate forecasting of solar energy produced by a solar thermal power plant can improve considerably the profitability at different phases of the project. Direct normal irradiance (DNI) is the fundamental input for energy production, and thus DNI forecasting is the basis for energy forecasting.

At the very outset of the project, when pre-feasibility assessments are being carried out, forecasting DNI during the plant lifetime is needed for the long-term project profitability characterization. Since it is not possible to know what is going to happen in the next 25 years, what usually is done is to evaluate the conditions of the past years assuming that the conditions of the last 25 years are very similar to those for same period in the future. Future scenarios can also be considered taking into account assessments in the context of international panels of climate changes [17]. These assessments type is called long-term forecasting.

When the power plant is already set up, there are different needs of forecasting that have to be characterized to improve the plant exploitation. Due to the dependency of the energy generation with meteorological variables (mainly solar radiation) the better the prediction of the solar radiation, the better the prediction of the energy output. For instance, this dependency has a strong impact on the determination of the starting production time and the establishment of the operation strategies for the next day, hours, and minutes. Power plant design depends on the solar radiation characteristics expected in the location during the exploitation years. Technical shutdowns are scheduled depending on the annual forecasting, and operation plans can be optimized to increase the production or to increase the profitability by matching the production with the expectations. Options are different depending on the plant technology, plant design, and main options as the storage hours.

Generally, forecasting tools are run by specialized companies or meteorological offices and cover wide areas of land. These forecasts are post-processed or locally adapted with local measurements in order to improve the forecasting behavior. In this chapter post-processing techniques are not considered separately because these processes are always applied at the final step of each technique taking into account the space−temporal characteristics of the expected forecasting, and traditionally are considered as statistical approaches.

Advances in Concentrating Solar Thermal Research and Technology
http://dx.doi.org/10.1016/B978-0-08-100516-3.00013-7

A relationship between the spatial resolution and the temporal frequency can be established. Thus, when dealing with areas at country or regional level, several kilometer spatial resolution and hourly temporal frequency are the more often used references. When dealing with a specific location, the objective has to be working with high-resolution data below 1 km, and up to 1 min of time frequency.

Solar radiation and specifically DNI has not been an objective in meteorological forecasting over the time. Weather forecasting tools were mainly developed with focus on temperature, pressure, and humidity [6]. However, with the renewable energies growth, when the installed power has a significance impact on the electricity grids, the accurate solar radiation and wind speed forecasting has an important role [53]. The role is important on both sides of the electricity production: on the utility side, the importance is related to the impact of the intermittency in the grid and the need to predict the total electricity in the network; on the promoter side, the importance is related to the plant operation for medium and long-term and with the dispatchability. Dispatchability, as the possibility of modulating the energy dumped to the grid, is a requested capability that aims to increase the economic return by delivering electricity when is most needed and valued. Thus, improvements in solar and wind forecasting can help in a better consideration of the energy provided by renewable power plants, due to the increasing interest in the development of methods for energy price forecasting, mainly after the energy market liberalization in many countries [50].

The previous deployment of wind energy forced the importance of wind forecasting in weather forecasting systems investigations before solar radiation [11]. However, as of 2016 the huge growth of photovoltaic (PV) and thermal electricity systems all over the world [21] are pushing to improve solar forecasting as well.

Although solar forecasting seems to be a general issue, forecasting needs will depend on the solar technology and the plant characteristics. In the context of technologies, PV technology is mainly interested in global radiation forecasting [40], but the needs of concentrated solar thermal (CST) electricity plants are focused on DNI [27]. In the case of CST, dispatchability has been demonstrated during summer days when large thermal storage option is included. In this case, forecasting needs change for the use of storage as an intermediate step before electricity production; and thus, short fluctuations have no impact on the electricity production. The thermal storage possibility in thermal electricity power plants is a significant difference of this technology that makes it the only renewable power system with real modulation options, and thus, capable of playing a real important role in the electricity generation markets in countries at medium and low latitudes [20]. Forecasting meteorological variables and modulation options are the key points that aim CST electricity plants to have a relevant position in the energy generation; it is possible to adjust generation to forecasted demand and then, increase the sold-electricity benefits (Fig. 13.1).

In this chapter, a summary of the solar forecasting techniques applied in concentrated solar power plants is presented, and starting from a classification of the possible techniques, each one of them is described. After this general presentation, the configuration of a solar thermal electricity plant forecasting system is defined and presented. At the end of the chapter the main challenges related to the forecasting solar radiation at solar power plants are summarized.

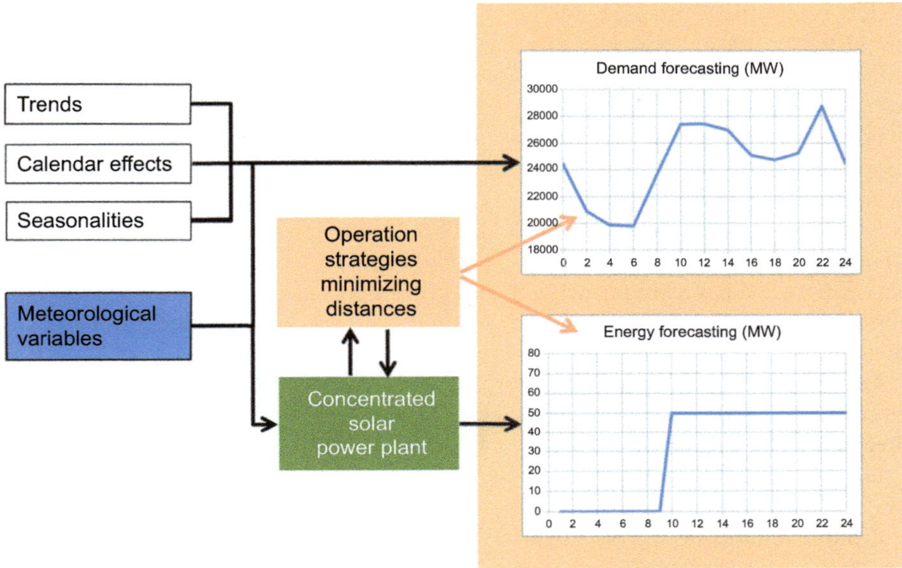

Figure 13.1 Importance of weather variables in forecasted demand and energy production.

13.2 Main forecasting techniques

Typical forecasting classifications are based on the time period forecasted or the treatment technique. In Fig. 13.2 the classical picture summarizing main characteristics is shown. Numerical weather prediction models (NWPMs) are the most suitable models for predicting solar radiation from 4 h up to several days. During the first hours, the

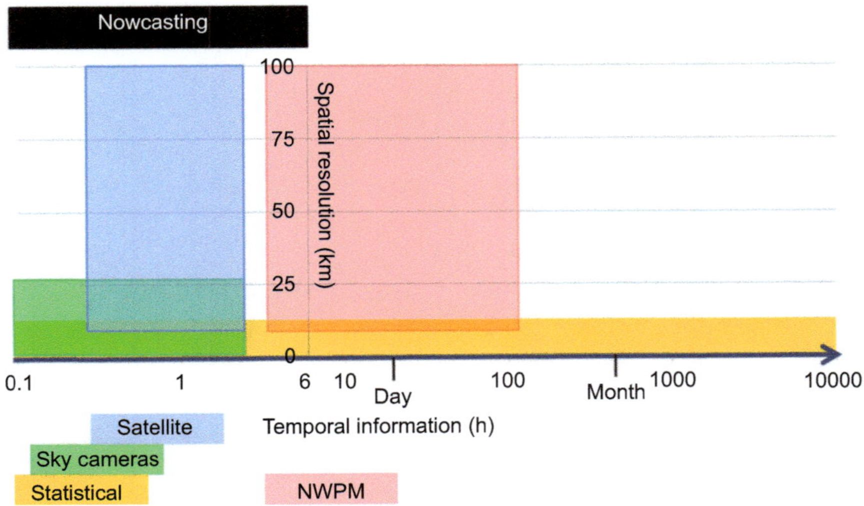

Figure 13.2 Main techniques used for different forecasting's horizon with different spatial resolutions. *NWPM*, numerical weather prediction models.

models need time to run and stabilize the outputs. Statistical forecasting is usually related to the use of local measurements and learning processes in order to derive future behavior. Depending on the time-frequency, this behavior can use data for years, months, days, hours, minutes, or seconds; then statistical models can cover all time resolutions and forecasting horizons. Since statistical models make use of local measurements, spatial resolution is related to less than 10 km around the measurements location and depends also on the topography.

Nowcasting is a common denomination for forecasting period up to 6 h. In the case of concentrated solar power plants, this period is mainly related with the period of plant exploitation, and has a big impact in the plant operation and on the electricity market. During this period, in addition to NWPM and statistical models, satellite- and sky cameras—derived forecasting are typically used.

This chapter focuses on forecasting from the point of view of a CST power plant. At the end of the chapter, some considerations related to single vs. aggregated or regional forecast are also presented.

In Table 13.1, the forecasting-related project moment and role are showed in relation with the forecasting name and technique. The most common classification is related to the forecasted time period. In the case of long-term and medium-term forecasting, statistical techniques are applied. In the forecasting level related to 10 days and 3 days ahead periods the treatment technique is NWPM in both cases, but typically

Table 13.1 Forecasting solar radiation: main techniques classification using time as main classification key

Project moment	Forecasting role	Typical forecasting name	Time period ahead	Main techniques	Typical time resolution
Before building the plant	• Project profitability	Long term	Years	Statistical models	Months
Power plant exploitation	• Maintenance planning	Medium term	Months		Days
	• Maintenance planning	10 days	Days	Numerical weather prediction models (NWPM)	Half days
	• Operation	3 days			Hourly
	• Electricity market	Nowcasting	(4–6 h)		30 min
			(0–6 h)	Statistical models	15 min
			(0–120 min)	Satellite images	15 min
			(0–60 min)	Sky cameras	1 min

the accuracy is much better in the case of 3 days ahead forecasting. Nowcasting is the most complex level of forecasting. In the nowcasting range, subsets of temporal windows can be established depending on the capabilities of each model. Sky cameras are able to predict solar radiation up to 60 min. In the case of satellite images, even when the models do not differ too much from the sky cameras model, because of the image resolution the forecasted horizon can achieve 2 h. Statistical models have then to address the time periods not covered by the rest of techniques.

Thus, different techniques deal with different forecasting ranges and time resolutions. Each one of these techniques has its own limitations and advantages, as are described briefly in next sections, but all of them have been mainly developed for global horizontal irradiance (GHI) forecasting. Due to the nature of solar radiation, with a bimodal probability density function, its estimation is always improved when the situations of cloudy sky and uncloudy sky moments are considered separately [39]. In uncloudy sky situations, solar radiation can be accurately modeled by clear sky models [18]. In the case of cloudy skies, due to the huge variability of cloud types, heights, and density, instead of using physical models solar radiation is modeled by statistical approaches using a cloudy-sky index as the main estimator [42].

Most of the forecasting techniques make use of these considerations to improve their results. They can also be applied as a post-process, or for downscaling purposes. It is important to mention that all the related techniques have not been developed specifically for direct normal radiation forecasting, and DNI was derived from GHI using global to beam models [55]. Thus, DNI had higher inaccuracies than global horizontal as is shown in a later section. Even so, DNI forecasting improves significantly when clear skies are under evaluation.

13.2.1 Numerical weather prediction models

NWPMs are models used and developed for weather forecasting purposes [6]. In a very simple way, they are supplied initial conditions and then, the differential equations describing the evolution of the atmosphere are solved. In order to test their behavior, these models can be used as well for "predicting the past," being run using ancillary initialization input data and assessing the forecasted horizons using ground measurements. Ground measurements can also be assimilated into the model or taken into account in a post-process treatment, considered as statistical approaches and described in a separate section. The option of running these models for "predicting the past" is called reanalysis of NWPMs. Table 13.2 depicts some examples of free time-series databases from reanalysis.

It is possible to classify the NWPMs into global models or mesoscale models. Typically, a global model has a global coverage and the most common one is the Global Forecast System—GFS (https://www.ncdc.noaa.gov/data-access/model-data/model-datasets/global-forcast-system-gfs). GFS uses as initialization input huge amount of data coming from automatic recording all over the world and from the satellite observation system. The spatial resolution is around 1 degree which matches to about 100 km at the earth's surface. The outputs from GFS are freely available and can be used as inputs for a mesoscale model. A mesoscale model

Table 13.2 Free databases time series reanalysis for research purposes

Name	Time range	Time frequency
ECMWF Interim Reanalysis (ERA Interim)	1979–present	3-hourly
ECMWF 40 year Reanalysis (ERA-40)	1958–2001	3-hourly
ERA-20C	1900–2010	3-hourly
NASA MERRA	1/1979–2/2016	Hourly
NASA MERRA-2	1/1980–present	Hourly
NCEP Climate Forecast System Reanalysis (CFSR)	1979–present	Hourly
NCEP/DOE Reanalysis AMIP-II (R2)	1979–present	4 times daily
NCEP/NCAR Reanalysis I (R1)	1948–present	4 times daily
NCEP North American Regional Reanalysis (NARR)	1979–present	4 times daily
NOAA-CIRES 20th Century Reanalysis (20CR)	1871–2012	4 times daily
NOAA-CIRES 20th Century Reanalysis (20CRV2c)	1851–2014	4 times daily

usually includes topographic and additional information as land cover information. The model from the European Centre for Medium-Range Weather Forecasts—ECMWF (http://www.ecmwf.int/en/forecasts) can be considered as being in between global and mesoscale models. It has products of different resolutions with different time steps depending on the product cost. The most extended mesoscale model is Weather Research and Forecasting—WRF (http://www.wrf-model.org/index.php). This model has been developed by the National Center for Atmospheric Research (NCAR), the National Center for Environmental Prediction (NCEP), and the Forecast Systems Laboratory (FSL). WRF is an open-source code run by a large number of universities and research centers. Outcomes are related to a volume and information at different levels from the ground level up to the stratosphere provided by all NWPMs. At each level, information is also provided at each point of a preconfigured grid resolution.

Development of NWPMs was focused on the improvement of the forecasted meteorological variables as temperature, precipitation, and humidity. But during the last decades, pushed by the wind energy sector, wind outputs have been significantly improved through the use and development of new parametrization in the context of open-source mesoscale models as MM5 or WRF [3].

During mid-2010s, some comparisons of solar radiation data from reanalysis products have demonstrated that in general, their results are far from the expected accuracy [4]. Even so, there are specific developments that show a relevant improvement when using new parametrizations [44] and post-processing treatments [25]. This new parametrization of the aerosol optical properties contributes to removing seasonal biases in the predicted GHI and also aim at the estimation of DNI as a new NWPM output.

NWPMs, in addition to the only pure forecasting technique, are the only way to obtain reliable information about the rest of meteorological variables needed for the solar-system simulation or the assessment of a solar plant, such as temperature, humidity, wind, and sometimes information related to clouds and atmospheric composition.

13.2.2 Image-derived solar radiation

While NWPMs provide multiple meteorological variables as outputs, image-derived models are focused on the estimation of solar radiation variables and these variables are commonly estimated with a better accuracy than in the case of NWPMs.

There are two different methodologies dealing with the treatment of images for solar radiation estimations: satellite-derived solar radiation and sky cameras–derived solar radiation estimations. Even though these methodologies use different input images of different nature (Fig. 13.3), the treatment is sometimes very close. Main differences of these techniques are related to the detectors point of view, which determines the image's field of view and the output resolution for each case.

In the case of satellite-derived models, detectors are placed at geostationary satellites facing the earth surface, seeing an earth part typically 60 degrees around the

Figure 13.3 Comparison of satellite view (facing the earth surface) and sky camera (facing the sky). Supported angles at a given location are also different. Satellite pixel is typically contained in 2 degrees from satellite position; sky camera has about 2π stereo-radians view.

sub-satellite point with a resolution in a range from 1 up to 10 km. Satellite images' time-frequency is typically between 15 and 30 min. In the case of sky cameras, detectors are placed at ground level facing the sky, seeing the sky equivalent to 2 km around the camera locations with a resolution between 1 and 10 m per grid point. Sky images' time-frequency is typically configured between 1 and 60 s.

But regardless of these differences, the first result in both cases of image treatments is a cloud index value at each image pixel. From this cloud index information, it is possible to assess cloud patterns motion vectors and thus, forecast a map of cloud indexes from the next moment. From this forecasted cloud image, GHI and DNI can be estimated as in the case of estimations models.

13.2.2.1 Satellite-derived cloud index

Satellite-derived solar radiation data sets are nowadays based on the use of geostationary satellite images. These images are the result of the reflection of sunlight on the earth's surface, so they have already suffered the topography effects and major atmospheric phenomena that take place when the sun's rays pass through the atmosphere. The geostationary orbit is a near-circular orbit placed 36,000 km far from the equator, managed by international bodies. Geostationary satellites are placed on this orbit spread from 45 up to 70 degrees and combining all of them a global earth observation is possible. Each satellite family has a nominal position related to a specific longitude and covers about 60 degrees from the sub-satellite point (point over the equator at 0 degree longitude in the case of Meteosat satellite family). Satellite families have changed over the years, and when a change in the onboard technology takes place, they are referred as a new satellite generation.

Satellite-derived solar radiation data sets have a gridded nature, where the main element is a pixel. Each pixel is a value that represents a specific area, depending on the characteristics of the satellite family, the generation, and its position over the earth's surface (Fig. 13.4). At the sub-satellite pixel, the usual spatial resolution is 2.5 km for old satellites and, as of 2016, 1 km for some of the current ones. The temporal resolution also varies from three times a day, up to 30 min, which is the most common as of 2016. It is important to take into account that pixels of a given image

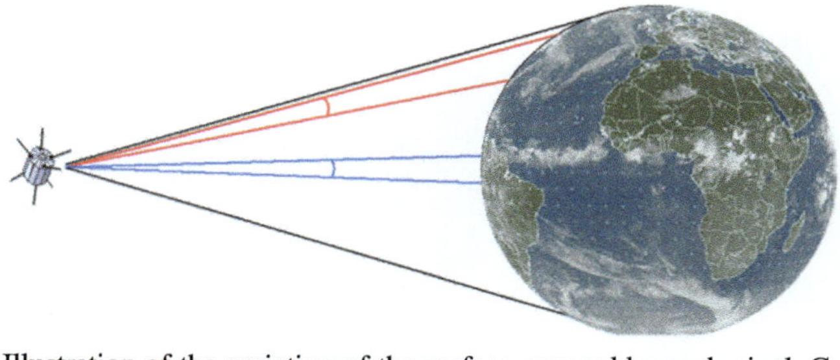

Figure 13.4 Illustration of the variation of the surface covered by each pixel. Covered surface increases when the distance to the equator and the distance to the sub-satellite point increase.

are related to different moments because the image recording has a scanning nature instead of being simultaneously recorded.

Each satellite family has onboard different types of devices for measuring the reflected radiation from the earth's surface, and then, the models vary from one set of satellite images to another. All satellite-derived solar radiation models can be divided mainly into statistical [33] and physical models [34]. In the case of statistical models, relationships between the satellite-derived cloud index and global radiation measured at the earth surface are established at several locations and later these relationships are applied to the whole image. In the case of physical models, these models are based on physical considerations taking into account the absorption and scattering behavior of the atmospheric components, the cloud reflectance and absorption coefficient, as well as the ground albedo among other parameters. Even as these models do not need ground measurements, they need detailed atmospheric information. Most of the new models have a mixture nature [39].

Satellite-derived solar radiation models are often operated by private companies or by research centers providing commercial products and not public databases in most of the cases. Some satellite-derived solar radiation data sets are free and have public access as the data sets shown in Table 13.3.

Cloud index is always computed for assessing satellite-derived solar radiation data. Fig. 13.5 shows the cloud index calculation based on the Heliosat model [8]. In this equation ρ_{ref} is the reference ground albedo for uncloudy skys; ρ_{cloud} is the estimated cloud albedo, and ρ_{ap} is the instantaneous apparent albedo. It can be observed that when a pixel is covered by clouds the relation is near 1 and 0 when the instantaneous apparent albedo is equal to the reference ground albedo.

Using the image of satellite-derived cloud index, cloud motion vector methodology discussed in the following section can be applied to assess future solar radiation values at each pixel. This is the basis of the solar forecasting from satellite images as can be seen in Refs. [1,14].

13.2.2.2 Cloud mapping using sky cameras

Sky cameras can be used for improvements in cloud detection and solar radiation characterization from the ground. There is a wide range of sky cameras that can be

Table 13.3 Free databases and services providing satellite-derived solar radiation time series

Name	Time frequency	Coverage
NASA SRB	3-hourly	World
NSRDB update	30-min	USA
CM-SAF	Hourly	Europe–Africa (Meteosat)
MACC RAD	15-min	Europe–Africa (Meteosat)
PVGIS	Hourly	Europe–Africa–Asia

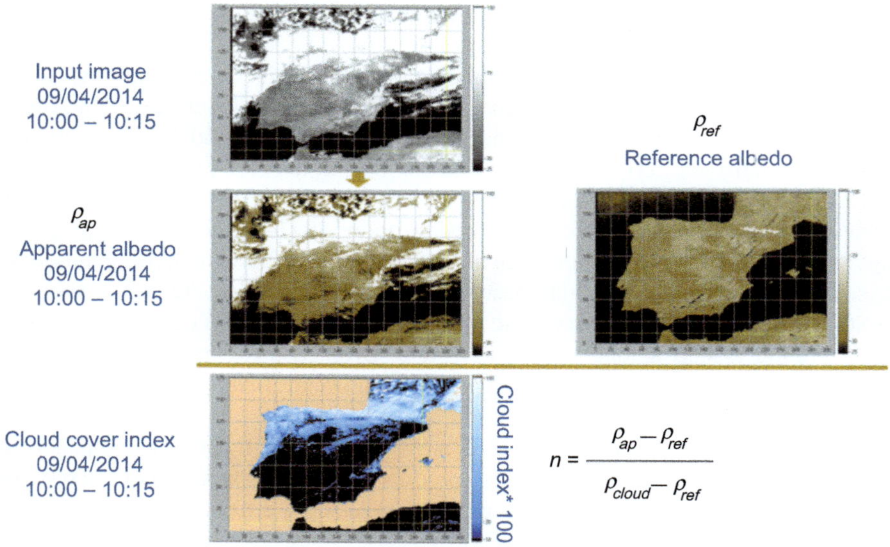

Figure 13.5 Satellite-derived cloud cover index calculation example.

classified into two main types. The first type is comprised of automatic sky cameras including real-time processing. These cameras have a hemispherical mirror and a shadow band that blocks the sun disk. The camera looking at the mirror captures the images. An internal algorithm aims to filter the clouds and calculate the cloud fraction from each image [48]. The second type looks at a similar option but uses commercial cameras with a fish-eye lens. These cameras have no sun-shadowing or processing options. These cameras are set up by research centers and techniques used differed from one developer to another.

The second type of sky cameras are being improved during the last years. Examples of this set of sky cameras can be reviewed in Refs. [16,23,46]. In Ref. [46], a method based on the use of two commercial cameras for deriving cloud maps, cloud hight, and cloud wind is proposed. In Ref. [16], in addition to the cloud fraction estimation, an automatic cloud classification algorithm is proposed based on statistical information from the RGB channels of the 30-bits color jpg images with a resolution of 3648×2736 pixel. The last example [23], adds new variables for the classification as the solar zenith angle or the visible fraction of the solar disk. In this work, authors used 8-bits color jpg images with spatial resolution 640×480 pixels.

In addition to the solar radiation estimation from cloud maps, research aimed at aerosol optical deep (AOD) estimation from sky cameras has been reported [36]. This work opens the door for DNI estimations directly from AOD estimation.

After the cloud map obtained from each sky-camera image, it is possible to apply cloud motion vectors for predicting the cloud position in the future and then, the solar radiation is derived from this. A detailed description of the prediction methodology from sky images can be found in Refs. [2,52].

13.2.2.3 Cloud motion vectors determination

Cloud motion vectors are the basis of satellite-derived forecasting and sky images–derived forecasting. The first reference of cloud motion vectors can be found in Ref. [28], but they are also described in all previous references of satellite and sky cameras forecasting. Fig. 13.6 presents a simplification of the cloud motion vector behavior. Two, or sometimes three consecutive images are needed and from each image, cloud maps are evaluated. Each cloud map is divided into a certain number of parts, which can be regular tiles, as in the case of Ref. [52], or irregular sectors as in Ref. [2]. Each image part is cross-correlated with all parts in the next image and the position with the highest correlation is identified. A vector movement can be identified from the initial part position to the next one (red small squares in Fig. 13.6 show the initial 1.5 image part, and the highest cross-correlated part in time 2 is identified at position 1.7). The vector movement is drawn in red when calculated and in green when it is applied. Green small square shows the forecasted position based on the application of the motion vector to the position 1.7. This procedure is repeated for each image part. Once the cloud map is forecasted, the solar radiation map is estimated.

13.2.3 Statistical forecasting

Statistical forecasting was the first set of techniques applied to the solar radiation forecasting [22]. This approach was based in the application of the model output statistics (MOS) methodology. In this case, solar radiation was not an output from NWPM but it was derived from the forecasted meteorological variables on daily basis. Forecasted daily global irradiation is calculated as an MOS of three different relationships: one relation of forecasted irradiation with the forecasted probability of precipitation; another relation of forecasted irradiation with the forecasted relative humidity, and the third one with the observed relative humidity.

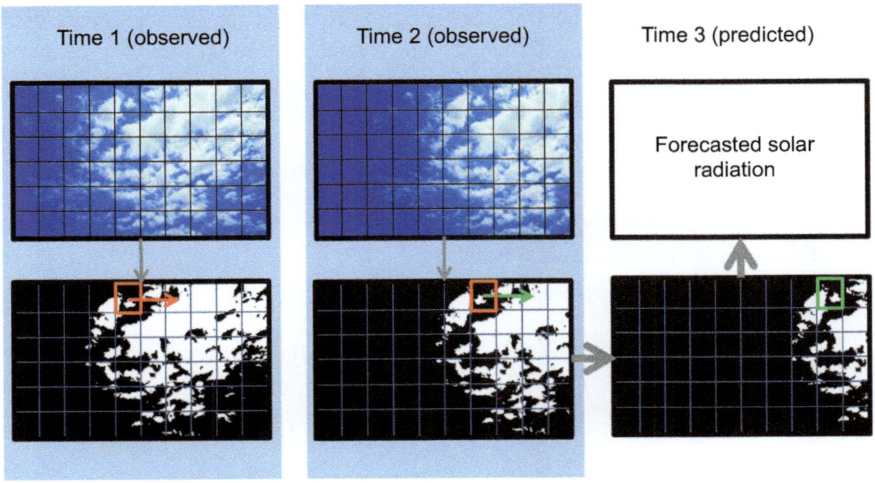

Figure 13.6 Cloud motion vector applied to a cloud map image.

After this work, many different groups started using series of solar radiation to develop univariate models using artificial intelligence techniques [47] as neuronal networks combined with wavelets [5]. Time-series assessment are sometimes proposed for this univariate type of models [41]. Univariate models are mainly the models applied for long-term forecasting. In this case, it is important to consider the need to take into account additional phenomena such as volcano eruption [43] or future scenarios [17] due to their well-known relationship with solar radiation.

Apart from univariate models, statistical models using local measurements are applied at the end of all other methodologies to remove biases, to improve spatial resolution, or to derive solar radiation from other variables. These statistical models are sometimes called post-processes to highlight the importance of considering this treatment apart from the general methodology and inform about the post-processes treatments of the models output.

13.3 Forecasting systems for CST power plants

A forecasting system for a CST power plant uses most of the described methodologies and real-time access to information from local measurements stations (Fig. 13.7) . As it has been mentioned, local measures are needed for statistical models application (linear models and time-series assessments); and for assimilation, kriging, and downscaling in the case of NWPM or satellite-derived forecasting (with spatial resolutions typically around 3 km). In the case of sky cameras local measures aim to remove bias and to check DNI consistency.

Ground measurements always have two different contributions in a forecasting system: the historical data bank to learn from the past and the current data to adjust model output in real time using Kalman filter techniques [49].

The main model that aims to predict the next days is the NWPM, to which statistical approaches are applied for the improvement of daily predictions using historical measurements. After the daily prediction, there is option for adjusting intraday predictions using real time measurements, and nowcasting from satellites and local sky cameras. Even so, if additional forecasting signals are available, combinations of all possible forecasted data will improve the final output.

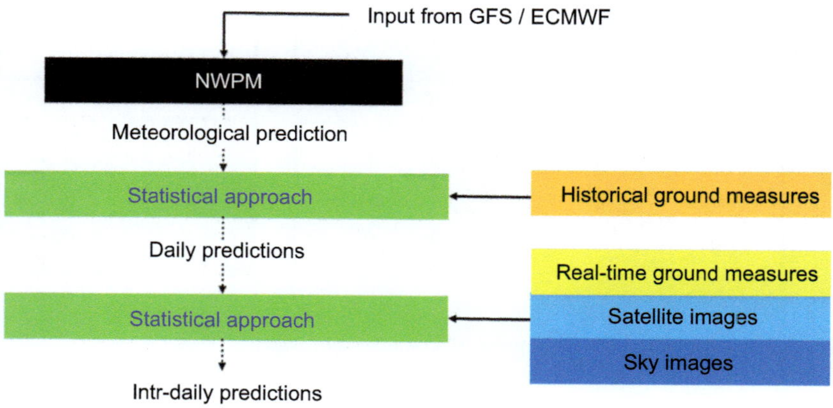

Figure 13.7 Classical solar forecasting system of a CST power plant.

13.4 Solar radiation forecasting baseline

In order to assess the main challenges related to forecasting needs in solar thermal power plants, first, a review of the solar forecasting evolution up to the current figure is presented.

Starting from nearly 300 of related references, a selection of them has been made. One of the first references fully related to solar forecasting is dated 1981 [22], but the real growth of solar forecasting started around 2000, and in 2006 can be seen one of the first reviews of solar radiation forecasting techniques [15]. Few years later it should be possible to find the first global benchmarking exercises and thus, the establishment of a baseline in solar forecasting is possible.

Some results from published comparison studies are shown in Refs. [4,26, 29,30,32,37,38,40]. After the revision of these publications, it can be seen that most of them deal with GHI instead of DNI, which is the main interest in our case of CST plants. Most of them are based on the evaluation of numerical weather prediction techniques, even when it is demonstrated that comparison of GHI coming from NWPM is far from the accuracy of GHI coming from satellite images. Due to this reason, statistical post-processing using local measurements for downscaling and biases reduction are included always as a step of the NWPM for solar radiation variables.

Main solar radiation forecasting review papers are included in Refs. [7,9,15,19,27,53], but only Ref. [27] is focused on DNI. In addition to this review, following works are also focused on DNI and CSP power plants forecasting: Refs. [12,25,27,31,35,56].

From Ref. [27] it is possible to derive that in the case of forecasted DNI in the range from 1 to 3 h, inaccuracies of hundreds of W/m^2 can be observed as a general mean. In the case of forecasted values from 1 up to 3 days, inaccuracies of thousands of Wh/m^2 can be observed in daily values. These results show a large opportunity for improvements in this field.

13.5 DNI and CST power plants forecasting: main challenges

Although up to now we have mainly discussed single forecasting, aggregated or regional forecast is also possible. Temporal variability of the non-dispatchable renewable power generation is minimized at national or utility level considering aggregated projects in large areas [45]. In this case, in addition, to minimize the temporal variability and then minimize the output inaccuracy because of the smooth effect of spatial aggregation, it is possible to assess optimal situations for renewable energy power plants. It is then possible to look for the location in a region or country that optimizes the energy balanced from all power plants. The determination of the adequate level of aggregation depending on the world situation and solar and wind resources is still under consideration. Related to long-term forecasting it is also important to improve the long-term forecasting including volcano eruptions [43] and future scenarios [17].

In the forecasting horizon of daily values, classification techniques can improve significantly the forecasting outputs. This forecasting horizon has an important role related to the planned operation strategies for the next day in that the plant's resources could be optimized in order to obtain the most profitable energy output from the power plant. As an example, CST power plants need a preheating time period before they start the energy production. Depending on the distribution of the expected radiation in the morning, the production can be delayed until the time stable conditions for productions are expected.

At the nowcasting level, the main challenge related to DNI forecasting and CST power output is related to the need of harmonization. As can be seen in Fig. 13.2, nowcasting methodologies are diverse in spatial and temporal resolution, and work is needed to provide a continuous series of data at the plant location. This is one of the challenges included in the DNICast project (http://www.dnicast-project.net/) under development. A related challenge is to provide reliable forecasts of high spatial–temporal resolution at plants level. This means to predict maps with the resolution of several meters and some seconds, where forecasts time gaps between image data and NWPM have to be filled through statistical methodologies. As prerequisites, improvements in aerosol knowledge and circumsolar radiation forecasting are also needed for the improvement of all methodologies [13,24,51] are examples of the importance of aerosol knowledge for an accurate DNI prediction. As regard to the circumsolar radiation, its impact is mainly related to central tower power plants and Ref. [54] shows the importance of an adequate determination as well as its effects on CSP plants. In central tower power plants, characterization of the horizontal attenuation at the surface level is also important for the output energy prediction [10].

13.6 Conclusions

In this chapter, a review related to DNI forecasting and nowcasting for CST systems has been presented. After a review of the main methodologies involved in solar radiation forecasting, specific considerations for CST plants have been taken into account. Before highlighting the main challenges in DNI forecasting, the baseline of solar radiation is presented showing a review of benchmarking exercises. Most of these comparison exercises are based on GHI. Some review papers, as well as selected research work, dealing specifically with DNI and solar thermal power plants have been also identified.

From Ref. [27] it is possible to derive that in the case of DNI forecasted in the range from 1 to 3 h, inaccuracies of hundreds of W/m^2 can be observed as a general mean. In the case of forecasted values from 1 up to 3 days, inaccuracies of thousands of Wh/m^2 can be observed in daily values. These results indicate a large opportunity for improvements in this field.

As regards the main advantages in this area, a relevant improvement when using new parametrizations [44] in NWPM can be emphasized. This new parametrization of the aerosol optical properties contributes to removing seasonal biases in the predicted GHI and also aims at the estimation of DNI as a new NWPM output. In addition, improved aerosol monitoring services [24,36] and the inclusion of them in CST forecasting systems will have a promising role in the improvements of DNI forecasting.

Related to the main challenges, different horizons can be considered. In long-term forecasting, inclusion of future events has a relevant impact in a fast-changing climate scenario. In this horizon, optimization tools for the optimal site location of renewable power plants, as also the sensitivity of the decision makers to push for a renewable scenario of a near to full renewable supply, are necessary. This will make possible searching for locations that best fit a balancing of the renewable resources [45].

In the forecasting horizon from 1 up to 3 days, improvements in daily classifications dealing with energy, temporal distribution, and temporal variability through each day are needed in order to optimize the operation strategies for the next day.

Finally, the forecasting horizon dealing with nowcasting, in the range from now up to 6 h, main challenges are examined. The first is related to the need of signals harmonization of outputs from all methodologies involved in this time period considering that several methodologies with different temporal resolutions cover different time periods. A related challenge is also to provide reliable forecasts of high spatial−temporal resolution at the plant level. This means, to predict maps with the resolution of several meters and some seconds, where forecasts time gaps between image data and NWPM have to be bridged by statistical methodologies. Advantages in aerosol knowledge, circumsolar radiation, and horizontal attenuation forecasting are also needed for improvements in all DNI-forecasting methodologies.

References

[1] Alonso-Montesinos J, Batlles FJ. Solar radiation forecasting in the short- and medium-term under all sky conditions. Energy 2015;83:387−93. http://dx.doi.org/10.1016/j.energy.2015.02.036.

[2] Alonso-Montesinos J, Batlles FJ, Portillo C. Solar irradiance forecasting at one-minute intervals for different sky conditions using sky camera images. Energy Conversion and Management 2015;105:1166−77. http://dx.doi.org/10.1016/j.enconman.2015.09.001.

[3] Al-Yahyai S, Charabi Y, Gastli A. Review of the use of numerical weather prediction (NWP) models for wind energy assessment. Renewable and Sustainable Energy Reviews 2010;14:3192−8. http://dx.doi.org/10.1016/j.rser.2010.07.001.

[4] Boilley A, Wald L. Comparison between meteorological re-analyses from ERA-Interim and MERRA and measurements of daily solar irradiation at surface. Renewable Energy 2015;75:135−43. http://dx.doi.org/10.1016/j.renene.2014.09.042.

[5] Cao S, Cao J. Forecast of solar irradiance using recurrent neural networks combined with wavelet analysis. Applied Thermal Engineering 2005;25:161−72. http://dx.doi.org/10.1016/j.applthermaleng.2004.06.017.

[6] Coiffier J. Fundamentals of numerical weather prediction. Cambridge: Cambridge University Press; 2011. http://dx.doi.org/10.1017/CBO9780511734458.

[7] Coimbra CFM, Kleissl J, Marquez R. Overview of solar-forecasting methods and a metric for accuracy evaluation. In: Solar energy forecasting and resource assessment. Elsevier; 2013. p. 171–94. http://dx.doi.org/10.1016/B978-0-12-397177-7.00008-5.
[8] Diabaté L, Moussu G, Wald L. Description of an operational tool for determining global solar radiation at ground using geostationary satellite images. Solar Energy 1989;42: 201–7. http://dx.doi.org/10.1016/0038-092X(89)90012-1.
[9] Diagne M, David M, Lauret P, Boland J, Schmutz N. Review of solar irradiance forecasting methods and a proposition for small-scale insular grids. Renewable and Sustainable Energy Reviews 2013;27:65–76. http://dx.doi.org/10.1016/j.rser.2013.06.042.
[10] Elias T, Ramon D, Dubus L, Bourdil C, Cuevas-Agulló E, Zaidouni T, et al. Aerosols attenuating the solar radiation collected by solar tower plants: the horizontal pathway at surface level. In: Solarpaces; 2016. p. 150004. http://dx.doi.org/10.1063/1.4949236.
[11] Foley AM, Leahy PG, Marvuglia A, McKeogh EJ. Current methods and advances in forecasting of wind power generation. Renewable Energy 2012;37:1–8. http://dx.doi.org/10.1016/j.renene.2011.05.033.
[12] Gastón M, Ramirez L, Pagola I, Fernandes PM, Lozano S, Nicuesa X. A system of direct radiation forecasting based on numerical weather predictions, satellite image and machine learning. 30th ISES Biennial Solar World Congress 2011, SWC 2011 2011;3: 2009–14.
[13] Gueymard CA. Temporal variability in direct and global irradiance at various time scales as affected by aerosols. Solar Energy 2012;86:3544–53. http://dx.doi.org/10.1016/j.solener.2012.01.013.
[14] Hammer A, Heinemann D, Lorenz E, Lückehe B. Short-term forecasting of solar radiation: a statistical approach using satellite data. Solar Energy 1999;67:139–50. http://dx.doi.org/10.1016/S0038-092X(00)00038-4.
[15] Heinemann D, Lorenz E, Girodo M. Forecasting of solar radiation. In: Solar energy resource management for electricity generation from local level to global scale; 2006. p. 83–94. http://dx.doi.org/10.1016/j.solener.2006.09.009.
[16] Heinle A, Macke A, Srivastav A. Automatic cloud classification of whole sky images. Atmospheric Measurement Techniques 2010;3:557–67. http://dx.doi.org/10.5194/amt-3-557-2010.
[17] ICCP. Climate change 2014: Synthesis report. Contribution of working Groups I, II and III to the fifth assessment report of the intergovernmental panel on climate change. Geneva, Switzerland: ICCP; 2014.
[18] Ineichen P. Validation of models that estimate the clear sky global and beam solar irradiance. Solar Energy 2016;132:332–44. http://dx.doi.org/10.1016/j.solener.2016.03.017.
[19] Inman RH, Pedro HTCC, Coimbra CFMM. Solar forecasting methods for renewable energy integration. Progress in Energy and Combustion Science 2013;39:535–76. http://dx.doi.org/10.1016/j.pecs.2013.06.002.
[20] International Energy Agency. Technology roadmap. In: IEA. Berlin/Heidelberg: Springer-Verlag; 2014. p. 52. http://dx.doi.org/10.1007/SpringerReference_7300.
[21] International Energy Agency. Solar energy perspectives. 2011. http://dx.doi.org/10.1787/9789264124585-en.
[22] Jensenius J. Insolation forecasting, solar resources (solar heat technologies), fundamentals and applications. Cambridge: MIT Press; 1981.
[23] Kazantzidis A, Tzoumanikas P, Bais AF, Fotopoulos S, Economou G. Cloud detection and classification with the use of whole-sky ground-based images. Atmospheric Research 2012;113:80–8. http://dx.doi.org/10.1016/j.atmosres.2012.05.005.

[24] Klüser L, Schepanski K. Deliverable Nr.: 2.3. Assessment report on aerosol monitoring as nowcasting scheme, Direct Normal Irradiance Nowcasting methods for optimized operation of concentrating solar technologies. 2015.

[25] Lara-Fanego V, Ruiz-Arias JA, Pozo-Vázquez D, Santos-Alamillos FJ, Tovar-Pescador J. Evaluation of the WRF model solar irradiance forecasts in Andalusia (southern Spain). Solar Energy 2012;86:2200–17. http://dx.doi.org/10.1016/j.solener.2011.02.014.

[26] Lauret P, Voyant C, Soubdhan T, David M, Poggi P. A benchmarking of machine learning techniques for solar radiation forecasting in an insular context. Solar Energy 2015;112: 446–57. http://dx.doi.org/10.1016/j.solener.2014.12.014.

[27] Law EW, Prasad AA, Kay M, Taylor RA. Direct normal irradiance forecasting and its application to concentrated solar thermal output forecasting – a review. Solar Energy 2014;108:287–307. http://dx.doi.org/10.1016/j.solener.2014.07.008.

[28] Leese JA, Novak CS, Ray Taylor V. The determination of cloud pattern motions from geosynchronous satellite image data. Pattern Recognition 1970;2:279–92. http://dx.doi.org/10.1016/0031-3203(70)90018-X.

[29] Lorenz E, Kühnert J, Heinemann D, Nielsem KP, Remund J, Müller SC. Comparison of irradiance forecasts based on numerical weather prediction. In: Proceedings of the European Photovoltaic Solar Energy Conference; 2015. p. 1524–37.

[30] Lorenz E, Remund J, Müller SC, Traunmüller W, Steinmaurer G, Pozo D, et al. Benchmarking of different approaches to forecast solar irradiance. In: Proceedings of the European Photovoltaic Solar Energy Conference. Berlin, Germany; 2009. p. 4199–208. http://dx.doi.org/10.4229/24thEUPVSEC2009-5BV.2.50.

[31] Marquez R, Coimbra CFM. Forecasting of global and direct solar irradiance using stochastic learning methods, ground experiments and the NWS database. Solar Energy 2011; 85:746–56. http://dx.doi.org/10.1016/j.solener.2011.01.007.

[32] Mathiesen P, Kleissl J. Evaluation of numerical weather prediction for intra-day solar forecasting in the continental United States. Solar Energy 2011;85:967–77. http://dx.doi.org/10.1016/j.solener.2011.02.013.

[33] Noia M, Ratto CF, Festa R. Solar irradiance estimation from geostationary satellite data: I. Statistical models. Solar Energy 1993;51:449–56. http://dx.doi.org/10.1016/0038-092X(93)90130-G.

[34] Noia M, Ratto CF, Festa R. Solar irradiance estimation from geostationary satellite data: II. Physical models. Solar Energy 1993;51:457–65. http://dx.doi.org/10.1016/0038-092X(93)90131-7.

[35] Nonnenmacher L, Kaur A, Coimbra CFM. Day-ahead resource forecasting for concentrated solar power integration. Renewable Energy 2016;86:866–76. http://dx.doi.org/10.1016/j.renene.2015.08.068.

[36] Olmo FJ, Cazorla A, Alados-Arboledas L, López-Alvarez MA, Hernández-Andrés J, Romero J. Retrieval of the optical depth using an all-sky CCD camera. Applied Optics 2008;47:H182–9. http://dx.doi.org/10.1364/AO.47.00H182.

[37] Pedro HTC, Coimbra CFM. Assessment of forecasting techniques for solar power production with no exogenous inputs. Solar Energy 2012;86:2017–28. http://dx.doi.org/10.1016/j.solener.2012.04.004.

[38] Perez R, Beauharnois M, Hemker Jr K, Kivalov S, Lorenz E, Pelland S, et al. Evaluation of numerical weather prediction solar irradiance forecasts in the US. In: American Solar Energy Society – Proc. ASES Annual Conference; 2011.

[39] Perez R, Cebecauer T, Šúri M. Semi-empirical satellite models. In: Solar energy forecasting and resource assessment. Elsevier; 2013. p. 21–48. http://dx.doi.org/10.1016/B978-0-12-397177-7.00002-4.

[40] Perez R, Lorenz E, Pelland S, Beauharnois M, Van Knowe G, Hemker K, et al. Comparison of numerical weather prediction solar irradiance forecasts in the US, Canada and Europe. Solar Energy 2013;94:305–26. http://dx.doi.org/10.1016/j.solener.2013.05.005.
[41] Reikard G. Predicting solar radiation at high resolutions: a comparison of time series forecasts. Solar Energy 2009;83:342–9. http://dx.doi.org/10.1016/j.solener.2008.08.007.
[42] Rigollier C, Lefèvre M, Wald L. The method Heliosat-2 for deriving shortwave solar radiation from satellite images. Solar Energy 2004;77:159–69. http://dx.doi.org/10.1016/j.solener.2004.04.017.
[43] Robock A. Volcanic eruptions and climate. Reviews of Geophysics 2000;38:191–219. http://dx.doi.org/10.1029/1998RG000054.
[44] Ruiz-Arias JA, Dudhia J, Gueymard CA. A simple parameterization of the short-wave aerosol optical properties for surface direct and diffuse irradiances assessment in a numerical weather model. Geoscientific Model Development 2014;7:1159–74. http://dx.doi.org/10.5194/gmd-7-1159-2014.
[45] Santos-Alamillos FJ, Pozo-Vázquez D, Ruiz-Arias JA, Lara-Fanego V, Tovar-Pescador J. Analysis of spatiotemporal balancing between wind and solar energy resources in the southern Iberian Peninsula. Journal of Applied Meteorology and Climatology 2012;51: 2005–24. http://dx.doi.org/10.1175/jamc-d-11-0189.1.
[46] Seiz G, Baltsavias EP, Gruen A. Cloud mapping from the ground: use of photogrammetric methods. Photogrammetric Engineering & Remote Sensing 2002:941–51.
[47] Sfetsos A, Coonick AHH. Univariate and multivariate forecasting of hourly solar radiation with artificial intelligence techniques. Solar Energy 2000;68:169–78. http://dx.doi.org/10.1016/S0038-092X(99)00064-X.
[48] Shields JE, Johnson RW, Koehler TL. Automated whole sky imaging systems for cloud field assessment. In: Fourth symposium on global change studies; 1993. p. 228–31.
[49] Soubdhan T, Ndong J, Ould-Baba H, Do M-T. A robust forecasting framework based on the Kalman filtering approach with a twofold parameter tuning procedure: application to solar and photovoltaic prediction. Solar Energy 2016;131:246–59. http://dx.doi.org/10.1016/j.solener.2016.02.036.
[50] Taylor JW, Espasa A. Energy forecasting. International Journal of Forecasting 2008;24: 561–5. http://dx.doi.org/10.1016/j.ijforecast.2008.08.001.
[51] Toll V, Gleeson E, Nielsen KP, Männik A, Mašek J, Rontu L, et al. Impacts of the direct radiative effect of aerosols in numerical weather prediction over Europe using the ALADIN-HIRLAM NWP system. Atmospheric Research 2016;172–173:163–73. http://dx.doi.org/10.1016/j.atmosres.2016.01.003.
[52] Urquhart B, Ghonima M, Nguyen DA, Kurtz B, Chow CW, Kleissl J. Sky-imaging systems for short-term forecasting. In: Solar energy forecasting and resource assessment. Elsevier; 2013. p. 195–232. http://dx.doi.org/10.1016/B978-0-12-397177-7.00009-7.
[53] Widén J, Carpman N, Castellucci V, Lingfors D, Olauson J, Remouit F, et al. Variability assessment and forecasting of renewables: a review for solar, wind, wave and tidal resources. Renewable and Sustainable Energy Reviews 2015;44:356–75. http://dx.doi.org/10.1016/j.rser.2014.12.019.
[54] Wilbert S. Determination of circumsolar radiation and its effect on concentrating solar power. 2014.
[55] Wong LT, Chow WK. Solar radiation model. Applied Energy 2001;69:191–224. http://dx.doi.org/10.1016/S0306-2619(01)00012-5.
[56] Yang D, Jirutitijaroen P, Walsh WM. Hourly solar irradiance time series forecasting using cloud cover index. Solar Energy 2012;86:3531–43. http://dx.doi.org/10.1016/j.solener.2012.07.029.

14 Advanced control strategies to maximize ROI and the value of the concentrating solar thermal (CST) plant to the grid

E.F. Camacho, A.J. Gallego
Universidad de Sevilla, Sevilla, Spain

14.1 Introduction

There are two main drawbacks to concentrating solar thermal energy systems: (1) the resulting energy costs are not yet competitive and (2) solar energy is not always available when needed. Considerable research efforts are being devoted to techniques that may help to overcome these drawbacks; control is one of those techniques [1].

One of the main challenges identified by the US National Academy of Engineering is to make solar energy economical [2]. This issue can be addressed by reducing investment and operating costs and increasing the solar plant performance [3]. Advanced control techniques can help to reduce operating costs and increase the solar plant performance. This chapter describes two examples of how advanced control techniques can help to optimize operation of solar plants and, in consequence, maximize the return of investment (ROI).

The first example focuses on solar trough plants. One of the first operative trough solar plants was at the Plataforma Solar de Almería (PSA), in southern Spain. This trough plant consisted of a field of solar collectors (ACUREX), a heat storage system, and an electrical conversion unit (0.5 MW Stal-Laval turbine). This plant has been operating since 1980, and many control strategies for solar systems have been tested there [4–6].

The nine SEGS trough plants (354 MW), which were commissioned between 1985 and 1990 in California, are considered to be the first commercial plants. Most commercial solar plants have been built and commissioned in the last decade. As examples, we can mention the three 50 MW parabolic trough plants Andasol 1, 2, and 3 [7] in Guadix (Spain); the solar tower plants of Abengoa PS10 and PS20; Gemasolar solar tower built by Torresol Energy; the three 50 MW Solnova and the two 50 MW Helioenergy parabolic trough plants of Abengoa in Spain; and the SOLANA and Mojave Solar parabolic trough plant constructed in Arizona and California, each of 280 MW power production [8].

A parabolic trough solar plant consists of a collector field, a power conversion system (PCS), a storage system, and auxiliary elements such as pumps, pipes, and valves [9]. The solar collector field collects solar radiation and focuses it onto a tube in which a heat transfer fluid, usually synthetic oils, circulates. The oil is heated up and is used

Advances in Concentrating Solar Thermal Research and Technology
http://dx.doi.org/10.1016/B978-0-08-100516-3.00014-9

by the PCS to produce electricity by means of a turbine. The storage system is necessary to cover the possible mismatch between the solar energy available and the demand; this is one of the advantages of solar thermal energy: storage of thermal energy is easier and cheaper than storage of electrical energy [10]. A three-layer nonlinear model predictive control strategy is developed to obtain the optimal set point. An estimation of the electrical production gain obtained by working at the optimal set point instead of working at the highest allowable temperature is shown.

The second example deals with the operation of solar tower plant with flat receivers. A solar tower plant consists of a field of mirrors (*heliostats*) arranged around a tower equipped with a solar irradiation receiver. By tracking the sun, the heliostats focus the solar irradiance on the receiver. The field can be composed of a high number of heliostats (more than 900 in recent commercial plants [8]), each of which is independently controlled [3]. The simplest (and most common) approach is to point at the center of the receiver to minimize energy losses by spillage of uncalibrated heliostats [11]. However, problems due to an improper thermal distribution on the receiver may appear, leading to the deterioration of the receiver and even its destruction [12]. In this chapter, a method to optimize the flux distribution and the solar radiation collected by the receiver is presented.

14.2 Optimal operation in solar trough plants

This section provides an example of how advanced control strategies, in particular model predictive control (MPC), can help in optimizing solar trough plant operation. The results described here are based on the previous work described in Refs. [13,14].

Solar trough plants heat up a heat transfer fluid, usually oil. The heated fluid is used to feed a steam generator producing superheated steam used to generate electricity or for any industrial applications. Superheated and pressurized steam has also been produced directly in the solar field [11,15].

The main goal of a control system for a parabolic trough field is to maintain the outlet temperature of the field close to a desired set point; because the primary energy source, solar energy, cannot be manipulated, the oil flow q is used as a control signal. The set point may change substantially throughout daily operation due to changes in the production requirements, solar radiation conditions, solar hour, etc. [16].

This goal is not an easy task due to the complex dynamics and multiple disturbances affecting the solar field. The transport delay is strongly dependent on the control signal (oil flow); at low flow the control of solar plants becomes trickier. The relation between control signal and transport delay is explained in Ref. [17]. To address this problem raised in the control of solar energy systems (not only in parabolic trough technology), the use of a dead time compensator is useful, as shown in Ref. [18].

Determining the best operating temperature (the outlet temperature set point) that maximizes the electrical production is an essential issue. An example using a mathematical model of the ACUREX solar trough plant is presented in this section.

The modeling of each component of the ACUREX solar trough field is presented subsequently. For a more detailed explanation of the modeling see Ref. [11].

14.2.1 Distributed parameter model

Each of the ACUREX loops consists of four 12-module collectors suitably connected in series. The loop is 172 m long and is divided into active parts (parts where the solar radiation reaches the tube) measuring 142 m and passive parts (joints and other parts not reached by concentrated radiation) measuring 30 m [19,20]. The model is described by the following system of partial differential equations (PDE) describing the energy balance:

$$\begin{aligned}\rho_m C_m A_m \frac{\partial T_m}{\partial t} &= IK_{opt} noG - H_l G(T_m - Ta) - LH_t(T_m - T_f) \\ \rho_f C_f A_f \frac{\partial T_f}{\partial t} + \rho_f C_f q \frac{\partial T_f}{\partial x} &= LH_t(T_m - T_f)\end{aligned} \tag{14.1}$$

where the subindex *m* refers to metal and *f* refers to the fluid. The model parameters and their units are shown in Table 14.1.

Density ρ, specific heat C, and the coefficient of heat transmission depend on fluid temperature. The coefficient of heat transmission depends on temperature and oil flow [20]. The geometric efficiency depends on hourly angle, solar hour, declination, Julian day, local latitude, and collector dimensions [20] and [19].

Table 14.1 Parameters description

Symbol	Description	Units
t	Time	s
x	Space	m
ρ	Density	kgm^{-3}
C	Specific heat capacity	$JK^{-1}kg^{-1}$
A	Cross-sectional area	m^2
$T(x,y)$	Temperature	K,°C
$q(t)$	Oil flow rate	m^3s^{-1}
$I(t)$	Solar radiation	Wm^{-2}
no	Geometric efficiency	Unitless
K_{opt}	Optical efficiency	Unitless
G	Collector aperture	m
$T_a(t)$	Ambient temperature	K,°C
H_l	Global coefficient of thermal loss	$Wm^{-2\circ}C^{-1}$
H_t	Coefficient of heat transmission metal–fluid	$Wm^{-2\circ}C^{-1}$
L	Length of pipeline	m

To solve this PDE system, a two-stage finite difference equation has been programmed, considering 1 m segments and solving Eq. (14.1).

Similar equations have been used to model the pipes. The main difference is that the isolated pipes do not collect solar radiation and the thermal losses are different.

14.2.2 Concentrated parameter model

The concentrated parameter model provides a lumped description of the whole field. The variation in the internal energy of the fluid can be described by the equation:

$$C\frac{dT_{out}}{dt} = K_{opt}noSI - qP_{cp}(T_{out} - T_{in}) - H_l(T_m - T_a) \tag{14.2}$$

where C is the thermal capacity of the loop, K_{opt} is the optical efficiency taking into account elements such as reflectivity, absorptance, and interception factor. The geometric efficiency (no) is determined by the position of the mirrors with respect to the beam radiation vector. Factor P_{cp} takes into account some geometrical and thermal properties of the loop. T_{in} is the inlet temperature, H_l is the coefficient of thermal losses, T_m is the mean temperature between inlet and outlet temperatures, and T_a is the ambient temperature.

Coefficient H_l can be approximated by 1.05 *kW*/°C, P_{cp} by 1.924e6 *kJ*/°C, C by 2267 *kW*/°C, and S takes the value of 2674 m^2. These values have been obtained by multiple regression analysis [20].

14.2.3 Power conversion cycle

Most thermal power conversion cycles in solar thermal plants are based on Rankine cycle and, to a lesser extent, Brayton cycle [21].

The Rankine efficiency can be modeled by Eq. (14.3):

$$\eta_{rank} = K\left(1 - \frac{T_{cf}}{T_{hf}}\right) \tag{14.3}$$

where T_{hf} and T_{cf} are the hot and cold focus temperatures, respectively, and K is a constant modeling the performance loss with respect to the ideal Carnot cycle. The T_{cf} is considered to be the condenser temperature. The evolution between points 2 and 3 is produced at a constant pressure, thus T_{hf} can be computed for different working temperatures using the concept of the mean thermodynamic temperature [22]. A relation between the outlet oil temperature of the solar field and the cycle efficiency can be obtained (Eq. (14.4)) using data supplied by thermodynamics tables.

$$\eta_{rank} = -4.98e - 6\ Tout^2 + 3.37e - 3\ Tout - 0.194 \tag{14.4}$$

The constant K is estimated using equations of the PCS system shown in Refs. [11] and [23]. The turbine installed in the plant (500 kW Stal-Laval) can only operate within a small range of temperature 277–292 °C.

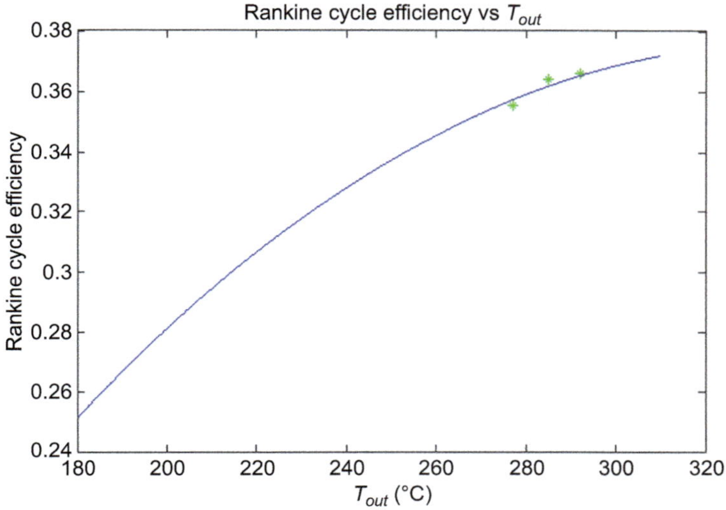

Figure 14.1 Rankine cycle efficiency, η_{rank}, versus outlet temperature of the solar field.

To approximate the Rankine cycle to the real one, three values of the real Rankine cycle have been used to obtain the constant K. Although the operation range of the turbine is quite narrow, adjusting the curve with three points can be considered acceptable. The constant K is a parameter to be tuned, ensuring the cycle efficiency fits the real one in the operating range of the real PCS. The overall efficiency of the solar field, η_{solar}, was considered as 0.52 [11].

Fig. 14.1 depicts the Rankine cycle efficiency, η_{rank}, versus the oil outlet temperature. As seen, the graphic follows the typical evolution of this kind of cycle [24].

14.2.4 Three-layer algorithm

Commercial plants are usually considered to operate optimally when working at the maximum reachable temperature allowed by environmental and security conditions. The reason is that the power cycle, typically a Rankine cycle, increases its efficiency at higher temperatures. However, thermal losses of the solar field increase with the operating temperature, and therefore, working at high temperatures does not necessarily increase the overall efficiency of the combination PCS + solar field [25]. Moreover, the electrical consumption of the pump increases at low temperatures because high temperatures imply lower oil flow and vice versa [26].

Some ideas have been proposed in the literature to increase the efficiency of solar plants [27]. A two-layer hierarchical control strategy was proposed in Ref. [28]. Two schemes were proposed for the reference governor layer, one based on fuzzy logic and one was suggested to maximize economic profit. The regulation layer consisted of an I + PD controller. Optimization is performed in steady state and the PCS is modeled as a constant, and therefore the dependence of the conversion cycle efficiency on the operating temperature is not taken into account.

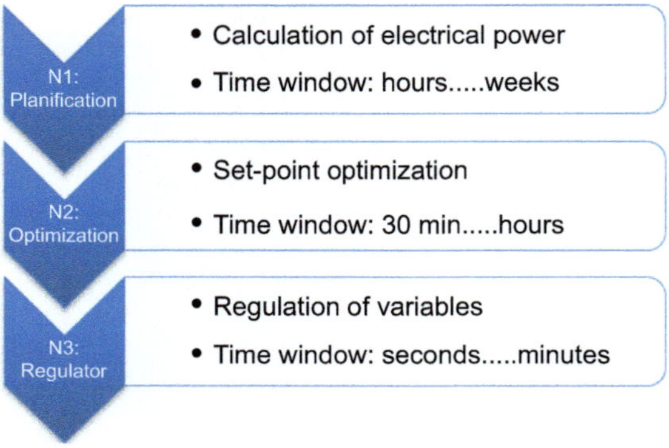

Figure 14.2 Three-layer algorithm.

The three-layer algorithm is proposed as a general hierarchical control strategy to cope with issues occurring at different timescales. The problem has a natural hierarchical structure because the task, goal, and the time windows for each layer are different. Fig. 14.2 shows the three layers and typical time windows used for them.

The first layer deals with the operational planning of the solar plant and it computes the electrical power to be produced and delivered. The second layer computes the optimal temperature set points for the solar plant, for the electrical power set points provided by the first. The third layer addresses the problem of tracking the temperature set points calculated by the second layer.

14.2.4.1 The first layer: calculation of electrical power

The operational scheduling is performed in this layer. The main decision variable is the electrical production profile throughout the scheduling horizon. This is computed taking into account the predicted price of electrical energy [1], the predicted electrical demand, the predicted solar radiation, and other factors. Suitable time windows for this layer range from hourly to weekly planning. Prediction models of all plant subsystems and electrical demand are fundamental for making decisions, as the electrical demand may change substantially depending on the date and environmental conditions. Accurate electrical energy price forecasts are very important for producers and consumers to negotiate good contracts and to establish appropriate bidding strategies on the short-term market [29].

Two cases can be considered: noncommitted production and committed production. In the noncommitted production case, prices are fixed at the time window considered with no commitments to the energy delivered. In the case of committed production, the plant managers have to decide the daily production and the energy to be delivered and stored at each time period.

For simplicity, only the noncommitted production is described in this section. In Ref. [11], a more complete description of the case of committed production is presented. Simulating the case of committed production would require a complex model of the market behavior.

If prices are constant, the best option would be to produce and deliver as much electricity as possible. Storage systems are useful when more solar energy can be collected than converted or delivered or when the energy conversion system is not fully operative and it is not possible to convert all the collected solar energy or when the prices change throughout the day.

For the sampling period k, the energy power balance is given by Eq. (14.5):

$$E_r(k) = \gamma_{te}(E_{solar}(k) - E_{s+}(k) + E_{s-}(k)) - E_{load}(k) \tag{14.5}$$

where $E_r(k)$ is the energy delivered to the grid during period k, $E_{solar}(k)$ is the solar energy generated, which depends on solar radiation and the operating conditions of the plant, $E_{load}(k)$ is the energy consumed by local systems, $E_{s+}(k)$ is the energy stored at interval $[k-1,k]$, $E_{s-}(k)$ is the energy extracted from the storage and delivered to the grid, and γ_{te} is the thermal to electrical power conversion factor. Both E_r and E_{stored} are subject to limits because the storage capacity is not infinite and the plant is not connected to an infinite grid where power can be injected or extracted as necessary.

Let β_{stor} be the efficiency of the storage system and α_{stor} models the stored energy lost due to thermal losses. The total energy stored $E_{stored}(k)$ is given by Eq. (14.6):

$$E_{stored}(k+1) = E_{stored}(k) + E_{s+}(k) - \beta_{stor}E_{s-}(k) - \alpha_{stor}E_{stored}(k) \tag{14.6}$$

Taking a prediction horizon of N samples, the objective is to determine the sequence $[E_r(k), E_r(k+1),\ldots, E_r(k+N-1)]$ that maximizes [11]:

$$\begin{aligned} J(N) = &\sum_{j=1}^{j=N} E_r(k+j)e_1(k+j) + E_{stored}(k) + \sum_{i=1}^{i=N}(E_{s+}(k+i) \\ &- \beta_{stor}E_{s-}(k+i))\alpha_{stor}^{N-i} \end{aligned} \tag{14.7}$$

subject to

$$\begin{aligned} &0 \le E_{stored}(k+j) \le E_{max} \\ &0 \le E_{s+}(k+j) \le E_{smax} \\ &0 \le E_{s-}(k+j) \le E_{smax} \\ &E_{rmin} \le E_r(k+j) \le E_{rmax} \\ &E_r(k) = \gamma_{te}(E_{solar}(k+j) - E_{s+}(k) + E_{s-}(k)) - E_{load}(k)) \end{aligned}$$

$E_{solar}(k+1)$ has to be estimated using models for forecasting solar radiation [30]. The decision variables are $E_r(k+j)$, $E_{s+}(k+j)$, and $E_{s-}(k+j)$ for $j=0,\ldots, N$. A receding horizon policy is implemented and only the first element of the sequence is used. Note that the optimization problem (Eq. 14.7) can be solved using a linear programming algorithm.

14.2.4.2 The second layer: set point optimization

In this subsection, the computation of the optimal set point is described. In Ref. [13], a more detailed description of this procedure can be found.

The second layer computes the optimal set point with a time window of minutes (20–60), trying to satisfy the electrical production scheduled by the first layer. A model of each subsystem and a model for forecasting solar radiation are required for predicting the evolution of the plant variables. The computation of the optimal set point is based on these models. In this paper, the set points are considered to be constant for periods of 30 min. Two assumptions for the computation of the optimal set point are considered:

- The turbine works at full power and the electricity produced can be delivered to the grid.
- The maximum temperature difference between inlet and outlet temperatures is 80 °C, when the turbine is working at full power.

The fundamental decision to be made is to determine the outlet temperature of the field, which is a function of the weather conditions, solar radiation, optical and geometrical efficiencies, and the general state of the whole plant. On one hand, when the operation temperature rises, the PCS efficiency increases and the electrical consumption of the oil pump decreases at low oil flow because the pressure drop is lower than at high oil flow [31]. On the other hand, thermal losses of the solar field increase and the efficiency of the solar field decreases [20]. The optimal set point for the operation temperature depends on these variables.

The total amount of solar energy collected by the solar field depends on solar radiation I, the solar field surface S, optical and geometrical efficiency, and shade factors:

$$E_{collected}(I, K_{opt}, no, S) = IK_{opt}noS \tag{14.8}$$

The overall efficiency of the collectors, the product $K_{opt}no$, is composed of two parts: geometrical efficiency, determined by the relative position between the Sun and the solar field, and optical efficiency depending on elements such as mirror reflectivity, metal absorptance, the interception factor, etc. Some of these variables are difficult to measure or estimate. The algorithm proposed in Ref. [32] can be used to obtain an estimation of the overall efficiency of the collectors.

The thermal losses of the ACUREX solar field are a function of the inlet and outlet temperatures and the ambient temperature (other variables such as humidity or wind speed may also have an effect but are not considered here). Let ΔT_m be the difference in temperature between the average inlet and outlet temperatures and the ambient temperature [19]:

$$\begin{aligned} E_{losses} &= 0.00667\Delta T_m^2 - 0.164\Delta T_m - 0.22(kW) \\ E_{losses} &= E_{losses}(T_{out}, T_{in}, T_{amb}) \end{aligned} \tag{14.9}$$

The evolution of the outlet temperature of the solar field is governed by the equations described in Section 14.2 (system of Eq. (14.1)). The heat power transferred to the heat transfer fluid, E_{solar}, is a function of the oil flow and inlet and outlet temperatures (Eq. (14.10)). T_{out} is, in turn, a function of the inlet temperature T_{in}, oil flow q, thermal losses E_{losses}, and the collected solar energy E_{solar} (Eq. (14.2)):

$$\begin{aligned} E_{solar} &= q\rho_f C_f (T_{out} - T_{in}) \\ E_{solar} &= E_{solar}(E_{collected}, E_{losses}, q, T_{out}, T_{in}) \end{aligned} \tag{14.10}$$

The overall efficiency of the solar field, η_{solar}, is the relation between the power delivered by the solar field to the heat transfer fluid, E_{solar}, and the power coming from the Sun, IS:

$$\eta_{solar} = \frac{E_{solar}}{I\,S} \tag{14.11}$$

The Rankine cycle efficiency, η_{rank}, depends on the outlet temperature of the oil flow coming from the solar field, as previously stated. The overall efficiency of the plant, $\eta_{plant}=\eta_{rankine}\eta_{solar}$, namely, the efficiency of the set solar field + Rankine cycle, depends on the inlet and outlet temperatures of the solar field, the oil flow, and the ambient temperature. Electrical production is computed by multiplying the thermal power coming from the solar field by the Rankine cycle efficiency.

$$E_{elect} = E_{solar}\eta_{rank} = IS\eta_{plant} \tag{14.12}$$

Finally, the electrical consumption of the pump depends on the oil flow and the inlet and outlet temperatures of the solar field. For computing the pressure drop and energy losses inside the pipes, the well-known Darcy equations are used [13,33].

Let E_{net} be the net electrical production:

$$E_{net} = E_{elect} - E_{pump} = F(T_{out}, T_{in}, T_{amb}, I, q, K_{opt}, no, S) \tag{14.13}$$

Function E_{net} depends on the temperature difference between T_{out} and T_{in}, the oil flow, the solar radiation, the overall efficiency of the solar field, and other variables. The main objective is to maximize the net electrical production, the difference between the electrical production and the electrical consumption of the oil pump, and other auxiliary elements.

The control objective is the regulation of the solar field outlet temperature using the oil flow as a control signal. The optimization algorithm must search for an optimal set point for the outlet temperature, subject to the maximum and minimum temperatures allowed and the maximum temperature difference allowed (in the ACUREX solar field this is about 80 °C [20]). Let ΔT_{max} be the maximum temperature difference between inlet and outlet temperatures, E_{max} the maximum energy that can be injected to the grid, and E_r the energy reference sent by the first layer. The optimization problem can be posed as follows:

$$\max_{T_{ref}} .E_{net}(T_{ref}, T_{in}, T_{amb}, I, q, K_{opt}, no, S)$$

subject to

$$
\begin{aligned}
&E_r \le E_{net} \le E_{max} \\
&T_{min} \le T_{ref} \le T_{max} \\
&T_{ref} - T_{in} \le \Delta T_{max}
\end{aligned}
\tag{14.14}
$$

14.2.4.3 The third layer: regulation

The goal of the third layer is to track the set point provided by the second layer using control algorithms with a sampling time of seconds.

As pointed out earlier, the main goal is to maintain the outlet temperature of the solar field as close as possible to a set point. The oil flow q is used as control signal because the primary energy source, solar radiation, cannot be manipulated. Many works have been carried out in the ACUREX field dealing with the control of the solar trough plant [4,5].

14.2.5 Simulation results

Simulation results are shown in this subsection. Fig. 14.3 shows the overall efficiency, the product $\eta_{field}\eta_{rank}$, for different radiation levels with a thermal jump between T_{in} and T of 80°C. As can be seen, maximum efficiency is not reached at the maximum allowed temperature of the ACUREX plant (about 300°C).

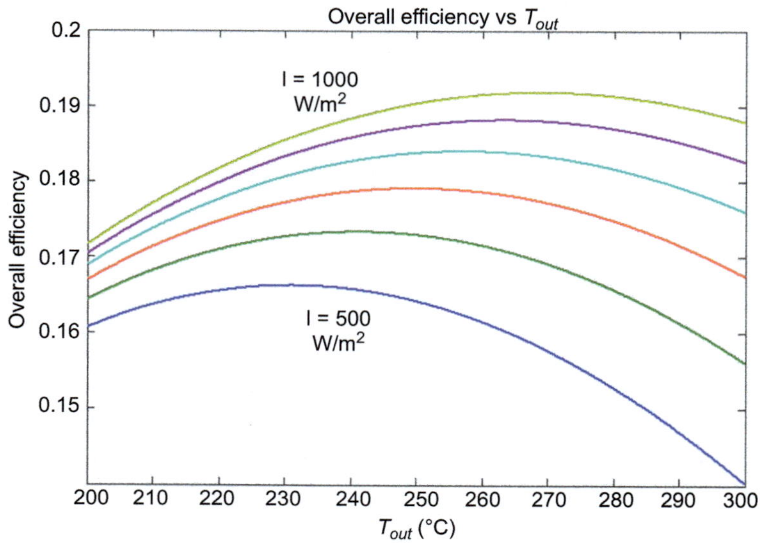

Figure 14.3 Efficiency evolution for different levels of solar radiation.

In the next simulation, an example of a complete operational day is simulated. As a control algorithm, a nonlinear MPC is used [14]. The MPC problem can be posed as follows:

$$\min_{\Delta u} J(\Delta u, y(t)) = \sum_{k=0}^{N_y} \left(y_{t+k|t} - w_{t+k}\right)^T Q\left(y_{t+k|t} - w_{t+k}\right) + \sum_{k=0}^{N_c} \Delta u_{t+k}^T R \Delta u_{t+k}$$

subject to

$$\begin{aligned}
&y_{min} \leq y_{t+k|t} \leq y_{max}, k = 1, \ldots, N_y \\
&\Delta u_{min} \leq \Delta u_{t+k} \leq \Delta u_{max}, k = 1, \ldots, N_c \\
&U_{min} \leq U(t+k|t) \leq U_{max}, k = 1, \ldots, N_c \\
&U(t+k|t) = U(t+k-1) + \Delta u(t+k-1), k = 1, \ldots, N_c \\
&x(t+k+1|t) = F(x(t+k), U(t+k)), k = 1, \ldots, N_y \\
&y(t+k) = H(x(k)), k = 1, \ldots, N_y
\end{aligned} \tag{14.15}$$

where w_{t+k} is the reference trajectory, Δu_{t+k} is the increment of the control signal, U is the control signal, x is the state vector, and y is the output. Q and R are weighting matrices that penalize the error between the output and the reference and the control signal, respectively. N_c is the control horizon and N_c is the prediction horizon. The functions F and H are nonlinear and assumed to be known.

In our case, the nonlinear model used in the MPC algorithm is a simplification of the system of Eq. (14.1). The loop is divided into 8 segments instead of the 172 used for simulation purposes. This simplification is required to alleviate the computational burden of the nonlinear programming problem. The optimization Matlab toolbox is used to solve it.

The power production obtained with the proposed three-layer algorithm is compared with that obtained when operating at the maximum allowable temperature, which is considered to be 295°C. The ACUREX solar collector field shuts down if the temperature surpasses 305°C [32]. The optimal set points are computed every 30 min.

The upper part of Fig. 14.4 shows the optimal set points computed by the first-layer algorithm. As can be seen, the best operating temperatures are not the maximum allowable and they are subject to change throughout the operation. The inlet temperature is that returned by the power conversion cycle. The nonlinear MPC algorithm achieves a good tracking performance with rise times between 4 and 6 m, with oscillations at the beginning of the day due to inlet temperature variations. The bottom part of Fig.14.4 plots the solar radiation corresponding to a clear day and the oil flow.

Fig. 14.5 shows net electrical production. It is worth pointing out that working at the optimal set points produces higher electrical production than working at the maximum allowable temperature. The production increment for the 6 h of simulation is about 8.058%, which constitutes an important gain.

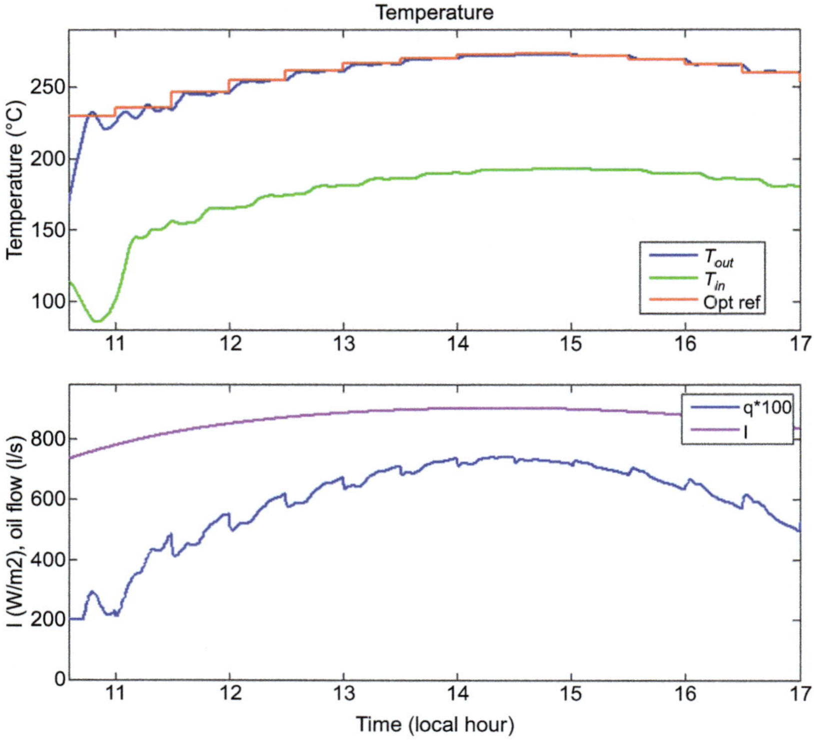

Figure 14.4 Optimal Set points tracking.

14.3 Optimization of flux distribution in solar tower plants

This section deals with the operation of a solar tower plant with flat receivers. A solar tower plant consists of a field of mirrors (*heliostats*) arranged around a tower equipped with a solar irradiation receiver. By tracking the sun, the heliostats focus the solar irradiance onto the receiver. The field can be composed of a large number of heliostats (more than 900 in recent commercial plants [8]), each of which is independently controlled [34].

Solar power towers are characterized by higher temperatures at the receiver (700−800°C with metal receivers and higher than 1000 °C with ceramic receivers) w.r.t. parabolic trough plants (ranging from 400 °C to 550 °C depending on the fluid used). In general, operation at higher temperatures results in larger throughputs and cheaper thermal energy storage [35].

When operating solar tower plants, the simplest (and most common) approach is to point at the center of the receiver to minimize energy losses by spillage of uncalibrated

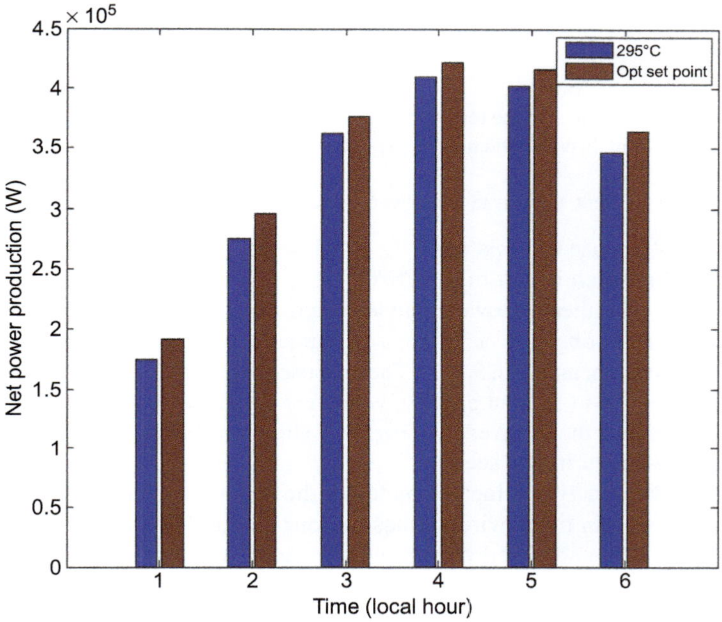

Figure 14.5 Hourly net electrical production.

heliostats [11]. However, problems due to an improper thermal distribution on the receiver may appear, leading to the deterioration of the receiver and even its destruction [12]. Optimizing the aiming points of the heliostats—to maximize the solar radiation collected and to avoid overheating problems—is an important task [11]. One possible solution—implemented in the CESA-1 plant at the PSA [36]—is to consider various focus points on the receiver. When multiple focuses are used, the control system should decide how many heliostats to "allocate" to each focus.

Some works related to the optimization of the flux distribution on solar receivers are available in the literature. For example, in Ref. [37] a heuristic method for the optimization of the temperature distribution was developed for a volumetric receiver. Real tests showed that the proposed strategy provides a more uniform heat distribution in the receiver. A method based on a genetic algorithm is proposed in Ref. [38] for the adjustment of the aim targets of parabolic heliostats in a small-scale tower plant, showing a reduction of undesired peaks in the flux distribution. The recent work of Belhomme et al. [39] presents a procedure for the optimization of the aim points based on the ant colony metaheuristic, whose effectiveness is demonstrated on a concentrated photovoltaic receiver test case.

Besides achieving an appropriate irradiance distribution—to minimize the thermal gradient over the receiver—the main objective is to maximize the collected energy, while taking into account the physical constraints on the receiver and the subsequent electricity generation system.

The method described in this section to optimize the flux distribution and the solar radiation collected by the receiver computes the aiming points for each heliostat, seeking three goals:

- Shave temperature peaks on the receiver, thus reducing the thermal stress on its components.
- Minimize the heat losses, which are proportional to the fourth power of the receiver's temperature.
- Maximize the incident solar irradiance over the receiver.

The plant model used in this paper is based on the CESA-1 solar thermal tower plant (Fig. 14.6), which is part of the PSA.

The CESA-1 solar thermal tower plant is composed of a field of 300 heliostats, each one providing a reflecting area of 39.6 m^2, a volumetric receiver, a steam generator, an energy storage system, and a PCS [40]. The proposed algorithm is tested on a subset of the available heliostats (180 out of 300), which are sufficient for the production of an adequate insolation at the receiver [11]. Fig. 14.7 shows the layout of the CESA-1 solar tower plant considered in this section.

The volumetric receiver is located on top of the tower, at a height of 86 m. It consists of a series of thin metal wire meshes (porous media) [41], with 2500/3000 kW of nominal/maximum absorbed power and a mean air temperature of 700°C. Solar radiation is concentrated by the heliostat field on the volumetric receiver surface, heating up the wire mesh. The heat is then transferred to the air circulating through the porous media.

To attain an appropriate flux distribution at the receiver, a five-point aiming strategy is currently implemented to cover the whole receiver surface [37]. For a more complete description of the plant and its operation, the reader is referred to Refs. [11] and [42].

Figure 14.6 CESA-1 tower plant (Copyright by PSA).

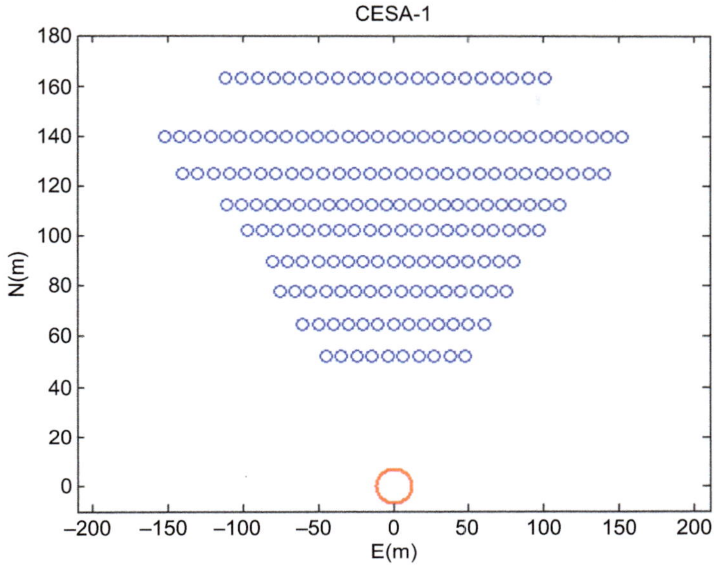

Figure 14.7 Layout of the CESA-1 solar tower plant.

14.3.1 Mathematical modeling

In this section, the mathematical model used to compute the flux density at the receiver is outlined. Two assumptions are made: (1) the surface of each mirror is perfectly flat and continuous and (2) the sun irradiance is modeled as a Gaussian distribution.

Several works dealing with the modeling of the heliostat field can be found in the literature. In Ref. [43], a real-time hybrid heliostat field simulator using Modelica as modeling language is presented.

In Ref. [44], various techniques for modeling the distribution of solar irradiance are presented. In this section, the sun is modeled as a circular source whose irradiance is normally distributed over its area [11]. As pointed out in Ref. [45], the total power reflected by a heliostat onto the receiver can be computed as follows:

$$P_h = I_n S_m cos(\theta) \varsigma_{at} \varsigma_{sp} \rho_m \tag{14.16}$$

where I_n is the direct normal irradiance, S_m is the mirror's area, θ is the incidence angle of the solar rays onto the heliostat, ς_{at} is the atmosphere transmittance, ς_{sp} is the spillage factor, and ρ_m denotes the mirror reflectivity.

To compute the atmosphere transmittance, a method proposed in Ref. [46] is used. The spillage factor is calculated by the formulas in Ref. [45], and the incidence angle θ is a function of the sun vector and the heliostat position [11]. The solar vector can be computed as follows:

$$\vec{S} = cos(\alpha_s)cos(\phi_s)\vec{i} + cos(\alpha_s)sin(\phi_s)\vec{j} + sin(\alpha_s)\vec{k} \tag{14.17}$$

where $\overrightarrow{S}$ is the solar vector, $\left(\overrightarrow{i},\overrightarrow{j},\overrightarrow{k}\right)$ are the vectors in the east, north, and zenith direction, respectively. The azimuth angle is represented by ϕ_s and α_s stands for the solar elevation, both computed using the model described in Ref. [21].

Let $P = (e_r, n_r, z_r)$ be an aiming point for a given heliostat located at (e_h, n_h, z_h). Where the coordinates are expressed in meters to the east, north and height of the tower center point (see figure 14.7) .The unit vector linking the target point P and the center of the heliostat is given by (Eq. 14.18):

$$\overrightarrow{R} = \frac{(e_r - e_h)\overrightarrow{i} + (n_r - n_h)\overrightarrow{j} + (z_r - z_h)\overrightarrow{k}}{\sqrt{(e_r - e_h)^2 + (n_r - n_h)^2 + (z_r - z_h)^2}} \tag{14.18}$$

The incidence angle can be obtained by computing the scalar product between $\overrightarrow{S}$ and $\overrightarrow{R}$:

$$cos(2\theta) = \overrightarrow{S}\cdot\overrightarrow{R} \tag{14.19}$$

The characterization of the flux density distribution produced by a heliostat when aiming at a desired point of the receiver has been tackled in different ways (e.g., [47]). In this chapter, the distribution is considered to be an elliptic Gaussian flux image [48]. The Sun can be modeled as a circular, normally distributed irradiance source described by Eq. (14.20) (in the sun coordinate reference system):

$$I(x_s, y_s) = \frac{P_h}{2\pi\sigma^2}\exp\left(-\frac{x_s^2 + y_s^2}{2\sigma^2}\right) \tag{14.20}$$

characterized by a standard deviation $\sigma = 2.325$ [11]. The goal is to compute the projection of the flux onto the receiver. Two coordinate transformations are needed: one from the sun to any given heliostat's reference, and another from any heliostat to the receiver reference. For any point P, the relation between its coordinates in the receiver reference system and the sun reference system is

$$P_r = R_h^r R_s^h P_s = MP_s \tag{14.21}$$

$$P_s = M^{-1} P_r \tag{14.22}$$

where R_h^r denotes the heliostat-to-receiver transformation matrix, and R_s^h the sun-to-heliostat transformation matrix. M is the coordinate transformation matrix between the sun and the receiver plane. For a comprehensive discussion of the computation of the sun image projection onto the receiver the reader is referred to Ref. [49]. Finally, the elliptic Gaussian flux distribution projected on the receiver is expressed by

$$f(x_r, y_r) = \frac{P_h}{2\pi\sigma^2}\exp\left(-\frac{k_1(x_r - \overline{x}_r)^2 + k_2(y_r - \overline{y}_r)^2}{2\sigma^2}\right) \tag{14.23}$$

where k_1 and k_2 can be obtained from the transformation matrix M^{-1}, which depends on complex trigonometric expressions [49].

14.3.2 Optimization problem

Basically, the proposed algorithm is intended to optimize the aim point assigned to each heliostat, depending on the position of the Sun and the heliostat field layout. Three (possibly conflicting) requirements have to be fulfilled by the optimization:

1. The aim points must lie on the receiver's surface.
2. The aim points should be chosen to achieve a uniform irradiance distribution over the receiver.
3. The energy collected by the receiver should be maximized.

To demonstrate the effectiveness of the proposed method, the receiver has been assumed to be a 4×4 m^2 rectangular surface for simulation purposes. To keep the problem at a manageable size, the receiver is discretized in squares of 0.25 × 0.25 m^2. Let F_r:$l \times m \rightarrow \mathbb{R}$ represent the flux density collected at each square of the receiver, i.e., the aggregate contribution of the whole heliostats field. Let $f_i(x_r, y_r; \overline{x}_i, \overline{y}_i)$ be the flux density—computed through Eq. (14.23)—over the square centered in (x_r,y_r), produced by heliostat i by aiming at the target point $(\overline{x}_i, \overline{y}_i)$. The total flux on the receiver can be computed as follows:

$$F_r(x_r, y_r) = \sum_{i=1}^{N_h} f_i(x_r, y_r; \overline{x}_i, \overline{y}_i) \tag{14.24}$$

where N_h is the total number of heliostats.

Let J designate the objective function to be minimized by the optimization problem:

$$J = \varrho_1 \sum_{j=1}^{l} \sum_{k=1}^{m} (F_r(j,k) - \max(F_r)) - \varrho_2 \sum_{j=1}^{l} \sum_{k=1}^{m} F_r(j,k) \tag{14.25}$$

where l and m denote the size of the discretized receiver's area, and max(F_r) is the peak value of the irradiance on the receiver. The first component of the objective function (Eq. 14.25) reveals how uneven the irradiance distribution is over the receiver's surface. The terms of the summation are the difference in the value of the flux density at each discretized point of the receiver's surface w.r.t. the peak value. The second part of Eq. (14.25) expresses the total amount of irradiance reaching the receiver. The weighting parameters ϱ_1 and ϱ_2 set the priority of achieving either a uniform flux distribution or the maximum aggregate energy.

Thus, the problem (Eq. 14.26) consists of determining the coordinates of the aim point $(\overline{x}_i, \overline{y}_i)$ for each heliostat i that minimize the cost function (Eq. 14.25):

$$\min_{\overline{x},\overline{y}} J(\overline{x}, \overline{y}; I_n, \theta, \ldots) \tag{14.26}$$

subject to

$$\bar{x}_{\min} \leq \bar{y} \leq \bar{x}_{\max}$$
$$\bar{y}_{\min} \leq \bar{y} \leq \bar{y}_{\max}$$

Notice that flux F_r over the receiver depends on the current aiming points and implicitly on several additional factors—such as the position (and orientation) of their relative heliostats, the time of the day, the day of the year, the transmittance of the atmosphere, and the mirrors' spillage factor. Hence, the optimization problem is highly nonlinear. Furthermore, even with just two decision variables per heliostat [i.e., $(\bar{x}_i, \bar{y}_i)$], the resulting global problem is characterized by a huge number of decision variables.

14.3.3 Simulation results

In the simulation results, the proposed method is compared to two approaches commonly implemented in solar tower plants:

1. The whole heliostat field aims at the center of the receiver (CS).
2. A multiaiming (MA) strategy, where five focuses are considered, similar to the CESA-1 plant [37]. For this simulation, focuses are located at points (x_r, y_r)={(−1.25,−1.25), (−1.25,1.25), (0,0), (1.25,1.25), (1.25,−1.25)} (m).

The nonlinear optimization algorithm is executed at intervals of 15 min, to achieve a sufficiently accurate solution. The problem (Eq. 14.26) is solved by an interior-point method [50], with the Matlab's Optimization Toolbox.

Fig. 14.8 shows a simulation of the flux obtained with CS and MA strategies on a day in June. The azimuth is $\phi_s = -0.45$ rad and the solar elevation $\alpha_s = 1.28$ rad. For the same day (and time of day), the flux distribution obtained by using the proposed algorithm is shown in Fig. 14.9. As expected, an improved distribution over the MA strategy is achieved.

Fig. 14.10 depicts the results of a simulation performed for a 3-h interval around noon on a May day (Julian day 130). Fig. 14.11 plots the values of direct solar irradiance used for this simulation; the value measured at the beginning of each interval is held constant throughout the optimization.

The top plot in Fig. 14.10 presents the total irradiance collected at the receiver, as obtained by the procedure proposed in this paper, and by the MA and center-aiming strategies. Indeed, the total flux projected by the heliostats controlled through the proposed algorithm is close to that obtained by pointing all the heliostats at the center of the receiver. From this point of view, the use of fixed aim points in the MA strategy results in being suboptimal. The bottom plot in Fig. 14.10 shows the difference between the maximum and minimum value of the flux over the receiver. As expected, the flux density distribution is greatly improved over the CS. Moreover, the evenness of the distribution is sensibly improved over the MA strategy.

Although in common plant operation, the heliostats are properly calibrated (see, e.g., the methodology presented in Ref. [51]), aiming inaccuracy may eventually arise.

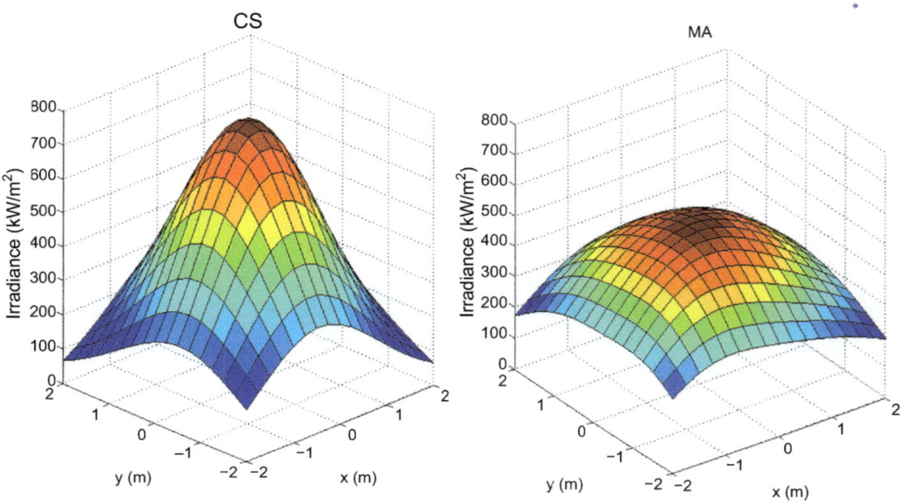

Figure 14.8 Irradiance distribution on the receiver: test on a June day (ϕ_s=−0.45 rad, α_s=1.28 rad) with center-aiming (CS) and multiaiming (MA) strategies.

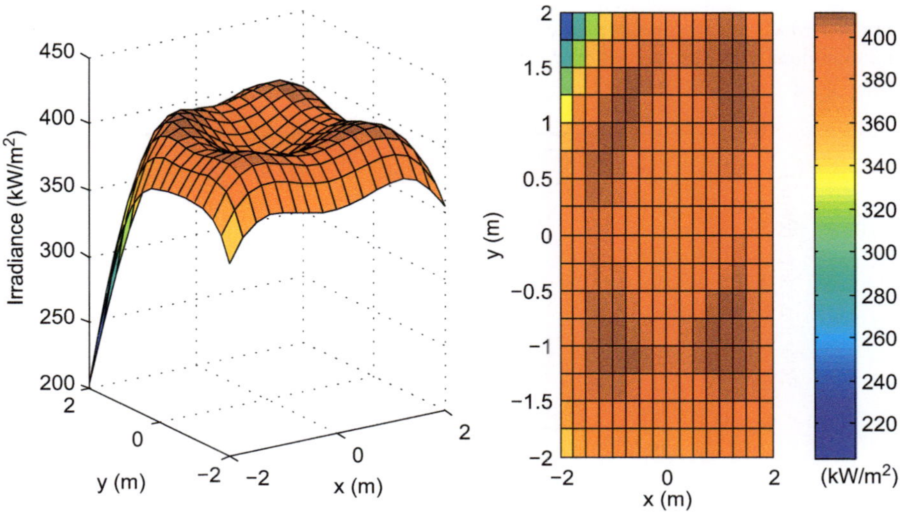

Figure 14.9 Same setting as in Fig. 14.8: irradiance distribution achieved by the proposed algorithm.

In particular, the error is more noticeable as the distance between the heliostats and the tower increases (a same error in the pointing angle would produce larger linear displacements on the receiver plane). For the purpose of analyzing the effect of calibration errors in the distribution of irradiance over the receiver, we carried out some simulations considering each heliostat affected by a random calibration error (uniformly distributed between 1 m and −1 m on both axes).

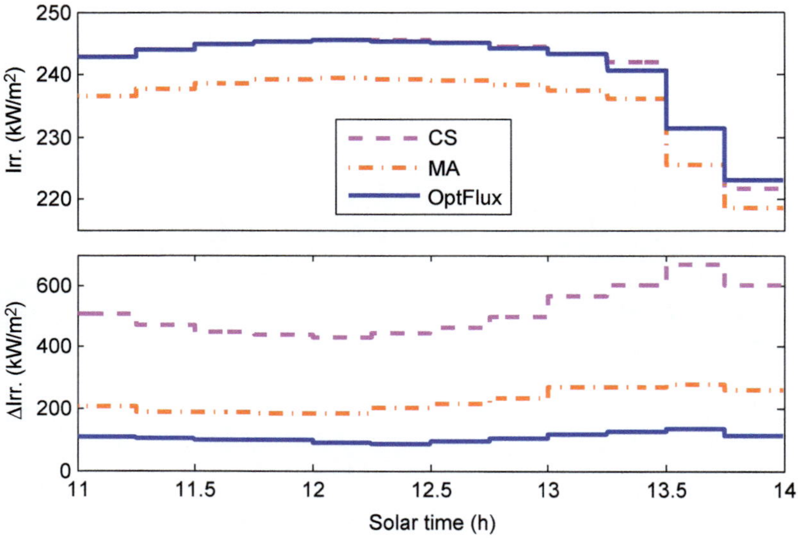

Figure 14.10 Day of the year 130 (May), time of the day (11−14 h). Top plot: comparison between the irradiance at the receiver for three different strategies: center-aiming (CS), multiaiming (MA), and the proposed one (OptFlux). Bottom plot: difference between maximum and minimum irradiance values over the receiver, for the three strategies.

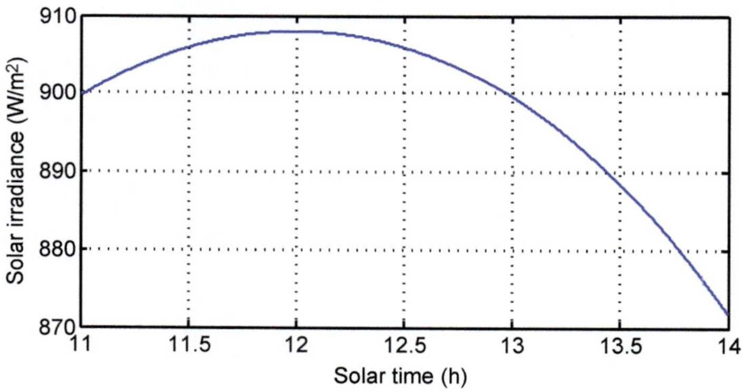

Figure 14.11 Evolution of the solar irradiance relative to Fig. 14.10. Values are considered constant through the optimization.

The example given in Fig. 14.12 compares the points on the receiver actually aimed at by all heliostats in ideal conditions (blue circles) and when affected by offset (red crosses). Notice that calibration errors may cause energy losses because some heliostats may aim outside the receiver surface.

Figs. 14.13 and 14.14 show the effect of calibration errors on the flux distributions obtained by the three different strategies (proposed algorithm, CS and MA). Even if the offset in the aim points deteriorates the evenness of the distribution obtained by the

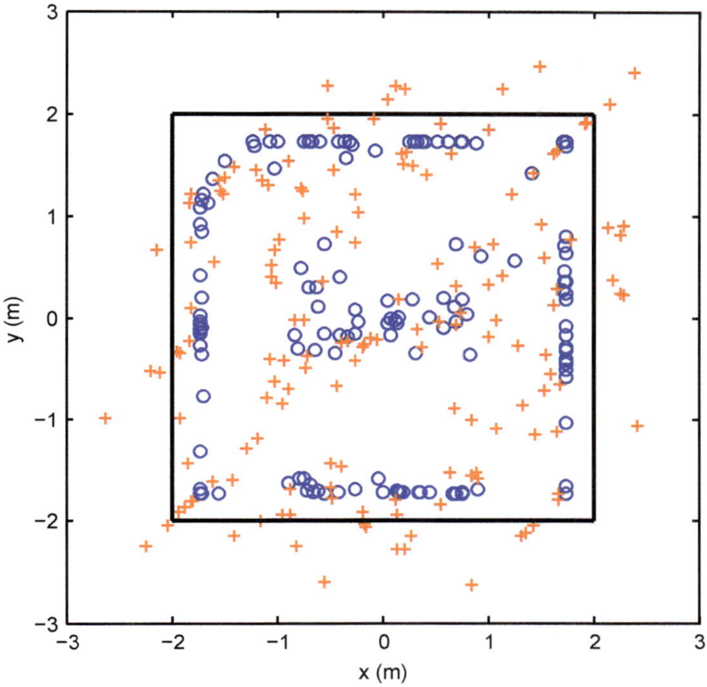

Figure 14.12 Points aimed at by the heliostats: *blue circles* (ideal pointing), *red crosses* (offset due to calibration errors).

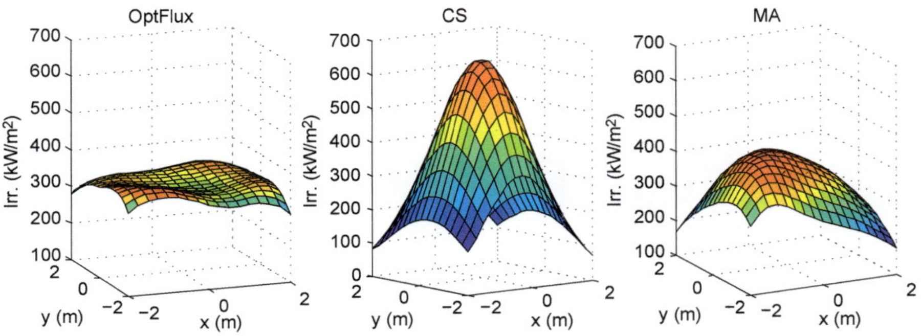

Figure 14.13 Flux density distributions affected by calibration errors on a summer day (17th July, 12 h). Left: proposed algorithm; center: center-aiming (CS); right: multiaiming (MA).

proposed algorithm, this is still capable of providing better uniformity w.r.t. the MA strategy.

Finally, the data in Table 14.2 refer to a simulation carried out in the same setting of the test in Fig. 14.10, with the additional effect of calibration errors. The average irradiance $\overline{F}_r$ collected at the receiver is equivalent to that achieved by the CS method. On

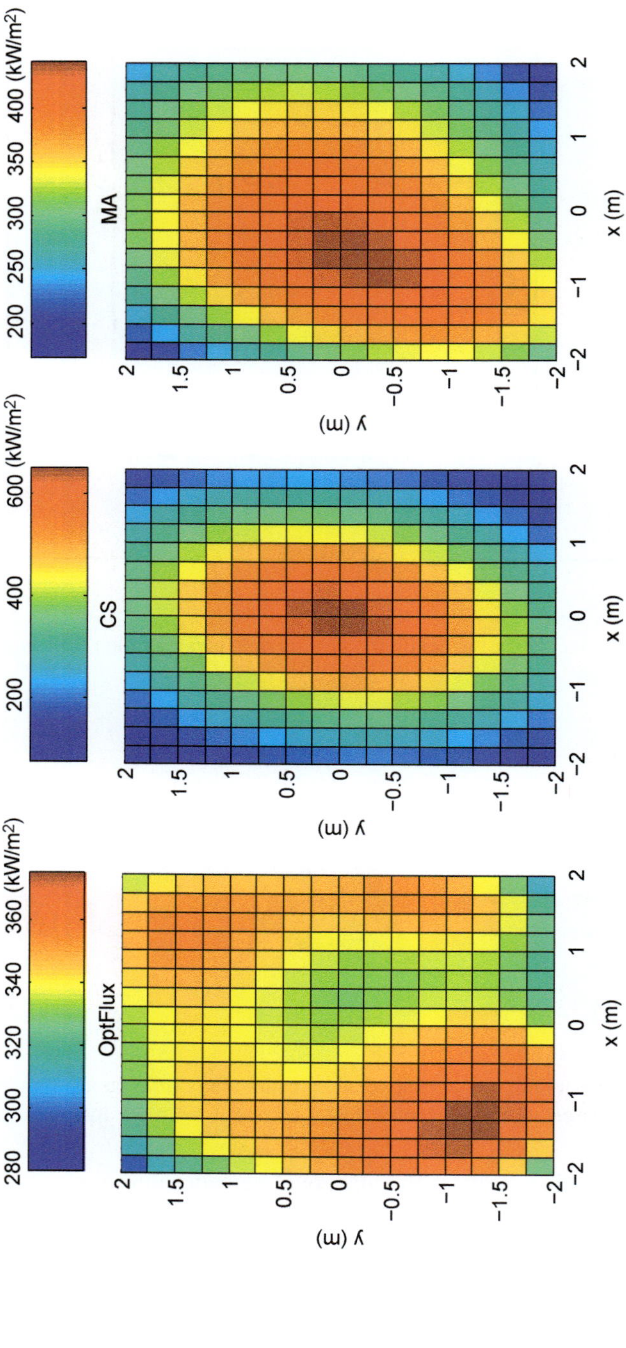

Figure 14.14 Alternative view of the irradiance distributions (w/calibration errors) in Fig. 14.13. Left: proposed algorithm; center: center-aiming (CS); right: multiaiming (MA).

Table 14.2 Average irradiance and maximum variation over the receiver's surface w/ calibration errors (day of year: 130)

	OptFlux	CS	MA
$\overline{F}_r\left[kW/m^2\right]$	229.66	229.70	223.97
$\Delta F_r[kW/m^2]$	95.02	576.64	274.25

CS, center-aiming; *MA*, multiaiming.

the other hand, the difference ΔF_r between the maximum and minimum irradiance value is substantially reduced by the proposed algorithm.

14.4 Conclusions and future works

This chapter has presented two cases demonstrating the effectiveness of the use of advanced control techniques when operating solar energy systems.

The first case describes a hierarchical control strategy to optimize the operation of solar trough plants. Commercial plants are usually considered to operate optimally when working at the maximum reachable temperature allowed by environmental and security conditions. Higher temperatures produce higher efficiencies in the power conversion cycle, but the thermal losses of the solar field are also increased. On the other hand, working at higher temperatures implies lower flow rates so that the oil pump consumption is reduced. The optimal set point (the working temperature of the field that maximizes the overall efficiency of the plant) has to take into account these factors. A model of the PSA solar trough plant is used to test the proposed algorithm. A simulation example has shown that important gains in electricity production are obtained by using the proposed hierarchical control scheme.

The second case deals with one of the most important issues in the operation of solar tower plants: achieving a uniform flux density distribution at the receiver and maximizing the energy collected by the receiver. Simulations have been carried out to compare the effectiveness of the proposed method with two commonly implemented approaches: the CS and the MA strategy. The results show that the proposed algorithm sensibly improves the flux density distribution obtained by an MA strategy. At the same time, it is capable of optimizing the total irradiance collected at the receiver, performing in this sense equivalently to the CS. Furthermore, the proposed method has demonstrated to be robust to calibration errors.

14.4.1 Future works

Since the first commercial solar trough plants were commissioned the size of the plants has grown considerably. They are usually divided into quadrants or sectors composed of several loops, which can be operated independently. Each sector can be affected by different levels of solar radiation and optical efficiency. This fact produces that the optimal set point may be different for each sector. Distributed control and optimization strategies are an open and interesting line of research.

As far as solar tower plants are concerned, two future lines of research can be mentioned. The first deals with the optimal scheduling of solar tower plants. The idea is similar to that presented here for solar trough plant: using a mathematical model to obtain the best operating point for maximizing the ROI.

The second line concerns the use of distributed or coalitional approaches for the optimization of flux distribution. The commercial solar tower plants can be composed of thousands of heliostats. The number of decision variables can be too high to solve the centralized optimization problem in real time. Distributed or coalitional optimization algorithms can be very useful in addressing the problem.

References

[1] Camacho EF, Berenguel M. Control of solar energy systems. In: 8th IFAC Symposium on advanced control of Chemical Processes; 2012. p. 848−55.

[2] N. A. Engineering, National academy of engineering, grand challenges for engineering. 2008. www.engineeringchallenges.org.

[3] Camacho EF, Samad T, Garcia-Sanz M, Hiskens I. Control for renewable energy and smart grids. Technical Report. IEEE Control Systems Society; 2011. www.ieeecss.org.

[4] Camacho E, Rubio F, Berenguel M, Valenzuela L. A survey on control schemes for distributed solar collector fields. part I: modeling and basic control approaches. Solar Energy 2007;81:1240−51.

[5] Camacho EF, Rubio FR, Berenguel M, Valenzuela L. A survey on control schemes for distributed solar collector fields. part II: advanced control approaches. Solar Energy 2007b; 81:1252−72.

[6] Limon D, Alvarado I, Alamo T, Ruíz M, Camacho E. Robust control of the distributed solar collector field ACUREX using MPC for tracking. In: Proceedings of the 17th World Congress the International Federation of Automatic control; 2008. p. 958−63.

[7] Solar Millennium AG, the construction of the Andasol power plants. 2011. http://www.solarmillennium.de/Technologie/Referenzprojekte/Andasol/Die_Andasol_Kraftwerke_entstehen_,lang2,109,155.html.

[8] Abengoa-Solar. Plataforma solucar: Planta ps20. 2009. http://www.abengoasolar.com/web/es/plantas_solares/plantas_propias/espana/.

[9] Duffie J, Beckman J. Solar engineering of thermal processes. 2nd ed. Wiley-Interscience; 1991.

[10] Herrmann U, Kearney DW. Survey of thermal energy storage for parabolic trough power plants. Journal of Solar Energy Engineering 2002;124:145−52.

[11] Camacho EF, Berenguel M, Rubio F, Martínez D. Control of solar energy systems. Springer-Verlag; 2012.

[12] Spirkl W, Ries H, Kribus A. Optimal parallel flow in solar collectors for nonuniform irradiance. Transactions of the ASME 1997;118:156−9.

[13] Camacho EF, Gallego AJ. Optimal operation in solar trough plants: a case study. Solar Energy 2013;95:106−17.

[14] Camacho EF, Gallego AJ. Model predictive control in solar trough plants: a review. In: 5th IFAC Conference on nonlinear MPC, September 17–20, Sevilla (Spain); 2015.
[15] Zarza E, Valenzuela L, Leon J, Hennecke K, Eck M, Weyers H, Eickhoff M. The DISS project: direct steam generation in parabolic troughs. In: Proceedings of ASME International solar energy; 2001.
[16] Gallego AJ, Camacho EF. Adaptative state-space model predictive control of a parabolic-trough field. Control Engineering Practice 2012;20(9):904–11.
[17] Brus L, Zambrano D. Black-box identification of solar collector dynamics with variant time delay. Control Engineering and Practice 2010;18:1133–46.
[18] Roca L, Guzmán JL, Normey-Rico JE, Berenguel M, Yebra L. Robust constrained predictive feedback linearization controller in a solar desalination plant collector field. Control Engineering and Practice 2009;17(9):1076–88.
[19] Carmona R. Análisis, modelado y control de un campo de colectores solares distribuidos con sistema de seguimiento en un eje. Ph.D. thesis. Universidad de Sevilla; 1985.
[20] Camacho EF, Rubio FR, Berenguel M. Advanced control of solar plants. Springer-Verlag; 1997.
[21] Goswami DY, Kreith F, Kreider JF. Principles of solar engineering. 2nd ed. 2000.
[22] Baerh HD. Tratado Moderno de Termodinámica. 1st ed. Springer-Verlag; 1965.
[23] ITET. The IEA/SPSS solar thermal power plants, vol. 2. Berlin: Springer; 1986.
[24] Kapooria RK, Kumar S, Kasana KS. An analysis of a thermal power plant working on a rankine cycle: a theoretical investigation. Journal of Energy in Southern Africa 2008;19(1):77–83.
[25] Stine WB, Geyer M. The power from the sun. 2001. http://www.powerfromthesun.net.
[26] Camacho EF, Berenguel M, Gallego AJ. Control of thermal solar energy plants. Journal of process control 24(2): 332–340.
[27] Berenguel M, Cirre CM, Klempous R, Maciejewski H, Nikodem M, Nikodem J, Rudas I, Valenzuela L. Hierarchical control of a distributed solar collector field. Computer Aided Systems Theory EUROCAST 2005 2005;3643:614–20.
[28] Cirre CM, Berenguel M, Valenzuela L, Klempous R. Reference governor optimization and control of a distributed solar collector field. European Journal of Operational Research 2009;193:709–17.
[29] Fosso OB, Gjelsvik A, Haugstad A, Mo B, Wangensteen I. Generation scheduling in a deregulated system the Norwegian case. IEEE Transaction on Power Systems 1999;14(1): 75–81.
[30] Kratzenberg MG, Colle S, Beyer HG. Solar radiation prediction based on the combination of a numerical weather prediction model and a time series prediction model. In: EUROSUN 2008, 1st International Congress on heating, Cooling and Buildings; 2008.
[31] Potter MC, Wiggert DC. Mecánica de Fluidos. 3rd ed. Paraninfo; 2003.
[32] Gallego AJ, Camacho EF. Estimation of effective solar radiation in a parabolic trough field. Solar Energy 2012;86:3512–8.
[33] Cengel Y, Cimbala J. Mecánica de los fluidos: fundamentos y aplicaciones. Mcgraw Hill; 2006.
[34] Gallego AJ, Fele F, Camacho EF. On the optimization of flux distribution in solar tower plants with flat receivers. In: ECC 2014. Strasbourg, 24–27 June; 2014.
[35] Romero M, Buck R, Pacheco JE. An update on solar central receiver systems, projects and technologies. Transactions on the ASME 2002;124:98–108.
[36] CIEMAT. Plataforma Solar de Almería. 2015. http://www.psa.es.
[37] García-Martín FJ, Berenguel M, Valverde A, Camacho EF. Heuristic knowledge-based heliostat field control for the optimization of the temperature distribution in a volumetric receiver. Solar Energy 1999;66(5):355–69.

[38] Hamza K, Saitou K, Gandhi U. Optimal scheduling of parabolic heliostats aim targets in a mini-tower solar concentrator system. In: Proceedings of the ASME 2011 International Design Engineering Technical Conferences and Computers and Information in Engineering Conference, Washington (USA); 2011.
[39] Belhomme B, Pitz-Paal R, Schwarzbozl P. Optimization of heliostat aim point selection for central receiver systems based on the ant colony optimization metaheuristic. Journal of Solar Energy Engineering 2013;136(1).
[40] Yu Q, Wang Z, Xu E, Zhang H, Lu Z, Wei X. Modeling and simulation of 1 MWe solar tower plant's solar flux distribution on the central cavity receiver. Simulation Modelling Practice and Theory 2012;29:123–36.
[41] Ávila-Marín AL. Volumetric receivers in solar thermal power plants with central receiver system technology: a review. Solar Energy 2001;85:891–910.
[42] Haeger M, Keller L, Monterreal R, Valverde A. Phoebus technology program solar air receiver (TSA). Technical Report. Madrid: CIEMAT; 1994.
[43] Bonilla J, Roca L, González L, Yebra LJ. Modelling and real-time simulation of heliostat field in central receiver plants. In: Proceedings of the 6th Int. Conf. On mathematical Modelling, Vienna (Austria); 2009. p. 2576–9.
[44] Buie D, Monger AG. The effect of circumsolar radiation on a solar concentrating system. Solar Energy 2004;76(1–3):181–5.
[45] Collado FJ, Turégano JA. Calculation of the annual thermal energy supplied by a defined heliostat field. Solar Energy 1989;42(2):149–65.
[46] Hottel HC. A simple model for estimating the transmittance of direct solar radiation through clear atmospheres. Solar Energy 1975;18:129–34.
[47] Collado FJ. One-point fitting of the flux density produced by a heliostat. Solar Energy 2010;84:673–84.
[48] Guo M, Wang Z. On the analysis of an elliptical Gaussian flux image and its equivalent circular Gaussian flux images. Solar Energy 2011;85:1144–63.
[49] Collado FJ, Gómez A, Turégano JA. An analytic function for the flux density due to sunlight reflected from a heliostat. Solar Energy 1986;37(3):215–34.
[50] Bertsekas DP. Convex analysis and optimization. 1st ed. Athena Scientific; April 2003.
[51] Berenguel M, Rubio FR, Valverde A, Lara PJ, Arahal MR, Camacho EF, et al. An artificial vision-based control system for automatic heliostat positioning offset correction in a central receiver solar power plant. Solar Energy 2004;76:563–75.

Part Six

Cost competitive CST plants concepts

Linear Fresnel reflector (LFR) plants using superheated steam, molten salts, and other heat transfer fluids

15

M. Collares-Pereira, D. Canavarro, L.L. Guerreiro
University of Evora, Evora, Portugal

15.1 Introduction (motivation)

As explained in Chapter 3 "Improved field design for linear Fresnel reflector systems," there are substantial advantages of linear Fresnel reflector (LFR) technology, when looking for low-cost energy delivery.

Low-lying flat mirrors, fixed receivers, and best ground cover ratio of all concentrated solar power (CSP) technologies are all good reasons for LFR to have inherently lower cost when compared with parabolic trough (PT); but conventional LFR suffers from a lower efficiency, mainly because it loses more incident power to the cosine-of-incidence-angle effect.

In Chapter 3 advanced LFR concepts were presented that uniquely resolve limitations of both PT and conventional LFR, offering the possibility of delivering much higher concentration with all the advantages explained there.

The new solutions presented must now be discussed from other complementary points of view, including heat transfer, delivery and conversion, system integration, and operation, for the complete picture to emerge, with all the potential advantages of these advanced concepts.

However few more components and system-related comments are in order.

15.2 Heat transfer fluids

The most common heat transfer fluids are water (under pressure), steam (saturated and superheated), and thermal oils. New fluids are emerging, such as, molten salts, air, and CO_2.

Conventional LFRs have been offered [1−3] to produce water or saturated steam for operating temperatures up to 270°C. They are typically associated with a Rankine cycle for electricity production [1,3] or with a direct thermal application such process heat, air conditioning/refrigeration, and desalination as in the case of [2].

Advances in Concentrating Solar Thermal Research and Technology
http://dx.doi.org/10.1016/B978-0-08-100516-3.00015-0

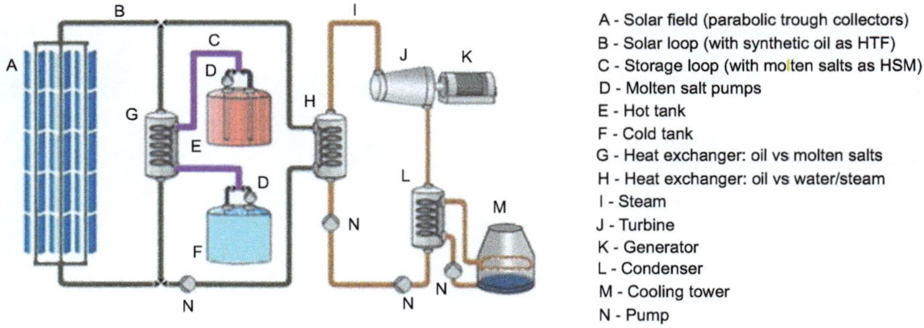

Figure 15.1 Schematics of a conventional solar-thermal electricity plant with storage.

PTs have been used for all these applications as well. In the case of electricity production the most common HTFs are thermal oils, for energy delivered up to 390°C, providing a much higher solar-to-electricity conversion efficiency [4]. In southern Iberian Peninsula locations these systems can have a yearly conversion efficiency of about 15%, while the LFRs referred to above would have about 8–9%.

Thermal oil as HTF usually requires a heat exchanger, as in solar-thermal electricity (STE) applications to produce steam for the Rankine cycle (steam turbine).

Fig. 15.1 shows a typical configuration of a complete system.

This figure also shows another characteristic [5] of many of the present solar power plants for STE: the thermal energy storage (in this case with molten salts in a two-tank configuration) connected to the thermal oil loop by means of a heat exchanger. Energy storage is a highly desirable feature and sets STE completely apart from PV electricity production. Many hours of storage can be provided in this fashion, allowing for nominal power production for many hours on demand, even with no sunshine: thus STE is said to be truly dispatchable.

Direct steam production has also been considered and amply studied for these PT systems [6], with the goal of eliminating the oil to steam heat exchanger and perhaps also to raise the operating temperature and pressure, for a higher conversion efficiency. But direct steam production with PTs has still many operational problems associated with it and does not help with the question of thermal energy storage, which is difficult to achieve, in particular when steam is in a superheated form, which leads to higher pressures and costs.

Therefore, why not use the molten salts directly as HTF and as means to store energy, transforming the scheme in Fig. 15.1 to that of Fig. 15.2?

This is exactly what is being done at the Central Receiver plant of Torressol/Gemasolar [7], as shown in Fig. 15.3.

The question is: can this also be done with linear concentrator technology—PT and LFR?

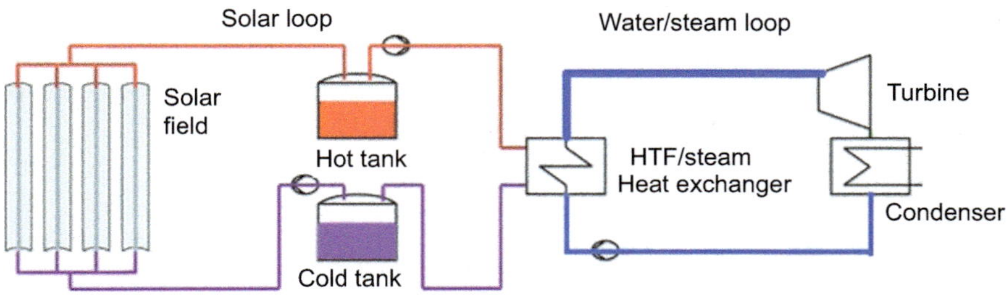

Figure 15.2 Solar system operating with molten salts as HTF and as heat storage medium.

Figure 15.3 Gemasolar plant: 19 MWe, 15 h of storage (Seville, Spain), with operating temperatures going up to 565°C on the molten salt receiver loop (540°C/110 bar for the steam produced). Source: SENER/Torresol Energy.

This chapter considers the use of molten salts in this fashion, using the latest optical concentration developments, the advanced LFR technology.

Before going further, a comment on other HTFs is in order, in particular, gases like air. The use of gases is potentially very interesting but brings an added problem, that of the substantially smaller heat transfer coefficient within the receiver, requiring higher pressures (as also in the case of CO_2) and higher temperatures The usefulness of the concept will only be apparent for much higher temperatures (i.e., from 600°C above). This likely means that they are better suited to be combined with central receiver (three-dimensional, 3D) type optics due to its capacity to deliver substantially higher concentration values. They constitute a topic of R&D today, in connection with suitable Brayton thermodynamic conversion cycles and also in search for solutions around the issue of suitable compatible energy storage.

Therefore, they will not be further considered in this chapter, dedicated to a linear (two-dimensional, 2D) technology, that of advanced LFR.

15.3 Higher temperatures: molten salts as HTF and thermal energy storage medium

Conventional PT plants already store heat in sensible form, transferred from the heated thermal oil. These plants use a binary molten salt mixture, a so-called solar salt, an eutectic mixture formed by 40% KNO_3 and 60% $NaNO_3$ by weight. It has a fusion point around 222°C and is used to fit within the operational limits of the thermal oil (up to ca. 393°C), meaning that it usually operates within the 290–393°C range, that is, providing a ΔT of 100°C.

It is simple to calculate that if storage is required for 7.7 h of peak power production (a common situation) as in [8], it is necessary to have about 26,500 ton of molten salts. If the salts could be heated to a higher temperature, as in the case of [7], the same storage capacity would require 2.65 times less salt [(565–290)/(390–290)], a substantial reduction on volume and costs.

Therefore, there is a strong incentive to operate at higher temperatures, from the perspective of storage size alone. But the other benefit from higher temperatures will come from higher Rankine cycle efficiency: available turbines range from efficiency values 0.36–0.39 with steam at 380°C to 0.42–0.44 for steam at 540°C [9].

A concern with salts is the freezing/fusion point. R&D is under way to study other mixtures with substantially lower freezing points, for future consideration in CSP systems.

Other questions involved in the use of molten salts in linear CSP plants have to be extensively dealt with before the concept can be used. In Europe demonstration plants such as that of PCS at Casaccia, Italy or the EMSP in Evora, Portugal, and the solar direct molten salts loop next to an existing power plant at Priolo Gargalo, Sicily, Italy are a part of that effort [10–12].

Specifically problems concerning freeze protection, start-up, and shut down in daily operation must be thoroughly addressed at R&D level before this solution becomes a commercial option. Currently, there is a broad ongoing research related to molten salts, namely:

- new mixture (ternary and quaternary) full characterization, presenting lower freezing points (many of them already studied at laboratory level);
- corrosion and degradation mechanisms, studying the compatibility of new salt mixtures with different metallic materials used for piping, tank, and circuit components;
- development of new components and associated materials for usage in direct contact with molten salts;

- new operational strategies, with an optimization of operation and maintenance when using molten salts directly as HTF and HSM (heat storage media).

A relevant fact is that the combination of molten salts and LFR technology simplifies some of the unsolved issues, because of the much simpler possible mechanical configurations and arrangements available, when compared with parabolic trough technology, as follows:

- LFR technology operates with fixed receivers in contrast with those in standard PT collectors, where potential problems arising from movable joints or flexible hoses must be addressed.
- LFR technology places the evacuated receivers high above the ground (usually 8 m or more) and that makes gravity assisted salts drainage a simpler matter.
- LFR technology reduces considerably the length of the piping loop needed, especially not just because of high concentration, as explained for advanced LFR technology, but also because of solutions like the one described in the following, where a double receiver is placed on a single tower.
- LFR technology can be implemented easily with a slight pending gradient, to facilitate draining operations.

For all the mentioned reasons advanced LFR technology has been gathering renewed interest, since it can supply both electricity (using molten salts as HTF) or process heat (using water/steam as HTF) at a lower cost than today's STE standard.

In terms of energy storage concepts, there will also be an opportunity for evolution; for instance, change from two-tank configurations to one-tank only and hybrid solutions with salts and different solid materials as storage media with extensive ongoing research to validate new materials. In fact, this area has been producing new interesting opportunities for future R&D+I (research and development plus innovation), for instance by incorporating residues from other industries to operate as new solid storage materials or by developing concepts related to safer operation and monitoring of solar plants, like prewarming and draining concepts (see Fig. 15.4) [13].

All these opportunities are being seized not only by R&D institutions but also by industry, recognizing that these are new opportunities for competitive STE production.

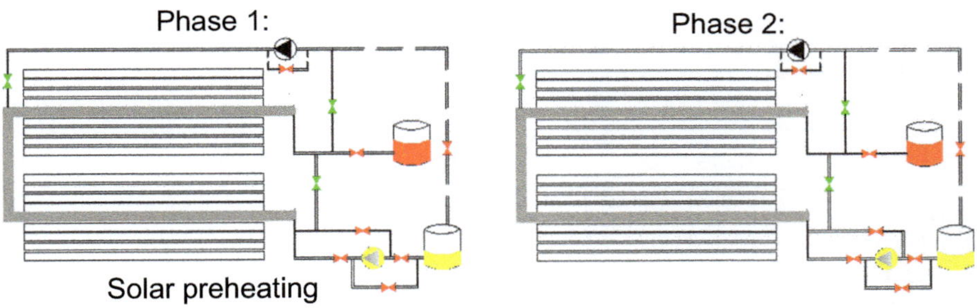

Figure 15.4 New operating concepts using molten salts as direct HTF.

15.4 Advanced LFR and molten salts: a new concept plant

In Chapter 3, advanced LFRs solutions were presented. Very high concentration values (about 50X or even above) were obtained. These will reduce thermal losses substantially, when compared with the typical 26X concentration value of PTs.[1]

A yearly expected performance simulation can be made for the collector described in Fig. 11, of Chapter 3, namely the dual asymmetric macrofocal CEC LFR. Rather than presenting here an optimized version of this concept with different widths and curvature radii for the primary mirrors, we note that the configuration shown starts with 1 m fixed width mirrors commercially produced [14], with different radius and relative spacing, in comparison with that of their commercial solution.

The evacuated tubular receiver used is the usual 70 mm inner tube as supplied by companies like Archimede Solar [15] or Schott Solar [16].

Table 15.1 summarizes the parameters of the final configuration, including the expected (calculated) optical efficiency at normal incidence through ray tracing.

15.5 Yearly performance

The results presented in the following paragraphs follow a analysis similar to the one made in [17,18]. They take into account a typical choice of materials and respective optical properties.[2] It should be noted that the intercept factor η_{int} was considered to be 0.98, given that the second stage concentrator and the wide acceptance angle can indeed compensate for eventual manufacture and for tracking inaccuracies.

The plant configuration for the calculation made below is that of Fig. 15.2. Inlet fluid temperature to the collector field will be 290°C and outlet temperature below 565°C.

Yearly performance can now be evaluated. To do so, the direct normal irradiance (DNI) hourly Meteonorm data for Faro (Portugal), 2234 kW h/m^2/year, latitude 37.03°N and for Hurghada (Egypt), 3043 kW h/m^2/year, latitude 27.26°N are chosen. The idea is to show results for a location that has one of the highest DNI values in Europe and also to show the impact that a lower latitude (and higher irradiation value)

[1] Concentration is defined to the perimeter of the receiver, not to its diameter, as it is sometimes done; in that case values are reported above 80X, but that is just wrong!

[2] Materials properties used for the collectors in the comparison.

	Reflectivity	Absorptivity	Transmissivity
Primary mirror	92% ([19])	—	—
Secondary mirror	92% ([19])	—	—
Receiver tube	—	96% ([16])	—
Glass cover	—	—	96% AR-coated glass tube ([16])

Table 15.1 **Optical and other parameters for the advanced LFR with an half-acceptance angle of 0.75 degrees**

Configuration	Primary width (m)	Total mirror aperture (m)	Receiver height (m)	Number of mirrors	Mirror length (m)	φ (°)	C (X)	θ (°)	CAP	η_{opt0}
Advanced LFR	25.7	22	10.8	22	1	48.57	45	0.75	0.58	0.70

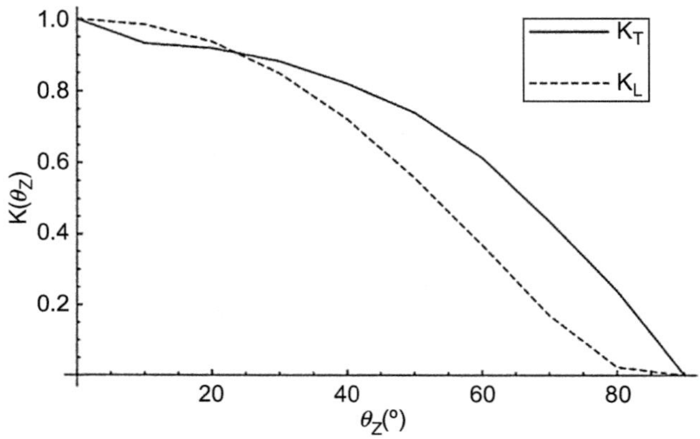

Figure 15.5 Transversal (K_T) and longitudinal (K_L) IAM curves of the advanced LFR concentrator.

has on the expected results. Lower latitudes are more favorable for LFR collectors, since they reduce the cosine-of-incidence-angle effect.

The calculation is made on yearly terms, with real solar radiation data and ray tracing, taking into account end effects, cosine-of-incidence-angle effects, and incidence-angle-modifiers effects for the incident rays, which are not limited to transversal, and at normal incidence (the design condition).

Fig. 15.5 shows the incidence-angle modifiers (longitudinal and transversal) for the advanced LFR concentrator (Fig. 10, Chapter 3).

The results of the yearly optical performance of the advanced LFR concentrator are shown in Table 15.2.

To complete the calculation, all the way to final solar-to-electricity conversion efficiency, it is important to include all thermal losses and expected performance from thermodynamic conversion from heat to electricity in a conventional turbine. In the present project the plan is to use as HTF a combination of molten salts (solar salt), to extract heat at 565°C. The very high concentration achieved and the evacuated tube technology, enable the system to operate with high efficiency at 565°C—very important to increase the power block overall efficiency.

For the calculations the so-called SCHOTTSOLAR receiver, tested for heat losses in different R&D institutions, was considered. Heat losses from the evacuated tubular receiver can be extrapolated from the data provided by SCHOTTSOLAR [16,20,21] and confirmed by NREL [22] from tests made up to 500°C.[3]

From [20] the following fit to the data was used:

$$H_T = 0.141 \times T_{abs} + 6.48 \times 10^{-9} \times T_{abs}^4 \tag{15.1}$$

[3] Archimede Solar confirms very similar (if not even better) behavior for their tubes and, therefore, the decision was made to use this published data.

Table 15.2 Energy delivery, per meter of tube length, at zero thermal losses

Optic	Location	DNI (kW h/m²/year)	Collected energy [same vacuum tube (kW h)][a]	Collected energy per m² of aperture area (kW h/m²)
Advanced LFR	Faro, Portugal	2359.01	19,795	899
	Hurghada, Egypt	3044.48	27,842	1265

[a]For receivers of 70 mm of diameter and 1 m of length.

yielding a value of 740 W/m (m of tubular length) at 565°C, with accuracy claimed to be of the order of ±10 W/m up to 500°C (T_{abs} in °C).

To these tubular losses, a further loss mechanism must be added, corresponding to connecting pipe losses, estimated to be between 65 W/m (inlet fluid temperature at 290°C) and 130 W/m (outlet fluid temperature at 565°C) of linear piping, from a simple model of pipes insulated with rock wool (10 cm thick) and carrying a hot fluid. With these values it is possible to simulate the performance of a specific collector field.

A plant (>60 MW peak power) size of 250 000 m² of mirrors was chosen (see Fig. 15.6) with 23 rows of 500 m length each. The double tubular receiver allows for a strategy of using one side with the HTF flowing in one direction and, on the other side, in the opposite direction with a U-turn at one extremity. This configuration reduces piping substantially and also easily accommodates receiver expansion at the U end.

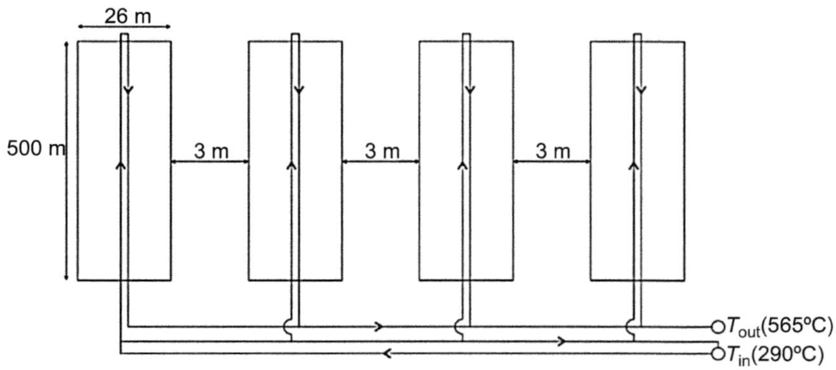

Figure 15.6 Schematic plant layout: not to scale for better viewing and representing just a portion of the total plant (4 of 23 rows).

Finally, the electrical energy delivered can be calculated according to Eq. (15.1). On an instantaneous basis the electricity produced can be calculated on an hourly basis according to:

$$q_{\mathrm{el,hi}} = \left[A_{\mathrm{col}} \times DNI \times \eta_{\mathrm{opt}} \times \eta_{\mathrm{int}} \times K_{\mathrm{L}} \times K_{\mathrm{T}} - (H_{\mathrm{T}}(T_{\mathrm{HTF}}) + H_{\mathrm{R}}(T_{\mathrm{HTF}}) + H_{\mathrm{P}}(T_{\mathrm{HTF}}))\right]^{+} \times \eta_{\mathrm{S}} \times \eta_{\mathrm{T}} \tag{15.2}$$

where A_{col} is the total mirror area of the collector, T_{HTF} is the temperature of the HTF fluid, H_{R} are the heat losses of the receiver, H_{P} are the heat losses of the connecting pipes, η_{S} is the efficiency of the steam generation, and η_{T} is the efficiency of the turbine. The superscript "+" means that only positive values of the quantity in straight brackets are to be considered, that is, from the energy available from the concentrator optical performance thermal losses are subtracted for each value of T_{HTF}. The result, if positive, is multiplied by the thermodynamic conversion efficiencies to yield electricity produced. The total inlet pipe length of 1100 m and outlet pipe length to be 550 m were assumed, and energy-weighted average operating temperature (with a result higher than the heat loss according to Eq. (15.1), at $T_{\mathrm{in}} + T_{\mathrm{out}}/2$) were used.

The yearly sum $Q_{\mathrm{el,yearly}}$ is then given by:

$$Q_{\mathrm{el,yearly}} = \sum_{i=1}^{8760} q_{\mathrm{el,hi}} \tag{15.3}$$

Note: Instead of the product $K_{\mathrm{L}}*K_{\mathrm{T}}$ a more rigorous approach was followed—that of using ray tracing to calculate the real incidence-angle modifier for each solar elevation angle and azimuth pair.

It was further assumed that overall thermodynamic conversion efficiency, from heat to electricity, is 0.42*0.98. Turbine efficiency (η_{T}) was chosen to be 0.42 [9], from commercial offers between 0.40 and 0.42 for steam at 540–545°C and 100–110 bar. The value 0.98 is the conversion efficiency (η_{S}) from molten salt–delivered heat at 565°C to steam production at 540°C and 110 bar as a result of practical plant operation experience at Gemasolar/Torresol [9]. Table 15.3 shows the results for the expected energy delivered to the turbine, for the electricity produced and also the yearly solar-to-electricity conversion efficiency for Faro and for Hurghada.

15.6 Discussion

These results compare very favorably with present day expected commercial efficiency from Fresnel and PT technologies, when installed in a place like Faro, with values like 8–9% for LFR and 14–16% for PT, respectively.[4]

[4] Conventional PT is also expected to improve its performance with larger troughs and other optical improvements, but not necessarily with the ease and low cost of the LFR solution proposed here.

Table 15.3 Yearly results, energy and final conversion efficiency, for the advanced LFR concept plant with a total mirror area of 250,000 m^2 (no storage)

Location	Thermal Energy delivered (kWh)	Electricity produced (kWh)	Total average yearly efficiency
Fora	201×11^8	8.56×10^7	0.145
Hurghada	3.02×10^8	1.24×10^8	0.163

It is also interesting to note the effect that lower latitude has on the final result in terms of overall efficiency.

At the time of writing of this chapter in 2016, these results are no more than a calculation exercise, with the goal of showing the potential behind advanced LFR technology. However a research project for building and testing a prototype with 2200 m^2 of mirror area has been undertaken by a consortium of partners from industry and research institutions from Europe and South Africa.

This project is planned for the Evora Molten Salt Test Platform in Portugal (University of Evora [10]); infrastructure is being built at present—see Fig. 15.7 (a) and (b)—with two molten salt tanks, solar field, and a complete water/steam loop associated.

All claims about optical and thermal performance of the concentrator and of the collector field need to be verified. The CLFR type described above has not yet been built, operated, and tested in practice. Specific molten salt operational problems in linear (2D) concentrating systems have to be addressed and solved, before a solution like this one can be considered ready for the market.

A final comment is due about the calculations presented. On purpose, the assumptions made about material and component properties were not overly optimistic. For instance, a slightly higher turbine efficiency could have been considered (for instance 0.44 instead of 0.42) and, in particular, a further optimized optics (to reduce the number of rays going more than once through the glass tube, to conserve etendue completely) could have been used and this could be expected to raise the optical efficiency from 0.70 to 0.71 or even 0.72. This would also benefit the incidence-angle modifiers, through further reduction of shading and blocking using more elaborate options for etendue conservation [23,24].

It should also be noted that the reflectivity values (0.92) assumed for the primary and secondary were also very conservative, since there are mirrors on the market with values of 0.93–0.94 or even higher.

On the other hand, no parasitic losses (pumping, for instance) were included in this calculation and thus, the potential improvements referred in the preceding paragraph will probably cover the negative impact of these losses (at least 5% of the electricity produced [4]) and the final result will be the same.

(a)

(b)

Figure 15.7 (a) Solar field and (b) details of the molten salt tanks and heat exchanger (molten salt to steam production).

Therefore, the safest way to view the previously referred potential improvements is to let them compensate for parasitic and other losses (intercept factor not exactly 0.99, manufacturing and tracking accuracies not exactly to the desired standard, etc.) that will occur, for an end result, which will hopefully not be too far from that of Table 15.3.

As stated, and beyond this excellent expected performance, an advanced LFR of this sort, offers further advantages for a complete collector field: (1) higher concentration for a fixed receiver results in a larger aperture and thus a substantial reduction of the number of necessary rows to achieve a given installed power (in this case, with a primary of about 26 m the reduction of rows in a comparable PT field would be higher than a factor of 4); (2) there is also a higher degree of compactness (occupied ground per peak watt). The first one will certainly impact on kWh production cost, and the second one also in applications where land or roof top occupation is at premium. The first one also means less heat loss and less parasitic losses (e.g., pumping power), all of them having a potential impact in cost reduction.

15.7 Conclusions

High concentration is a very important feature for potential energy delivery cost reduction as explained in the Chapter 3.

Advanced LFR concentrators are uniquely placed to deliver high concentration and thus stand as a very interesting way to produce low-cost electricity or simple low-cost heat for direct use.

LFR technology has a much smaller presence in the commercial plant market than its rival, linear PT technology. One of the reasons, as explained, is that LFR has been perceived as potentially cheap, but associated with low efficiency conversion from solar to electricity. This seems to have relegated LFR to applications like industrial process heat or other niche markets. These new ideas around advanced LFR technologies are a breakthrough and stand to completely change that perception, showing the true potential of LFR with advanced optics to operate at much higher temperatures as well and in conditions that are much easier to install, operate, and control. The impact of this perception is that LFR technology will have recovered its competitive edge for CSP and many companies will seek to visit or even revisit this new opportunity, quite besides the fact that these ideas will also create improvements (cost reduction) in other applications as well, such as in industrial process heat.

A renewed interest in the technology will thus have a serious impact on the CSP market. Advanced LFR technologies with molten salts will become a contender for large-scale competitive electricity production.

A final note a comment on industrial process heat is in order given the new degrees of freedom introduced by advanced LFR technology:

- For this application, high concentration is also potentially quite important, not so much because of temperature performance (as in the case of CSP) but because it can provide very large primaries (twice the typical LFR sizes) reducing the number of rows or increasing the degree of compactness.
- Another consideration is the possibility of keeping concentration to the usual values, but adopting non-imaging optics with much larger acceptance angles, easing tolerances on manufacture, tracking requirements, and so on, for ease of operation and lower cost.

References

[1] Novatec. www.novatecsolar.com/.

[2] Industrial Solar. www.industrial-solar.de/.

[3] Ausra/Areva. www.solar.areva.com/.

[4] Morin G, et al. Comparison of linear Fresnel and parabolic trough collector power plants. Solar Energy 2012;86:1−12.

[5] Price H, et al. Advances in parabolic trough solar power technology. Journal of Solar Energy Engineering May 2002;124.

[6] Eck M, Zarza E. Saturated steam process with direct steam generating parabolic troughs. Sol Energy 2006;80:1424−33.

[7] Torresol/Gemasolar. www.torresolenergy.com/TORRESOL/gemasolar-plant/en.

[8] Andasol 1 e 2. https://en.wikipedia.org/wiki/Andasol_Solar_Power_Station.
[9] Torresol/Gemasolar. Private Communication.
[10] EMSP Facility, Portugal. http://www.en.catedraer.uevora.pt/.
[11] ENEA PCS Facility, Italy. www.enea.it/en/video/the-pcs-plant.
[12] Priolo Gargalo Facilit, Italy. https://en.wikipedia.org/wiki/Archimede_solar_power_plant.
[13] MSALFR Project. Call: H2020-LCE-07-2015-2017.
[14] SUNCNIM. www.cnim.fr.
[15] Archimede Solar. www.archimedesolarenergy.it/en_prodotti.htm.
[16] Schott Solar. www.schott.com/csp/english/.
[17] Chaves J, Canavarro D, Collares-Pereira M. Dual asymmetric macrofocal CEC LFR solar concentrator (patent pending).
[18] Canavarro D, Chaves J, Collares-Pereira M. A novel compound elliptical-type concentrator for parabolic primaries with tubular receiver, Submitted to Solar Energy (Elsevier).
[19] Alanod Solar. http://www.bluetec.eu/de.
[20] Burkholder F, Kutscher C. Heat loss testing of Schott's 2008 PTR70 parabolic trough receiver. Technical Report NREL/TP-550-45633. May 2009.
[21] Dreyer S, Eichel P, Guardig T, Heckers Z, Jonker S, Kuckelkhorn T, et al. Heat loss measurements on parabolic trough receiver. SolarPACES; 2010.
[22] NREL. www.nrel.gov.
[23] Chaves J. Introduction to nonimaging optics. second ed. CRC Press, Taylor and Francis Group; 2015.
[24] (contributions by Shatz, N., Bortz, J., C.) Winston R, Miñano JC, Benítez P. Nonimaging optics. Amsterdam: Elsevier Academic Press; 2005.

16 Central tower systems using the Brayton cycle

R. Buck, S. Giuliano, R. Uhlig
Deutsches Zentrum für Luft-und Raumfahrt (DLR), Stuttgart, Germany

16.1 Introduction and history

In conventional power plant technology, gas turbine systems are known as cost-effective and reliable components. In combination with a bottoming steam cycle ("combined cycle," CC) very high conversion efficiencies can be achieved, exceeding 60% with the most modern technology. This is significantly higher than the efficiencies that steam cycles in state-of-the-art concentrated solar power (CSP) systems can achieve (about 42% in molten salt solar tower systems). For this reason, solar gas turbine systems are considered an attractive option for increasing the solar-to-electric conversion efficiency of CSP plants, and thus reducing the levelized cost of electricity (LCOE).

A solar gas turbine system consists of the following main components:

- solar concentrator: heliostat field with tower or parabolic dish;
- solar receiver for high temperatures;
- gas turbine unit with combustor;
- optional: recuperator, intercooling or bottoming steam cycle;
- generator.

Fig. 16.1 shows a scheme of a typical solar-hybrid gas-turbine (SHGT) system in a CC configuration. The solar radiation is concentrated by a huge number of heliostats, all focused on the receiver on top of a tower. In the receiver, the concentrated solar radiation is absorbed heating the compressed air downstream the gas turbine's compressor. Depending on the actual solar power input and the selected configuration, the air is heated to a certain temperature before entering the combustor section. In the combustor, fuel is supplied to perform heating to the required turbine inlet temperature (TIT). The hot gas is then expanded in the turbine section and drives both the compressor and the generator for power production. In the shown CC configuration, the hot exhaust gas from the turbine is then used in a heat recovery steam generator to provide steam to a bottoming steam cycle, generating additional power and thus increasing overall conversion efficiency.

16.1.1 Advantages of solar-hybrid gas-turbine systems

SHGT systems offer several advantages, compared with other CSP systems. The most important are:

High conversion efficiency systems: SHGT systems can achieve thermal cycle efficiencies up to 50%, depending on system configuration and power level. These cycle

Advances in Concentrating Solar Thermal Research and Technology
http://dx.doi.org/10.1016/B978-0-08-100516-3.00016-2

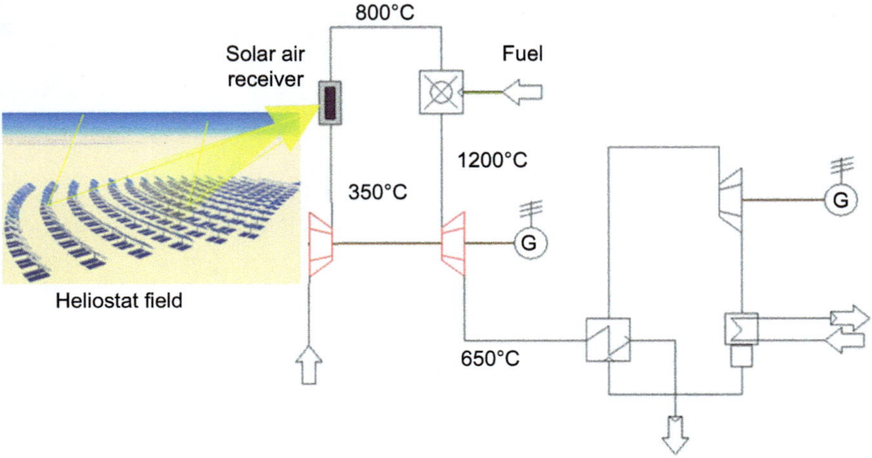

Figure 16.1 Scheme of a solar-hybrid combined cycle system.
Copyright: DLR.

efficiencies are significantly higher than efficiencies of state-of-the-art solar steam cycles.

Reduction of LCOE: Higher conversion efficiency results in a smaller heliostat field with lower cost. With the heliostat field being the most significant cost contribution in a solar tower system, this will result in reduced investment and field maintenance cost and lower LCOE.

Ensured grid capacity with hybridization: Hybridization of the SHGT plant using fossil fuel is a key asset. Keeping an adapted combustor as part of the gas turbine unit makes the plant capacity fully available whenever required. Thus, even in longer periods without sunshine no additional power plants must be kept as standby, and grid stability can be supported.

Less expensive power block: The specific power-block cost of gas turbine systems or CC systems are lower than comparable steam cycles.

Reduced water consumption: SHGT cycles without bottoming steam cycle do not require any water for the power cycle. Even for systems with bottoming steam cycle, the water consumption is significantly lower than for a conventional steam cycle of comparable power level. This is due to the fact that the bottoming cycle in a CC plant represents only about one third of the total power generation.

Fast response time: Fuel flow to the gas turbine combustor can be controlled much faster than transients from varying solar input (i.e., temperature variations) that occur at the combustor inlet. Solar transients are significantly delayed by the thermal inertia in the receiver and the associated connection piping.

Simple plant control: The hybrid operation capability ensures stable and controllable operation, allowing delivery of any desired power output independent of level of solar power availability.

Storage integration possible: High-temperature storage systems using pressurized regenerator-type units can be integrated in parallel with the receiver. This allows increase of the solar share, for example, by using the stored solar energy during night time.

16.1.2 Challenges in solar gas turbine systems

Conventional gas turbines are usually designed as very compact units, with multiple and complex air flow paths especially for internal cooling purposes. For the integration of solar energy into gas turbines, several technical issues have to be solved before commercial deployment. The most important ones are:

High receiver temperature required: Gas turbines require high TITs. In order to replace as much fuel as possible, the receiver temperature must be close to the TIT. This temperature must be also provided at high receiver efficiency.

Combustor integration/modification: Depending on the configuration, the combustor must be adapted to accept either higher air inlet temperatures (serial mode), or operation over a wide range of air mass flows (parallel mode).

Modification of air flow paths: The main air flow must be extracted after the compressor section to allow external heating. The heated air must then be reintroduced before the combustor or the turbine section. Additionally, the internal cooling air paths might be affected. Design and cost of the required high-temperature piping and auxiliary equipment is also important.

Gas turbine control: Due to the high thermal inertia and additional volume in the solar subsystem, the dynamic response of the system must be adapted. Start-up and shutdown sequences also need to be adapted.

Emergency procedures: Due to the high thermal inertia and additional volume in the solar subsystem, turbine overspeed situations will occur rapidly when the load is removed (e.g., in case of loss of grid). Special countermeasures need to be taken to avoid damage to the gas turbine and the receiver systems.

Low pressure drop in external heating: Additional pressure drop between compressor and turbine results in decreased efficiency of the gas turbine cycle. When the additional pressure drop is too high, the surge limit of the compressor may be reached, leading to operation failures.

High-temperature storage: Storage units for SHGT systems need to operate at high temperatures and pressure, introducing high requirements for the inventory, the storage containment, and the thermal insulation.

16.1.3 Overview of solar gas turbine projects

Gas turbine systems require high TITs in the range of 900−1500°C. In solar systems, such temperatures can only be achieved by point-focusing solar systems; that is, solar tower systems or parabolic dishes. These systems provide average concentration of the solar insolation by a factor of at least 600.

Early work on solarized gas turbine systems focused on parabolic dish applications. Following the development of a suitable receiver, a first test setup was built and operated by Sanders based on a microturbine with about 5 kW_e nominal power [14]. Analysis of hybrid solar gas turbines was performed by Scheuerer [42]. followed by nonsolar laboratory testing using a modified ALLISON 250-C20B helicopter engine [43].

In 1978 the gas-cooled solar tower project GAST [8] was started for design and development of a pilot plant with open-cycle gas turbine and for testing the solar-specific components (heliostat, receiver, and hot gas pipes). On the basis of a 20 MW_e lead concept, technical investigations and laboratory tests were carried out. The receiver outlet temperature from the cavity receiver with metal tubes was 800°C.

Led by the company Cummins, a 25 kW_e solar gas turbine system was under development as backup solution within the USJVP program [23]. All relevant components of that system were completed, but the project stopped before integration to the system.

In 2001, a European project named "SOLGATE" started with the goal of a demonstration of an SHGT system on a solar tower [47]. Again, a modified ALLISON 250 helicopter engine was used, driving a generator. The combustor was modified to accept air inlet temperatures up to 800°C, using kerosene to provide the remaining heat input to achieve the nominal TIT. The solar receiver unit consisted of a serial connection of three receiver modules, each equipped with a secondary concentrator with hexagonal entrance aperture. The first receiver module used metallic tubes as the absorber, while the next two modules were pressurized volumetric receivers. The system was installed at the solar tower test facility PSA (Plataforma Solar de Almería, Spain). A controllable air bypass around the receiver was foreseen to enable operation of the receiver at higher temperatures, without exceeding the combustor inlet temperature limit of 800°C. During solar testing, electric power output of 230 kW was demonstrated. In a subsequent test campaign, receiver air exit temperatures up to 1030°C were achieved using the air bypass with a modified receiver unit [10].

A first step toward commercialization was taken in the European project named "SOLHYCO" [48]. A 100 kW industrial microturbine system was modified for integration with a 200 kW_{th} metallic tube receiver. Solar-hybrid system tests were again performed at the PSA. These tests demonstrated system operation with receiver outlet temperatures up to 803°C [3]. Unintendedly, the system was operated for a certain time in solar-only mode, when a combustor flame-out was not detected by the control system. However, system operation continued smoothly at reduced power output until the test operators recognized the situation.

Upscaling of SHGT technology was the goal of the following European project named "SOLUGAS" [40]. In the course of this project a complete SHGT system was built at the Sanlucar solar platform near Seville, Spain, including heliostat field and tower (Fig. 16.2). The gas turbine (Solar Turbines Mercury-50) was modified to accept solar preheating up to 650°C at the combustor inlet. Again, a bypass around the receiver allowed operation of the receiver at temperatures up to 800°C. In total, about 1000 solar test hours were accumulated in different load conditions.

Led by the French research institution CNRS, a number of R&D projects regarding SHGT systems were initiated under the name "PEGASE" [19,27]. The goal was the development of an SHGT system with 1.4 MW_e as next step in technology upscaling. Several receiver concepts for high temperatures were developed and tested, accompanied by system simulation and optimization including thermal storage.

In 2013 a European project named "OMSOP" was started, aiming for the development and demonstration of a small dish-based SHGT system [33]. The intended system will be modular and capable of producing electricity in the range of 3–10 kW per module.

Currently, AORA [4] is the only company offering SHGT systems on a commercial basis. The AORA system is based on a 100 kW_e microturbine which is adapted for

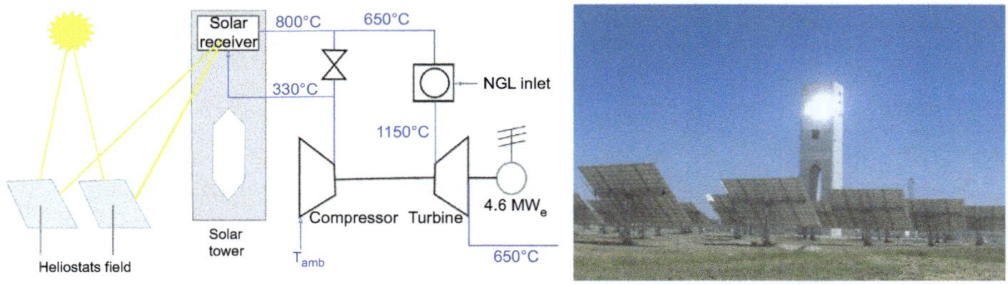

Figure 16.2 Left: SOLUGAS cycle scheme (inside *dashed line*). Right: plant in operation [40].

solar operation. The microturbine and the solar receiver are mounted on a tower, with a small heliostat field providing the concentrated solar radiation. Besides electricity production, up to 170 kW of process heat can be supplied from the hot turbine exhaust gas. Two prototype systems were operational in 2015.

The company Wilson Solarpower [56] is planning for mass production of small modular SHGT systems. The intended power level is about 400 kW_e, and storage is foreseen to ensure a high solar share.

Another solar-assisted gas turbine technology is the so-called Integrated Solar Combined Cycle System (ISCCS). Such a system is basically a conventional CC plant, with the gas turbine operated only on fossil fuel. Solar-produced steam can be introduced into the bottoming steam cycle, either to boost the output of the steam cycle or to reduce the fuel consumption by operating the gas turbine in part load during solar-assisted operation. However, the annual solar share of such plants is usually well below 10%. ISCCS plants are operational in Egypt and Morocco [13]. Another special system configuration, integrating compressed air energy storage with a solar gas turbine system to provide load shifting capability, was developed by Kesseli [32]. Table 16.1 gives an overview over the main projects.

Most of the hardware projects were accompanied by techno-economic studies to verify the cost reduction potential of SHGT systems. Several techno-economic studies were dealing with the optimized layout of SHGT systems in different configurations. A more detailed discussion of these studies follows.

16.2 Solarization of gas turbines

Solarization of gas turbine units generally refers to heating of the compressed air upstream the combustor or parallel to the combustor, by means of solar energy. Basically, two options exist:

- *Direct heating*: This is the preferred solution, with the solar receiver directly integrated in the compressed airflow path.
- *Indirect heating*: A high-temperature heat exchanger is integrated in the compressed air flow path; thermal energy from any suitable heat transfer medium is transferred by the heat exchanger to the compressed air. Due to the high temperatures, such heat exchangers are quite expensive.

Table 16.1 Overview over solar-hybrid gas-turbine systems

Project/ company	Year	Power level	Main achievements	Remarks
SANDERS	1983–84		Demonstration of a dish system with microturbine	System demonstration, sub-atmospheric Brayton cycle, atmospheric solar receiver
SOLGATE/ HST	1998–2002/ 2004	250 kW_e	First solar tower system with solar-hybrid gas turbine, receiver temperature up to 1030°C	Modified helicopter gas turbine, efficiency about 18%
SOLHYCO	2006–10	100 kW_e	First solar-hybrid system with commercial microturbine, biofuel capability	Solar-hybrid system operation, with receiver temperature 800°C
SO LUGAS	2008–12	4600 kW_e	Upscaling of solar-hybrid gas-turbine technology, demonstration of cavity receiver	Low solar share (combustor inlet temperature limited to 650°C)
PEGASE	2009–13	1400 kW_e	Technology development, several receiver concepts	Receiver tests up to 750°C
AORA		100 kW_e	Commercial solar-hybrid microturbine system, cogeneration option	Two systems in operation [4]

Point-focusing solar concentrator systems are required to achieve the high receiver temperatures for the operation of SHGT system. Dish-based solutions allow for low power levels per unit (<100 kW_e); higher power levels are obtained by installing multiple units. However, the efficiency of the gas turbine is limited at low power level, and the integration of storage is hardly feasible. Solar tower systems allow for high power levels with highly efficient gas turbine cycles, and storage can be integrated and installed at the bottom of the solar tower. A modular approach using standardized

SHGT systems with solar tower might be a suitable solution to distribute the development cost over a large number of units.

Several solar gas-turbine system concepts were developed the most important ones are described in the following section.

In SHGTs modified state-of-the-art gas turbines are used. Of interest in this application are open-cycle gas turbines having high electric efficiencies. The higher the cycle efficiency the smaller the solar field to be designed and thus, the investment on it decreases. There are several options to increase the cycle efficiency. Fig. 16.3 shows the most important solar-hybrid gas turbine configurations. Beside increasing the TIT it is, for example, possible to use recuperated gas turbine cycles, (multiple) reheated and/or (multiple) intercooled cycles, steam injection, inlet air cooling, CCs (with

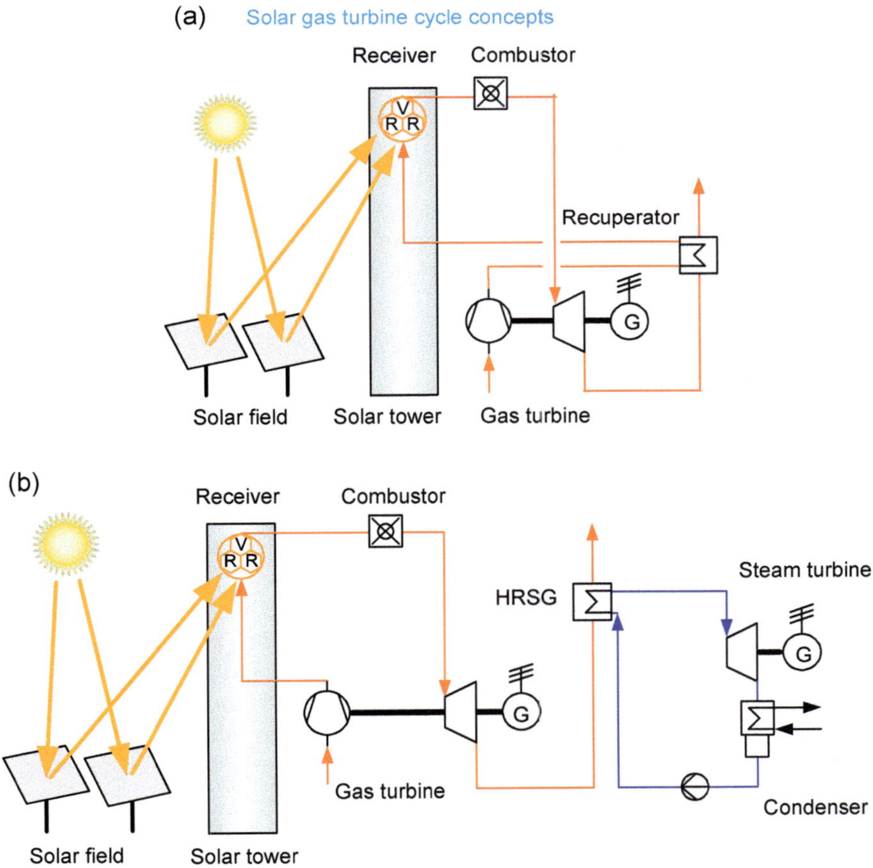

Figure 16.3 Scheme of solar gas turbine cycles. (a) Solar-hybrid recuperated Brayton cycle. (b) Solar-hybrid combined Brayton and Rankine cycle.
Copyright: DLR.

bottoming Rankine steam cycle or ORC), and others. However, to achieve high solar shares, and as the receiver outlet temperature is limited depending on the receiver technology used, the option of increasing the TIT further is not a favorable option for SHGTs.

As of 2016 two development paths are followed:

1. SHGT with high TIT: low solar share
   A gas turbine cycle with high TIT and thus high thermal efficiency is used. The solar gas turbine plant has therefore the highest possible efficiency, but because of the limited receiver outlet temperature more fuel has to be used and thus the solar share is low.
2. SHGT with low TIT (e.g., 950°C): high solar share
   The gas turbine cycle is modified to reach high efficiency with low TIT (e.g., 950°C) [31]. The TIT is adapted to the receiver outlet temperature and thus at design point the plant reaches up to 100% solar share.

Gas turbine ratings are specified with ISO conditions to be comparable (ambient temperature 15°C, elevation 0 m above sea level, relative humidity 60%). The performance of a gas turbine is strongly affected by the ambient conditions. With higher ambient temperatures and high elevation (low pressure) the density of air decreases and so do the power and efficiency of a gas turbine. Furthermore, the part load operation of gas turbines has a strong influence on the efficiency. The SHGTs are operated in countries with high solar irradiation, as this usually also means high ambient temperatures, the reduction in power and efficiency has to be considered. One of the common measures to improve the performance is to use air inlet cooling that is mainly done by evaporative cooling, fogging, and cooling with chillers.

16.3 Solar gas turbine cycle concepts

16.3.1 Solar-based heat input to solar gas turbine systems

Solar heating means solar power input ranging from 0% to 100% during the day and often fast temperature transients, for example, due to cloud passages. For power generation, from 5 kW_e small-scale microturbines to large-scale gas turbines up to 400 MW_e are available. Common to all SHGT systems is the external heating of the air. Current gas turbine systems operate at TITs ranging from 950 to about 1500°C, depending on the layout and power level of the unit. Modern high-efficiency gas turbines usually operate with a TIT close to the upper temperature range, while, for example, microturbines with uncooled turbine wheels operate at the lower temperature range.

In principle, there are two ways for solar heating of solar gas turbines: direct heating and indirect heating (Fig. 16.4). Both concepts have advantages and disadvantages [45]. The indirect system is challenging because of the needed high-temperature heat exchanger. Another important difference is the arrangement of receiver and combustion chamber: serial or parallel (Fig. 16.4).

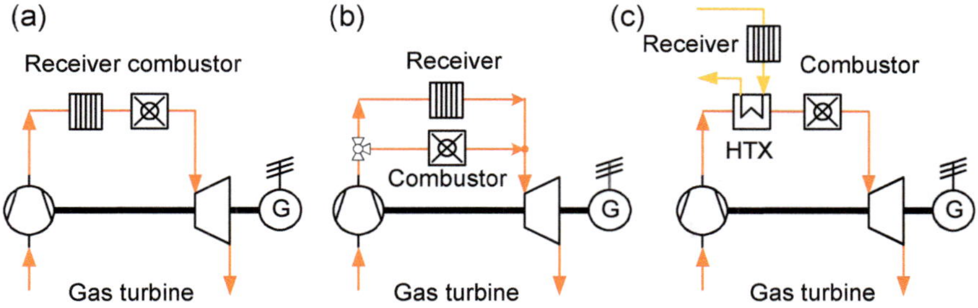

Figure 16.4 Solar based heat input to gas turbine systems. (a) Direct serial. (b) Direct parallel. (c) Indirect serial.
Copyright: DLR.

16.3.2 Open-cycle Brayton systems

16.3.2.1 Simple cycle

Simple cycle Brayton systems are state of the art. These cycles run as open thermodynamic cycle and consist only of a compressor, a combustor system, and the turbine section with generator. The energy of the exhaust gas is wasted. Commercial simple cycle gas turbines are robust, easy to operate, and reach efficiencies up to 40% at large power levels. Fig. 16.5 shows a solar gas turbine process with simple cycle. Because of the limited efficiency this option is not favorable for solar power generation. However, if the waste heat (usually about 400–600°C) can be used for process heat (e.g., heating or cooling applications) the economics can be improved.

16.3.2.2 Recuperated cycles

Recuperated gas turbine cycles are mainly realized at smaller scale (power levels up to 20 MW_e). At those power levels a bottoming steam cycle is quite expensive and reaches only moderate efficiencies. Microturbines usually have a recuperator to increase efficiency. Other gas turbines such as the Mercury-50 from Solar Turbines Inc. in the range of 4.5 MW_e (Fig. 16.6) and the THM1304 GT from MAN Diesel & Turbo SE in the 10 MW_e range use also a recuperator.

In a recuperated cycle the hot exhaust gas is used to preheat the compressed inlet gas (Fig. 16.7). The recuperator (i.e., a heat exchanger) is used to transfer thermal energy from the exhaust to the already compressed gas, before it enters the receiver. As the thermal energy of the exhaust is reused in the cycle, the efficiency increases. Recuperation is more effective in gas turbine cycles with a low pressure ratio. Therefore, those cycles are more sensitive against additional pressure losses (e.g., from a solar receiver).

As the air is preheated before entering the solar receiver the inlet temperature to the receiver is higher. This has a negative influence of the receiver efficiency, as well as a reduced temperature spread for a storage system, if used.

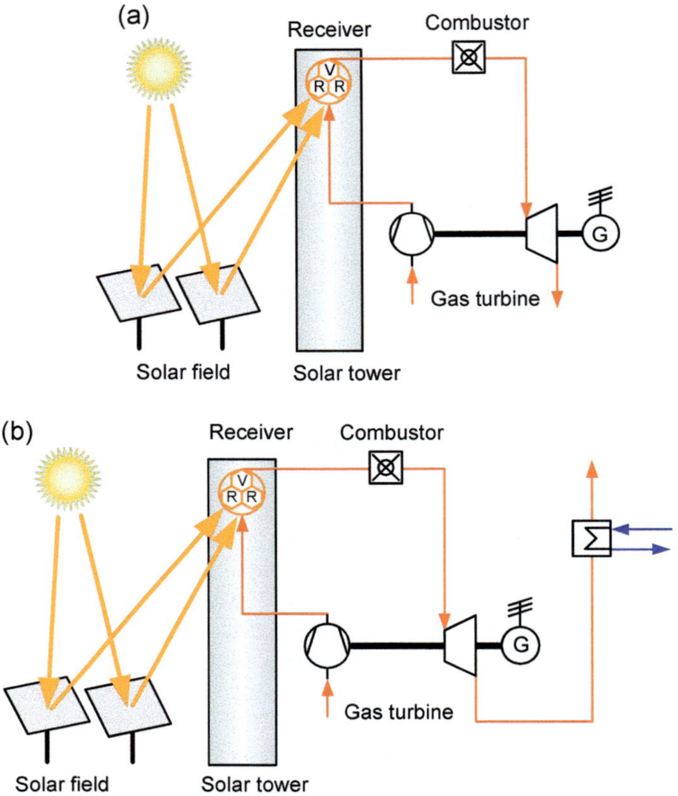

Figure 16.5 Scheme of solar gas turbine cycles. (a) Simple cycle solar gas turbine. (b) Simple cycle solar gas turbine with cogeneration.
Copyright: DLR.

The reported nominal efficiency for the small-scale Mercury-50 is 38.5% with a pressure ratio at about 10 [46].

16.3.2.3 Combined cycle

A CC combines a simple cycle gas turbine with a bottoming steam cycle (Fig. 16.8). Organic Rankine cycles (ORC) can also be applied. As the waste heat from the simple cycle is efficiently used, the efficiency of the cycle increases. In general the modifications to the gas turbine cycle are equal to the simple cycle. The HRSG and the steam cycle can be regarded as conventional system. Here all known modifications to the cycle can be used (e.g., dual pressure). For a 16-MW_e SHGT with CC an efficiency of 44.7% is reported in [44]. Puppe et al. [39] reported an efficiency of 43.7% for a 50 MW_e unit for an SHGT with CC at reduced TIT of 970°C. For an SHGT with bottoming ORC cycle the efficiency of up to 48.1% is reported.

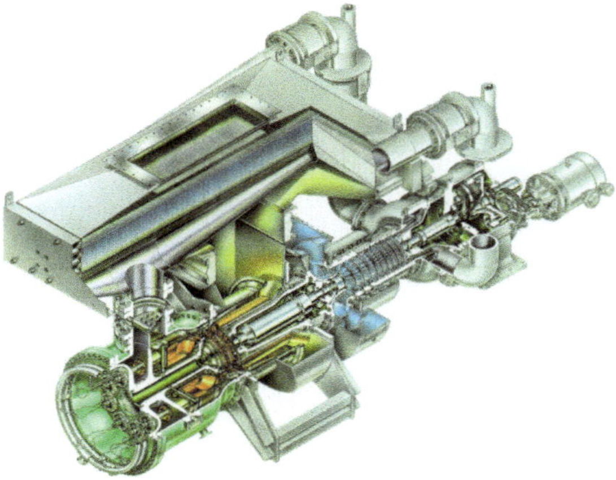

Figure 16.6 Standard Mercury-50 gas turbine.
Copyright: Solar Turbines.

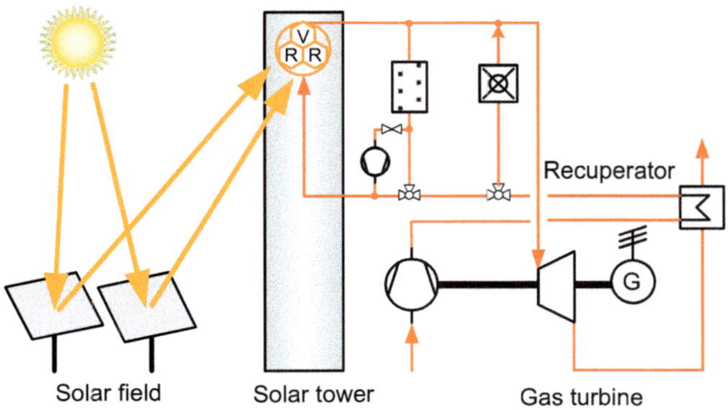

Figure 16.7 Scheme of SHGT with recuperation, parallel combustor, and storage.
Copyright: DLR.

16.3.2.4 Intercooled cycles

With the intercooling of the compressed air (Fig. 16.9) a reduced power consumption for compression and thus, more useable power from the gas turbine plant can be achieved. With higher pressure ratios those cycles can improve efficiency compared to the simple cycle. Currently only one commercial gas turbine with intercooling is available for power generation: LMS100 from GE Energy, with an efficiency of about 44% [36]. The part load efficiency is higher compared to simple cycle and thus, it has advantages when used often in part load.

A more sophisticated approach for intercooling was studied in the HYGATE project. This configuration is based on a multistage compressor with internal downstream

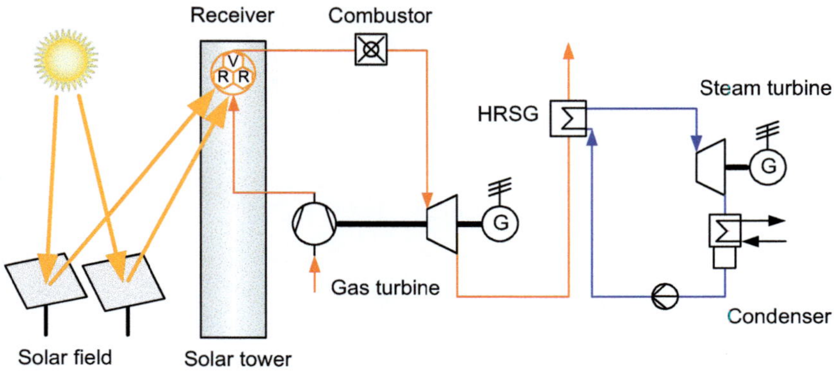

Figure 16.8 Scheme of combined Brayton and (steam) Rankine cycle.
Copyright: DLR.

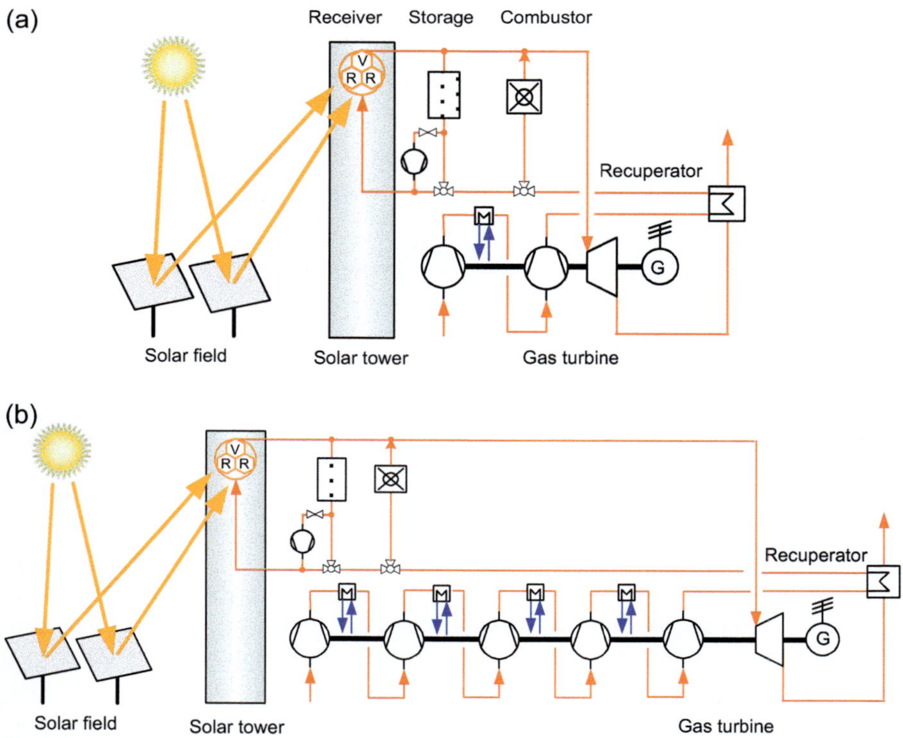

Figure 16.9 Schemes of intercooled solar gas turbine cycles. (a) Intercooled solar gas turbine. (b) With quasi isothermal compression and recuperation.
Copyright: DLR.

cooling of all but the last compressor stage to provide a high efficiency due to low compression temperatures. This compressor concept is adapted from the commercially available MAN isothermal compressors. The cooling medium is water, which is cooled by a dry cooling system. Instead of a bottoming cycle a recuperator is included to use the exhaust heat from the gas turbine, which is at a temperature of 420°C. The compressed air is heated in the recuperator from around 130 to 400°C. The design-point gross electric efficiency of the power block is 45.7% at reduced TIT of 970°C [31].

16.3.3 Closed Brayton cycle systems

In a closed-cycle gas turbine the heat is supplied from an external source. Gases such as air, helium, carbon dioxide, and nitrogen are used as working fluid. This type of recirculating turbine follows the Brayton cycle. Those turbines are interesting for high-temperature solar- and nuclear-power generation [22]. The advantage of closed-cycle gas turbines compared to open-cycle gas turbines is that they allow higher efficiencies at lower temperatures and have a better part load behavior. They allow an operation at higher pressure and the used gases have better thermodynamic properties. With this, advantages in the receiver design are possible.

The disadvantages are the sensitivity against internal pressure losses, the additional cost of the required heat exchanger at the cold end, the higher efforts for sealing the cycle, and that there are no commercially turbines available.

Bammert [6] studied in 1980 a closed-cycle gas turbine plant with air as well as helium as working fluid. For a 20-MW_e solar tower plant at a TIT of 800°C an electrical efficiency of 38.3% with helium and 37.1% with air was reported. For comparison, the efficiency for the open-cycle gas turbine was given as 34.1% (unfortunately a small-scale 250 kW_e gas turbine was considered). The optimal turbine inlet pressure was 31 bar for helium and 44 bar for air.

Closed-cycle gas turbine with collecting pipes from parabolic dish collectors has been discussed [35]. The problem was that no suitable gas turbines were available and that the feasibility of such collecting pipes for the working fluid was critical.

Closed-cycle gas turbines with CO_2 as working fluid are considered in [11]. With a bottoming ORC cycle an efficiency of 47.5% is reported at a TIT of 827°C.

16.4 System components

16.4.1 Gas turbine

Operating a gas turbine with solar heat input requires the possibility to externally heat the compressed air before entering the turbine section. Conventional gas turbines are usually not prepared for this operation mode, but must be modified in several aspects to allow this external air heating. These modifications include:

- air path: compressed air extraction and preheated air re-introduction;
- "solarized" combustor, in parallel or serial connection;
- optimization for solar operation (reduced TIT);

- adapted safety and emergency measures;
- adapted control system.

Air path

External air heating of the compressed air requires interface ducting between the gas turbine and the external components (solar receiver, storage system, connection piping, and so on). In gas turbines equipped with external can combustors, the interface to the combustor can be adapted for this purpose with little modifications to the gas turbine. Several gas turbine types from different manufactures exist with suitable combustor configurations.

However, most of the modern gas turbine units use annular combustor systems that are highly integrated with all other components. In these gas turbine configurations significant redesign of the casing and the air flow path is required to allow external air preheating.

In a solarized gas turbine system, a number of external components are included in the flow path between compressor and turbine, namely solar receiver, storage system (if used), connection piping, and flow control equipment. All these components result in both additional volume and additional thermal inertia.

Combustor

A standard gas turbine combustor is designed for stable operation with the compressor discharge air, that is, at an air inlet temperature of about 400°C, depending on the gas turbine configuration. In recuperated gas turbines, combustor systems for air inlet temperatures up to 600°C exist.

In solarized gas turbines, the combustor can be integrated with the solar components in two ways:

1. *Serial connection*: The combustor is installed downstream of the receiver (and storage, if available), upstream the turbine section.
2. *Parallel connection*: The combustor is installed parallel to the receiver (and storage, if available), upstream the turbine section.

In serial connection, the combustor receives high air inlet temperature (up to 1000°C), which is varying dependent on the actual solar power input. Due to the serial connection, the pressure drop is increased, as it is the sum of all components in series. In this configuration, the pressure drop is the sum of all components in series. Therefore, the pressure drop is higher than in parallel connection, with negative impact on turbine performance. With increased air inlet temperature, the fuel mass flow is reduced accordingly, resulting in different combustion regimes and flame stability issues. The cooling situation of the combustor components is changed significantly, as the cooling capability of the hotter air is reduced. In addition, the fuel/air mixture tends to self-ignition at higher air inlet temperatures, resulting in uncontrolled combustion with the risk of component damage. Also, combustion characteristics under elevated air temperatures and low fuel ratios vary significantly.

In parallel connection, the compressor mass flow is split between receiver/storage and the combustor, with the split ratio depending on the solar power input. The combustor always receives air at the low compressor discharge temperature (or recuperator exit temperature, in case of a recuperated cycle), but at a varying mass flow. To compensate for lower temperature in the receiver/storage flow path, the combustor might need some overheating to obtain the desired TIT, according to the mass flow split ratio. Due to the parallel connection, the resulting pressure drop of receiver, storage, and combustor equals the maximum value of all separate flow paths. Appropriate control valves are used to obtain this pressure drop in all flow paths with the desired mass flow split.

Component cooling and material issues are critical in both configurations; special designs are under development for solarized gas turbines [12,31].

Optimization for solar operation: reduction of TIT

In hybrid CSP systems it is often desirable to reduce the amount of fuel consumed annually, that is, to achieve a high solar share. It is assumed that reasonable annual solar efficiencies can be achieved at receiver exit temperatures up to around 1000°C. With this limitation, the desired high solar shares can only be realized when the TIT is also reduced to temperatures close to the maximum receiver outlet temperature. Then, the solar share is mainly determined by the solar resources, storage integration, and operation schedules. Solar-only operation is then possible directly from the receiver's output or from the stored solar energy, or from a combination of both.

Control system

In normal operation mode, the control of an SHGT is relatively simple as long as the combustor is within its stable operation regime. Fuel flow to the combustor can be controlled very fast and effectively, while the inertia effects of the receiver and the piping result in significantly longer time constants of the solar subsystem. Analysis of a control concept is described in [15].

Safety and emergency measures

The external solar components (receiver, piping, storage) represent a high thermal inertia and a large additional pressurized volume. In case of emergency situation (e.g., load shedding) this can lead within sub-second time periods to overspeed in the turbine section. Special measures need to be taken to control this critical situation. Several methods have been proposed and tested for this purpose, such as:

- addition of blow-off valves;
- shut-off valves;
- mechanical brakes;
- shunt resistors loading the generator.

In the SOLGATE project [47] a blow-off valve was added between the receiver and the combustor. Upstream the valve, a water cooler was installed to enable use of an inexpensive low-temperature valve. An orifice plate was used to limit the blow-off mass flow to the desired value.

Felsmann [15] developed a dynamic simulation tool and investigated the response to critical operation conditions, for example, load shedding. He concluded that appropriate and fast-reacting hardware is necessary to avoid damage to the SHGT system (Fig. 16.10).

Other issues

An SHGT system always has additional components in the flow path between the compressor and the turbine sections, mainly, the additional receiver or heat exchanger and the piping. This results in increased pressure drop between compressor and turbine. As a consequence, the compressor must be designed for enough safety to stay away from surge limit. Also, the additional pressure drop results in reduced cycle efficiency. The impact of this pressure drop is very dependent on the gas turbine type.

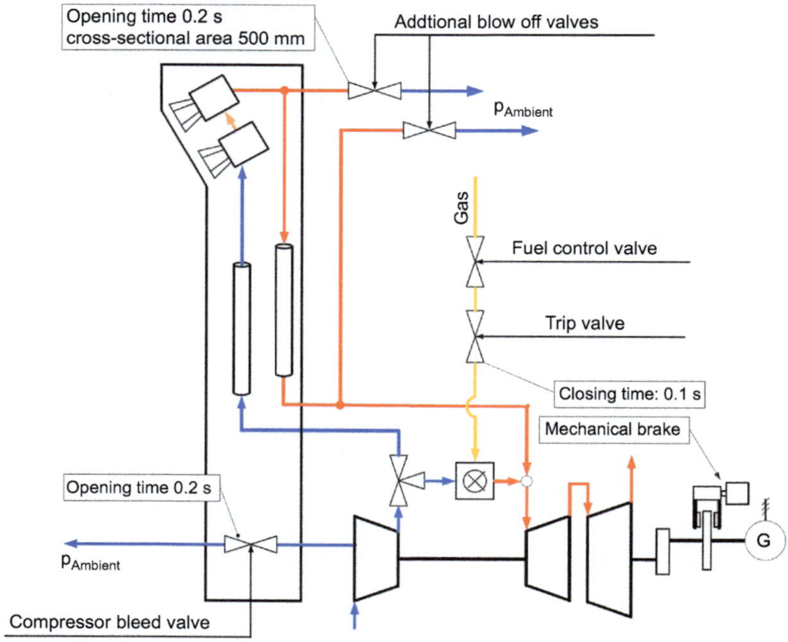

Figure 16.10 Measures for control of emergency situations [15].

16.4.2 Receivers for SHGT systems

The receivers convert the concentrated solar radiation to high-temperature heat and transfer the heat to the working medium. The receivers can have the typical tubular receiver configurations, or more complex designs, such as pressurized volumetric receivers.

As the receivers are exposed to high solar flux densities in combination with high temperatures, the requirements are quite stringent. The receiver should

- convert and transfer the heat with high efficiency;
- accept high and inhomogeneous heat fluxes (locally and in time);
- achieve long lifetime at acceptable cost.

To fulfill these needs the design of receivers has to take care of several thermal and mechanical boundaries. Heat transfer by convection, conduction, and radiation exchange has to be considered simultaneously. Receivers for central tower solar plants work at high solar heat fluxes and at high temperatures. The load situations are quite complex. The solar heat flux varies over time of day and year leading to different load situations. Fast transients can occur by clouds blocking the sunlight totally or partially. This leads to high and alternating stresses in the materials that affect the lifetime of the receiver. The thermal efficiency of the receiver is strongly influenced by the temperature and the average flux density on the receiver. The (over) temperature of the receiver is mainly influenced by the heat transfer medium used. For air at medium pressure levels (as in SHGT systems) the heat transfer capability is low in comparison to liquid media. As the pressure drop of the receiver is limited by the gas turbine, heat transfer coefficient enhancement is limited (higher fluid velocity and/or higher turbulence lead to higher pressure losses). Increasing the heat transfer area is another option to enhance the heat transfer capability, but it results in higher cost.

16.4.2.1 Metallic tubular receivers

As of 2016, all commercial receivers (for steam or molten salt) are built as *metallic tube receivers*. Multiple panels, each consisting of a number of parallel absorber tubes, are interconnected in serial or parallel configuration. Header sections collect and distribute the fluid (Fig. 16.11). The absorber tubes are normally coated with a black paint. The fluid is passed through the tubes and is convectively heated by the tube walls having absorbed the solar radiation on the outside.

Tube receivers can be built as external or cavity receivers. In external receivers the panels are arranged in a cylindrical or polygonal configuration. In cavity receivers the tubes are arranged along the walls of an insulated cavity. Due to the high operational temperatures and the relatively large absorber areas external receivers would lead to high thermal losses by radiation and convection. Therefore cavity-type receivers are usually the preferred solution for SHGT systems.

The SOLGATE project used a modular and multistage approach combining three different receiver modules, connected in series. The receivers used a hexagonal secondary concentrator to cover the focal spot like a honeycomb. The first stage used a

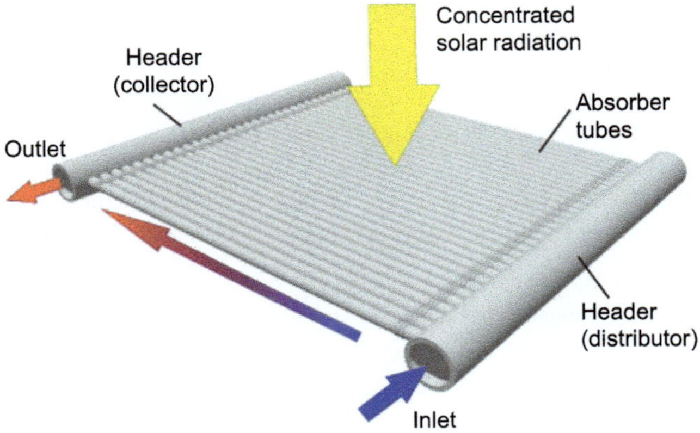

Figure 16.11 Principle schematic of tubular receiver.
Copyright: DLR.

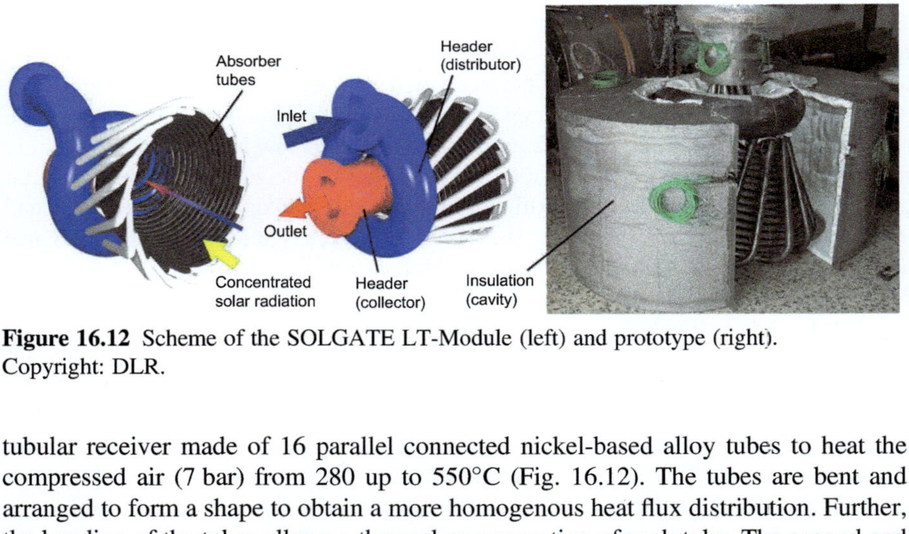

Figure 16.12 Scheme of the SOLGATE LT-Module (left) and prototype (right).
Copyright: DLR.

tubular receiver made of 16 parallel connected nickel-based alloy tubes to heat the compressed air (7 bar) from 280 up to 550°C (Fig. 16.12). The tubes are bent and arranged to form a shape to obtain a more homogenous heat flux distribution. Further, the bending of the tubes allows a thermal compensation of each tube. The second and the third stage are used to heat the air up to more than 1000°C using volumetric receivers. The receiver setup was tested successfully in 2003 [9,47].

A tubular receiver with a thermal power of 280 kW, preheating the air of a recuperated microturbine with 100 kW_e, was developed and tested within the *SOLHYCO* project [1]. The receiver was built using 60 nickel-based alloy tubes connected in parallel (Fig. 16.13). The irradiated length of the tubes was 2.5 m. To enhance the heat transfer a wire-coil insert was used within the absorber tubes. A torus built from bent tubes acted as the distributor. To compensate the different thermal expansions of the absorber tubes, every absorber tube was connected to the distributor using a metallic bellow. The mass flow of each absorber tube was then collected at an internally

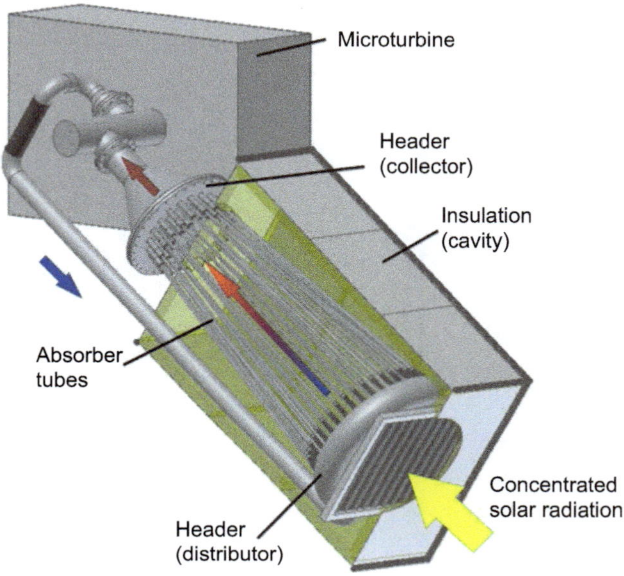

Figure 16.13 SOLHYCO receiver setup.
Copyright: DLR.

insulated collector. The system was tested at the PSA, Spain, in 2006, reaching the design outlet temperature of 800°C. However, due to problems with the cavity insulation (enabling cold air streams through the cavity) the convection losses were quite high. As a result, the measured receiver efficiency at design point reached only 39% instead of the expected 77.7% [3].

Under the European *SOLUGAS* project, a tubular receiver with a thermal power of 3 MW was developed and tested. The receiver was designed to heat the compressed air (10 bar_{abs}) of a gas turbine from 330 up to 800°C with a maximum pressure drop of 250 mbar. Due to limitations of the test setup the design receiver efficiency was limited to 73%. To allow flexible testing of the receiver only a fraction of the compressor mass flow was heated in the receiver. The outlet mass flow of the receiver was mixed with the other fraction bypassing the receiver. The receiver design follows a modular approach using 10 circularly arranged panels. Fig. 16.14 shows the design of the receiver (insulation is illustrated transparent). The solar radiation is absorbed and transferred to the fluid by 170 parallel absorber tubes with an inner diameter of 19.6 mm and an irradiated length of 5 m. The absorber tubes as well as the collector are made of a nickel-based alloy to withstand the high material temperatures while the distributor and the support frame are made of stainless steel. To compensate the different thermal expansions of the absorber tubes, every absorber tube was connected to the collector tube using a metallic bellow [53].

The receiver was tested at Abengoa's Solucar Platform and reached all design goals. More than 1000 h of solar operations have been accomplished [40].

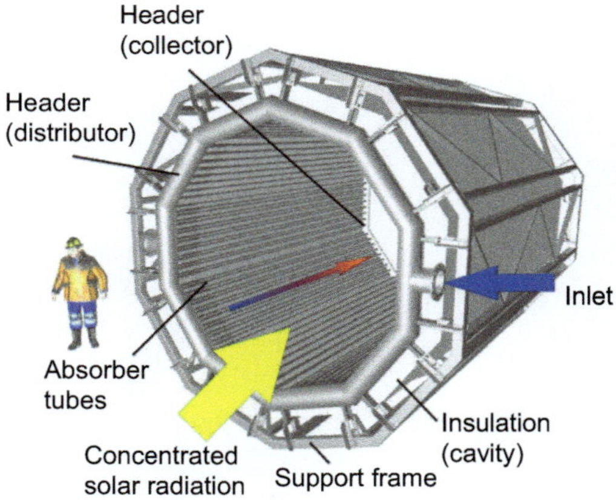

Figure 16.14 Scheme of the SOLUGAS receiver.
Copyright: DLR.

Cavity receivers show high heat losses by convection to ambient. The reason is that cold air enters the aperture at the bottom of the aperture, is heated by the hot walls, and leaves the cavity at the top. Wind can drastically increase these losses [20]. One way to reduce these losses is to close the aperture using a transparent window made of fused silica. As the manufacturing of such parts is limited in size a segmented approach can be used. It is also possible to cover only a part of the aperture (Fig. 16.15) to increase the stagnation zone with no heat exchange between hot walls and air [21]. A Prototype of such a window was tested at the SOLHYCO receiver tests [2]. A larger version of the window using 4.5 m long halved fused silica tubes was developed under the SOLTREC project. A system study showed a receiver efficiency enhancement and lower LCOE when such a window is used [29,54].

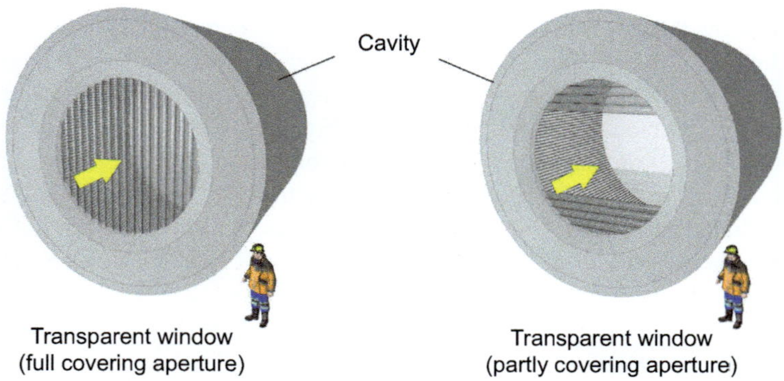

Figure 16.15 Transparent window.
Copyright: DLR.

Absorber tubes of such receivers have to accept inhomogeneously absorbed solar radiation. Due to the pressure drop limitations of the gas turbine the air flow velocity is relatively low, resulting in low heat transfer coefficients. This leads to high thermal gradients and as a consequence to high stresses in the tube walls. To reduce the gradients a so-called "profiled multilayer tube" (PML) was developed [52]. The PML tube consists of a profiled inner tube made of nickel-based alloy, a copper intermediate tube, and a nickel-based alloy outer tube. The PML tubes where manufactured using a hydroforming process, followed by an annealing step to bond the different materials by diffusion welding. The profiling allows an enhancement of the heat transfer performance by increasing the turbulence of the fluid flow. The copper layer reduces the thermal gradients due to the high thermal conductivity of copper. Simulations showed a possible decrease of thermal gradients from 220 K down to 65 K, resulting in significantly lower stresses (30 MPa instead 100 MPa). The maximum temperature of the tube is also reduced (812°C instead of 885°C). The copper layer is protected against oxidation by the inner and outer nickel-based alloy layers.

Using corrugated tubes adds a design space to the thermohydraulic layout of tubular receivers. The turbulence and the following heat transfer can be influenced by the surface structure. This means that the heat transfer performance can be more independent of the tube dimensions [55].

Under the *PEGASE* project a 4-MW_{th} receiver heating compressed air from 350 up to 750°C was developed. The receiver concept uses several metallic (nickel-based alloy) tubes located in a copper filled box of nickel-based alloy (Fig. 16.16). This approach combines the advantage of the high thermal conductivity of copper with the high temperature resistance of nickel-based alloys. The receiver is irradiated only at one side of the box, with the opposite side being insulated. Thus, only the irradiated side contributes to the radiation and convection losses of the receiver. The heat is transferred by conduction to the tubes where the air is heated up. A twisted tape insert was used to enhance the heat transfer coefficient. A prototype with a thermal power of 360 kW_{th} was tested in solar operation. An outlet temperature of 784°C was reached with an efficiency of 81.3% and a pressure drop of 300 mbar [26].

Volumetric receivers use highly porous structures for solar absorption and heat transfer. These structures offer a huge internal surface area, allowing effective heat

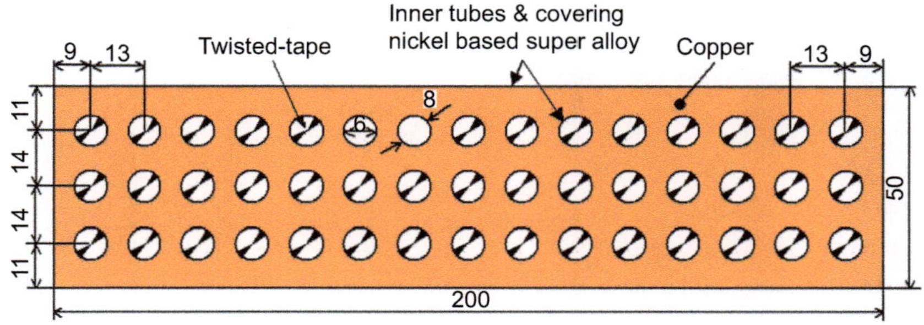

Figure 16.16 Scheme of the PEGASE receiver [26].

transfer. The concentrated solar radiation is absorbed in the volume of the absorber, as the porous structure allows for penetration of the radiation. Air or other gas passing through the structure is heated by forced convection. The porous absorber is usually a matrix or a foam structure made of SiC ceramic. There are two main directions of volumetric receiver concepts under development.

An open volumetric receiver uses ambient air as heat transfer fluid [5,16]. The hot air can then be used to heat up the compressed air of the gas turbine by a heat exchanger. But such heat exchangers would be costly due to the large heat transfer areas needed and the necessary high-temperature alloys.

If the fluid is under pressure a so-called pressurized volumetric receiver (PVR) can be used. Such a receiver consists of an internally insulated pressure vessel, the porous absorber, and a transparent quartz window covering the vessel's aperture. As the quartz window is limited in size, it is necessary for high power levels to connect several receivers to a cluster. For a complete coverage of the focal spot, secondary concentrators with hexagonal entry apertures in front of the receivers are used [28,51] (Fig. 16.17). Under the REFOS, and later in the SOLGATE project, volumetric receivers had been designed and tested. The receiver of the REFOS project used a metallic mesh absorber to reach outlet temperatures of about 800°C at a power level of about 400 kW_{th} and a pressure of 16 bar. The absorber was located within an inner insulated pressure vessel closed by a dome-shaped quartz window. This receiver was later on used in the SOLGATE project as the middle stage of a modular receiver set up. The first stage, a tubular receiver, passed the fluid with a temperature of 550°C to the metallic volumetric absorber module for heating up the air to 800°C. Within the last stage a volumetric receiver using a ceramic foam absorber the air was finally heated up to 1000°C before entering the combustor of the turbine. Several hundred hours of operation were performed under different solar load conditions [9,28,51].

The directly irradiated annular pressurized receiver (DIAPR) is a volumetric receiver designed to heat compressed air (10–30 bar) up to 1300°C (Fig. 16.18). The design uses a large number of ceramic pins arranged within a cavity closed by a window. The layout of the pins is optimized to reach good radiative exchange between the pins. The inhomogeneous heat flux distribution is homogenized by the

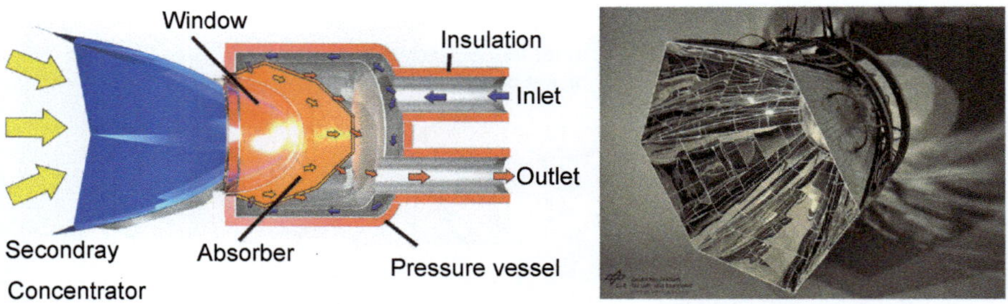

Figure 16.17 Scheme of the pressurized volumetric receiver (PVR) (left); secondary concentrator (right).
Copyright: DLR.

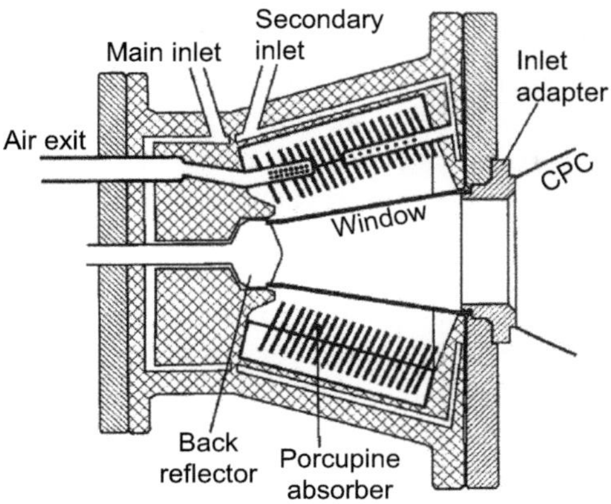

Figure 16.18 Scheme of the pressurized volumetric receiver (DIAPR) [34].

radiative exchange within the receiver. The air is heated by convective heat transfer directly from the heated surface of the pins. A prototype was tested at the Weizmann Institute's Solar Furnace reaching outlet temperatures of 1200°C with receiver efficiencies of 70–90% at a pressure level of 17–20 bar [34].

16.4.2.2 Other concepts

A pressurized modular air receiver avoiding pressurized windows was proposed in [37] with air at 5 bar. The concept is foreseen to operate up to outlet temperatures of 1200°C and 30 bar. So far a prototype of 3 kW_{th} has been tested in a solar furnace reaching an efficiency of 77% at an outlet temperature of 1000°C at 5 bar.

A volumetric particle receiver was proposed to reach high temperatures (1300K) and high efficiencies. In comparison to the pressurized volumetric receiver concept an air-particle mixture is used to absorb the solar radiation [17,18].

In direct absorption receivers (DAR) the heat transfer medium also acts as the absorber. As the solar radiation does not need to heat an absorbing structure first, and then transfer the heat based on a temperature difference, the DAR concept promises lower over temperatures and reduced requirements for the receiver structural materials. The use of ceramic particles is a promising option, allowing temperatures up to 1000°C [30,41]. The heat transfer to the gas turbine compressed air can be done using a moving bed particle heat exchanger. As the heat transfer coefficient of the particle flow to a tube bundle, for example, is very low, large heat transfer areas are needed. In addition, the high temperatures and the abrasion of the ceramic particles can lead to high cost due to expensive materials (high-temperature alloys or ceramics).

Several receiver concepts are in the R&D phase [57]. As of 2016, particle-to-compressed air heat exchangers at high temperatures (1000°C) are not available commercially, but designs have already been presented [38].

16.4.3 Storage options

In order to achieve high solar shares, thermal storage systems for SHGT have to provide high-temperature heat, preferably close to the TIT. Since storage of hot air is practically not feasible, a separate storage medium has to be used instead. While indirect integration using high-temperature heat exchangers is very expensive, direct contact heat exchange in regenerator-type storage units is the preferred option. These units are similar in concept to commercial regenerator storage units (e.g., Cowper stoves used with blast furnaces). Several options exist for the storage inventory: fixed bed with stacked material, or packed beds from regular or irregular materials. Due to the required high temperatures, ceramic materials must be used. Compared to Cowper stoves with charge/discharge cycle periods less than 1 h, units for storage of solar energy have usually one cycle per day.

Storage systems are integrated into the system parallel to the receiver. Fig. 16.19 shows a solar-hybrid system with storage, with the combustor connected in parallel. For charging the storage, the solar subsystem must be oversized. A blower can generate an additional air mass flow through the receiver and the storage. As the storage loop with the blower is practically a closed loop, the additional air mass flow does not alter the mass flow through the compressor and the turbine section. The operation of the system is then as follows:

- *Charge mode*: when the receiver delivers more power than the gas turbine can take, the excess power is taken by an increased mass flow, obtained from the compressor and an additional mass flow introduced by a blower downstream the storage. Thus, a certain fraction of the receiver outlet air passes through the storage. The receiver outlet temperature is maintained at the design outlet temperature.
- *Discharge mode*: when the receiver delivers less power than the gas turbine requires, the mass flow from the compressor is split into a fraction toward the receiver and another through the storage (with flow then reversed compared to charge mode).
- In both modes fluctuations in the receiver/storage outlet temperature can be compensated by the combustor connected in parallel.

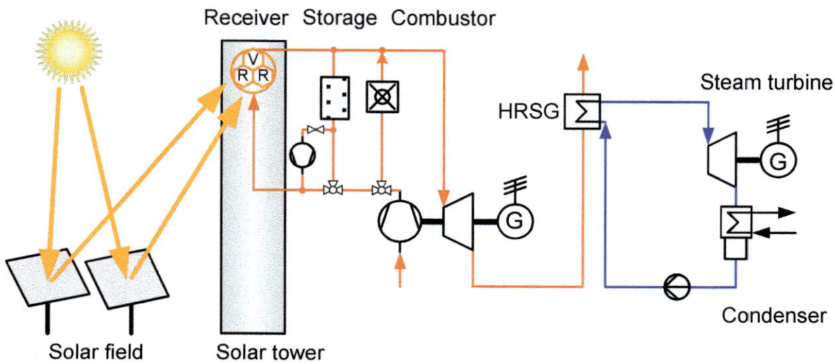

Figure 16.19 Scheme of a solar-hybrid combined cycle system with storage.
Copyright: DLR.

16.5 System studies

A large number of system studies were performed by groups to evaluate the cost and performance potentials of SHGT systems.

A comprehensive study on SHGT with CC at 100 MW_e was performed 1979 by Bechtel National, Inc [7]. The thermodynamic analyses were based on a Westinghouse gas turbine. For gas turbine inlet temperatures of 1093°C and 1316°C net efficiencies of 43.5% and 47.7% were reached. A major result of the study was that the hybrid plant offers a cost-effective alternative to conventional gas turbine and CC plants. As outlook the authors noted: "Preparation and assessment of system design with gas TITs matching the peak receiver temperatures is possible with heat pipe and with ceramic receivers."

Cost-optimized design and performance prediction for SHGT plants in the power levels 1.4, 4.2, and 16.1 MW_e for two different locations were shown in [44].

An annual average solar to net electric efficiency of up to 19% was calculated, among the highest conversion efficiencies for solar electric technologies. Solar LCOE between about €0.13/kWh and €0.90/kWh were calculated. Using the cost reduction potential that lies in combined design, construction and operation of multiple distributed plants leads to solar LCOE of below €0.10/kWh for a power level of 16.1 MW_e.

In the SOLHYCO project a 100 kW industrial microturbine system was modified for integration with a 280 kW_{th} metallic tube receiver (800°C) (SOLHYCO final report). Among other activities in this project different design variations for cogeneration systems and CC systems were analyzed and the cost for electricity generation evaluated. For a good solar site in Algeria the LCOE for the 100 kW_e unit is €0.101/kWh in an operating scheme with 25% solar share. For the large CC systems a 21 MW_e system was selected and evaluated. The LCOE of such system is €0.078/kWh being operated in base load configuration with a solar share of 18%. For solar shares of 35% the LCOE will be €0.094/kWh. This plant would be operated on mid-load for 4000 h per year. It could be shown that the solar-hybrid CC system can provide dispatchable power at a high conversion efficiency of 48%.

In the USYHNE project the technical feasibility and the economic benefit of SHGT cogeneration units with absorption chillers (power level 5–7.5 MW_e) were investigated [25]. The project served as preparation for a demonstration plant. The main result of the study shows, that in most cases it is more advantageous to generate electricity with the SHGT as much as possible (bottoming steam cycle) and then to use some of the generated electricity to drive a compression chiller to generate cooling energy.

During the projects SHCC and HYGATE several system studies were carried out. The projects are part of a technology program for the development of a commercial demonstrator of an SHGT in the 20-MW_e range. In the first phase (SHCC—solar hybrid combined cycle) several concepts of SHGT plants were identified and studied [45]. A major result was that the LCOE of such systems are competitive to other CSP reference plants (molten salt tower, parabolic trough with thermal oil) [24]. Additionally it was identified that a technological development should focus in future on further reducing the CO_2 emissions by increasing the solar share. This was studied in the following second phase of the program in the HYGATE project (Hybrid High Solar Share Gas Turbine Systems) [31]. Key improvements are the integration of thermal energy storage and the reduction of the operating temperature of the gas turbine to 950°C. As a result, the solar receiver can provide the necessary temperature for solar-only operation of the plant at design point, without using the auxiliary burner. Compared to the defined reference molten salt solar tower the SHGT plants have higher plant efficiencies, but have a slightly lower potential for CO_2 reduction. The LCOE are comparable and therefore the SHGT can be considered as an interesting alternative to molten salt tower plants [39].

A multiobjective optimization was used by [49,50] to identify pareto-optimal designs and highlight the trade-offs between minimizing investment costs and minimizing specific CO_2 emissions of SHGT with thermal energy storage and combined-cycle power blocks. A major result was that "advanced solar-hybrid combined-cycle power plants provide a 60% reduction in electricity costs compared to parabolic trough power plants. Furthermore, a 22% reduction in costs and a 32% reduction in CO_2 emissions are achieved relative to a combination of parabolic trough and combined-cycle power plants."

16.6 Conclusions

SHGT systems offer several advantages; most of all, reduced cost of solar electricity, full dispatchability, simplicity of operation, and reduced water consumption. Several hardware test systems were built and operated, accompanied by numerous studies on system optimization and techno-economic analysis.

However, such systems are not mature yet. Further development is required to obtain an efficient and reliable technology. The main challenges toward market introduction of solar gas turbine systems are as follows:

- Development of gas turbines for solar operation
  Highly efficient gas turbines adapted to solar conditions are a prerequisite. Suitable gas turbines are available as the basis, but gas turbines adapted to solar operation are not readily available. The required modifications are known and considered feasible, but significant effort must be made to provide technically proven units. The challenges are mainly associated with interfacing to external air heating, modification of the combustor system, component cooling, and system control. Currently, the gas turbine manufacturers hesitate to undertake these developments due to an unclear market perspective.
- Development of high-temperature receivers
  SHGT systems require high-temperature receivers with outlet temperatures in the range of 1000°C. The corresponding receiver technology is not mature yet. Several concepts are developed to different levels of technical readiness, but upscaling and long-term operation experience are lacking.
- General aspects
  All in all, further development effort is required to bring SHGT systems to the market. Modular system designs might be the right approach to limit the development cost on the gas turbine and receiver side, distributing these cost over a larger number of identical units. Selecting and developing the appropriate modular configuration needs to be done.

Abbreviations

CC system	Combined-cycle system
DAR	Direct absorption receiver
GT	Gas turbine
LCOE	Levelized cost of electricity
ORC	Organic Rankine cycle
PML	Profiled multilayer tube
PSA	Plataforma solar de Almería, Spain
SHGT	Solar-hybrid gas turbine
SiC	Silicon carbide
TIT	Turbine inlet temperature

References

[1] Amsbeck L, Buck R, Heller P, Jedamski J, Uhlig R. Development of a tube receiver for a solar-hybrid microturbine system. In: Proceedings of the 14th Biennial CSP SolarPACES Symposium 2008. 14th Biennial CSP SolarPACES Symposium 2008, 2008-03-04–2008-03-07, Las Vegas, NV (USA); 2008.

[2] Amsbeck L, Heisch G, Röger M, Uhlig R. Development of a Broadband Antireflection coated transparent silica window for a solar-hybrid microturbine system. In: SolarPACES 2009, 15–18, September 2009, Berlin, Germany, ISBN 978-3-00-028755-8.

[3] Amsbeck L, Denk T, Ebert M, Gertig C, Heller P, Herrman P, et al. Test of a solar-hybrid microturbine system and evaluation of storage deployment. In: Proc. SolarPACES 2010, September 21–24, Perpignan, France; 2010.

[4] AORA Website, http://aora-solar.com/.

[5] Ávila-Marín A. Volumetric receivers in solar thermal power plants with central receiver system technology: a review. Solar Energy May 2011;85(5):891–910.

[6] Bammert K. Layout of gas cycles for solar power generation. Electric Power Systems Research 1980;3:163–83. Elsevier Sequoia S.A.

[7] Bechtel National, 1779. Foster Wheeler Development Corporation; Northrup, Incorporated; Westinghouse Electric Corporation, Combustion Turbine Systems Division; Public Service Company of New Mexico; United States, Dept. of Energy; San Francisco Operations. 1979. Combined cycle solar central receiver hybrid power system study. Final technical report, prepared by Bechtel National Inc. for the United States Department of Energy, San Francisco Operations Office; Contract no. DE-AC03–78ET21050, Bechtel job no. 13007. DOE/ET-21050-1/1; 1979.

[8] Becker M. GAST: the gas cooled solar tower technology program. In: Becker M, Böhmer M, editors. Proceedings of the final presentation May 30–31, Lahnstein, Federal Republic of Germany/Deutsche Forschungs-und Versuchsanstalt für Luft-und Raumfahrt e.V. (DFVLR). Springer; 1989.

[9] Buck R, Bräuning T, Denk T, Pfänder M, Schwarzbözl P, Tellez F. Solar-hybrid gas turbine-based power tower systems (REFOS). Journal of Solar Energy Engineering (ASME) 2002;124(1):2−9.
[10] Buck R. Hocheffiziente Solarturm-Technologie (HST). Schlussbericht, Förderkennzeichen Z II 6 (D)-46040−1/3.3. February 2005.
[11] Chacartegui R, et al. Alternative cycles based on Carbon dioxide for central receiver solar power plants. Applied Thermal Engineering 2010. http://dx.doi.org/10.1016/j.applthermaleng.2010.11.008. Elsevier.
[12] Coogan S, Brun K, Teraji D. Micromix combustor for high temperature hybrid gas turbine concentrated solar power systems. Energy Procedía 2014;49. ISSN: 1876-6102:1298−307.
[13] CSP Today Global Tracker Website, http://social.csptoday.com/tracker/proiects/table.
[14] Parabolic dish module experiment, NASA STI/Recon Technical Report N March 1986. http://adsabs.harvard.edu/abs/1986STIN...8631987.
[15] Felsmann C, Gampe U, Freimark M. Dynamic behavior of a solar hybrid gas turbine system, GT2015−42437. In: Proc ASME Turbo Expo 2015, June 15−19, Montreal, Canada; 2015.
[16] Fend T, Hoffschmidt B, Pitz-Paal R, Reutter O, Rietbrock P. Porous materials as open volumetric solar receivers: experimental determination of thermophysical and heat transfer properties. Energy April−May 2004;29(5−6):823−33.
[17] Fernández P, Miller F. Assessment of the overall efficiency of gas turbine-driven CSP plants using small particle solar receivers. In: Solar Paces2013, Las Vegas. USA; 2013.
[18] Fernández P, Miller F, McDowell M, Hunt A. Design space exploration of a 5 MWth small particle solar receiver. In: Solar Paces2013, Las Vegas. USA; 2013.
[19] Ferrière A. The PEGASE project. International Scientific Journal for Alternative Energy and Ecology 2008;6(62). ISSN: 1608-8298.
[20] Flesch R, Stadler H, Uhlig R, Pitz-Paal R. Numerical analysis of the influence of inclination angle and wind on the heat losses of cavity receivers for solar thermal power towers. Solar Energy 2014;110. ISSN: 0038-092X:427−37. http://dx.doi.org/10.1016/j.solener.2014.09.045.
[21] Flesch R, Grobbel J, Stadler H, Uhlig R, Hoffschmidt B. Reducing the convective losses of cavity receivers. In: SolarPACES Conference 2015, 13−16. October 2015, Capetown; 2015.
[22] Frutschi HU. Closed-cycle gas turbines: operating experience and future potential. New York: ASME Press; 2005. London: Professional Engineering.
[23] Gallup DR, Kesseli JB. A solarized Brayton engine based on turbo-charger technology and the DLR receiver. In: 29. Intersociety energy conversion engineering conference, Monterey, USA, 7−12 August 1994; 1994.
[24] Giuliano S, Buck R, Eguiguren S. Analysis of solar-thermal power plants with thermal energy storage and solar-hybrid operation strategy. Journal of Solar Energy Engineering. ASME 2011 ISSN: 0199-6231. http://dx.doi.org/10.1115/1.4004246.
[25] Giuliano S, Schillings C, Hoyer-Klick C, Al Nuaimi S, Al Obaidli A. Upscaling of solar-hybrid gas turbine cogeneration units (USHYNE), Deutsches Zentrum für Luft-und Raumfahrt, Stuttgart, Schlussbericht, BMU-Projekt FZK 03UM0088. 2008.
[26] Grange B. Thermal performances of a high temperature air solar absorber based on compact heat exchange technology. Journal of Solar Energy Engineering 2001;133: 031004.

[27] Grange B, Dalet C, Falcoz Q, Siros F, Ferrière A. Simulation of a hybrid solar gas-turbine cycle with storage integration. Energy Procedía 2014;49. ISSN: 1876-6102:1147–56.

[28] Heller P, Pfänder M, Denk T, Tellez F, Valverde A, Fernandez J, et al. Test and evaluation of a solar powered gas turbine system. Solar Energy 2006;80(10):1225–30. Proc. 14th SolarPACES Symposium, Las Vegas, 2008.

[29] Hertel J, Uhlig R, Söhn M, Schenk C, Heisch G, Bornhöft H. Fused silica windows for solar receiver applications. In: SolarPACES conference 2015, 13–16. October 2015, Capetown, South Africa; 2015.

[30] Ho C, Christian J, Gill D, Moya A, Jeter S, Abdel-Khalik S, et al. Technology advancements for next generation falling particle receivers. Energy Procedía 2014;49:398–407. 19th SolarPACES conference, Las Vegas, 2013.

[31] HYGATE. MAN Diesel & Turbo SE, Deutsches Zentrum für Luft-und Raumfahrt e.V., Technische Universität Dresden, VGB Powertech e.V.: HYGATE -Hybrid High Solar Share Gas Turbine Systems, 2015. 2015. Schlussbericht, BMWi/BMU-Projekt FZK 03ET2041 A-C/0325382 A-B.

[32] Kesseli J. Brayton power conversion system. United States. 2012. http://dx.doi.org/10.2172/1045668. http://www.osti.gov/scitech/servlets/purl/1045668.

[33] Wang W, Ragnolo G, Aichmayer L, Strand T, Laumert B. Integrated design of a hybrid gas turbine-receiver unit for a solar dish system. Energy Procedia 2015;69:583–92.

[34] Kribus A, Doron P, Rubin R, Reuven R, Taragan E, Duchan S, et al. Performance of the directly-irradiated annular pressurized receiver (DIAPR) operating at 20 bar and 1,200 degrees C. ASME Journal of Solar Energy Engineering 2001;123:10–7.

[35] McDonald CF. A hybrid solar closed-cycle gas turbine combined heat and power plant concept to meet the continuous total energy needs of a small community. Heat Recovery Systems 1986;6(5):399–419. Pergamon Journals Ltd.

[36] Payrhuber K, Rayome D. Experience and Applications with the innovative LMS100® Gas Turbine. VGB Fachtagung Gasturbinen und Gasturbinenbetrieb. Mannheim: VGB PowerTech e.V; 2009.

[37] Pozivil P, Aga V, Zagorskiy A, Steinfeld A. A pressurized air receiver for solar-driven gas turbines. In: SolarPACES 2013, Las Vegas, USA; 2013.

[38] Prosin T, Pryor T, Creagh C, Amsbeck L, Uhlig R. Solar gas turbine systems with centrifugal particle receivers, for remote power generation. In: SolarPACES 2014; 2014.

[39] Puppe M, et al. Hybrid high solar share gas turbine systems with innovative gas turbine cycles. In: Energy Procedía 2015. International Conference on Concentrating Solar Power and Chemical Energy Systems, SolarPACES 2014, 16–19, September 2014, Peking, China; 2015.

[40] Korzynietz R, Brioso JA, del Río A, Quero M, Gallas M, Uhlig R, Ebert M, Buck R, Teraji D. Solugas – Comprehensive analysis of the solar hybrid Brayton plant. Solar Energy October 2016;135:578–89. http://dx.doi.org/10.1016/j.solener.2016.06.020. ISSN 0038-092X.

[41] Röger M, Amsbeck L, Gobereit B, Buck R. Face-down solid particle receiver using recirculation. Journal of Solar Energy Engineering ASME 2011 ISSN: 0199-6231:031009-1–031009-8. http://dx.doi.org/10.1115/1.4004269.

[42] Scheuerer K. Berechnung des stationären und instationären Betriebsverhaltens einer Solar-Kraftanlage mit Gasturbine (Ph.D. thesis). Technische Universität München; 1986.

[43] Schmuttermair H. Experimentelle Simulation und Analyse des Betriebsverhaltens einer Solar-Kraftanlage mit Gasturbine (Ph.D. thesis). Technische Universität München; 1992.

[44] Schwarzbözl P, Buck R, Sugarmen C, Ring A, Crespo MJM, Altwegg P, et al. Solar gas turbine systems: design, cost and perspectives. Solar Energy October 2006;80(10).

[45] SHCC. MAN Diesel & Turbo SE, Deutsches Zentrum für Luft-und Raumfahrt e.V. Technische Universität Dresden, VGB Powertech e.V; 2011. Entwicklung eines Referenzkonzepts für eine Solarhybrid-GuD-Anlage (SHCC) der Leistungsklasse bis 20 MW, 2011, Schlussbericht, BMU-Projekt FZK 0325086.

[46] Solar turbines, Mercury 50 brochure. 2013. http://s7d2.scene7.com/is/content/Caterpillar/C10553041.

[47] Final Publishable Report SOLGATE-solar hybrid gas turbine electric power system, ISBN 92-894-4592-0. EUR 21615, 47 pp.

[48] SOLHYCO. Final Public Report SOLHYCO: solar-hybrid power and cogeneration plants. 2011. Available at: http://cordis.europa.eu/publication/rcn/13318 en.html.

[49] Spelling J. Hybrid solar gas-turbine power plants (Ph.D. thesis), ISBN 978-91-7501-704-4.

[50] Spelling J, Laumert B, Fransson T. Advanced hybrid solar tower combined-cycle power plants. Energy Procedía 2014;49. ISSN: 1876-6102:1207−17. http://dx.doi.Org/10.1016/j.egypro.2014.03.130.

[51] Sugarmen C, Ring A, Buck R, Heller P. Solar-hybrid gas turbine power plants-test results and market perspective. In: Proceedings ISES Solar World Congress, Sweden; 2003.

[52] Uhlig R, Buck R. Development of a "multi-layer tube" for high temperature solar receivers; heatSET 2005 heat transfer in components and systems for Sustainable energy Technologies 5−7 April 2005, Grenoble, France. 2005.

[53] Uhlig R. Transient stresses at metallic solar tube receivers. In: 17th SolarPACES 2011 Conference, 20−23. September 2011, Granada, Spain; 2011.

[54] Uhlig R, Flesch R, Gobereit B, Giuliano S, Liedke P. Strategies enhancing efficiency of cavity receivers. In: SolarPaces 2013, 17.09.2013−20.09.2013, Las Vegas; 2013.

[55] Uhlig R, Gobereit B, Rheinländer J. Advancing tube receiver performance by using corrugated tubes. Energy Procedía 2015;69. ISSN: 1876-6102:563−72.

[56] Wilson Solarpower Website, http://wilsonsolarpower.com/.

[57] Wu W, Trebing D, Amsbeck L, Buck R, Pitz-Paal R. Prototype testing of a centrifugal particle receiver for high-temperature concentrating solar applications. Journal of Solar Energy Engineering 2015;137. ISSN: 0199-6231. http://dx.doi.org/10.1115/1.4030657.

Solar power towers using supercritical CO_2 and supercritical steam cycles, and decoupled combined cycles

17

M.A. Silva-Pérez
University of Sevilla, Seville, Spain

17.1 Introduction

Solar power-tower systems (also known as *central receiver systems*) can efficiently achieve high temperatures because of the high concentration ratios they can achieve using different configurations of the collector field and receiver. The combination of solar power towers with high-temperature cycles permits to increase the global efficiency in the conversion of solar radiation to electricity with respect to today's CSP plants based on the sub-critical Rankine cycle and could result in LCOE reduction, as far as the increase in efficiency outweighs the increased costs associated to the use of more expensive equipment and materials.

Although operating temperatures higher than 1000°C can be achieved with solar power towers, there are still significant technological barriers that must be overcome before CSP plants operating at these elevated temperatures reach the market. The use of power cycles operating at temperatures in the range 600–800°C, only moderately higher than those of the current CSP plants, has been identified as a promising path to increase the efficiency and reduce the LCOE of the next generation of CSP plants that would require technological developments achievable in a relatively short term.

The technological developments required to improve the technology readiness level of these options seem to be achievable in the short term at least in some cases, although, as of 2016, none of these concepts has been demonstrated beyond a pilot scale. In some cases, the adaptation of the state-of-the-art equipment would be sufficient, while in other cases further research is required.

In any case, this path has been recognized in different programs and initiatives, like DOEs Sunshot [16], the European Solar Thermal Electricity Industry Association, ESTELA, [12] or the Australia Solar Thermal Research Initiative, ASTRI [26].

In this chapter we present a review of three options to follow this path:

- power towers with supercritical steam Rankine cycles (SSRCs);
- power towers with supercritical CO_2 cycles;
- decoupled solar combined cycles (DSCCs).

Advances in Concentrating Solar Thermal Research and Technology
http://dx.doi.org/10.1016/B978-0-08-100516-3.00017-4

17.2 Solar power towers with supercritical cycles

In this section, the different options for the integration of solar power towers and supercritical steam and CO_2 cycles are described and discussed. Both supercritical cycles operate in temperature ranges only moderately superior to state-of-the-art solar power towers, although the different physical and chemical properties of the cycle working fluid involve specific requirements for the solar power tower in terms of integration schemes, heat transfer media (HTMs), materials, and so on.

17.2.1 Supercritical steam Rankine cycles

SSRCs operate at pressures greater than 22.1 MPa in a range of temperatures only moderately higher (600–720°C, up to 760°C in the case of advanced ultra-supercritical cycles) than those of the state-of-the-art solar power towers and have been identified as a promising option to increase the efficiency and reduce the cost of the electricity generated by CSP plants in the near or medium term.

The Sunshot Initiative of the US Department of Energy [16] identified the possibility to adapt current molten salt and direct steam generation (DSG) solar power towers to supercritical steam cycles operating in the range 600–700°C. Years before, the long-term scenario for 2020 analyzed by [37] already considered a solar power tower plant using an SSRC. Kolb [23] explored the benefits of a new generation of molten salt CSP plants with SSRCs operating in the range 600–650°C, finding potential LCOE reductions of up to 8% compared to current sub-critical molten salt plants. [39] studied the cost reduction potential of ultra-supercritical steam cycles—350 bar and up to 720°C with a thermal efficiency of 55%—coupled with a solar tower using tubular receivers with different liquid metals and salt mixtures as HTMs. According to their estimates, the LCOE could be reduced up to 15% compared to the current molten salt CSP plants. The same authors [40,41] compared the potential of two innovative receiver concepts—internal direct absorption, beam down—with tubular receivers using different HTMs and ultra-supercritical steam cycles, estimating potential LCOE reductions from 7.2% (direct absorption receiver with chloride salt) to 0.5% (beam down with molten nitrate salts) with respect to molten salt power towers of 2016. Peterseim and Veeraragavan [34] compared three solar power towers using an advanced steam cycle with sub-critical parameters and two supercritical solar power towers: the first, a hybrid solar–natural gas configuration with state-of-the-art molten salts and steam parameters of 280 bar and 620°C and the second, a solar-only plant with a precommercial, advanced molten salt and steam parameters of 620°C at 280 bar. They found that the LCOE could be reduced about 4.3% with the third option, that requires the development of a commercial molten salt mixture, stable at about 700°C. [38] analyzed the requirements, in terms of materials technology, for the use of advanced ultra-supercritical steam cycles operating at steam conditions up to 760°C with up to 35% improved efficiency compared to superheated steam cycles of 2016.

However, working at high temperature and pressure increases the costs of materials and equipment. In addition, there are some challenges associated to the development of SSRC solar power towers:

- *Size of the plant.* A major challenge for the immediate deployment of SSRC solar power towers is the upscaling of current power tower plants or downsizing the SSRC. The largest solar power tower built today is Crescent Dunes (Nevada, USA) with a rated power of 110 MW and 10 full-hour molten salt thermal energy storage (TES); on the other hand, the smallest commercially available supercritical steam turbines are in the range of the 250 MW. Hybrid solar–fossil fuels options may provide a path to overcome the current gap [34].
- *High-temperature receivers.* State-of-the-art molten salt receivers operate at temperatures of about 560°C. Increasing this temperature requires the development of new salt mixtures which are stable at the temperatures required by SSRC plants (600–700°C) or the use of other HTMs like solid particles, liquid metals, air, or supercritical steam. On the other hand, materials that can withstand the demanding conditions during the lifetime of the plant will be required for the receiver, heat exchangers [35], and other equipment. Further technology developments are required in all cases.

17.2.1.1 Integration of solar power towers and supercritical steam Rankine cycles

There are multiple possible configurations of SSRCs solar power towers: direct or indirect supercritical steam generation, direct, indirect, or thermochemical storage system, solar-only or hybrid plants, and so on. In the following, we provide an overview of the main variants referenced in the literature.

Direct steam generation

In the DSG configuration, water enters the solar receiver at low temperature and supercritical pressure and is heated to temperatures above 550°C. No additional heat exchanger or steam generator is required unless the plant includes an indirect-type TES. The main concern for the development of direct supercritical steam receivers is the demanding requirements in terms of materials and design to withstand the high-pressure and high-temperature conditions of the steam [19]. An early 2010s project in CSIRO [3,15] has demonstrated the feasibility of such receivers, with a pilot project (309 kW thermal power) generating supercritical steam at pressures from 22.5 to 23.5 MPa and temperatures of 570°C (Fig. 17.1).

Indirect steam generation

Another option to integrate solar power towers with supercritical steam is to use a high-temperature HTM. In this case, the concentrated solar flux is used to increase the temperature of the HTM and a high-pressure HTM steam generator is used to generate the supercritical steam feeding the turbine. The overall configuration of the plant can be very similar to current molten salt plants (Fig. 17.2). Potential HTMs are molten salt mixtures, liquid metals, solid particles, and so on.

Figure 17.1 A picture of the CSIRO supercritical steam receiver [15].

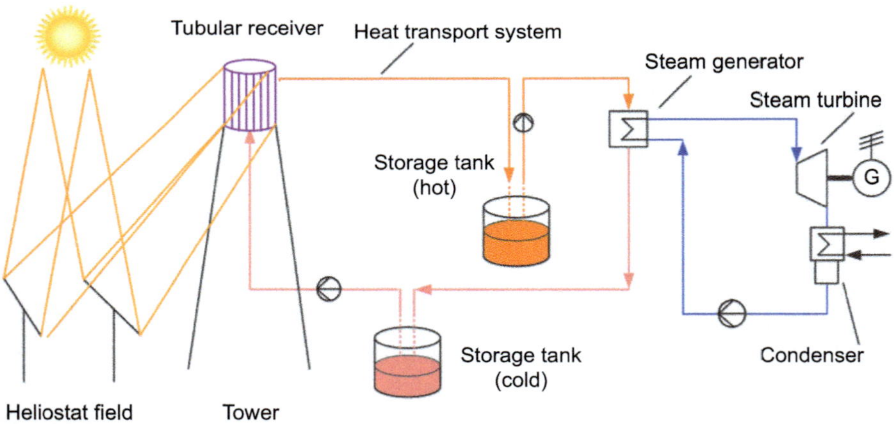

Figure 17.2 Sketch of a supercritical steam plant with liquid HTM and TES [41].

Molten salt mixtures appear today as the most immediate option for SSRC solar power towers. A hybrid configuration based on the current molten salt plants producing supercritical steam at 280 bar and 545°C with additional natural gas superheating to reach a steam temperature of 620°C was analyzed by Peterseim and Veeraragavan [34] (Fig. 17.3).This design would not require any major technological development since only the molten salt—steam generator would need to be modified to operate at supercritical pressure.

The use of other salt mixtures that have stability beyond 650°C would allow to increase the steam temperature without the use of natural gas superheating. Chlorides, carbonates, and fluorides have been identified as potential candidates [19,40,41].

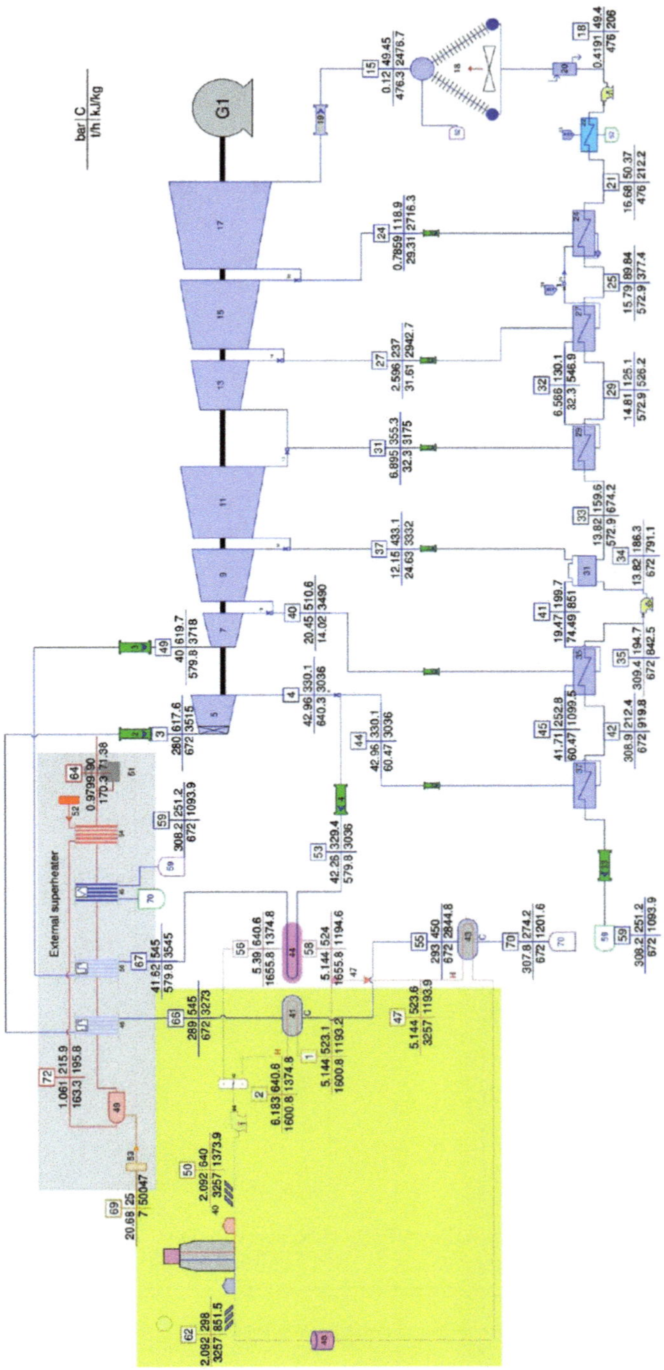

Figure 17.3 Process diagram of the supercritical solar tower with natural gas superheating.

From Peterseim JH, Veeraragavan A. Solar towers with supercritical steam parameters—is the efficiency Gain worth the Effort? Energy Procedia 2015a;69:1123–32. http://dx.doi.org/10.1016/j.egypro.2015.03.181.

However, operating beyond 650°C will require the development of new receiver materials and, in some cases, the use of oxygen blankets to avoid or reduce the rate of corrosion in the hot storage tank and other equipment [23,37].

Liquid metals and metal eutectics can operate at the high temperatures required to generate supercritical steam. They exhibit excellent heat transfer characteristics allowing for high efficiency and high flux density on the receiver. The use of alkali metals (mainly sodium) and lead–bismuth eutectic (LBE) has been analyzed by different authors [19,31]. Table 17.1 presents a comparison of the physical properties of solar salt (state-of-the-art molten nitrate salt mixture), sodium, and LBE. While the upper temperature limits of both liquid metals are well above the current range of supercritical and ultra-supercritical steam plants, sodium has the drawback of its extreme reactivity with air and water and LBE has a low thermal capacity, preventing their application in direct TES configurations.

Solid particles (see Chapter 10 of this book) can achieve high temperatures too. They can be used in different types of receivers and configurations as HTM and TES medium—direct absorption falling particles [43], fluidized bed [25], particles in tube [13], and so on—and have the potential to be relatively cheap. In principle, the integration of solid-particles solar power towers with supercritical steam cycles seems straightforward, with an overall scheme similar to that of molten salts solar power towers. However, the solid-particle technology in its different variants is still at a relatively early stage of development [19].

Table 17.1 Comparison of solar salt and two candidate liquid metals

Physical property	Solar salt	Liquid Na	Liquid LBE
Lower temperature limit,°C	220	98	125
Upper temperature limit,°C	600	883	1670
Heat capacity c_p, $kJkg^{-1} k^{-1}$	1.52	1.27	0.143
Thermal conductivity λ, $Wm^{-1} k^{-1}$	0.53	69.8	13.7
Density ρ, kg m^{-3}	1804	850	10,139
Dynamic viscosity μ, MPa s	1.69	0.27	1.44
Prandtl number	4.85	0.005	0.015
Other characteristics	Low-cost direct TES is possible	React with air and water	Larger density, lower c_p

From Pacio J, Singer C, Wetzel T, Uhlig R. Thermodynamic evaluation of liquid metals as heat transfer fluids in concentrated solar power plants. Applied Thermal Engineering 2013;60:295–302. http://dx.doi.org/10.1016/j.applthermaleng.2013.07.010.

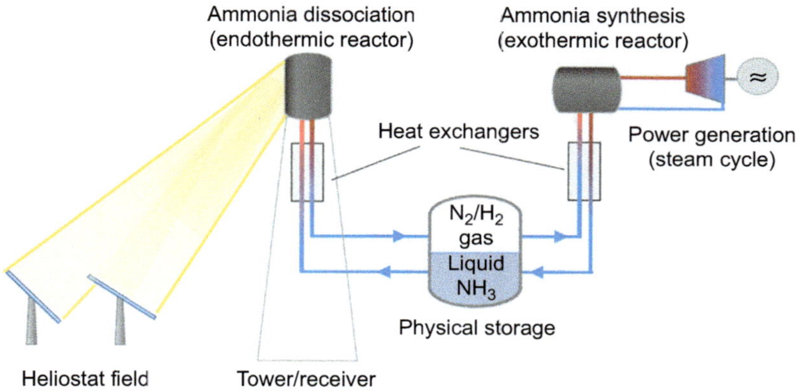

Figure 17.4 Schematic of ammonia-based thermochemical energy storage system [24].

Ammonia-based thermochemical energy storage

Ammonia-based thermochemical energy storage (TCES), Fig. 17.4, is based on the realization of the endothermic dissociation of NH_3 at the solar receiver–reactor absorbing solar energy. The products of the reaction, N_2 and H_2, are stored in a physical tank (charging process). The thermochemical energy stored in the tank can be used by performing the reverse, exothermic reaction at the discharge reactor (steam generator). A prototype ammonia-based TCES coupled to a parabolic dish was demonstrated at the Australian National University, ANU, achieving temperatures of 475°C.

The use of ammonia TCES in SSRCs has been explored by several researchers [9,24], exploring alternatives for the physical gas containment (salt caverns, drilled underground shafts), supercritical steam generator–reactor, and the solar reactor. The authors expect to develop a TCES with a cost of US $15/kW h thermal.

17.2.2 Supercritical carbon dioxide Brayton cycles

Supercritical carbon dioxide (sCO_2) closed-loop Brayton cycles have been introduced and described in Chapter 11 of this book. sCO_2 cycles have the potential to provide higher efficiency than the sub-critical steam cycles used in CST power plants and equivalent or higher than supercritical steam cycles operating on the same temperature range, with the advantage of lower operating pressure and greater compactness. In addition, the heat rejection temperature ranges of sCO_2 cycles make them appropriate for the use of dry-cooling. Sensible heat TES can be easily integrated because the CO_2 presents a single phase in all the cycle processes. These characteristics reveal a potential to develop highly efficient and compact CST plants (Fig. 17.5).

Different sCO_2 cycle configurations have been discussed in Chapter 11. The selection of the most appropriate one for integration in CST towers depends on the cycle efficiency, the temperature difference, and the complexity—which is directly associated to its cost.

Chacartegui et al. [7] compared two configurations of sCO_2 Brayton cycles—simple and recompression—and an sCO_2–ORC combined cycle, integrated with central

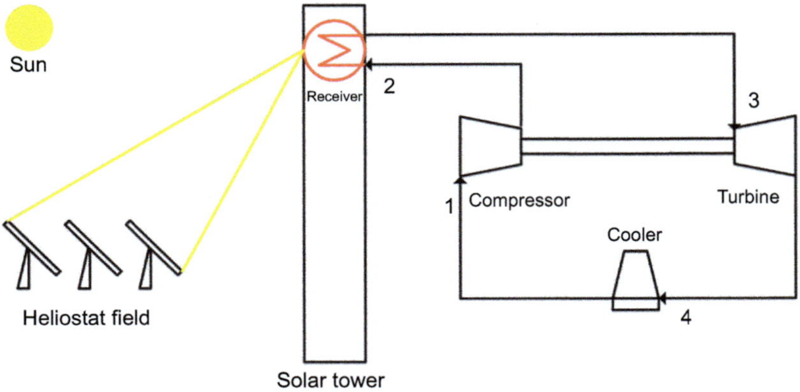

Figure 17.5 Simple closed-loop supercritical carbon dioxide Brayton cycle [2].

receiver systems. Their analysis, which used a simple solar tower model, concluded that the recompression cycle provides the best performance. Other analyses using an effectiveness model for the recuperator [42] suggested that the performance of the partial cooling and recompression cycles were similar. A later study using a conductance model for the recuperators [27] showed that the partial cooling cycle provides higher efficiency than the recompression cycle up to high values of the recuperator conductance. Assuming that the cost of the equipment increases with the recuperator conductance, the partial cooling configuration is better for low- and medium-conductance values. In addition, they also found advantages for the partial cooling cycle regarding the design and operation of the solar receiver. The simple sCO_2 Brayton cycle has a lower efficiency than the partial cooling and the recompression cycle, but has the advantage of its simplicity for near-term implementation.

Iverson et al. [21] modeled and validated—within the limits of the experimental setup available—an sCO_2 cycle for CST plants. The results include the transient part-load response of the cycle and the identification of necessary research for successful implementation of sCO_2 Brayton cycles in CST power plants, including the development of turbines, bearings, seals, heat exchanger designs, and materials.

Padilla et al. [32,33] compared the thermal and exergetic performance of four cycle configurations: simple, recompression, partial cooling, and recompression with main compression intercooling configurations, finding that the latter has the best performance with a thermal efficiency of about 47% at temperature greater than 700°C.

17.2.2.1 Integration of solar power towers and supercritical CO_2 cycles

As in the case of SSRCs, the options for the integration of sCO_2 cycles and solar power towers are multiple: direct or indirect sCO_2 generation, direct, indirect, or thermochemical storage system, solar-only or hybrid plants, and so on. In the following, we provide an overview of the main variants referenced in the literature.

Direct sCO_2 receivers

The main advantage of direct receivers, where the working fluid is the same at the solar receiver and power cycle, is that it eliminates the need for intermediate heat exchangers, thus avoiding the thermal and exergy losses and the cost associated to this equipment. In the case of solar powers with sCO_2 cycles, there are three main options for a direct sCO_2 receiver: tubular, pressurized volumetric, and fluidized bed, small particle-gas receiver. Of these three options, only the first seems to be sufficiently mature for its commercial deployment in the near to medium term.

Tubular receivers would operate at pressures of up to 30 MPa. Ortega and Christian [30] established the design requirements for tubular CO_2 receivers operating at supercritical conditions. They have also developed a coupled optical–thermal-fluid model [29] and performed a structural and creep–fatigue evaluation of such receiver [28]. Their results show that thermal efficiencies close to 85% can be achieved at the receiver when using appropriate aiming-point strategies to obtain adequate flux profiles on the receiver surface and flow patterns with high recirculation of the working fluid.

Besarati et al. [4] proposed a direct CO_2 solar receiver based on compact heat exchanger (CHE) technology. The receiver consists of Inconel 625 plates with square shaped channels bonded together, operating at pressures close to 20 MPa, and temperatures between 530°C and 707°C with a solar flux density of 500 kW/m^2.

Indirect receivers for sCO_2

The integration of sCO_2 receivers with storage is today a major, unsolved challenge; thermal [18] storage of supercritical fluids is not viable [22] and the use of a different TES medium does not seem to be an efficient option. In this context, the use of different HTMs as working fluid appears as the best option for integration of solar power towers with storage and sCO_2 cycles.

As in the case of the SSRCs, the concentrated solar flux is used to increase the temperature of the HTM and a high-pressure HTM—CO_2 heat exchanger is used to generate the high-temperature supercritical CO_2 feeding the sCO_2 turbine. The overall configuration of the plant can be very similar to molten salt plants (Fig. 17.6). Potential HTMs are, again, molten salt mixtures, liquid metals, and solid particles in different receiver configurations. Ho et al. [20] reviewed several high-temperature designs for sCO_2 Brayton cycles, concluding that the most viable option today for indirect CO_2 heating and TES is the use of falling particles receivers.

Liquid metals—mainly sodium and LBE—can operate at the high temperatures required by sCO_2, but, again, sodium has the problem of its extreme reactivity with air and water and LBE has a low thermal capacity, preventing their immediate application in direct TES configurations.

Thermochemical energy storage for sCO_2 cycles

Several thermochemical cycles for TES coupled to sCO_2 cycles have been proposed. Buckingham et al. [5] propose redox transitions in metal oxides and sulfur-based cycles, where energy is stored inexpensively in the form of elemental sulfur (Fig. 17.7). These

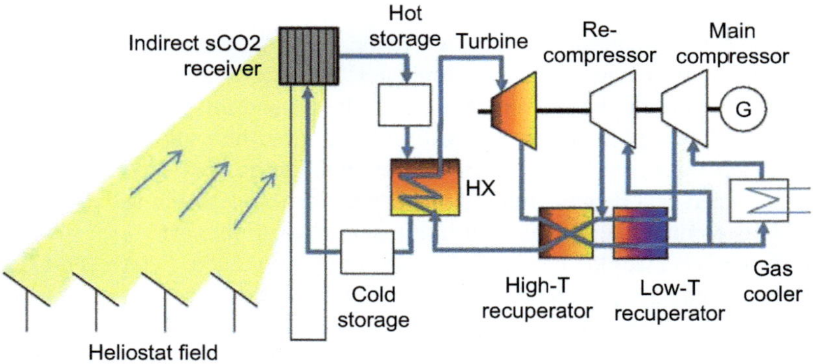

Figure 17.6 Schematic of a solar-driven, indirectly heated, closed-loop supercritical CO_2 Brayton power cycle.
From Ho CK, Carlson M, Garg P, Kumar P. Cost and performance Tradeoffs of alternative solar-driven S-CO2 Brayton cycle configurations. In: Proceedings of the ASME 2015 power and energy conversion Conference. San Diego, California, 2015. p. 1−10.

reactions that occur at temperatures between 500°C and 1000°C take place at the solar receiver (moving-bed reactor). The authors estimate that these configurations have the potential to increase the efficiency of the plant with respect to the state-of-the-art molten salt Rankine cycle plants, however, with significant uncertainties related to cost.

Calcium looping with sCO_2 cycle is a promising option too, to be integrated with solar power towers [6]. The calcium looping process is based on the calcination–carbonation reaction of $CaCO_3$. It has a high volumetric energy density and can produce temperatures close to 900°C, well coupled to sCO_2 cycles and solar power towers.

17.2.3 Comparison of supercritical steam and carbon dioxide Brayton cycles

A review of high-efficiency thermodynamic cycles and their applicability to CSP systems performed by Dunham and Iverson [11] concluded that steam Rankine systems may offer higher thermal efficiencies up to temperatures about 600°C, material limits for steam components as of 2016 while an sCO_2 recompression Brayton cycle may be the best candidate for higher temperature, with potential efficiency of about 60% at 30 MPa and above 1000°C and wet-cooling.

The benchmarking exercise realized by [8] concluded that a superheated steam Rankine cycle is both more efficient and more cost-effective than the three sCO_2 power cycle concepts and the SSRC considered in their analysis (Figs. 17.8 and 17.9). However, the validity of these conclusions is limited by the fact that they considered a constant thermal power input of 213.7 MW—relatively small for state-of-the-art supercritical and combined cycles—and that the comparison was mainly based on the normalized cost ($/kW) rather than the LCOE.

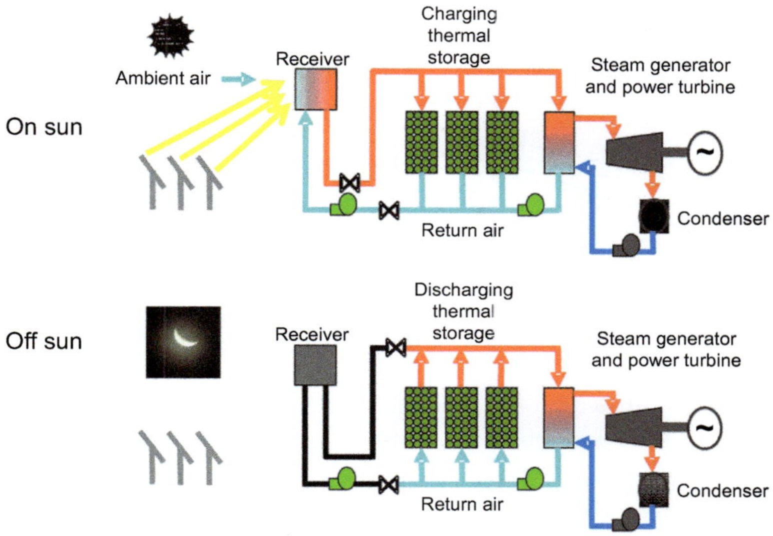

Figure 17.7 Metal oxide TES configuration.
From Buckingham R, Wong B, Brown L. Thermochemical energy storage for concentrated solar power—coupling to a high efficiency supercritical CO2 Power cycle. In: Proceedings of the SolarPaces Conference 2011. p. 0–5.

In any case, any conclusions as of 2016 can be only considered of a temporary character, since there is a vast activity in the field of high-temperature materials for advanced power generation and other related fields that could modify the outcomes of these analyses.

17.3 Decoupled solar combined cycles

DSCCs have been identified as a promising option for LCOE reduction that takes advantage of the high temperature achievable by means of CSP systems and the use of TES. A DSCC is the combination of a high-temperature cycle where the heat is provided by a CSP system and a lower temperature, bottoming cycle. The heat rejected during the operation of the high-temperature cycle is used to charge the TES. The energy stored in the TES can be asynchronously used to feed the bottoming cycle, thus decoupling the operation of both cycles and allowing for great operation flexibility.

The DSCC concept seems "naturally" linked to the solar tower technology because of its capacity to efficiently achieve high concentration ratios and high temperature, thus taking advantage of the high exergy of the solar radiation.

The DSCC concept provides great flexibility in the design of the plant. Despite its relatively recent development—the first reference to DSCCs dates from 2012 [36]—several different configurations can be found in the literature.

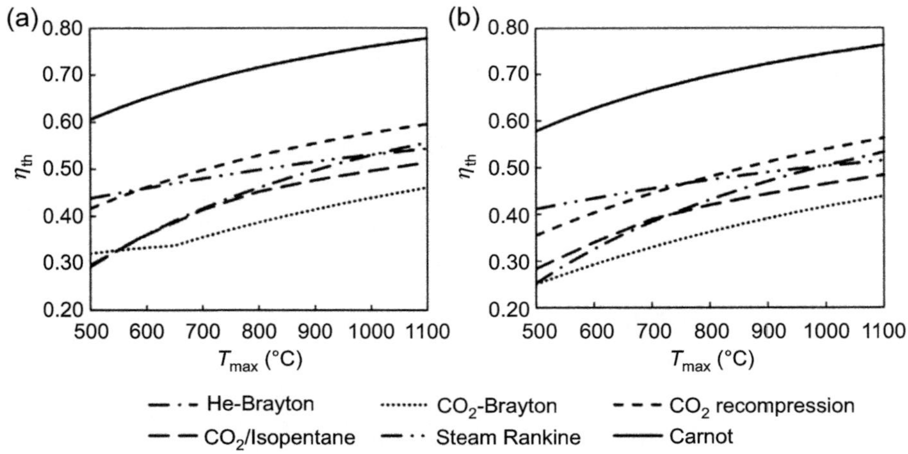

Figure 17.8 Comparison of cycle performance at 20 MPa maximum pressure under (a) wet-cooling and (b) dry-cooling conditions.
From Dunham MT, Iverson BD. High-efficiency thermodynamic power cycles for concentrated solar power systems. Renewable and Sustainable Energy Reviews 2014;30:758–70. http://dx.doi.org/10.1016/j.rser.2013.11.010.

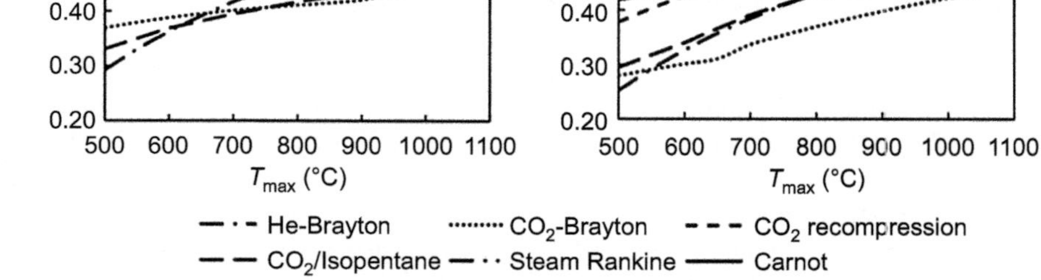

Figure 17.9 Comparison of cycle performance at 30 MPa maximum pressure under (a) wet-cooling and (b) dry-cooling conditions.
From Dunham MT, Iverson BD. High-efficiency thermodynamic power cycles for concentrated solar power systems. Renewable and Sustainable Energy Reviews 2014;30:758–70. http://dx.doi.org/10.1016/j.rser.2013.11.010.

Researchers of CENER (National Renewable Energy Center, Spain) identified and analyzed two configurations [36], both of them based on multitower solar fields, each tower having its own gas turbine (Brayton cycle). Both the multitower solar energy collection and the Brayton cycle have the potential to achieve high efficiency.

In the first configuration (concept A of the authors) the heat rejected from every gas turbine is used to charge a common, single medium-temperature thermal storage system that provides thermal energy to the bottoming cycle, a single low-temperature organic Rankine cycle (ORC). The potential of this configuration to achieve competitive LCOE lies in the use of the relatively low-cost TES (based on a thermocline tank with mineral oil) and the ORC. Because of the characteristics of both the TES and the ORC, this configuration could be used in relatively small CSP plants.

The second configuration (concept B) combines the multitower system with a single high-temperature molten salts TES and a superheated Rankine cycle. The TES is charged trough air–molten salt heat exchangers (one per tower). The emphasis of this configuration is on the efficiency achievable by the combination of the Brayton and superheated Rankine cycles and the high TES capacity, allowing for large CSP plants.

In both cases, the authors consider a solar energy collection system with a heliostat field, a beam-down reflector, and a secondary concentrator coupled to a ground-based high-temperature air receiver (Fig.17.10).

The authors analyze two cases ("conservative" and "optimistic") for each configuration:

- *Concept A*. Both the optimistic and the conservative Brayton cycles have a rated power of 13.06 MW and the ORC of 1.7 MW, with four towers and storage capacities of 13 (conservative) and 14 h (optimistic). The estimated LCOE are 11.7 c€/kWh and 9.9 c€/kWh (relative reductions of 26% and 41% with respect to the base case—Gemasolar CSP plant located near Seville, Spain—respectively).
- *Concept B*. The optimistic case has a Brayton cycle of 44.8 MW and 10 solar towers, while the conservative leads to a Brayton cycle of 53.7 MW and 12 towers. The Rankine cycle has a power of 19.9 MW in both cases. The preliminary results show potential LCOE reductions of 26–41% with respect to the reference case.

A variant of the first configuration is proposed and analyzed in greater detail by the same group [14]. This variant, based on already existing components, uses a medium-temperature superheated steam Rankine cycle (40 bar, 274°C) instead of the ORC. In this case, the system consists of 32 towers and a TES capacity of 9 full hours of the Rankine cycle of the solar field is composed of small, single-facet heliostats with a biomimetic layout arranged in a hexagonal shape. The pressurized air receiver with advanced cavity configuration operates at 800°C. This relatively low temperature is considered to minimize the risks and the efficiency penalties associated with higher temperature. The rated power of each Brayton cycle is 3.34 MW, and the common Rankine cycle has 10.8 MW. The authors use a probabilistic approach to assess the performance and estimate the achievable cost reduction, concluding that the LCOE could be reduced up to 25% from the reference case, based on the Gemasolar CSP plant, with a probability of 90% (Fig.17.11).

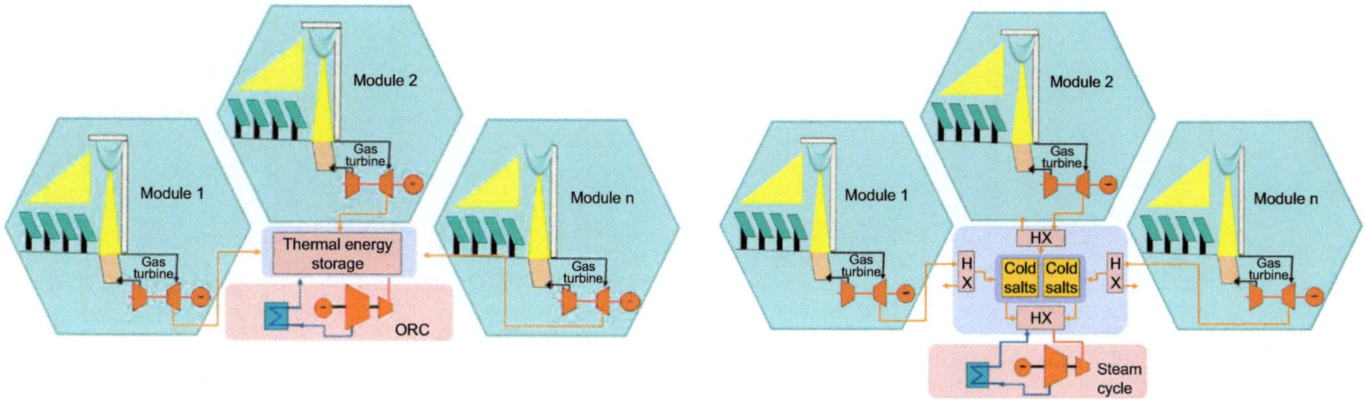

Figure 17.10 Schemes of the decoupled solar combined cycle, concept A (left) and concept B (right) [36].

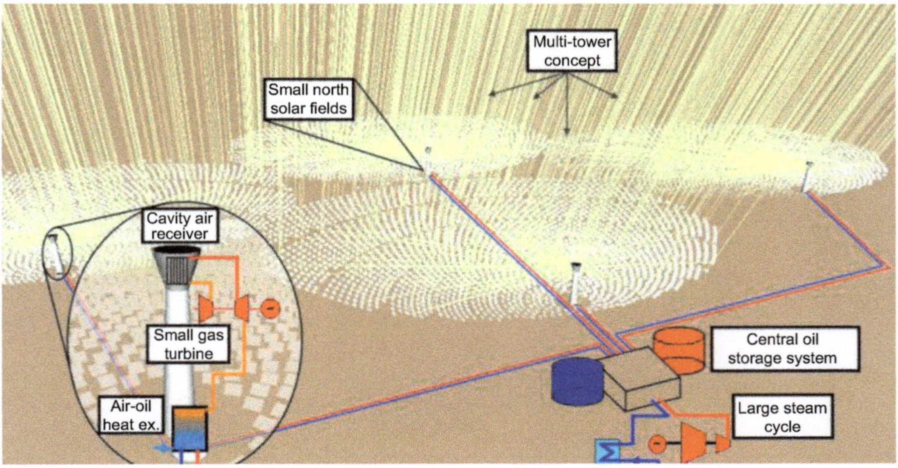

Figure 17.11 Scheme of the DSCC concept proposed by [14] based on a multitower system with Brayton cycles and a central oil storage systems feeding a bottoming medium-temperature Rankine cycle.

Agalit et al. [1] propose a decoupled configuration with a Brayton cycle and a superheated-steam Rankine cycle, and two TES systems (Fig. 17.12). The solar field with a high-temperature air receiver (800−1200°C) is connected to the gas turbine and a high-pressure regenerator/packed-bed storage system using natural (quartzite) rocks. The system includes a fuel combustor which can be used to complement or replace the solar energy input. The combustor and the TES provide operational flexibility to the Brayton cycle. The exhausts from the gas turbine feed the second, low-pressure TES, also based on natural rocks, which in turn feeds the Rankine cycle through an air-steam heat exchanger. The authors focus their work in the simulation of the thermal storage systems, not providing any LCOE estimates.

This configuration is a modification of the SUNSPOT concept [17] developed at the Stellenbosch University of South Africa, which includes only the low-pressure TES and is conceived for partially decoupled operation (the Rankine cycle would operate during nigh time). The authors elaborate a relatively simple model which is used to calculate the LCOE, estimating values of €0.11−0.18/kWh.

Crespo [10] proposes a DSCC hybrid configuration, with a large fuel-driven Brayton cycle and a molten salt heat exchanger to recover the exhaust thermal energy of the air turbine, combined with a molten salt CSP plant operating at about 560°C (Fig. 17.13). The storage system can be charged either from the solar receiver or from the gas turbine exhausts at the same conditions. The generation of electricity from the steam cycle is completely decoupled either from the gas turbine or from the solar part. According to the author's estimates, the efficiency in the conversion of the thermal energy from the fuel to electricity is similar to that of conventional combined cycles, with solar shares (fraction of the electricity generated from solar radiation to total electricity generated) ranging from 50% (in base load operation) to 80%.

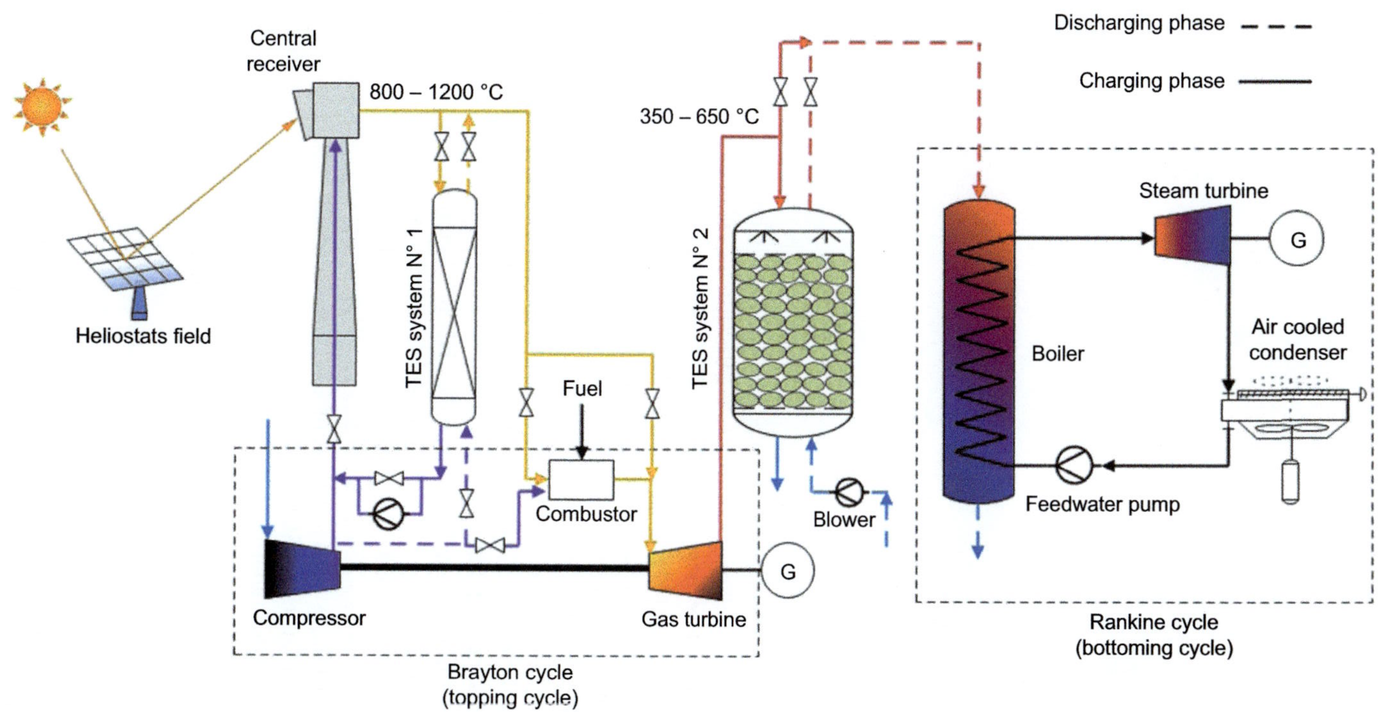

Figure 17.12 Scheme of the DSCC proposed by [1]. This concept includes a high-temperature regenerator/TES for the topping Brayton cycle and a second, low-pressure TES which feeds the bottoming Rankine cycle.

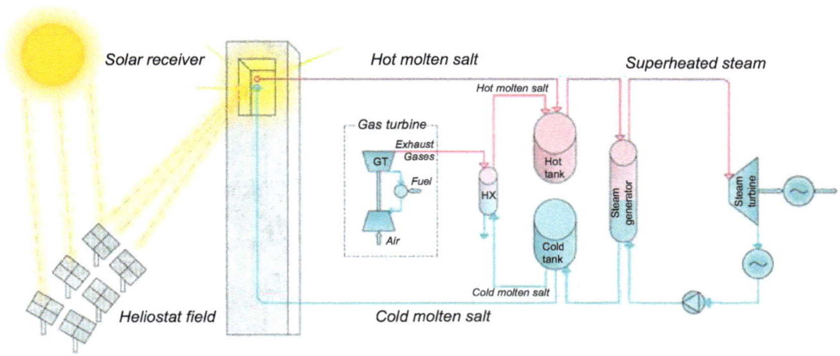

Figure 17.13 Hybrid DSCC concept.
Proposed by Crespo L. Ste plants: beyond Dispatchability Firmness of Supply and integration with VRE. Energy Procedia 2015;69:1241–1248. http://dx.doi.org/10.1016/j.egypro.2015.03.161.

As a summary, the combination of a high-temperature power cycle with a bottoming cycle using a TES system in place of the conventional heat recovery steam generator not only allows to a very large extent the decoupling of the operation of the bottoming cycle from that of the high-temperature cycle, but it also introduces additional degrees of flexibility in the design and operation of solar tower systems that can be used to create more flexible, reliable, and cost-effective systems.

17.4 Summary and conclusions

In this chapter, the integration of SRCCs and supercritical CO_2 Brayton cycles with solar power towers is reviewed. Both power cycles operate in temperature ranges only moderately superior to molten salt plants, being thus well suited to the temperature levels achievable with central receiver systems. The integration of these power systems poses, however, significant challenges both on the power cycle and on the solar energy collection system, especially in terms of the identification or development of new materials that can withstand the demanding operating conditions of these systems without increasing the costs so much that the potential efficiency increase is negatively counterweighed.

According to the literature, SRRCs can be a good option for temperatures up to 600–650°C, while sCO_2 cycles would be more competitive for higher temperatures. However, there is no complete consensus about this conclusion, since most of the possible configurations require significant technology developments, especially in the solar energy collection and TES systems, whose outcomes are difficult to estimate in terms of performance and costs.

Thermochemical storage based on ammonia dissociation, metal oxides or sulfur redox reactions, or *calcium looping* coupled to supercritical plants also seem to be promising long-term options.

This chapter also includes a section on DSCCs, where the heat rejected by a high-temperature cycle integrated with a solar power tower is used to charge a TES system which, in turn, feeds a bottoming cycle. Several DSCC configurations are described and discussed. DSCCs have the potential to achieve high efficiency without the need of significant technology development, being an excellent candidate for the next generation of CST plants.

References

[1] Agalit H, Zari N, Maalmi M, Maaroufi M. Numerical investigations of high temperature packed bed TES systems used in hybrid solar tower power plants. Solar Energy 2015;122: 603–16. http://dx.doi.org/10.1016/j.solener.2015.09.032.

[2] Al-Sulaiman FA, Atif M. Performance comparison of different supercritical carbon dioxide Brayton cycles integrated with a solar power tower. Energy 2015;82:61–71. http://dx.doi.org/10.1016/j.energy.2014.12.070.

[3] ARENA. Advanced steam generating receivers for high-concentration solar collectors final report : project results. 2014.

[4] Besarati SM, Yogi Goswami D, Stefanakos EK. Development of a solar receiver based on compact heat exchanger technology for supercritical carbon dioxide power cycles. Journal of Solar Energy Engineering 2015;137:031018. http://dx.doi.org/10.1115/1.4029861.

[5] Buckingham R, Wong B, Brown L. Thermochemical energy storage for concentrated solar power - coupling to a high efficiency supercritical CO_2 Power cycle. In: Proceedings of the SolarPaces Conference; 2011. p. 0–5.

[6] Chacartegui R, Alovisio A, Ortiz C, Valverde JM, Verda V, Becerra JA. Thermochemical energy storage of concentrated solar power by integration of the calcium looping process and a CO_2 power cycle. Applied Energy 2016;173:589–605. http://dx.doi.org/10.1016/j.apenergy.2016.04.053.

[7] Chacartegui R, Muñoz De Escalona JM, Sanchez D, Monje B, Sanchez T. Alternative cycles based on carbon dioxide for central receiver solar power plants. Applied Thermal Engineering 2011;31:872–9. http://dx.doi.org/10.1016/j.applthermaleng.2010.11.008.

[8] Cheang VT, Hedderwick RA, McGregor C. Benchmarking supercritical carbon dioxide cycles against steam Rankine cycles for concentrated solar power. Solar Energy 2015;113: 199–211. http://dx.doi.org/10.1016/j.solener.2014.12.016.

[9] Chen C, Aryafar H, Warrier G, Lovegrove KM, Lavine AS. Ammonia synthesis for producing supercritical steam in the context of solar thermochemical energy storage. SolarPACES 2015. http://dx.doi.org/10.1115/POWER201549190. 2015 050010, 0–9.

[10] Crespo L. Ste plants: beyond dispatchability firmness of supply and integration with VRE. Energy Procedia 2015;69:1241–8. http://dx.doi.org/10.1016/j.egypro.2015.03.161.

[11] Dunham MT, Iverson BD. High-efficiency thermodynamic power cycles for concentrated solar power systems. Renewable and Sustainable Energy Reviews 2014;30:758–70. http://dx.doi.org/10.1016/j.rser.2013.11.010.

[12] Estela. Strategic research agenda. 2012. p. 2020–5.

[13] Flamant G, Gauthier D, Benoit H, Sans J, Boissière B, Ansart R, et al. A new heat transfer fluid for concentrating solar systems: particle flow in tubes. In: Energy Procedia. Elsevier B.V.; 2014. p. 617–26. http://dx.doi.org/10.1016/j.egypro.2014.03.067.

[14] García-Barberena J, Monreal A, Mutuberria A, Sánchez M. Towards cost-competitive solar towers - energy cost reductions based on decoupled solar combined cycles (DSCC). Energy Procedia 2013;49:1350–60. http://dx.doi.org/10.1016/j.egypro.2014.03.144.

[15] Gardner W, Mcnaughton R, Kim J, Barrett S. Development of a solar thermal supercritical steam generator. In: Proceedings of the 50th Annual Conference, Australian solar energy Society; 2012.

[16] Gary J, Turchi C, Siegel N. CSP and the DOE Sunshot initiative. System 2011;1:8.

[17] Heller L, Gauché P. Modeling of the rock bed thermal energy storage system of a combined cycle solar thermal power plant in South Africa. Solar Energy 2013;93:345–56. http://dx.doi.org/10.1016/j.solener.2013.04.018.

[18] Ho CK, Carlson M, Garg P, Kumar P. Cost and performance tradeoffs of alternative solar-driven S-CO2 Brayton cycle configurations. In: Proceedings of the ASME 2015 power and energy conversion Conference. San Diego, California; 2015. p. 1–10.

[19] Ho CK, Iverson BD. Review of high-temperature central receiver designs for concentrating solar power. Renewable and Sustainable Energy Reviews 2014;29:835–46. http://dx.doi.org/10.1016/j.rser.2013.08.099.

[20] Ho CK, Conboy T, Ortega J, Afrin S, Gray A, Christian JM, et al. High-temperature receiver designs for supercritical CO 2 closed-loop Brayton cycles. In: Asme 2014 8th International Conference on energy Sustainability. Boston (MA): ASME; 2014. http://dx.doi.org/10.1115/ES2014-6328.

[21] Iverson BD, Conboy TM, Pasch JJ, Kruizenga AM. Supercritical CO_2 Brayton cycles for solar-thermal energy. Applied Energy 2013;111:957–70. http://dx.doi.org/10.1016/j.apenergy.2013.06.020.

[22] Kelly BD, Izygon M, Vant-Hull L. Advanced thermal energy storage for central receivers with supercritical coolants. DE-FG36–08GO18149. 2010. http://dx.doi.org/10.2172/981926.

[23] Kolb GJ. An evaluation of possible next generation high-temperature molten-salt power towers. 2011.

[24] Lavine AS, Lovegrove KM, Jordan J, Bran G, Chen C, Aryafar H, et al. Thermochemical energy storage with Ammonia: aiming for the SunShot cost Target. SolarPACES 2015. http://dx.doi.org/10.1063/1.4949126. 2015 050028, 2–9.

[25] Ma Z, Mehos M, Glatzmaier G, Sakadjian BB. Development of a concentrating solar power system using fluidized-bed technology for thermal energy conversion and solid particles for thermal energy storage. Energy Procedia 2015;69:1349–59. http://dx.doi.org/10.1016/j.egypro.2015.03.136.

[26] Manuel J, Blanco C. Astri overview. 2015.

[27] Neises T, Turchi C. A comparison of supercritical carbon dioxide power cycle configurations with an emphasis on CSP applications. Energy Procedia 2013;49:1187–96. http://dx.doi.org/10.1016/j.egypro.2014.03.128.

[28] Ortega J, Khivsara S, Christian J, Ho C, Dutta P. Coupled modeling of a directly heated tubular solar receiver for supercritical carbon dioxide Brayton Cycle : structural and creep— fatigue evaluation. Applied Thermal Engineering 2016. http://dx.doi.org/10.1016/j.applthermaleng.2016.06.031.

[29] Ortega J, Khivsara S, Christian J, Ho C, Yellowhair J, Dutta P. Coupled modeling of a directly heated tubular solar receiver for supercritical carbon dioxide Brayton cycle: optical and thermal-fluid evaluation. Applied Thermal Engineering 2016. http://dx.doi.org/10.1016/j.applthermaleng.2016.05.178.

[30] Ortega JD, Christian JM. Design requirements for direct supercritical carbon dioxide. In: Proceedings of the 9th International Conference on energy Sustainability; 2015. p. 1–6.

[31] Pacio J, Singer C, Wetzel T, Uhlig R. Thermodynamic evaluation of liquid metals as heat transfer fluids in concentrated solar power plants. Applied Thermal Engineering 2013;60: 295–302. http://dx.doi.org/10.1016/j.applthermaleng.2013.07.010.

[32] Padilla RV, Soo Too YC, Benito R, Stein W. Exergetic analysis of supercritical CO_2 Brayton cycles integrated with solar central receivers. Applied Energy 2015;148:348–65. http://dx.doi.org/10.1016/j.apenergy.2015.03.090.

[33] Padilla RV, Too YCS, Beath A, McNaughton R, Stein W. Effect of pressure drop and reheating on thermal and exergetic performance of supercritical carbon dioxide Brayton cycles integrated with a solar central receiver. Journal of Solar Energy Engineering 2015; 137:051012. http://dx.doi.org/10.1115/1.4031215.

[34] Peterseim JH, Veeraragavan A. Solar towers with supercritical steam parameters—is the efficiency gain worth the effort? Energy Procedia 2015;69:1123–32. http://dx.doi.org/10.1016/j.egypro.2015.03.181.

[35] Sabharwall P, Clark D, Glazoff M, Zheng G, Sridharan K, Anderson M. Advanced heat exchanger development for molten salts. Nuclear Engineering and Design 2014;280: 42–56. http://dx.doi.org/10.1016/j.nucengdes.2014.09.026.

[36] Sánchez M, Blanco MJ, García-Barberena J, Monreal A. The potential for cost reduction of solar towers with decoupled combined cycles. In: SolarPaces Conf; 2012.

[37] Sargent, Lundy. Assesment of Parabolic trough and power tower solar technology cost and performance forecast. 2003. p. 347.

[38] Siefert JA, Libby C, Shingledecker J. Concentrating solar power (CSP) power cycle improvements through application of advanced materials. 2016. p. 070030. http://dx.doi.org/10.1063/1.4949177.

[39] Singer C, Buck R, Pitz-Paal R, Müller-Steinhagen H. Assessment of Solar power tower driven ultrasupercritical steam cycles applying tubular central receivers with varied heat transfer media. Journal of Solar Energy Engineering 2010;132:041010. http://dx.doi.org/10.1115/1.4002137.

[40] Singer C, Buck R, Pitz-Paal R, Müller-Steinhagen H. Economic potential of innovative receiver concepts with different solar field configurations for supercritical steam cycles. Journal of Solar Energy Engineering 2013;136:21009. http://dx.doi.org/10.1115/1.4024740.

[41] Singer C, Giuliano S, Buck R. Assessment of improved molten salt solar tower plants. In: Energy Procedia. Elsevier B.V; 2013. p. 1553–62. http://dx.doi.org/10.1016/j.egypro.2014.03.164.

[42] Turchi CS, Ma Z, Neises TW, Wagner MJ. Thermodynamic study of advanced supercritical carbon dioxide power cycles for concentrating solar power systems. Journal of Solar Energy Engineering 2013;135:041007. http://dx.doi.org/10.1115/1.4024030.

[43] Wu W, Gobereit B, Amsbeck L, Pitz-paal R. Direct absorption receivers for high temperatures. SolarPACES 2011:1–8. 2011.

Solar thermal processing

18

R. Bader, W. Lipiński
The Australian National University, Canberra, ACT, Australia

18.1 Introduction

The sun is a virtually unlimited source of radiant heat at approximately 5800K. Hence, solar radiation can be used as a source of renewable process heat to drive energy-intensive chemical processes. The engineering challenge lies in introducing this intermittent and spatially dilute energy source to processes in an industrially useful, technically feasible, and cost-effective manner.

In 2013, 18% of the global end-use energy was supplied in the form of electricity, while two-thirds were supplied as fossil-derived hydrocarbon fuels (coal, oil, and natural gas) [1]; 40% of the energy supplied as fossil fuels is used for transportation and 27% is used for industrial applications [1]. Hence, mitigating the majority of the global greenhouse gas emissions requires a substitution of fossil fuels with renewable energy not only in the electricity sector but also in the transportation and industrial processing sectors. Concentrated solar energy can be used to provide a large portion of the energy demand for transportation and industrial processes in the form of fuels, process heat, and high-flux radiation.

Fig. 18.1 gives an overview of processes other than electricity production that can utilize concentrated solar energy, including reactants, products, and process types. Additional processes are likely to emerge as research into novel applications of concentrated solar energy continues. Processes are organized according to their application into three groups: H_2/CO production; material processing and chemical commodity production; and other thermal processes. A large body of research focuses on the production of H_2 and CO. The mixture of H_2 and CO, referred to as synthesis gas (or syngas), is the feedstock for the production of synthetic fuels (e.g., diesel, kerosene, gasoline, methanol, and dimethyl ether) via the Fischer–Tropsch process and other catalytic processes. In addition, H_2 plays an important role in many industrial applications, most notably in the production of ammonia via the Haber–Bosch process. Ammonia is used as the main source of nitrogen in fertilization and could be used in the future as a vector for hydrogen as a fuel [14]. H_2/CO production processes can be further divided according to the feedstock. In H_2O/CO_2 splitting processes with CO_2 either recovered from flue gases or captured from the atmosphere the heating value of the produced syngas corresponds to the amount of solar energy stored in the process, resulting in a carbon-neutral fuel. On

Advances in Concentrating Solar Thermal Research and Technology
http://dx.doi.org/10.1016/B978-0-08-100516-3.00018-6

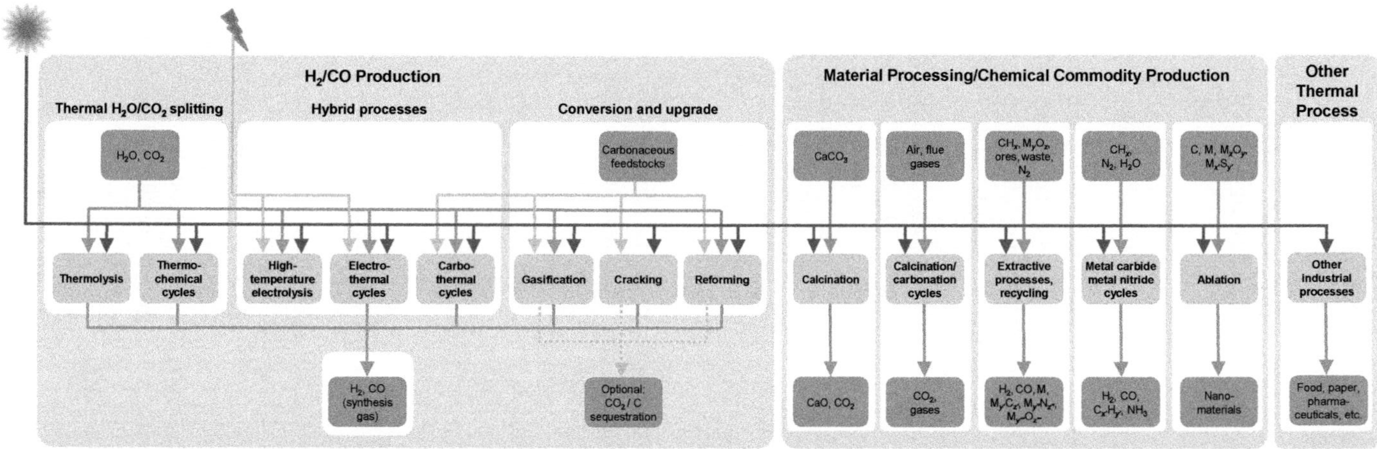

Figure 18.1 Applications of concentrated solar thermal energy for thermal and thermochemical processing: H_2/CO production processes, material processing and chemical commodity production, and other thermal industrial processes. M refers to metal. This figure expands upon the schematics in Refs. [2–4], with additions from Refs. [5–13].

the other hand, solar conversion of carbonaceous feedstocks, such as coal, petcoke, natural gas, biomass, and carbonaceous waste materials, to syngas leads to a calorific upgrade of the feedstock and offers the possibility for decarbonization. While the solar processing of fossil feedstocks does not result in carbon-neutral products and hence is not a sustainable long-term solution, it may provide a viable transition path from today's largely fossil fuel-based to a carbon-neutral economy. In addition, a group of hybrid H_2/CO production processes can be defined that either combine multiple feedstocks or multiple energy sources.

Besides syngas production, many other industrial applications for concentrated solar energy have been proposed. The calcination of limestone is the main energy-intensive step in the production of cement and, as such, responsible for about 4.5% of the total global anthropogenic CO_2 emissions [9]. Replacing fossil fuels with concentrated solar energy in the production of cement has the potential to reduce the CO_2 emissions from the process by up to 50% [9]. The cyclic calcination of $CaCO_3$ and carbonation of CaO can also be used to separate pure CO_2 from flue gases and atmospheric air.

Other proposed important process applications of concentrated solar energy are the thermochemical or carbothermal reduction of metal and semiconductor oxides in the processing of mineral ores and waste to produce pure metals and semiconductors, lower valence metal oxides, metal carbides, and metal nitrides. Metal carbides and nitrides can be used either as high-performance ceramics, or in solar-driven thermochemical cycles to produce renewable commodity materials such as ammonia, syngas, and hydrocarbons.

Besides these specific chemical process applications, concentrated solar energy can also be used as a generic source of process heat for other industrial production processes. Major process heat applications are found in the agricultural, food, chemical, pharmaceutical, and paper and pulp industries and for water desalination. A key aspect for the industrial use of process heat is its dispatchability via the use of thermal storage. Finally, in thermal processes that require extreme heat fluxes, concentrated solar radiation can also be introduced directly to the process. Examples include high-flux solar-driven heat treatment, cutting, surgery, and ablation.

18.2 H_2/CO production

The processes discussed in this section aim at using concentrated solar energy to produce syngas (H_2 and CO). Syngas is a universal feedstock for processes that produce a range of synthetic fuels and solid organic materials. Syngas can be produced from various fossil and non-fossil feedstocks and via several solar process routes. Solar processes are grouped into thermochemical H_2O/CO_2 splitting processes, hybrid processes using a combination of thermal energy and electricity or a carbonaceous reducing agent, and processes for the conversion and upgrade of carbonaceous feedstocks.

18.2.1 Thermochemical H_2O/CO_2 splitting

18.2.1.1 Thermolysis

The direct thermal dissociation of H_2O and CO_2 according to:

$$H_2O(l) \rightarrow H_2 + \frac{1}{2}O_2, \Delta H^{\circ}_{298K} = 286 \text{ kJ/mol}, \tag{18.1}$$

$$CO_2 \rightarrow CO + \frac{1}{2}O_2, \Delta H^{\circ}_{298K} = 283 \text{ kJ/mol}, \tag{18.2}$$

is conceptually the simplest way to produce syngas from H_2O and CO_2 [15,16]. However, at ambient pressure, reactions (18.1) and (18.2) occur only at extremely high temperatures. For example, 20% dissociation of H_2O and CO_2 requires about 2800K and 2600K, respectively [15,17]. In addition, unwanted gaseous species (e.g., H, O, and OH) are coproduced at these high temperatures [15,17]. Furthermore, in order to avoid the recombination of the product gases upon cooling, they must be separated at the extremely high dissociation temperatures required by the process, which is technologically challenging. Effusional [15,18] and electrolytic [19] separation and quenching [16,20] have been proposed to mitigate the recombination of the products. Finally, the availability of construction materials that can be operated at temperatures in excess of 2500K over extended periods of time is limited. These drawbacks and challenges with the direct thermal dissociation of H_2O and CO_2 have been the motivation for the development of alternative thermochemical processes to split H_2O and CO_2.

18.2.1.2 Thermochemical cycles

Thermochemical cycles use multiple chemical reactions to dissociate H_2O and/or CO_2. In the simplest case of a two-step redox process, an intermediate oxygen exchange material, M, with sufficient reducing power is used in the first process step to split H_2O and/or CO_2 into H_2 and/or CO according to:

$$M + xH_2O \rightarrow MO_x + xH_2 \tag{18.3}$$

and/or:

$$M + xCO_2 \rightarrow MO_x + xCO \tag{18.4}$$

In the second process step, the oxidized form of M, MO_x, is regenerated in an endothermic thermal reduction step, operated with concentrated solar energy as the source of high-temperature process heat:

$$MO_x \rightarrow M + \frac{x}{2}O_2 \tag{18.5}$$

The main advantages of splitting H_2O/CO_2 via multistep thermochemical cycles compared to their direct thermolysis are (1) the need for high-temperature separation or quenching of H_2/CO and O_2 is avoided because the two gases are produced in separate process steps, and (2) the maximum process temperature can be reduced. Numerous candidate cycles have been scanned for their suitability to split H_2O using concentrated solar thermal energy [21–23], but only a few have been investigated in-depth. Table 18.1 gives an overview of the main H_2O/CO_2 splitting thermochemical cycles that have been investigated to date. Several additional H_2O/CO_2 splitting thermochemical cycles have been considered. Comprehensive reviews of H_2O/CO_2 splitting via thermochemical cycles can be found in Refs. [2,23,34,35].

Among the cycles that have been studied so far, those that comprise more than two chemical process steps tend to be operable at lower temperatures than the two-step cycles, which makes the reactor design and material selection less challenging and reduces the required solar flux concentration ratio. On the other hand, increasing the number of cycle steps increases irreversibilities in the process due to internal heat transfer, heat losses to the environment, and the number of transport and separation processes of reactants and products, leading to an increase in the overall complexity and costs of the process [21]. Consequently, a large part of the research since around 2000 has focused on two-step cycles using metal oxide redox materials, and in particular non-volatile materials that remain solid throughout the process, which simplifies the separation of products.

Research since around 2000 has been dedicated to thermodynamic, kinetic, and heat transfer analyses, as well as material and solar reactor developments, with the overarching goals to improve process efficiency and stability. Doping of ferrite redox materials has been considered in order to prevent sintering during redox cycling. Co- and Ni-ferrites show promise as they reduce at about 1700K and hence about 100K below their sintering temperatures [36]. Doped ferrites supported on zirconia support structures resulted in both improved reaction kinetics and resistance to sintering [34]. Non-stoichiometric redox cycling of ceria at temperatures up to about 1773K leads to limited sintering but results in relatively low H_2/CO yields. Numerous dopants have been considered to increase the oxygen exchange capacity of ceria. A trend of increasing reducibility with decreasing ionic radius of the dopant was observed [37]. For Zr-doped ceria, both the oxygen yield and the fuel (H_2) productivity increase significantly with increasing dopant concentration [38,39]. Different porous ceria structures have been tested, in order to improve volumetric radiation absorption, mass transfer, and reaction kinetics, including monolithic reticulate and particulate porous ceramics with single- and dual-scale porosity [27,40–42], porous ceramic felts [27], and three-dimensionally ordered macroporous structures [43].

Since 2013 metal oxide materials with the perovskite structure (ABO_3) have been proposed for solar thermochemical H_2O/CO_2 splitting. This class of redox materials has long been used as electrolyte material in solid oxide fuel cells and shows great promise for H_2O/CO_2 splitting due to their increased H_2/CO yields and lower reduction temperatures compared to ceria. An overview of perovskite material compositions investigated so far is provided in Table 18.2. Strontium-doped lanthanum manganites ($Sr_xLa_{1-x}MnO_{3-\delta}$) were investigated thermodynamically and experimentally

Table 18.1 Overview of the main H_2O/CO_2 splitting thermochemical cycles

Name	Categorization	Chemical reactions	References
Ferrite cycles	Two-step, non-volatile, stoichiometric	$M_xFe_{3-x}O_4(s) \rightarrow xMO(s)+(3-x)FeO(s)+1/2O_2(g)$ $xMO(s)+(3-x)FeO(s)+H_2O/CO_2(g) \rightarrow M_xFe_{3-x}O_4(s)+H_2/CO(g)$ M: Co, Ni, Zn, Mg, Mn	[17,24]
Zinc oxide cycle	Two-step, volatile, stoichiometric	$ZnO(s) \rightarrow Zn(g)+1/2O_2(g)$ $Zn(s/l)+H_2O/CO_2(g) \rightarrow ZnO(s)+H_2/CO(g)$	[17,25]
Ceria cycles	Two-step, non-volatile, non-stoichiometric	$M_xCe_{1-x}O_2(s) \rightarrow M_xCe_{1-x}O_{2-\delta}(s)+\delta/2O_2(g)$ $M_xCe_{1-x}O_{2-\delta}(s)+\delta H_2O/CO_2(g) \rightarrow M_xCe_{1-x}O_2(s)+\delta H_2/CO(g)$ M: 1+: Li; 2+: Mg, Ca, Sr; 3+: Sc, Y, Pr, Sm, Gd, Dy, La; 4+: Zr, Hf; 5+: Ta; 6+: Cr	[26,27]
Hercynite cycle	Two-step, non-volatile, stoichiometric	$CoFe_2O_4(s)+3Al_2O_3(s) \rightarrow CoAl_2O_4(s)+2FeAl_2O_4(s)+1/2O_2(g)$ $CoAl_2O_4(s)+2FeAl_2O_4(s)+H_2O(g) \rightarrow CoFe_2O_4+3Al_2O_3(s)+H_2(g)$	[28]
Perovskite cycles	Two-step, non-volatile, non-stoichiometric	$ABO_3(s) \rightarrow ABO_{3-\delta}(s)+\delta/2O_2(g)$ $ABO_{3-\delta}(s)+\delta H_2O/2CO_2(g) \rightarrow ABO_3(s)+\delta H_2/CO(g)$ A: Ba, Ca, Ce, Dy, Gd, La, Nd, Sm, Sr, Y; B: Al, Co, Cr, Fe, Mg, Mn	[29]
Sulfur–iodine cycle	Three-step, volatile, stoichiometric	$SO_2(g)+I_2(l)+2H_2O \rightarrow 2HI+H_2SO_4(aq,inf)$ $2HI(aq,inf) \rightarrow H_2(g)+I_2(g)$ $H_2SO_4(aq,inf) \rightarrow H_2O(g)+SO_2(g)+1/2O_2(g)$	[30]
Manganese oxide cycle	Three-step, non-volatile, stoichiometric	$1/2Mn_2O_3(s) \rightarrow MnO(s)+1/4O_2(g)$ $MnO(s)+NaOH(1) \rightarrow NaMnO_2(s)+1/2H_2(g)$ $NaMnO_2(s)+1/2H_2O(l) \rightarrow 1/2Mn_2O_3(s)+NaOH$	[31]
UT-3 cycle	Four-step, volatile, stoichiometric	$CaBr_2(s)+H_2O(g) \rightarrow CaO(s)+2HBr(g)$ $CaO(s)+Br_2(g) \rightarrow CaBr_2(s)+1/2O_2(g)$ $Fe_3O_4(s)+8HBr(g) \rightarrow 3FeBr_2(s)+4H_2O(g)+Br_2(g)$ $3FeBr_2(s)+4H_2O(g) \rightarrow Fe_3O_4(s)+6HBr(g)+H_2(g)$	[32]

δ refers to the non-stoichiometric change in the oxygen content of non-stoichiometric redox materials.
Adapted from [33], with additions from [23], Copyright, with permission from Elsevier.

Table 18.2 Overview of perovskite material compositions investigated for H_2O/CO_2 splitting

A-site atoms	B-site atoms	References
La, Sr	Mn	Scheffe et al. [29]
La, Sr	Al, Mn	McDaniel et al. [44]
La, Sr	Mn	Demont and Abanades [45]
Ba, La, Sr	Co, Fe, Mn	Demont et al. [46]
La, Sr	Mn	Yang [47]
Ce, La, Sr	Co, Fe, Mn	Jiang et al. [48]
Ba, Ca, La, Sr, Y	Al, Mg, Mn	Demont and Abanades [49]
Ca, Dy, Gd, La, Nd, Sm, Sr, Y	Mn	Dey et al. [50]
La, Sr	Cr, Co	Bork et al. [51]
Ca, La, Sr	Mn, Al	Galvez et al. [52]

[29]. These perovskites exhibit higher oxygen non-stoichiometry than ceria at identical conditions (1773K, oxygen partial pressure of 1 Pa), but lower reactivity with H_2O/CO_2. Nevertheless, higher fuel productivity was predicted compared to ceria. Strontium- and manganese-doped lanthanum aluminates ($Sr_xLa_{1-x}Mn_yAl_{1-y}O_{3-\delta}$) were shown experimentally to release several times more oxygen than ceria under the same conditions (1623K, He sweep gas flow with O_2 partial pressure of 0.2 mbar) [44]. Subsequent re-oxidation of the perovskites with 40 vol% H_2O and CO_2 at 1273K yielded up to 9 times and 6 times more H_2 and CO, respectively, compared to ceria. The time to complete the re-oxidation step was comparable for the two material classes. Redox cycling between 1623K and 1273K over 80 cycles resulted in stable CO production of 125−150 μmol/g per cycle. Several additional perovskite material compositions have been studied [46]. Most of these materials exhibit higher oxygen exchange capacity compared to ceria at temperatures in the range of 1273−1673K and noticeable access to metallic valence transitions during re-oxidation in steam [46].

Further developments of advanced redox materials are expected to be promoted by first-principle material analyses [53−55], direct pore-level numerical simulations [56−59], and micro- and nanoscale material design, engineering, and manufacturing [60−62]. Recent reviews of material developments for solar thermochemical H_2O/CO_2 splitting can be found in Refs. [23,33,35].

The performance of thermochemical H_2O/CO_2 splitting cycles can be quantified in terms of the thermal efficiency:

$$\eta_{\text{thermal}} = \frac{\dot{W}}{\dot{Q}_{\text{solar}}} \tag{18.6}$$

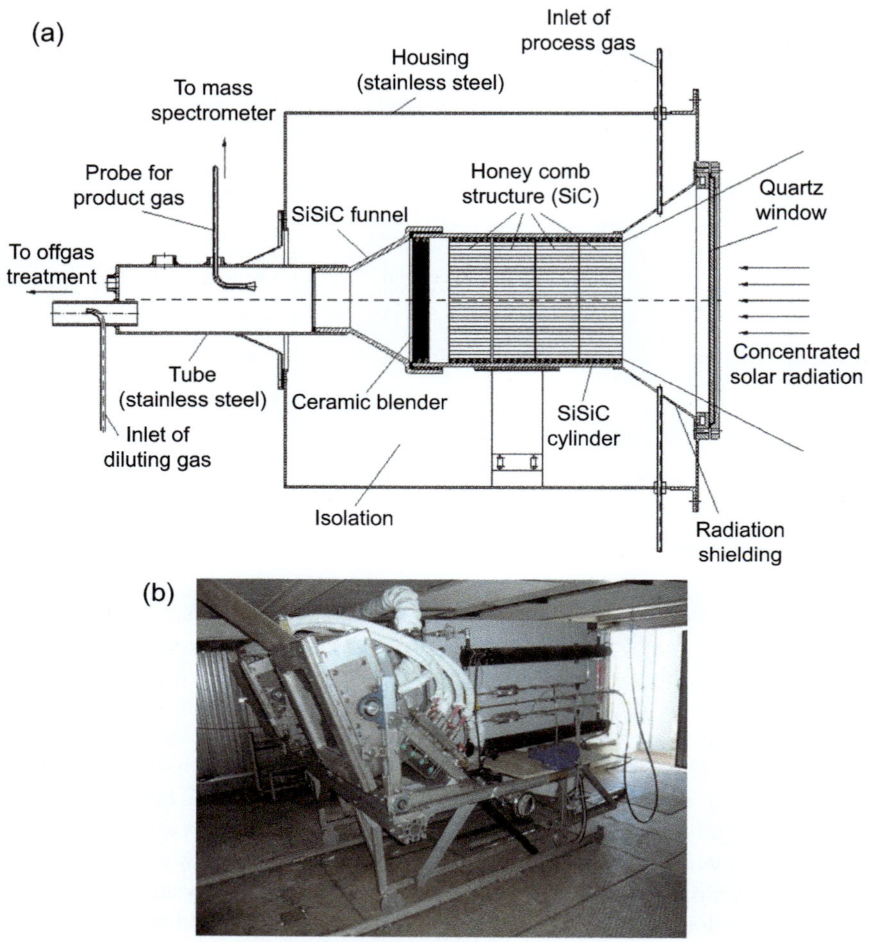

Figure 18.2 Solar reactor technology to realize ferrite cycles: (a) schematic and (b) 100 kW pilot-scale dual-chamber solar receiver-reactor based on honeycomb absorber structure coated with ferrite-based redox materials; (c) prototype of a counter-rotating ring reactor with integrated solid-phase heat recuperation and (d) prototype circulating fluidized-bed reactor. (a) Reprinted from [72], Copyright with permission from the American Society of Mechanical Engineers. (b) Reprinted from [73], Copyright with permission from Elsevier. (c) Reprinted from [74], Copyright with permission from the American Society of Mechanical Engineers. (d) Reprinted from [75], Copyright with permission from Hydrogen Energy Publications, LLC.

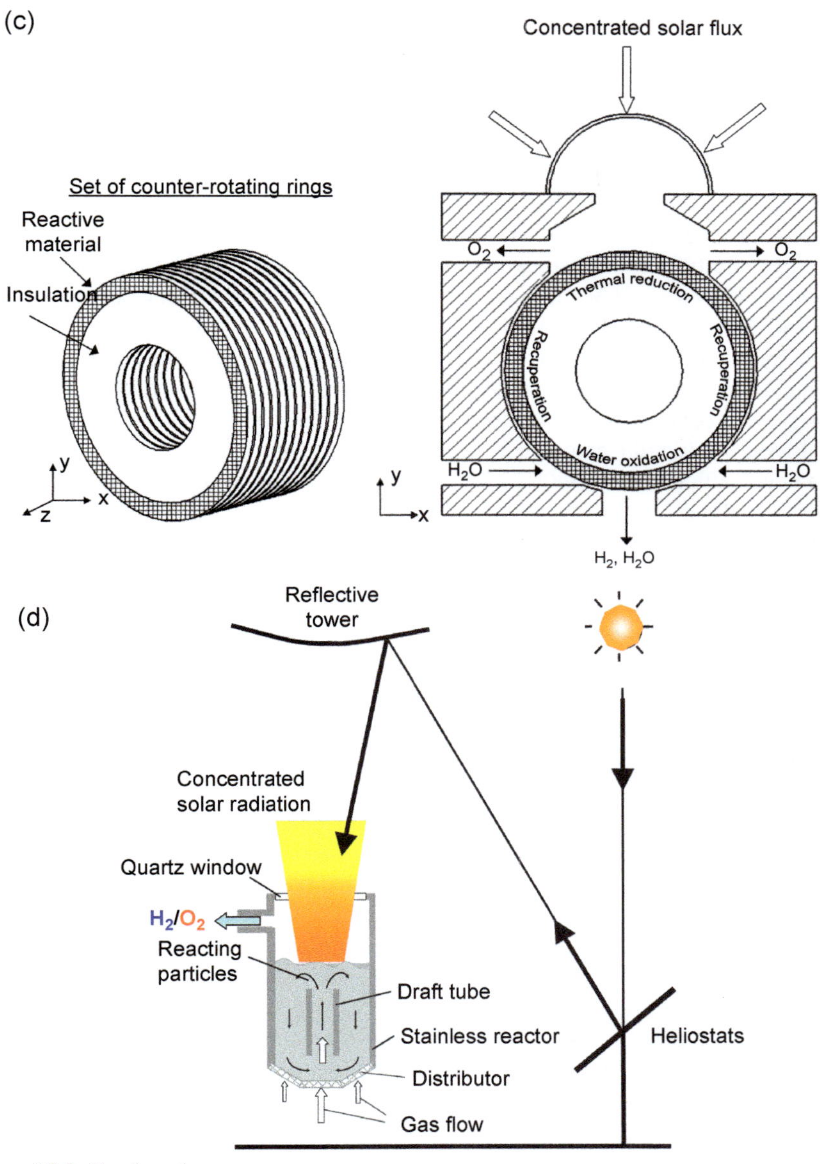

Figure 18.2 Continued.

or the solar-to-fuel energy conversion efficiency:

$$\eta_{\text{solar-to-fuel}} = \frac{\Delta \dot{H}_{\text{fuel}}}{\dot{Q}_{\text{solar}}} \tag{18.7}$$

depending on whether the intended use of the produced fuel is for work and/or heat. In Eqs. (18.6) and (18.7), $\dot{Q}_{solar}$ is the solar thermal power input to the entire solar process. The numerators in Eqs. (18.6) and (18.7) are the net mechanical power and the rate of heat stored in the solar process, respectively. First- and second-law analyses were conducted to determine the theoretical efficiencies of two-step thermochemical H_2O/CO_2 splitting cycles. For the pure ferrite cycle, the predicted thermal efficiency is <8% (solar-to-fuel efficiency <10% [34]) without any heat recovery [63]. Assuming complete conversion of the sensible and latent heat rejected by the hot products to work via an ideal Carnot engine, predicted thermal efficiencies are up to 61%. In a similar analysis, the thermal efficiency of the zinc oxide cycle is predicted to be 50–55% with complete heat recovery from the process units and 30–34% without any heat recovery. Corresponding solar-to-fuel efficiencies are 10–24% higher than the thermal efficiencies [64]. A first-law analysis was conducted for temperature-swing non-stoichiometric ceria cycles with thermal reduction enabled by an inert sweep gas flow [65]. The predicted solar-to-fuel efficiency is >50% with complete solid- and gas-phase heat recovery and <6% without any heat recovery, illustrating the importance of heat recovery in non-stoichiometric ceria cycles, due to the low conversions of reactants and oxygen exchange material. Operating the cycle isothermally eliminates the need for solid-phase heat recovery, but requires very high sweep gas flow rates and, consequently, a high degree of gas-phase heat recovery [66,67]. The possibilities of solid–solid heat recuperation to reduce the heat losses in temperature-swing operation, and vacuum pumping as an alternative to sweep gas flow to lower the oxygen partial pressure in the reduction step have been considered [68–71].

The Carnot efficiency for solar energy emitted at 5780K is around 95%. Irreversibilities in solar thermochemical H_2O/CO_2 splitting cycles include material-related temperature limitations of about 2000K in air (~3000K in vacuum), radiative and conductive heat losses from the hot solar reactor, unrecovered enthalpy of hot reaction products (including the oxygen exchange material), and internal heat transfer resistances. Further efficiency penalties are due to the auxiliary energy demand by separation processes, vacuum pumps, flow circulation pumps, solar system electronics and controls, and upstream and downstream processes. Additional entropy generation takes place in the conversion of fuel to mechanical work.

Reactor developments since around 2000 for solar thermochemical H_2O/CO_2 splitting via the ferrite, zinc oxide, ceria, and hercynite cycles are depicted in Figs. 18.2–18.4. Reactor concepts include those containing fluidized beds [75] and honeycomb support structures for redox cycling of ferrites [83], moving-bed [78], rotating-cavity [77], and aerosol flow reactors [84,85] for zinc oxide dissociation, and directly [26] and indirectly [80] irradiated stationary cavity receiver-reactors containing porous structures, a dual-chamber packed-bed reactor [81], and dual-zone reactors with internal rotary cylinders [74,86–88] for redox cycling of ferrites and/or ceria. The development stage of these reactors varies greatly and ranges from concept stage, to initial prototype testing, to pilot-scale testing, and to demonstration-scale plant design.

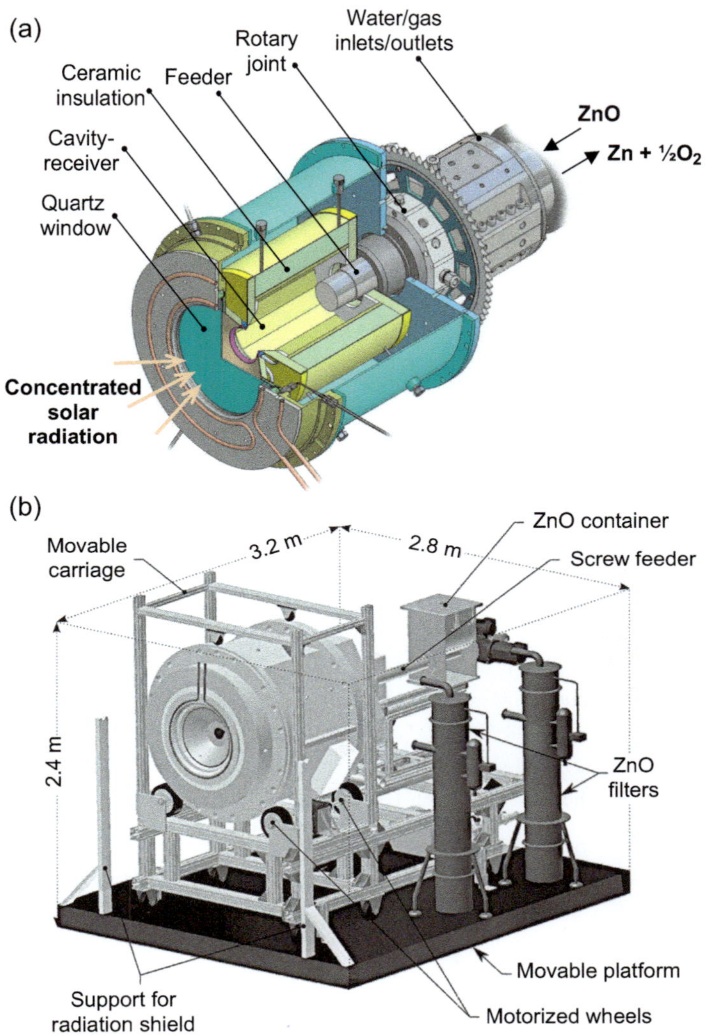

Figure 18.3 Solar reactor technology to realize the zinc oxide cycle: (a) prototype and (b) 100-kW pilot-scale rotating-cavity solar reactor for the thermal dissociation of ZnO particles; (c) gravity-fed moving-bed reactor for the thermal dissociation of ZnO particles and (d) aerosol flow reactor for the in situ formation and hydrolysis of zinc nanoparticles. (a) Reprinted from [76]. (b) Reprinted from [77], Copyright with permission from the American Society of Mechanical Engineers. (c) Reprinted from [78], Copyright with permission from Hydrogen Energy Publications, LLC. (d) Reprinted from [79].

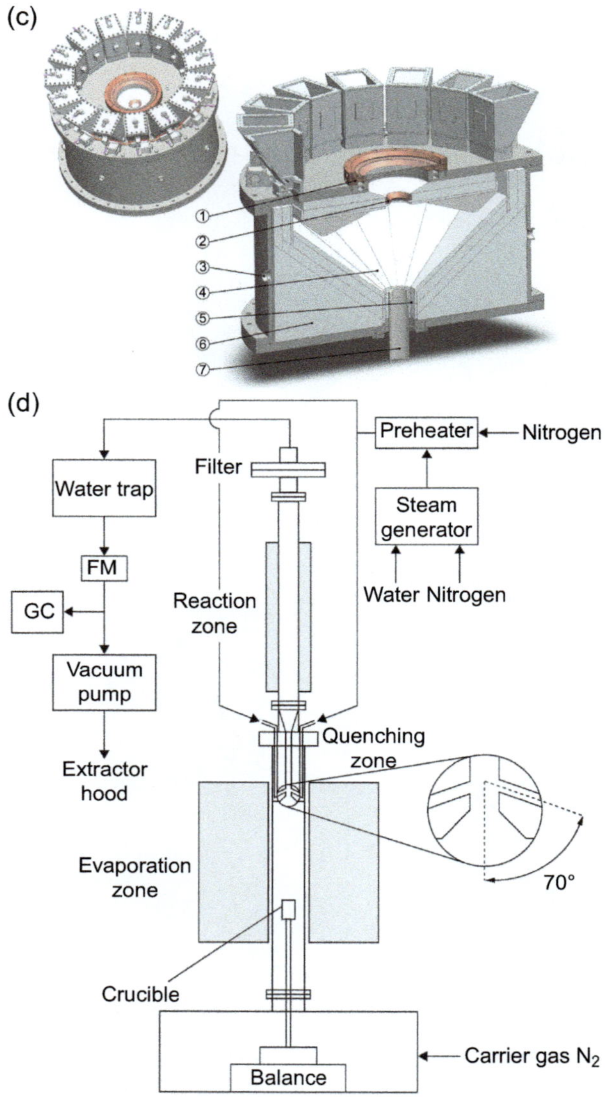

Figure 18.3 Continued.

A 100-kW dual-chamber pilot-scale solar reactor for ferrite-based redox cycling was developed and tested on sun (Fig. 18.2(a)–(b)) [72,73,89]. The redox cycle is conducted sequentially in a single reactor. The redox material is coated on a monolithic silicon carbide honeycomb structure contained in a solar receiver and exposed to direct concentrated solar radiation. The aperture of the receiver is covered with a quartz glass

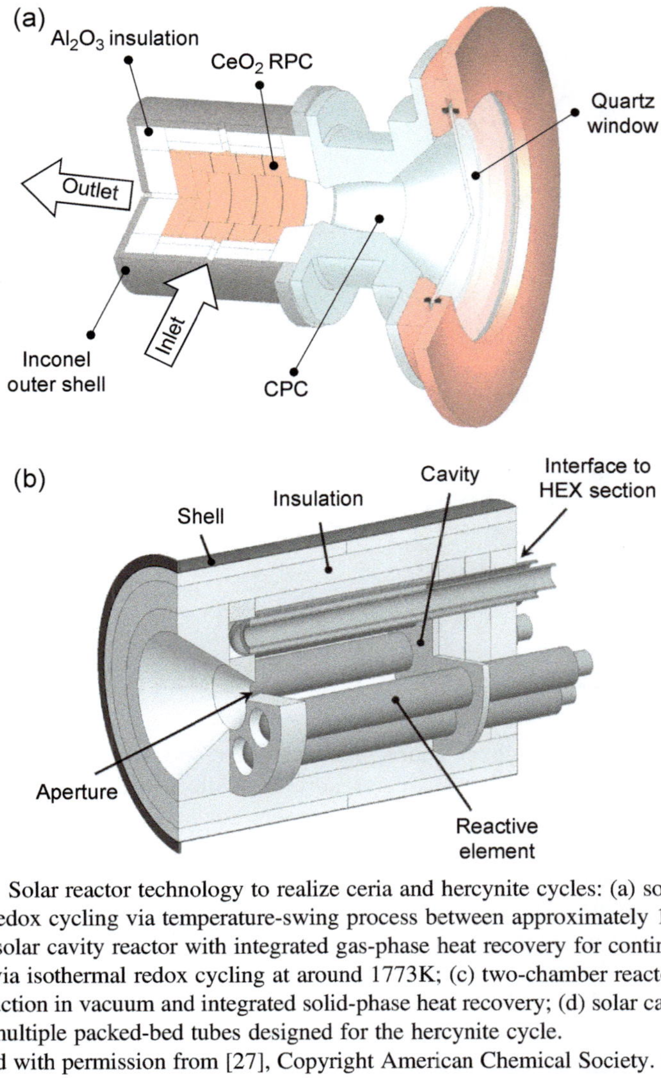

Figure 18.4 Solar reactor technology to realize ceria and hercynite cycles: (a) solar cavity reactor for redox cycling via temperature-swing process between approximately 1073 and 1873K; (b) solar cavity reactor with integrated gas-phase heat recovery for continuous fuel production via isothermal redox cycling at around 1773K; (c) two-chamber reactor with thermal reduction in vacuum and integrated solid-phase heat recovery; (d) solar cavity reactor containing multiple packed-bed tubes designed for the hercynite cycle.
(a) Reprinted with permission from [27], Copyright American Chemical Society. (b) Reprinted from [80], Copyright with permission from the American Society of Mechanical Engineers. (c) Reprinted from [81], Copyright with permission from Elsevier. (d) Reprinted from [82], Copyright with permission from Elsevier.

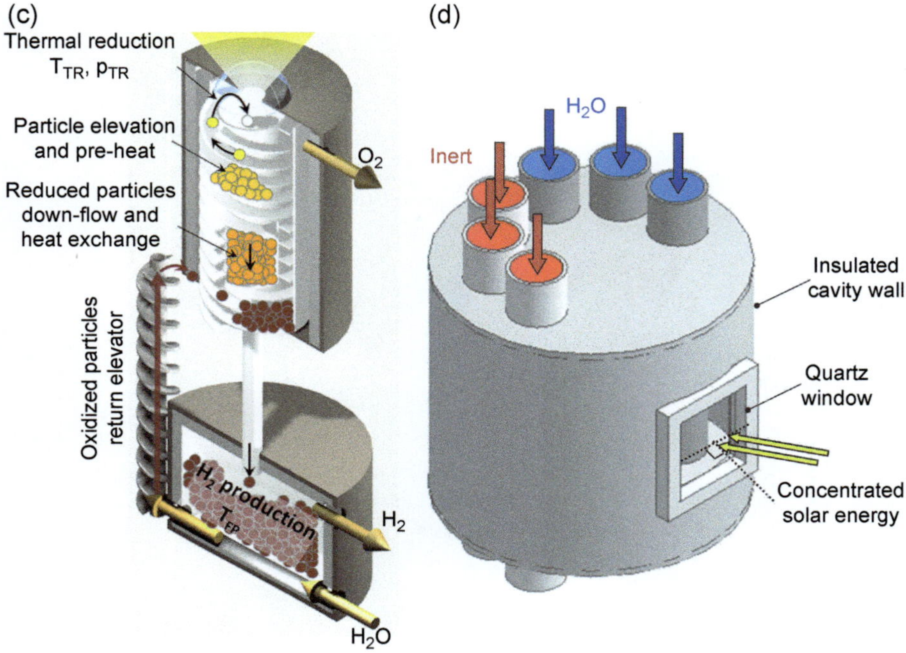

Figure 18.4 Continued.

window. Gases enter through inlet ports in the frustum of the receiver, flow through the axial channels of the honeycomb structure, and are collected in a funnel at the back of the reactor. Two full H_2O splitting cycles were conducted with H_2O conversion of up to 30%, resulting in about 35 g of H_2 produced [73].

A 100-kW pilot-scale solar reactor was developed to conduct the thermal dissociation of ZnO particles (Fig. 18.3(a)–(b)) [77,90–92]. The reactor consists of a rotating cylindrical cavity receiver with a windowed aperture to control the gas atmosphere inside the reactor. Particles are fed into the reactor by a screw feeder and are distributed over the cylindrical wall of the cavity by its rotating motion, where the particles are directly exposed to concentrated solar radiation. Argon sweep gas is injected through radial and tangential nozzles near the window and around the frustum, generating a vortex flow that protects the window from zinc deposits and carries the gaseous products out of the reactor. Product gases exit the reactor through an annular outlet at the back of the reactor that forms a water-cooled quench unit in which argon is injected to quench the products to avoid their recombination. The reactor was tested in the 1 MW solar furnace in Odeillo, France. The total amount of dissociated ZnO in a typical experiment was 157 g [77].

Thermochemical H_2O and CO_2 splitting via non-stoichiometric redox cycling of ceria in a single prototype solar reactor, shown in Fig. 18.4(a), has been demonstrated [26]. The reactor consists of an insulated cavity containing cylindrical porous ceria structures [27,41]. The measured reactor's solar-to-fuel efficiency is in the order of

1–2%. Main thermal losses are attributed to heat radiation and conduction. A prototype solar thermochemical reactor to realize isothermal redox cycling of ceria and gas-phase heat recovery has been developed and tested (Fig. 18.4(b)) [80,93]. The reactor consists of an array of six reactive element tubes, filled with a packed bed of porous ceria pellets, which can be operated individually. This allows for simultaneous reduction of ceria in one section of the reactor and oxidation of ceria in another section, thus allowing for a constant solar power input to the reactor and quasi-steady fuel production. During 20 h of operation, the 4-kW reactor showed a stable rate of CO production of about 2 mL/s over 25 cycles.

18.2.2 Hybrid processes

18.2.2.1 High-temperature electrolysis

Hydrogen can be produced with solar energy via conventional water electrolysis using electricity generated with solar photovoltaic (PV) cells or concentrating solar power. Syngas can be obtained from H_2 and CO_2 via reverse water-gas shift reaction. However, with typical solar-to-electricity energy conversion efficiencies of single-junction PV cells and concentrating solar power systems limited to about 25% and hydrolysis cell efficiencies of about 70% [2], this process route is limited to maximum solar-to-fuel efficiencies of about 18%. Higher efficiencies may be obtained by reducing the required electric power input to the electrolysis cell by increasing the cell-operating temperature. For example, at 1000K the theoretically required cell voltage to hydrolyze H_2O is reduced to about 1 V, compared to about 1.23 V at ambient temperature. The heat input to the cell can be provided by a combination of ohmic losses occurring in the cell and concentrated solar energy. High-temperature electrolysis cells typically operate at about 973–1173K, suitable for concentrated solar energy input. High-temperature electrolysis is based on solid oxide fuel cell technology and materials, and offers the possibility of co-electrolyzing H_2O and CO_2 to produce syngas [94]. Reviews of solid oxide electrolysis technology can be found in Refs. [5,95].

18.2.2.2 Electro-thermochemical cycles

Thermodynamic analyses indicate that below 1000K, H_2O/CO_2 splitting thermochemical cycles require at least three process steps [96]. In fact, there are currently no known pure thermochemical two-step H_2O/CO_2 splitting cycles that operate at temperatures below 1373K [21]. However, the operating temperature of two-step H_2O/CO_2 splitting cycles can be lowered if at least one cycle step is hybridized electrochemically. An example is the two-step H_2O splitting hybrid sulfur cycle.

The first step is the thermochemical decomposition of sulfuric acid [34]:

$$H_2SO_4(aq) \rightarrow H_2O(g) + SO_2(g) + 1/2O_2(g) \tag{18.8}$$

The second step is the electrolysis of water:

$$SO_2(aq) + 2H_2O(l) \rightarrow H_2SO_4(aq) + H_2(g),\ E^\circ = -0.156\ V \qquad (18.9)$$

The first process step, Eq. (18.8), proceeds in three stages. The sulfuric acid is first vaporized, followed by its decomposition to H_2O and SO_3 at 623−673K. In the third stage, SO_3 is dissociated catalytically into SO_2 and $1/2O_2$ in the temperature range of 973−1273K [6]. The second process step, Eq. (18.9), is conducted at 353−393K in an electrolysis cell [97,98]. While the hybrid sulfur cycle does require electric energy input, it has the potential to be more efficient than direct water electrolysis due to the strongly reduced cell potential required in the electrolysis step compared to conventional water electrolysis.

The thermal decomposition of sulfuric acid has been conducted in a 2-kW two-chamber solar cavity reactor (Fig. 18.5). Both reaction chambers are made of AISI 316Ti steel, have individual apertures that are closed with a quartz window, and contain SiSiC (siliconized silicon carbide) volumetric absorber structures. In the lower-temperature chamber, sulfuric acid is vaporized and dissociated at 673K. The SO_3-H_2O vapor is then fed to the second chamber where the catalytic decomposition to SO_2 and $1/2O_2$ is conducted at 923−1223K. Iron and mixed metal oxides (Fe_2O_3 and $CuFe_2O_4$) are used as catalytic coatings on the absorber structure. Conversions of SO_3 of more than 90% were obtained at 1223K. Highest measured net efficiency of converting solar energy to enthalpy change in the gas flow was 28% [97].

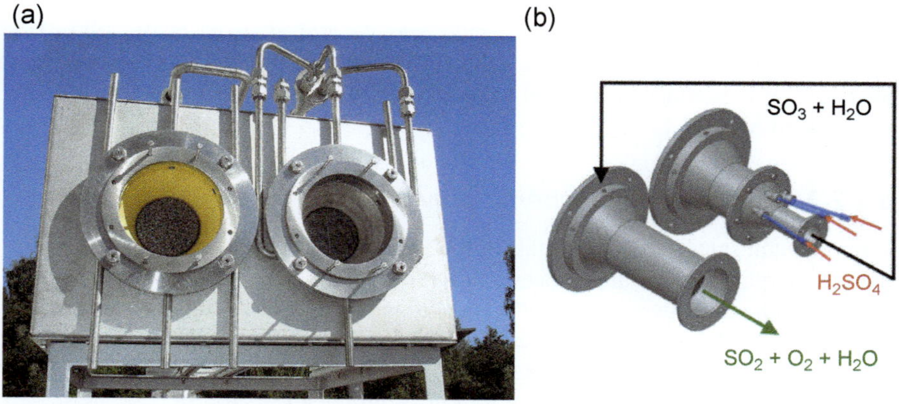

Figure 18.5 Two-chamber solar reactor to conduct the thermochemical decomposition of sulfuric acid.

Reprinted from [97], Copyright with permission from Hydrogen Energy Publications, LLC.

In an alternative two-step hybrid cycle, a metal (Fe, Mn, or Co) oxide is reduced with concentrated solar energy in air and subsequently used as an anode or solute to split H_2O electrolytically at a reduced cell potential [7].

18.2.2.3 Carbothermal cycles

The high temperatures required to thermally reduce metal oxide redox materials in two-step thermochemical H_2O/CO_2 splitting cycles lead to challenges associated with material stability and durability, and significant heat losses. Instead of including an electrochemical step, the required temperatures of H_2O/CO_2 splitting cycles can also be significantly reduced by hybridizing thermochemical cycles with a carbothermal reduction step, that is, by replacing reaction (18.5) by reaction (18.25) (see Section 18.3.3). In carbothermal H_2O/CO_2 splitting cycles, H_2 and/or CO are produced in both cycle steps. The process can be represented by [8]:

Carbothermal reduction step:

$$CH_x + \frac{1}{z}M_yO_z \rightarrow \frac{x}{2}H_2 + CO + \frac{y}{z}M \qquad (18.10)$$

H_2O/CO_2 splitting step:

$$\frac{y}{z}M + H_2O \rightarrow \frac{1}{z}M_yO_z + H_2 \qquad (18.11)$$

and/or:

$$\frac{y}{z}M + CO_2 \rightarrow \frac{1}{z}M_yO_z + CO \qquad (18.12)$$

A thermodynamic process analysis estimated the theoretical solar-to-fuel energy conversion efficiency of the isothermal carbothermal redox cycle, Eqs. 18.10–18.12, with non-stoichiometric ceria as the redox material and methane as the reducing agent [99]. If ceria is reduced carbothermally with methane at 1273K, the maximum reachable oxygen non-stoichiometry of ceria within the cubic fluorite phase of 0.25 can be reached and methane can be completely converted to syngas [99,100]. The results of the analysis show that with an increased oxidation non-stoichiometry of 0.1 nearly complete conversion of H_2O/CO_2 can be achieved in the H_2O/CO_2 splitting step. Operating the cycle at 1273K with a solar concentration ratio of 1000 and without any gas-phase heat recovery, the predicted solar-to-fuel efficiency (Eq. 18.7) is 38% for complete methane conversion. With a more realistic methane conversion of 25%, the predicted efficiency is 27% without heat recovery and 37% with 90% heat recovery [99].

Thermodynamic and experimental studies of the carbothermal reduction of metal oxides are further discussed in Section 18.3.3.

18.2.3 Conversion and upgrade of carbonaceous feedstocks

High-temperature solar thermal processing of carbonaceous feedstocks has been proposed as a hybrid solar–fossil and near-term route to production of H_2/CO as a feedstock for solar fuel synthesis and as commodity materials. Solar energy is used as the exclusive source of process heat, and the carbonaceous feedstocks are the source of carbon and hydrogen. The main processes are gasification, cracking, and reforming [101]. Chemical reaction equations presented in this section are written for carbonaceous feedstocks containing exclusively carbon, hydrogen, and oxygen. Impurities and other elements are typically found in fossil fuels and biomass, for example, sulfur and nitrogen, and lead to formation of additional product compounds as compared to those reported in this section.

18.2.3.1 Gasification

Solar thermal gasification has been studied for a variety of carbonaceous feedstocks including coal, coke, biomass, bitumen, and carbonaceous wastes [102]. While the use of biomass as the feedstock promises a carbon-neutral fuel, the availability of biomass for energetic uses is limited. Coal remains the most abundant fossil fuel resource in the world, with the total recoverable reserves estimated in year 2008 to be 909 billion tons. The United States share the largest portion of these reserves, which are 262.7 billion tons (29%). The main approach to coal utilization is the direct combustion for electricity generation, which produces about 0.92 kg of CO_2 per 1 kWh_e of electricity generated. In spite of environmental concerns, coal is likely to remain one of the leading energy sources in the near future. Coal gasification technologies can substantially reduce greenhouse gas emissions through increased conversion efficiency. Further emission reductions are possible, if the process heat required to drive the gasification process is obtained from a CO_2-free energy source such as concentrated solar energy, eliminating the need for combustion of a significant portion of the feedstock. In addition, elimination of internal combustion for supplying the process heat mitigates contamination of the synthesis gas with the combustion by-products [101,103]. The produced synthesis gas can then be used for efficient power generation in combined-cycle power plants or for synthesis of hydrocarbon fuels.

Thermal gasification of carbonaceous materials is a complex chemical process that involves several different intermediate reactions. Steam and carbon dioxide are the common oxidizing agents. A simplified net reaction of steam gasification was proposed [104]:

$$CH_xO_y + (1-y)H_2O \rightarrow \left(\frac{x}{2} + 1 - y\right)H_2 + CO, \tag{18.13}$$

where x and y are the molar ratios of hydrogen to carbon (H/C) and oxygen to carbon (O/C), respectively. Analogously, a simplified net reaction of dry gasification can be written as [105]:

$$CH_xO_y + (1 - y)CO_2 \rightarrow \frac{x}{2}H_2 + (2 - y)CO. \quad (18.14)$$

Reactions (18.13) and (18.14) proceed endothermically under atmospheric pressure at temperatures above 850K and 900K, respectively. The simplified reactions given by Eqs. (18.13) and (18.14) are the net result of the following intermediate reactions:

Boudouard reaction:

$$C + CO_2 = 2CO, \quad (18.15)$$

Methane cracking/hydrogasification:

$$C + 2H_2 = CH_4, \quad (18.16)$$

Reforming/methanation:

$$CH_4 + H_2O = CO + 3H_2, \quad (18.17)$$

Water–gas shift:

$$CO + H_2O = CO_2 + H_2, \quad (18.18)$$

Reactions (18.15)–(18.18) are strongly dependent on temperature, pressure, and the C/O ratio. The composition of syngas, and thus its suitability for production of synthetic fuels is highly dependent on the outcome of reactions (18.15)–(18.18). Thermodynamic analyses of solar thermal gasification of four types of coal—anthracite, bituminous, lignite, and peat—were studied [104,105]. Fig. 18.6 shows the equilibrium mole fractions of the main species for the steam and dry gasification, respectively, of anthracite over the temperature range 700–1500K. For temperatures below 700K, the intermediate reactions do not take place. Thus, components such as C, CH_4, CO_2, and H_2O are thermodynamically stable at temperatures below 700K. As the temperature increases from 700K to about 1200K, rates of the intermediate reactions increase and the fractions of C, CH_4, CO_2, and H_2O decrease. At temperatures higher than 1200K, both steam and dry gasification produce syngas that consists of only H_2 and CO with a constant H_2/CO ratio.

Kinetic models of solar thermal gasification processes have been proposed in several studies [106–108]. The models are typically based on the Langmuir–Hinshelwood type of kinetic rate expressions for production and consumption of each gas species [102]. A detailed discussion of the kinetic analyses and models is omitted from this work, and the reader is referred to the cited literature references.

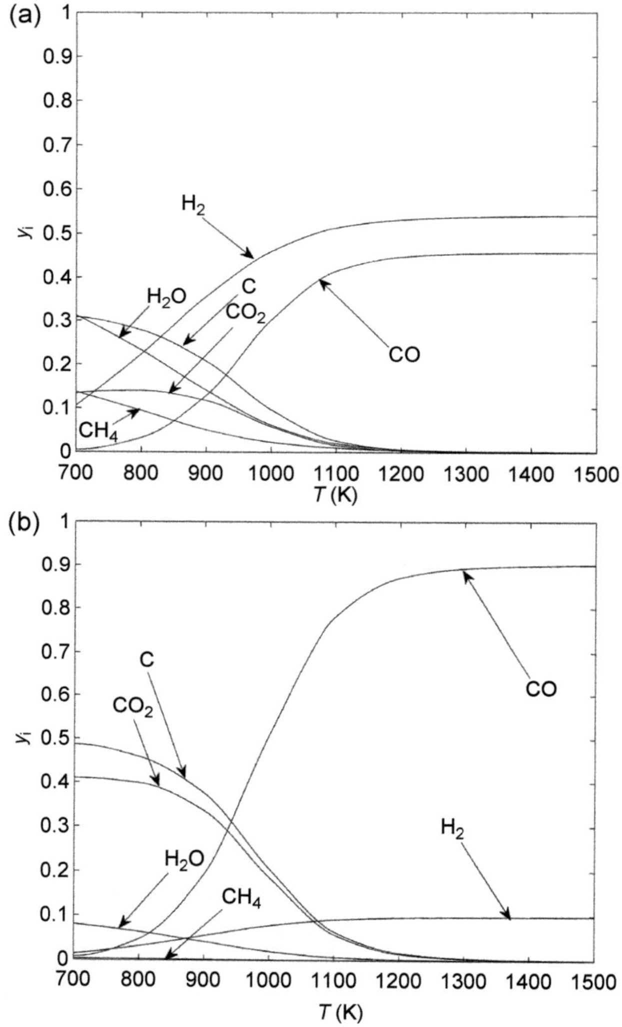

Figure 18.6 Equilibrium composition for (a) steam and (b) dry gasification of anthracite as a function of temperature.
Reprinted from [105], Copyright with permission from Elsevier.

Solar reactors for gasification of carbonaceous feedstocks are among the most mature reactor concepts in solar thermochemistry. Both directly and indirectly irradiated reactors have been developed [102,109]. The configurations of the reacting two-phase solid–gas media include fluidized beds [110], entrained particle flows [111], vortex flows [112], and packed-bed reactors [113]. Example solar reactors containing reacting media of such configurations are depicted in Fig. 18.7. Since 2010, reactors implementing more complex approaches to solar thermal gasification have been proposed, which include gasification of biomass in molten salts [114] and with supercritical water [115]. The solar reactor from the latter study is shown in Fig. 18.8.

18.2.3.2 Cracking

Solar-driven cracking of hydrocarbons to coproduce hydrogen and carbon black can be represented by the simplified reaction:

$$C_xH_y \rightarrow xC(gr) + \frac{y}{2}H_2 \tag{18.19}$$

The process is typically realized in the form of a directly irradiated gas flow laden with carbon particles [116–118], or an indirectly irradiated flow in tubes [119–122]. The particles serve as radiant absorbers and provide surface to the heterogeneous reaction. Fig. 18.9 shows an example directly irradiated solar thermochemical reactor featuring a horizontal cylindrical reaction cavity containing the reactive mixture of carbon particles and methane [116,117].

18.2.3.3 Reforming

Steam and dry reforming of hydrocarbons are described by the simplified reactions:

$$C_xH_y + xH_2O \rightarrow \left(x + \frac{y}{2}\right)H_2 + xCO, \tag{18.20}$$

$$C_xH_y + xCO_2 \rightarrow \frac{y}{2}H_2 + 2xCO. \tag{18.21}$$

A directly irradiated solar reactor for the steam and dry reforming of methane was developed, scaled up to 300 kW, and tested on the solar tower facility at the Weizmann Institute of Science, Israel (Fig. 18.10) [123–125]. The reactor uses an Al_2O_3 or SiC reticulate porous ceramic structure, supporting the γ-Al_2O_3 washcoat and rhodium catalyst and acting as both the radiation absorber and reaction site. On-sun operation was conducted with average absorber temperature between 700 and 860°C and total pressure of 3.5 bar, resulting in methane conversions >80% [124].

Selected design aspects of a solar reactor for methane decomposition in both cracking and reforming processes based on thermodynamic and kinetic analyses are discussed in Ref. [126]. A comprehensive review of solar methane reforming is given in Ref. [127].

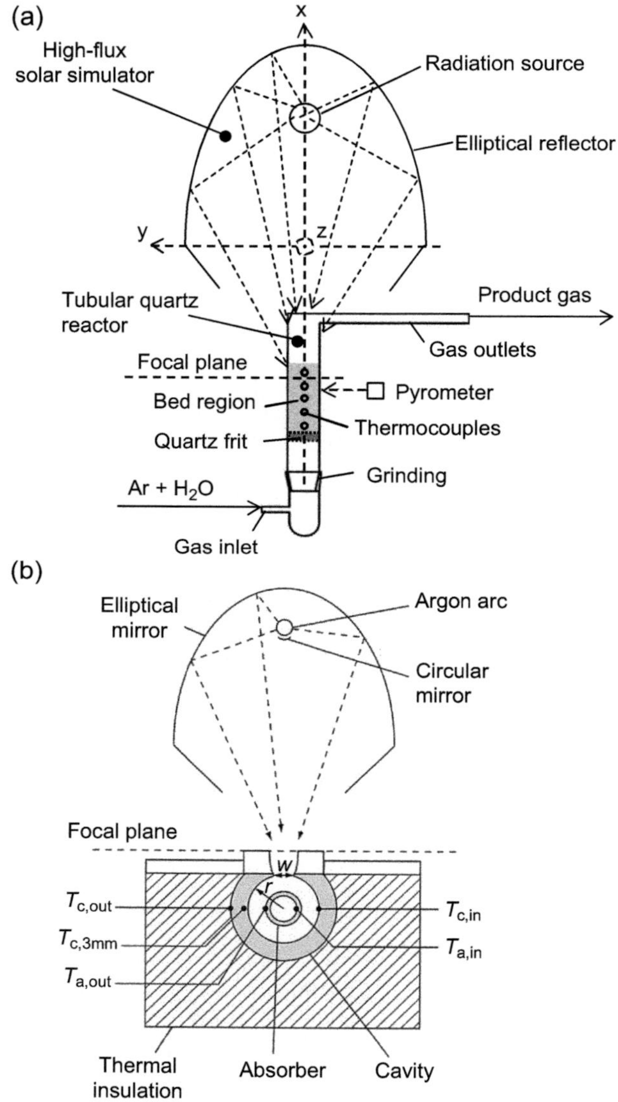

Figure 18.7 Examples of solar thermochemical reactors for thermal gasification of carbonaceous materials: (a) directly irradiated fluidized bed reactor; (b) indirectly irradiated entrained particle flow reactor; (c) directly irradiated vortex-flow reactor; and (d) indirectly irradiated packed-bed reactor.

(a) Reprinted from [110], Copyright with permission from Elsevier. (b) Reprinted from [111], Copyright with permission from Elsevier. (c) Reprinted from [112], Copyright with permission from the International Association for Hydrogen Energy. (d) Reprinted from [113], Copyright with permission from Elsevier.

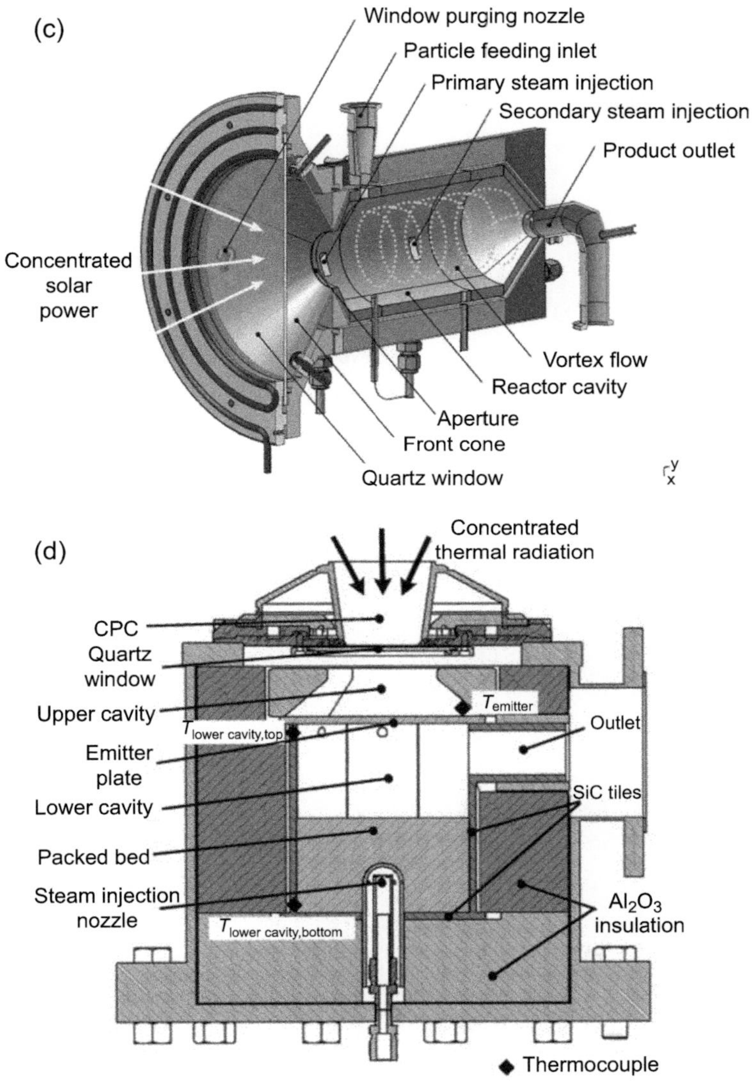

Figure 18.7 Continued.

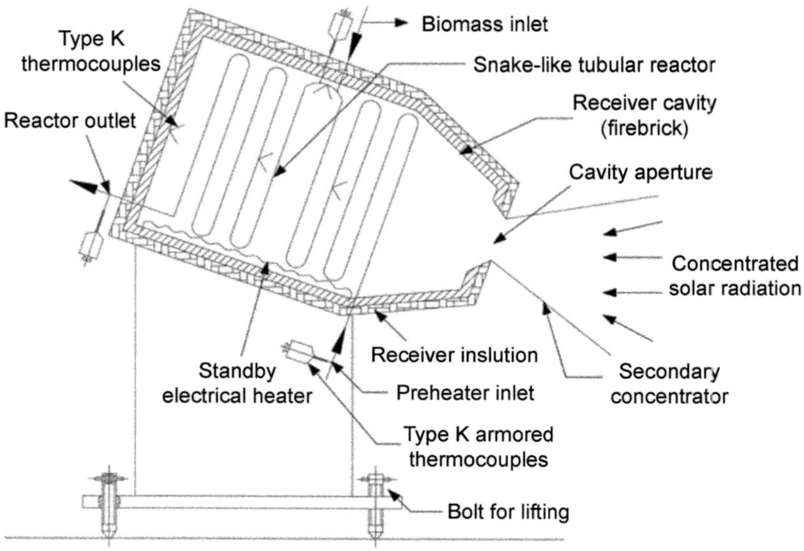

Figure 18.8 Solar thermochemical reactor for supercritical water gasification of biomass. Reprinted from [115], Copyright with permission from Elsevier.

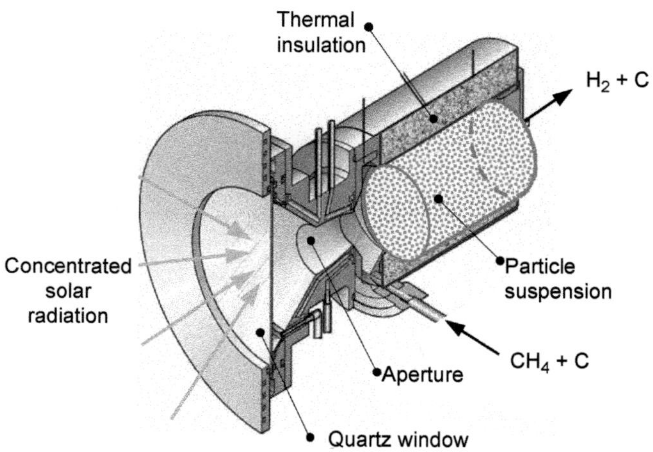

Figure 18.9 Schematic of a reactor prototype for solar methane cracking, featuring a flow of methane laden with carbon particles and directly exposed to concentrated solar radiation. Reprinted from [116], Copyright with permission from Elsevier.

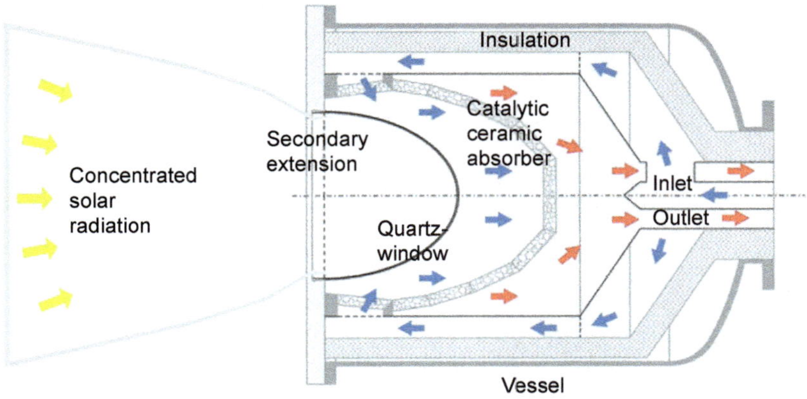

Figure 18.10 Directly irradiated 300-kW solar reactor for the solar reforming of methane. Reprinted from [123], Copyright with permission from the American Society of Mechanical Engineers.

18.3 Material processing and chemical commodity production

The high-temperature process heat derived from concentrated solar radiation is suitable for driving highly endothermic processes to process materials and produce chemical commodities. Examples include production of lime and cement, chemical-looping processes for gas separation, extractive metallurgy, recycling of hazardous waste materials, ammonia production via nitride-based cycles, nanomaterial production via ablation, and more.

18.3.1 Calcination of $CaCO_3$

The calcination of limestone ($CaCO_3$) to lime (CaO) according to:

$$CaCO_3(s) \rightarrow CaO(s) + CO_2(g), \ \Delta H^{\circ}_{298K} = 178 \text{ kJ/mol}, \tag{18.22}$$

is the main endothermic step in the production of lime and cement, and is operated above 1173K [2]. In most lime kilns, process heat is supplied by combustion of fossil fuels. The CO_2 emissions from the lime industry are about 1% of the global anthropogenic CO_2 emissions [9]. The CO_2 emissions from the process can be mitigated by operating the process with high-temperature solar process heat and sequestering the CO_2 evolving from the calcination reaction (18.22). The CO_2 can be stored or used as a feedstock in the production of solar fuels.

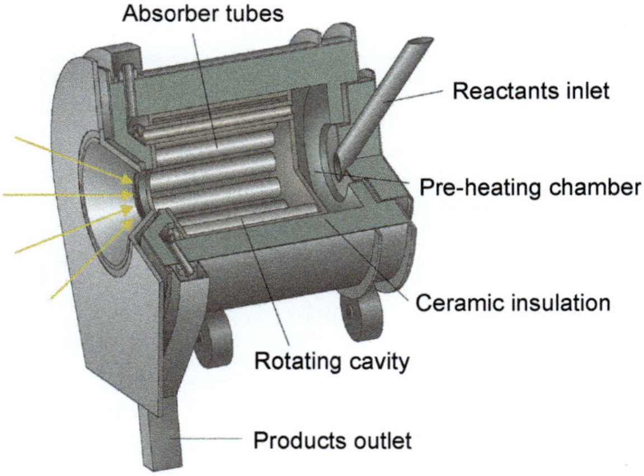

Figure 18.11 10-kW prototype solar rotary kiln for the calcination of limestone. Reprinted from [128], Copyright with permission from Elsevier.

A 10-kW prototype solar-heated rotary kiln was developed and tested (Fig. 18.11) [2,9,129]. It consists of a 225 mm × 252 mm indirectly irradiated cylindrical cavity receiver made from recrystallized silicon carbide (maximum operating temperature 1873K). Solar radiation at a concentration ratio of about 2000, provided by a solar furnace, enters the reactor through a 9 cm-diameter aperture in the water-cooled aluminum front. 1–5-mm large limestone particles are fed into the preheating chamber at the back of the reactor. Due to the inclination and the rotation of the reactor, the particles are moved into the 16 silicon carbide absorber tubes that line the solar cavity. Calcined CaO particles are collected in a funnel at the front of the reactor. The reactor was tested on-sun for over 100 h, producing up to 4 kg/h of lime at solar energy conversion efficiencies of up to 35%. The cost of lime produced in a 20-MW solar calcination plant was estimated to amount to 131–158 $/t, corresponding to two to three times the current selling price (cost estimation from 2006) [128].

In order to mitigate both the CO_2 emissions resulting from the combustion of fossil fuels to provide the process heat and the CO_2 emissions resulting from the calcination reaction itself, the combination of $CaCO_3$ calcination with methane reformation was explored in a directly irradiated solar cavity reactor [130]. Up to 83% conversion of $CaCO_3$ and 38% conversion of CH_4 were achieved at a nominal temperature of 1223K [130].

18.3.2 $CaCO_3$/CaO cycle

Operated in a closed cycle, the calcination of limestone followed by the carbonation of lime can be utilized to capture CO_2 from CO_2-rich flue gas streams or from air, or as a high-temperature thermochemical storage system. The process is described by:

$$CaCO_3(s) \rightarrow CaO(s) + CO_2, \Delta H^{\circ}_{298K} = 178 \text{ kJ/mol} \quad (18.23)$$

$$CaO(s) + CO_2 \rightarrow CaCO_3(s) \ \ \Delta H^{\circ}_{298K} = -178 \text{ kJ/mol} \quad (18.24)$$

To obtain a product stream of pure CO_2 during the calcination step, the reaction needs to be operated in a sweep gas flow of pure CO_2.

A thermodynamic process analysis was conducted to calculate the theoretical energy input required by the process (18.23)–(18.24) as a function of CO_2 concentration in the flue gas and heat recovery from the hot products [131]. The energy requirement of the process strongly depends on the CO_2 concentration in the inlet gas and on gas-phase heat recovery, while solid-phase heat recovery is relatively unimportant. With 100% gas- and solid-phase heat recovery, 207 kJ of thermal energy is required per mole of captured CO_2, irrespective of the CO_2 concentration in the flue gas stream. In the absence of gas-phase and solid-phase heat recovery, the energy requirement is between 45 MJ and 283 kJ per mole of captured CO_2 for initial CO_2 concentrations spanning those encountered in air (~0.03%) to those encountered in flue gas (15%). For comparison, the theoretical minimum work to separate CO_2 from a binary gas mixture is between 20 and 30 kJ/mol of captured CO_2 at atmospheric CO_2 concentrations and 7 kJ/mol of captured CO_2 for an initial CO_2 concentration of 15%.

A packed-bed solar reactor has been designed to conduct both process steps (Fig. 18.12) [132,133]. The reactor has a beam-up configuration. This downward-facing orientation of the solar cavity receiver reduces convective heat losses from the open receiver aperture. Calcium carbonate particles are contained in an annular

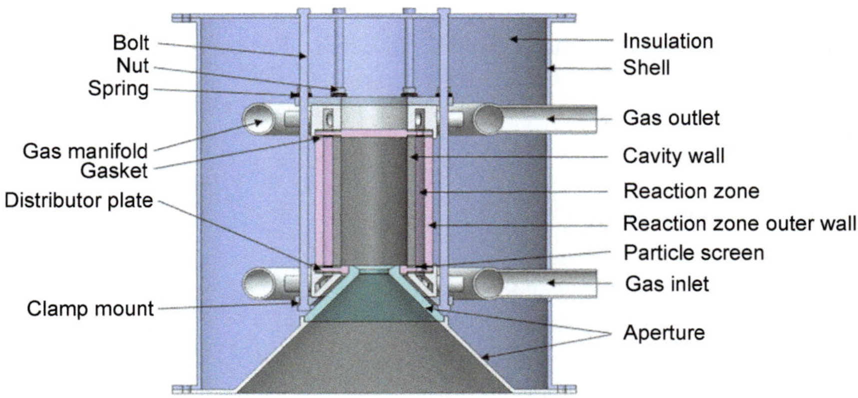

Figure 18.12 Packed-bed solar reactor to capture CO_2 via process (18.23)–(18.24). Reprinted from [132], Copyright with permission from the American Society of Mechanical Engineers.

space surrounding the cavity receiver. The particles are calcined in a pure stream of CO_2 with solar heat input. During the carbonation step, flue gases or atmospheric air flow through the particle bed. Due to the exothermicity of the chemical reaction, solar heat input to the reactor is not required during the carbonation step. Hence, two or more reactors could be operated in parallel and the solar energy input be switched between the reactors, for example, by changing the orientation of the heliostats of a solar tower plant [73].

Other reactors that have been examined to conduct one or both of the process steps (18.23)–(18.24) include rotary kilns (including the reactor shown in Fig. 18.11) [9,134], fluidized beds [134,135], and cyclone and vortex-flow reactors [130,136,137].

18.3.3 Extractive processes and recycling

The extraction of metals from mineral ores involves very energy-intensive process steps. For example, extraction of aluminum from alumina (Al_2O_3) via the electrolytic Hall–Héroult process requires an electric energy input of about 45 MJ per kg of aluminum, resulting in emissions of 4.9–7.4 kg CO_2-eq. per kg of aluminum [138]. Overall, the total energy consumption by the metallurgical industry is estimated to contribute 10% of the total global anthropogenic greenhouse gas emissions [2].

Solar thermochemical and carbothermal processes could replace some of the most energy-intensive extractive metallurgical processes and strongly reduce the CO_2 emissions associated with this industry. In addition, through the solar carbothermic conversion of waste materials, such as electric arc furnace dust (EAFD) and automobile shredder residue (ASR), hazardous and valuable elements can be recovered. Depending on the reactants and process conditions, solar carbothermal reduction of metal (and Si) oxide-containing feedstocks yields pure metals, lower valence metal oxides, metal carbides, or metal nitrides. In all cases, CO or syngas is coproduced, which can be further processed to synthetic fuels.

The overall chemical reactions for the carbothermal reduction of metal oxides, such as Fe_3O_4, ZnO, MgO, and Al_2O_3, to the metals with carbonaceous reducing agents (e.g. coal, coke, and natural gas), can be represented by (M stands for metal or Si) [8]:

$$CH_x + \frac{1}{z}M_yO_z \rightarrow \frac{x}{2}H_2 + CO + \frac{y}{z}M \tag{18.25}$$

For pure carbon (represented by graphite, $x = 0$) and methane ($x = 4$) as the reducing agents, Eq. (18.25) yields, respectively [11,139]:

$$C(gr) + \frac{1}{z}M_yO_z \rightarrow CO + \frac{y}{z}M \tag{18.26}$$

$$CH_4 + \frac{1}{z}M_yO_z \rightarrow 2H_2 + CO + \frac{y}{z}M \tag{18.27}$$

The purified metals can either be used as a commodity material for steel production and manufacturing or as an energy vector. Combustion of metals yields high-temperature heat, oxidation in fuel cells yields electricity, and oxidation with H_2O/ CO_2 yields syngas (Section 18.2) [140].

Thermodynamic equilibrium calculations for the carbothermal reduction of a range of metal oxides show that the onset temperature for the formation of the free metals can be considerably lowered by decreasing the total pressure in the system [141]. Carbothermal reduction in vacuum further enables vacuum distillation of the products and suppresses the formation of undesired by-products, such as Al-carbides and -oxycarbides [141].

For certain metal oxides, such as TiO_2, ZrO_2, and SiO_2, carbothermal reduction in an inert environment or in N_2 at ambient pressure can lead to the formation of carbides, nitrides, and lower valence metal oxides rather than pure metals [11]. These processes can be summarized as [2]:

$$C(gr) + M_yO_z\{+ N_2\} \rightarrow \{M_{y'}C_{z'}, M_{y''}N_{z''}, M_{y'''}O_{z'''}\} + CO \tag{18.28}$$

Carbides and nitrides, such as TiC, SiC, TiN, and Si_3N_4, are high-performance ceramic materials with exceptionally high mechanical hardness, chemical inertness, and heat resistance. This makes them valuable for a range of applications, for example as protective coatings of cutting and sliding surfaces, as non-toxic coatings of medical implants, as heat shields for spacecraft surfaces, and as base materials to manufacture high-temperature parts of cars, rockets, and solar reactors. Hydrolysis of other metal carbides and nitrides can be used for the production of commodity materials and/or fuels (Section 18.3.4) [2]. In the following sections, exemplary studies of carbothermal reduction of alumina, zinc oxide, and solid waste materials are described. Other studies investigated the carbothermal reduction of other metal and semiconductor oxides, including MgO, SiO_2, SnO_2, and GeO_2 [142–144], and the silicothermal and the electro-thermochemical reduction of MgO [145,146].

18.3.3.1 Aluminum

At ambient pressure, the formation of a pure mixture of Al and CO via reaction (18.26) requires excessively high temperatures of more than 2800K (Fig. 18.13(a)) [138]. Reducing the gas pressure in the process can substantially reduce the required temperature and the formation of undesired by-products, such as Al_2O and Al_4C_3. For example, at 10^{-6} bar the reaction goes to completion at less than 1500K (Fig. 18.13(b)) [138]. Exploratory experiments of vacuum carbothermal reduction of alumina were conducted in a directly solar-heated quartz tube reactor operated at pressures in the range of 3.5–12 mbar and temperatures in the range of 1300–2000K with biocharcoal as the reducing agent. Measured aluminum mass fraction in the condensed products was up to 19%, accompanied by the formation of Al_4C_3 and Al_4O_4C within the crucible [138].

The equilibrium composition of reaction (18.27) with Al_2O_3 and of the carbothermal reduction of bauxite components has been studied theoretically

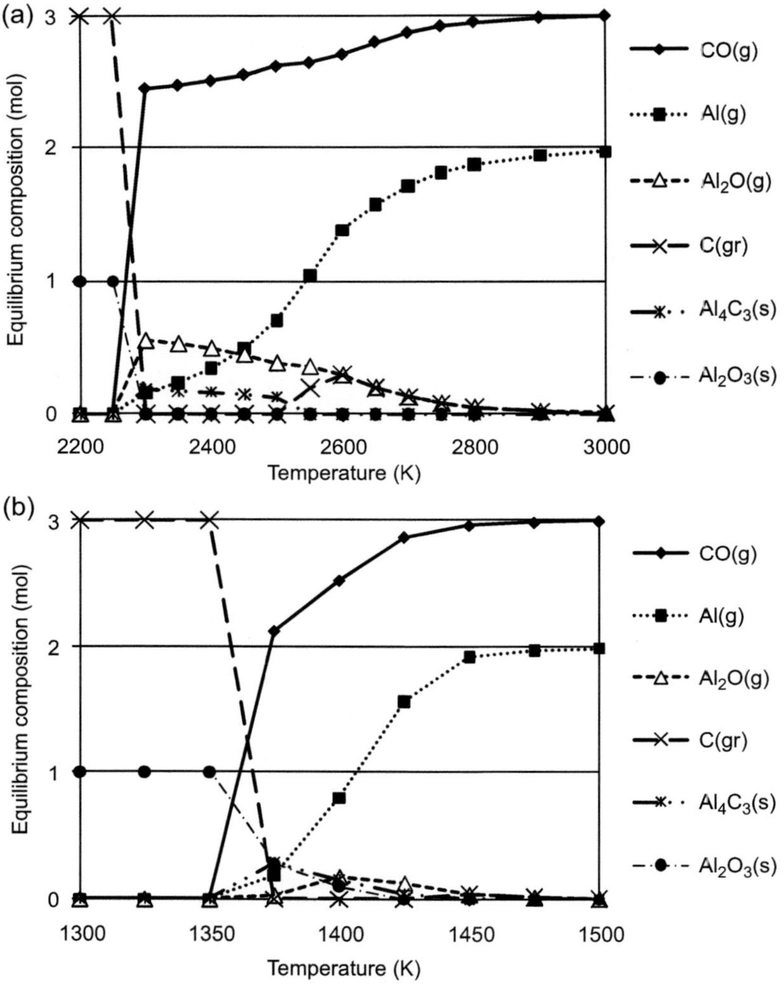

Figure 18.13 Equilibrium composition of the system $Al_2O_3 + 3C$ (a) at 1 bar and (b) at 10^{-6} bar.
Reprinted from [138], Copyright with permission from Springer.

[147,148]. It was predicted that carbothermal reduction of Al_2O_3 with methane at above 2573K and ambient pressure yields a mixture of Al and syngas with only limited amounts of unwanted by-products (Al_2O, AlH, and H). In a separate study, the beneficial role of gaseous oxygen in the carbothermal reduction of Al_2O_3 with methane was shown [149].

An experimental study demonstrated the production of AlN via carbothermal reduction of Al_2O_3 with carbon and methane as reducing agents in an N_2 flow under direct concentrated irradiation [150].

18.3.3.2 Zinc

The reduction of zinc oxide to zinc via the thermochemical, carbothermal, and electrothermal routes has been studied extensively for the production of zinc as an important commodity (annual production of 13.4 million tons in 2015 [151]), as a potential future energy vector used in zinc–air batteries and fuel cells, as well as in the context of the H_2O/CO_2 splitting zinc oxide cycle (see Section 18.2.1) [25,77,84 152–157]. The thermal dissociation of ZnO (Eq. 18.5) is thermodynamically favored above 2235K [158]. Somewhat lower reduction temperatures can be achieved if the product gases are diluted with inert gas to shift the reaction equilibrium to the right [64]. A major challenge in the thermal dissociation of ZnO is to avoid the recombination of the gaseous products either by quenching or by high-temperature gas separation [159,160].

The thermal reduction of ZnO was conducted in a solar-driven thermogravimeter to measure the thermal dissociation rate and study the chemical reaction kinetics under heat and mass transfer conditions closely approximating those in a directly irradiated solar reactor [161,162]. A directly irradiated rotating solar cavity reactor for the thermal dissociation of ZnO was developed, modeled, and tested at up to 100-kW solar power input (Fig. 18.3(a) and (b)) [77,90–92,163,164]. Other reactor concepts for the thermal reduction of ZnO can be found in Refs. [84,85,156,165].

In the presence of carbon or CH_4, ZnO can be completely reduced above approximately 1200K [11,153]. The carbothermal reduction of zinc was conducted in solar reactors at up to 300 kW [8,166–170]. A 300 kW pilot-scale two-cavity beam-down solar reactor, shown schematically in Fig. 18.14, was tested at temperatures of 1300–1500K, yielding up to 50 kg/h of 95% purity Zn with a solar-to-fuel efficiency (Eq. 18.7) of 30% [169].

Other processes to extract Zn that have been explored include the thermochemical reduction of willemite (Zn_2SiO_4), yielding Zn(g), O_2, and SiO_2(l) [171], the combined ZnO reduction and biomass pyrolysis [2], and the "solar clinkering" followed by the carbothermal reduction with biocharcoal of Waelz oxide [172].

18.3.3.3 Extraction of metals from solid waste materials

High-temperature carbothermal conversion of solid waste materials using concentrated solar energy as the source of process heat is an efficient way to extract metals from metal-rich feedstocks [173,174]. Two important sources of waste are electric arc furnace dust (EAFD) and automobile shredder residue (ASR) [173]. Both types of waste occur in millions of tons annually and contain large amounts of metals, mostly in oxidized form. Their typical elemental compositions are listed in Table 18.3. Solid carbon and methane are considered as reducing agents for EAFD, while the carbon contained within the waste material is used as the reducing agent for ASR [173].

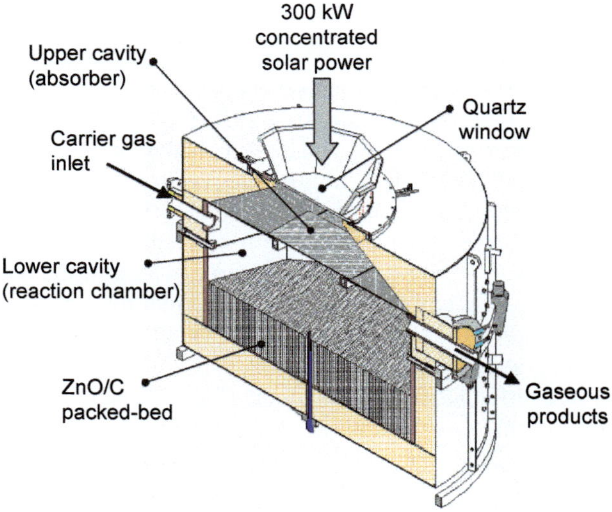

Figure 18.14 300-kW pilot-scale two-cavity beam-down solar reactor for the carbothermal reduction of ZnO.
Reprinted from [169], Copyright with permission from the American Society of Mechanical Engineers.

Table 18.3 Main metal elemental compositions of electric arc furnace dust (EAFD) and automobile shredder residue (ASR). Oxygen is complementary to 100%

Element	EAFD (mass %)	ASR (mass %)
Zn	37.8	2
Fe	13.5	14
Pb	10.1	0.8
Cu	0.23	2.5
Cd	0.09	0.02
Cr	0.12	0.18
Cl	4.8	1.6
S	0.6	0.4
Si	1.7	5
Alkaline earth elements	0–5	0–10
C	<2	38

Reprinted with permission from [173]. Copyright American Chemical Society.

Thermodynamic equilibrium calculations for the carbothermal reduction of EAFD and ASR indicate that above 1300K the main metallic constituents (Zn(g), Fe, Pb for EAFD; Zn(g), Fe, Pb, Cu for ASR) are present in elemental form, along with CO (and H_2), as well as SiO_2 added as vitrification agent (Fig. 18.15). The theoretical minimum process heat required to convert the feedstocks at 1500K (including heating from ambient temperature) is in the range of 2500−4100 kJ/kg, depending on feedstock and reducing agent used [173].

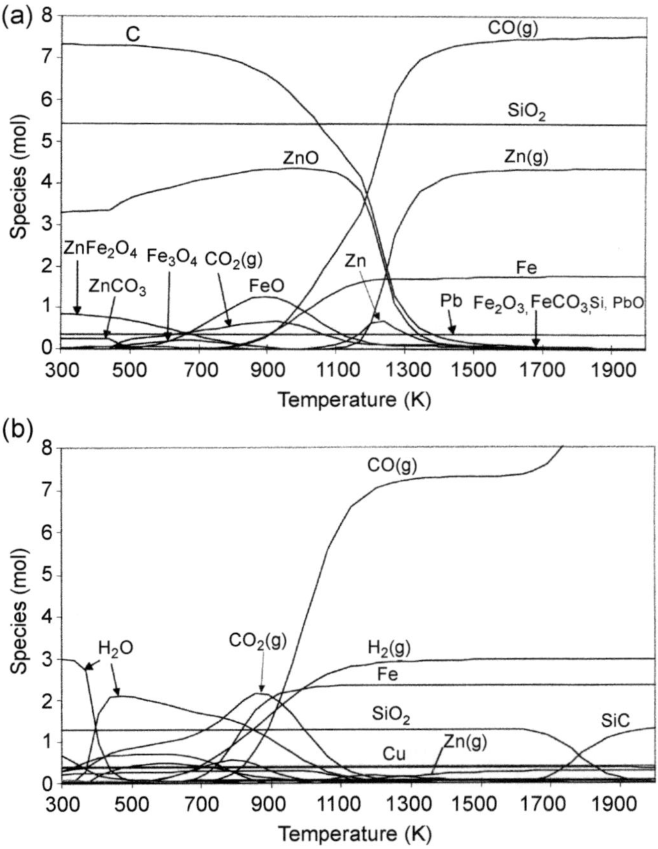

Figure 18.15 Equilibrium compositions for carbothermal reduction of waste materials: (a) EAFD, initial composition 3.46 mol ZnO, 0.91 mol $ZnFe_2O_4$, 0.12 mol Pb_3O_4, 7.56 mol C, and 5.45 mol SiO_2; (b) ASR, initial composition 29 mol C, 3 mol H_2O, 1.66 mol SiO_2, 1.18 mol Fe_2O_3, 0.39 mol CuO, 0.37 mol Al_2O_3, 0.31 mol ZnO, 0.05 mol Cl_2, and 0.02 mol Pb_3O_4 (Pb not shown in graph due to low concentration) [173].

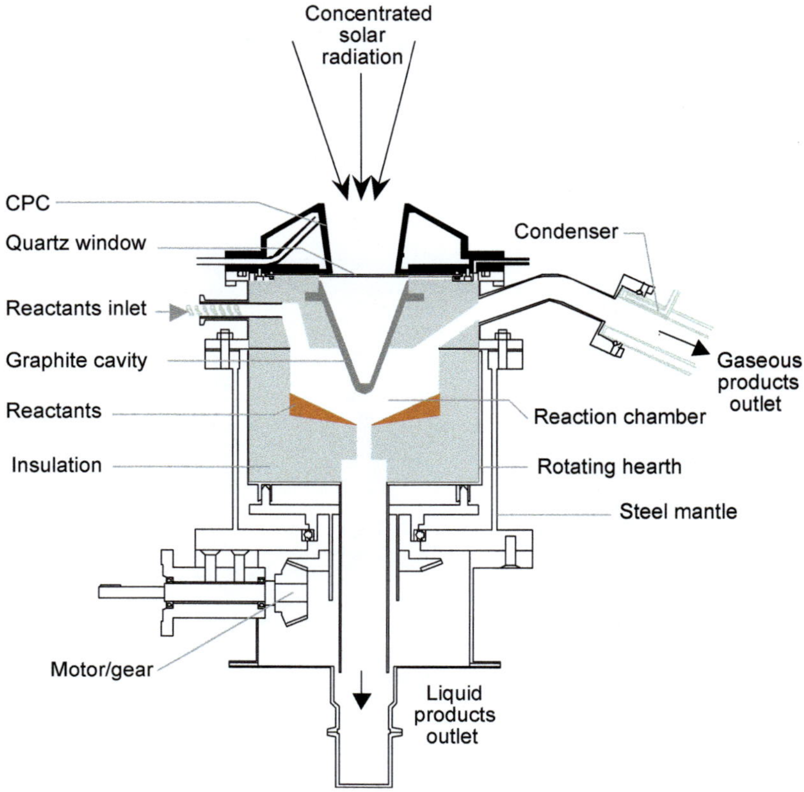

Figure 18.16 10-kW two-cavity prototype solar reactor used for the carbothermal conversion of EAFD.

Reprinted with permission from [174]. Copyright American Chemical Society.

The conversion of EAFD was conducted in a beam-down solar reactor, shown in Fig. 18.16. The reactor consists of a windowed inner cavity to absorb concentrated solar radiation and reemit it into the outer rotary reaction chamber. The feedstock, 87% EAFD and 13% activated charcoal, was either placed into the reactor as a batch of 100–200 g, or continuously fed with a screw feeder at a rate of 2–24 g/min. The reactor was purged with a nitrogen flow of 8–25 L_n/min. Gaseous products were continuously pumped out of the reactor via a water-cooled quench tube and a battery of filters to condense and collect solid products. Experiments were conducted with reactor temperatures in the range of 1140–1400K. More than 99% of Zn and Pd initially in the EAFD were extracted in batch mode operation at 1400K; in continuous mode, the corresponding values were 90% and 80% at 1250K. No ZnO was detected when the O_2 concentration in the outlet stream remained below 2% [174].

18.3.4 Metal carbide/nitride cycles

Metal carbides and nitrides formed in the carbothermal reduction of metal oxides according to Eq. (18.28) can be used in the synthesis of chemicals. For example, hydrolysis of CaC_2 yields acetylene, hydrolysis of manganese and iron carbides yields H_2 and hydrocarbons, and hydrolysis of AlN yields ammonia [2,175,176]. Hence, metal carbides and nitrides can be used to design cyclic processes for the production of commodity materials and fuels with concentrated solar energy as the source of high-temperature process heat.

18.3.4.1 Ammonia

Conventional ammonia production via the Haber—Bosch process uses H_2 obtained via methane reforming. About 100 million metric tons of ammonia are produced per year, using about 5% of the globally produced natural gas [14,177]. Thermodynamic and thermogravimetric analyses have been conducted to study the production of ammonia with concentrated solar energy via the two-step Al_2O_3/AlN thermochemical cycle according to [178—180]:

Carbothermal reduction:

$$Al_2O_3 + 3C + N_2 \rightarrow 2AlN + 3CO, \ \Delta H^{\circ}_{298K} = 708.1 \text{ kJ/mol} \tag{18.29}$$

Ammonia production:

$$2AlN + 3H_2O \rightarrow Al_2O_3 + 2NH_3, \ \Delta H^{\circ}_{298K} = -274.1 \text{ kJ/mol} \tag{18.30}$$

Results of thermodynamic equilibrium calculations for both process steps are shown in Fig. 18.17 [178]. The first endothermic process step, Eq. (18.29), goes to completion at about 2273K (Fig. 18.17(a)). In the second process step, Eq. (18.30), oxidation of AlN is favored at all temperatures between 273 and 1773K (Fig. 18.17(b)). In contrast, NH_3 is thermodynamically unstable at temperatures above 273K. However, NH_3 may exist in a metastable state in the absence of a catalyst [178]. These findings were confirmed with thermogravimetric experiments at 1 bar and in absence of any catalyst. Carbothermal reduction was conducted at 1773—1973K with activated carbon as the reducing agent and a flow of pure N_2. Ammonia production was conducted between 1173 and 1473K with steam concentrations of 20—80% in argon. At 1973K, carbothermal reduction yielded reaction extents (percentage of Al_2O_3 reacted) >80% after 30 min. Highest NH_3 yield (number of NH_3 moles in the gas products relative to initial number of moles of AlN) of 88% was obtained at 1273K with 80% steam concentration, and with a reaction extent of AlN of 93% [179]. Cycling experiments over four consecutive cycles showed trends of decreasing particle size and increasing specific surface area, resulting in increasing reaction rate in the first and increasing ammonia yield in the second process step [180].

An alternative process for NH_3 production was proposed with CrN instead of AlN [177]. Cr_2O_3 is reduced to Cr with both CO and H_2 as reducing agents. Nitridation of

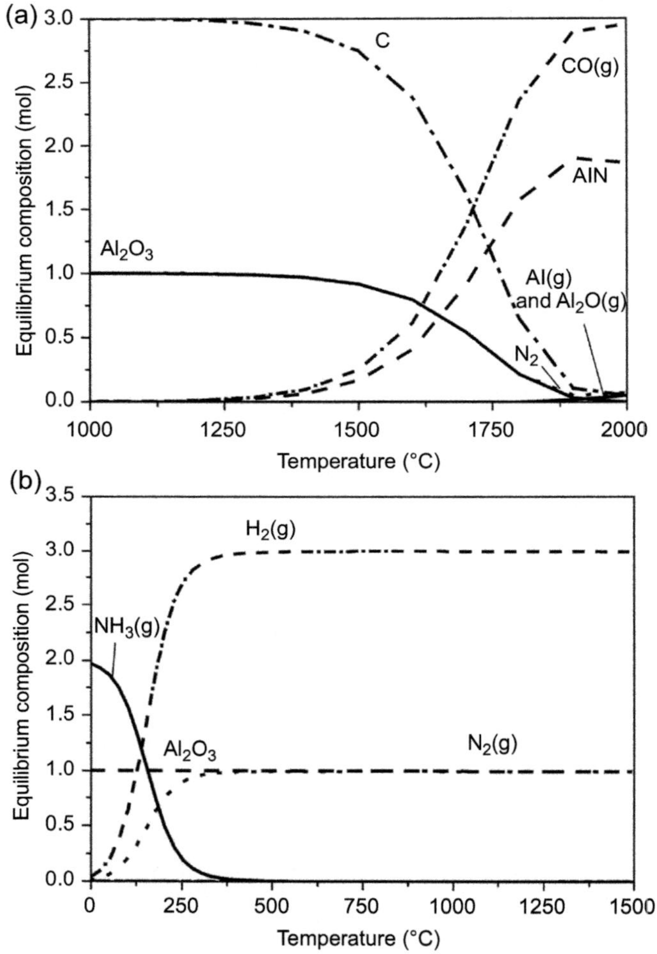

Figure 18.17 Equilibrium compositions for the systems $Al_2O_3 + 3C + N_2$ (a) and $2AlN + 3H_2O$ (b).
Reprinted with permission from [178]. Copyright American Chemical Society.

Cr_2O_3 to CrN occurs at 773−1273K without reducing agent, followed by hydrolysis of Cr_2N in the same temperature range to produce NH_3. Experimentally determined NH_3 yield (relative to the N in the nitride) in the hydrolysis step was below 1 mol% [177]. Another proposed process is based on molybdenum nitride [181]. Thermodynamic and electronic structure analyses of metal reactants for ammonia production can be found in Refs. [14] and [182], respectively.

18.3.5 Ablation

Ablation of solid materials with pulsed lasers can be used to fabricate high-purity nanomaterials suitable for a wide range of potential applications, for example, in nanocomposite materials, as surface coatings, and for medical diagnostics and targeted drug delivery [183]. Pulsed lasers can deliver light pulses at heat fluxes in the order of about 10^6 MW/m^2 and can be used to reach a local plasma near the surface of the ablated material, which enhances the ablation process [184,185]. Heat fluxes theoretically attainable by concentrating solar radiation in air or vacuum are limited to about 46 MW/m^2. Nevertheless, ablation with concentrated solar radiation has been successfully applied to produce different nanostructures.

18.3.5.1 Carbon nanostructures

Carbon fullerenes (C_{60} and C_{70}) can be formed by condensation of carbon vapor. Fullerenes were first produced by vaporizing solid graphite disk samples in a high-density helium flow via laser ablation [186] and later by evaporating graphite electrodes via electric arc discharge into helium at subatmospheric pressure [187,188]. However, scale-up of these processes has proven difficult [2,189].

Fullerenes synthesis via ablation of a graphite sample with high-flux solar radiation in a solar furnace was first demonstrated in 1993 [190,191] and has since been upscaled to a 50-kW solar reactor, producing gram quantities of fullerenes [189,192,193]. A schematic of the solar reactor is shown in Fig. 18.18. It comprises a metallic tube (40-cm

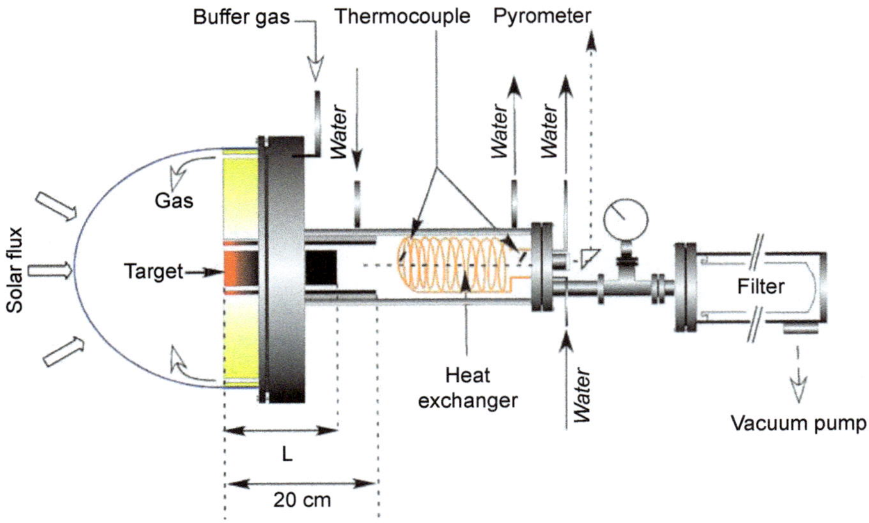

Figure 18.18 50-kW solar reactor for the production of carbon fullerenes.
Reprinted from [189], Copyright with permission from Elsevier.

long, 13-cm i.d.) closed at the front by a quartz dome. The sample is a 10–15-cm long, 6-cm o.d. graphite rod, surrounded by a protective insulated graphite tube. The front end of the carbon rod is vaporized by direct exposure to concentrated solar radiation with a radiative flux of about 8 MW/m^2. Evaporated carbon is entrained by the buffer gas (helium) flow and cooled in a dark water-cooled zone behind the rod to form fullerenes. The process was operated at 0.13–0.5 bar total pressure. Condensed carbon samples were collected from the reactor, the heat exchanger, and the filter. The carbon vaporization rate was 21 g/h, with a mean fullerenes yield of 13.5%. The same reactor setup was also used to produce single-walled carbon nanotubes [194].

18.3.5.2 Inorganic nanostructures

Ablation with high-flux solar radiation has also been used to produce inorganic single- and multi-walled fullerene-like nanostructures and nanotubes, as well as nanospheres, nanofibers, nanowires, and graphene-like nanosheets [12,195–198]. Inorganic materials studied include Cs_2O, Si, $SiO_{2\text{-}x}$, SiC, MoS_2, $MoSe_2$, WS_2, and WSe_2 [197–199]. Initial on-sun experiments with Cs_2O crystallites were conducted in an evacuated quartz ampoule, directly exposed to a radiative flux of 2.5–9.8 MW/m^2 delivered by an optical fiber connected to a solar mini-dish concentrator, shown in Fig. 18.19 (see also Section 18.3.6). TEM images confirmed the presence of nested closed-cage fullerene-like structures with a diameter of about 20–40 nm [195]. Following these initial experiments, solar production of single- and multi-walled MoS_2 nanotubes and closed-cage structures, and amorphous SiO_2 and crystalline Si nanostructures was demonstrated with the same experimental setup (Fig. 18.19). TEM images of the resulting nanostructures are shown in Fig. 18.20 [12].

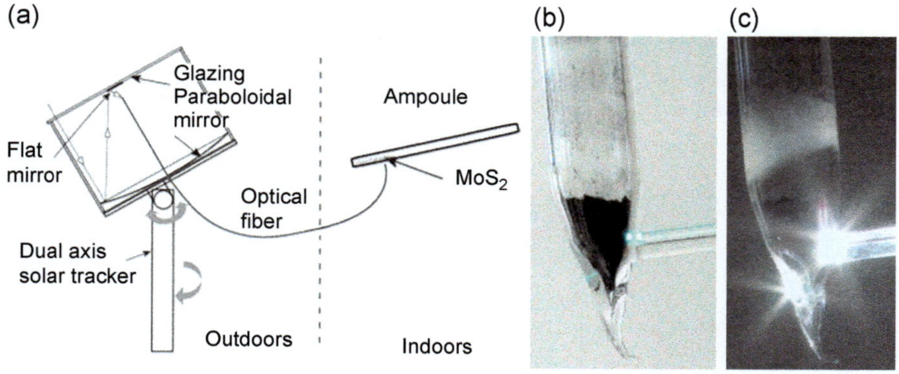

Figure 18.19 Experimental setup used to produce MoS_2, SiO_2, and Si nanostructures: (a) solar dish concentrator connected with an optical fiber to the evacuated quartz ampoule; quartz ampoule containing MoS_2 before (b) and during (c) a solar experiment.
Reproduced from [12], with permission from The Royal Society of Chemistry.

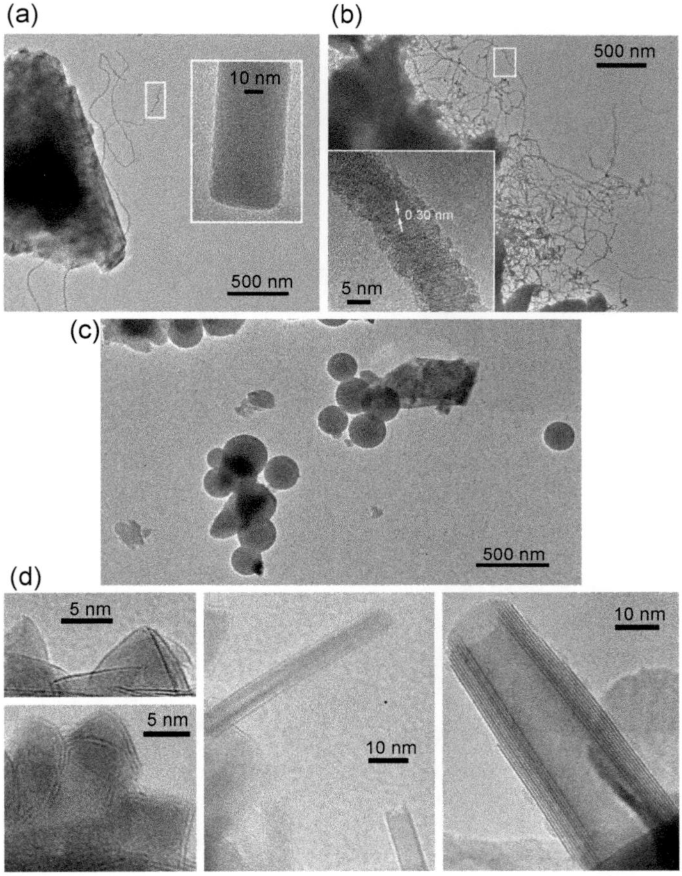

Figure 18.20 TEM images, showing nanofibers (a), nanowires (b), nanospheres (c), and single- and multi-walled closed-cage and tubular nanostructures (d) produced via high-flux ablation with concentrated solar radiation.
Reproduced from [12], with permission from The Royal Society of Chemistry.

18.3.6 Surgery

An optical system, shown in Fig. 18.21, to bundle and transport concentrated solar radiation was designed to provide high-flux solar radiation for solar surgery [200–203]. It uses a 200-mm diameter mini-dish solar concentrator. Concentrated radiation is coupled into a 20-m long 1 mm diameter fused silica optical fiber via a flat beam-down mirror. The system delivers a radiative heat flux of 11–12 MW/m^2 at the distal tip of the fiber. Another optical fiber-based solar energy transmission system was designed, consisting of a bundle of 19 silica optical fibers coupled to a two-stage concentrating system, capable of transmitting up to 60 W of radiative power with an efficiency of 60% [204].

Figure 18.21 Optical system for the transmission of high-flux concentrated solar radiation. Reprinted from [200], Copyright with permission from Elsevier.

18.4 Other thermal processes

In 2013, about 28% of the global final energy consumption was used by the transport sector, while about 29% was used by the industry sector (most energy-intensive industrial sectors: iron and steel, chemical and petrochemical, non-metallic minerals, food, and tobacco), and the remaining 43% were used by other energy and non-energy-related sectors, including the residential sector (23%) [205]. In particular, the industrial and residential sectors include a vast number of low- to medium-temperature process heat applications where solar thermal energy could be used. In order to serve as a generic source of process heat that can be used on demand, solar thermal energy needs to be transferred between collector and application with a heat transfer fluid and stored for periods of low or no solar energy collection and/or high process heat demand.

Table 18.4 provides an overview of solar collector technologies for solar process heat generation, including their typical solar concentration ratios and operating temperature ranges. Stationary and one-axis tracking solar collectors typically use air, water, water/ethylene glycol solutions, thermo-oils, or a refrigerant as the heat transfer medium [13]. Two-axis tracking systems can operate at higher temperatures and typically use molten nitrate salts, water/steam, liquid sodium, or pressurized air as the heat transfer fluid. Low-cost heat transfer fluids such as water or molten salt can be directly stored in an insulated thermal storage tank. Alternatively, inexpensive storage media, such as pebble stones, concrete, phase-change materials (e.g., KNO_3-NaNO_3 mixtures), or reversible chemical reactions can be used [3,13,208,209].

Table 18.5 gives an overview of typical low- to medium-temperature industrial process heat applications along with their temperature requirements. Other applications not listed in Table 18.5 include refrigeration, sea water desalination, extraction operations in metallurgical and chemical processing, evaporation, distillation, dehydration, hydrolyzing, polymerization, cleaning, and more [210,211]. In addition, low- to medium-temperature heat can be used for industrial and domestic water heating and space heating and cooling. The required temperatures for these applications range from 60 to 260°C and hence are best met with low-concentration solar systems, such as flat-plate collectors with stationary (compound parabolic) concentrators, or one-axis tracking systems, such as parabolic trough and linear Fresnel concentrators.

In addition, there exists a range of potential high-temperature applications for solar process heat. For example in Australia, currently large amounts of natural gas are used to generate process heat at temperatures >800°C for the metal processing industry, basic chemicals and ammonia industries, and non-metallic minerals and cement industries [207]. These sectors offer additional large opportunities for the substitution of high-temperature solar process heat generated with two-axis tracking solar collectors.

Table 18.4 Overview of stationary and one- and two-axis sun-tracking solar collectors suitable for process heat generation

Tracking	Collector type	Concentration ratio	Operating temperature (°C)
None	Flat-plate collector	1	30–80
	Evacuated tube collector	1	50–200
	Compound parabolic collector	1–5	60–240
One-axis	Linear Fresnel collector	10–40	60–450
	Parabolic trough collector	10–50	60–450
Two-axis	Solar tower collector	100–1500	100–2000
	Solar dish collector	100–2000	100–2000

Adapted from [206] with modifications based on [207], Copyright with permission from Elsevier.

Table 18.5 Low- to medium-temperature applications of industrial process heat

Industry	Process	Temperature (°C)
Dairy	Pressurization	60–80
	Sterilization	100–120
	Drying	120–180
	Concentration	60–80
	Boiler feed water	60–90
Tinned food	Sterilization	110–120
	Pasteurization	60–80
	Cooking	60–90
	Bleaching	60–90
Textile	Bleaching, dyeing	60–90
	Drying, degreasing	100–130
	Fixing	160–180
	Pressing	80–100
Paper	Cooking, drying	60–80
	Boiler feed water	60–90
	Bleaching	130–150
Chemical	Soaps	200–260
	Synthetic rubber	150–200
	Processing heat	120–180
	Preheating water	60–90
Meat	Washing, sterilization	60–90
	Cooking	90–100
Beverages	Washing, sterilization	60–80
	Pasteurization	60–70
Flours and by-products	Sterilization	60–80
Timber by-products	Thermodiffusion	80–100
	Drying	60–100
	Preheating water	60–90
	Preparation pulp	120–170
Bricks and blocks	Curing	60–140

Table 18.5 Continued

Industry	Process	Temperature (°C)
Plastics	Preparation	120–140
	Distillation	140–150
	Separation	200–220
	Extension	140–160
	Drying	180–200
	Blending	120–140

Adapted from [206], Copyright with permission from Elsevier.

18.5 Other solar processes

Other technologies and applications of concentrated solar radiation, including alternative solar electricity technologies, such as concentrated PV [212] or solar thermoelectrics [213], PV/thermal hybrid solar technologies [214], solar thermal seawater desalination [215], photochemical and photocatalytic water treatment [216,217], solar cooking [218], solar daylighting [219], solar-pumped laser [220], as well as new emerging solar technologies, such as photo-electrochemical H_2O splitting and CO_2 conversion [221], are beyond the scope of this chapter. The reader is referred to the literature cited above and references cited therein.

18.6 Conclusions and future trends

Solar energy is by far the most abundant source of renewable energy and, as such, will naturally play a major role in the future global energy system. Rapidly decreasing costs of PV as well as concentrated solar thermal electricity have resulted in a rapid expansion of solar electric power generation. As a result, to date, solar energy has been mainly associated with electricity production. However, the majority of today's energy supplies come in the form of hydrocarbon fuels rather than electricity. Major energy-intensive sectors and applications include transportation, metal and steel production, cement production, and ammonia production. Concentrated solar energy is a vast but—until today—largely untapped source of highly exergetic energy, capable of providing the required energy input to a broad range of energy-intensive sectors and applications.

This chapter has provided an overview of applications of concentrated solar radiation as a source of high-temperature process heat besides electricity generation. The pertinent scientific literature has been reviewed, with a focus on advances in solar process technology research and development. Technologies and applications have been divided broadly into three main groups, *H_2/CO production, material processing and chemical commodity production*, and *other thermal processes*.

Several processes have been described for the production of syngas (H_2 and CO) as a universal feedstock for the production of synthetic fuels. Concentrated solar energy can be used to produce syngas either via solar upgrading of carbonaceous feedstocks, such as coal, petcoke, methane, and biomass, or via thermochemical, electro-thermochemical, or carbothermal processes that split H_2O and CO_2. While processes involving fossil feedstocks are not CO_2-neutral, they do significantly reduce the CO_2 emissions involved with the use of fossil fuels and provide a viable transition path from today's fossil fuels to tomorrow's CO_2-neutral solar fuels produced from CO_2 and H_2O or other sustainable feedstocks. Two-step thermochemical cycles to split H_2O and CO_2 have been studied intensively in the literature as a long-term technology to produce CO_2-neutral synthetic fuels, due to their relative simplicity and their ability to reduce the temperatures required to dissociate H_2O and CO_2 compared to their direct thermolysis. Non-volatile metal oxides, including ferrites, ceria, and different perovskites, appear as promising oxygen exchange materials due to their ability to release oxygen and split H_2O/CO_2 without changing their solid phase, which simplifies the separation of the gaseous products. In addition, the electro-thermochemical hybrid sulfur cycle appears to be a promising technology to split H_2O, due to its strongly reduced electric energy demand compared to pure water electrolysis.

Concentrated solar energy can further be used in the production of a broad range of commodity materials. Calcination of limestone, the main energy-intensive process in the production of lime (quicklime) and cement, was demonstrated in a 10-kW solar-driven rotary kiln reactor with solar energy conversion efficiencies of up to 35%. The calcination/carbonation cycle with $CaCO_3/CaO$ has further been shown to offer a method to remove CO_2 from flue gases or ambient air and to store solar energy thermochemically. Coproduction of CO and aluminum via the solar carbothermal reduction of aluminum oxide has been discussed as a cleaner alternative to the energy-intensive electrochemical route via the Hall–Héroult process. Other elements that can be produced via the solar carbothermal route include zinc, magnesium, and silicon, as well as hazardous elements, such as lead, chlorine, and cadmium extracted from waste materials. Solar carbothermal reduction of metal oxides can also be conducted to produce metal carbides and nitrides. These products can either be used as high-performance materials or as intermediates to produce fuels and commodity materials. For example, AlN has been used in a solar carbothermal two-step Al_2O_3/AlN cycle to produce ammonia from molecular nitrogen and water. Further, concentrated solar radiation can also be used as a source of high-flux radiation for processes requiring extreme heat fluxes, such as cutting, ablation, and heat treatment. Finally, there is a large potential to integrate solar process heat into low- to medium-temperature industrial processes across a range of industrial sectors, and for domestic water and space heating and cooling.

The research results to date demonstrate the usefulness of concentrated solar radiation as a universal, high-quality heat source, capable of providing virtually unlimited renewable energy for the production of clean electricity as well as fuels and materials.

References

[1] International Energy Agency. Key world energy statistics. 2015.

[2] Meier A, Steinfeld A. Solar energy in thermochemical processing. Encyclopedia of Sustainability Science and Technology 2012:9588−619.

[3] Romero M, Steinfeld A. Concentrating solar thermal power and thermochemical fuels. Energy & Environmental Science 2012;5(11):9234−45.

[4] Bader R, Lipiński W. Solar thermochemical processes. In: Solar energy, Crawley GM, editor, World Scientific Publishing. New Jersey 2016:345−94.

[5] Graves C, Ebbesen SD, Mogensen M. Co-electrolysis of CO_2 and H_2O in solid oxide cells: performance and durability. Solid State Ionics 2011;192(1):398−403.

[6] Roeb M, Thomey D, De Oliveira L, Sattler C, Fleury G, Pra F, Tochon P, Brevet A, Roux G, Gruet N, Mansilla C, Lenaour F, Poitou S, Allen RWK, Elder R, Kargiannakis G, Agrafiotis C, Zygogianni A, Pagkoura C, Konstandopoulos AG, Giaconia A, Sau S, Tarquini P, Haussener S, Steinfeld A, Canadas I, Orden A, Ferrato M. Sulphur based thermochemical cycles: development and assessment of key components of the process. International Journal of Hydrogen Energy 2013;38(14):6197−204.

[7] Palumbo R, Diver RB, Larson C, Coker E, Miller JE, Guertin J, Schoer J, Meyer M, Siegel NP. Solar thermal decoupled water electrolysis process I: proof of concept. Chemical Engineering Science 2012;84:372−80.

[8] Tsuji M, Wada Y, Tamaura Y, Steinfeld A, Kuhn P, Palumbo R. Coal gasification using the ZnO/Zn redox system. Energy & Fuels 1996;10:225−8.

[9] Meier A, Gremaud N, Steinfeld A. Economic evaluation of the industrial solar production of lime. Energy Conversion and Management 2005;46(6):905−26.

[10] Nikulshina V, Hirsch D, Mazzotti M, Steinfeld A. CO_2 capture from air and co-production of H_2 via the $Ca(OH)_2-CaCO_3$ cycle using concentrated solar power−Thermodynamic analysis. Energy 2006;31(12):1715−25.

[11] Murray JP, Steinfeld A, Fletcher EA. Metals, nitrides, and carbides via solar carbothermal reduction of metal oxides. Energy 1995;20(7):695−704.

[12] Gordon JM, Katz EA, Feuermann D, Albu-Yaron A, Levy M, Tenne R. Singular MoS_2, SiO_2 and Si nanostructures—synthesis by solar ablation. Journal of Materials Chemistry 2008;18(4):458−62.

[13] Kalogirou SA. Solar thermal collectors and applications. Progress in Energy and Combustion Science 2004;30(3):231−95.

[14] Michalsky R, Pfromm PH. Thermodynamics of metal reactants for ammonia synthesis from steam, nitrogen and biomass at atmospheric pressure. AIChE Journal 2012;58(10):3203−13.

[15] Fletcher EA, Moen RL. Hydrogen and oxygen from water. Science 1977;197(4308):1050−6.

[16] Traynor AJ, Jensen RJ. Direct solar reduction of CO_2 to fuel: first prototype results. Industrial & Engineering Chemistry Research 2002;41(8):1935−9.

[17] Gálvez ME, Loutzenhiser PG, Hischier I, Steinfeld A. CO_2 splitting via two-step solar thermochemical cycles with Zn/ZnO and FeO/Fe_3O_4 redox reactions: thermodynamic analysis. Energy & Fuels 2008;22(5):3544−50.

[18] Kogan A. Direct solar thermal splitting of water and on-site separation of the products - II. Experimental feasibility study. International Journal of Hydrogen Energy 1998;23(2):89−98.

[19] Ihara S. On the study of hydrogen production from water using solar thermal energy. International Journal of Hydrogen Energy 1980;5(5):527–34.

[20] Diver RB, Pederson S, Kappauf T, Fletcher EA. Hydrogen and oxygen from water—VI. Quenching the effluent from a solar furnace. Energy 1983;8(12):947–55.

[21] Abanades S, Charvin P, Flamant G, Neveu P. Screening of water-splitting thermochemical cycles potentially attractive for hydrogen production by concentrated solar energy. Energy 2006;31:2469–86.

[22] Perret R. Solar thermochemical hydrogen production research (STCH) thermochemical cycle selection and investment priority. Sandia Report 2011. SAND2011–3622.

[23] Agrafiotis C, Roeb M, Sattler C. A review on solar thermal syngas production via redox pair-based water/carbon dioxide splitting thermochemical cycles. Renewable & Sustainable Energy Reviews 2015;42:254–85.

[24] Nakamura T. Hydrogen production from water utilizing solar heat at high temperatures. Solar Energy 1977;19(5):467–75.

[25] Palumbo R, Lédé J, Boutin O, Ricart EE, Steinfeld A, Möller S, Weidenkaff A, Fletcher EA, Bielicki J. The production of Zn from ZnO in a high-temperature solar decomposition quench process — I. The scientific framework for the process. Chemical Engineering Science 1998;53(14):2503–17.

[26] Chueh WC, Falter C, Abbott M, Scipio D, Furler P, Haile SM, Steinfeld A. High-flux solar-driven thermochemical dissociation of CO_2 and H_2O using nonstoichiometric ceria. Science 2010;330(6012):1797–801.

[27] Furler P, Scheffe J, Gorbar M, Moes L, Vogt U, Steinfeld A. Solar thermochemical CO_2 splitting utilizing a reticulated porous ceria redox system. Energy & Fuels 2012;26(11): 7051–9.

[28] Scheffe JR, Li J, Weimer AW. A spinel ferrite/hercynite water-splitting redox cycle. International Journal of Hydrogen Energy 2010;35(8):3333–40.

[29] Scheffe JR, Weibel D, Steinfeld A. Lanthanum – strontium – manganese perovskites as redox materials for solar thermochemical splitting of H_2O and CO_2. Energy & Fuels 2013;27(8):4250–7.

[30] Kubo S, Nakajima H, Kasahara S, Higashi S, Masaki T, Abe H, Onuki K. A demonstration study on a closed-cycle hydrogen production by the thermochemical water-splitting iodine - sulfur process. Nuclear Engineering and Design 2004;233(1–3): 347–54.

[31] Sturzenegger M, Nüesch P. Efficiency analysis for a manganese-oxide-based thermochemical cycle. Energy 1999;24(11):959–70.

[32] Sakurai M, Bilgen E, Tsutsumi A, Yoshida K. Solar UT-3 thermochemical cycle for hydrogen production. Solar Energy 1996;57(1):51–8.

[33] Scheffe JR, Steinfeld A. Oxygen exchange materials for solar thermochemical splitting of H_2O and CO_2: a review. Materials Today 2014;17(7):341–8.

[34] Kodama T, Gokon N. Thermochemical cycles for high-temperature solar hydrogen production. Chemical Reviews 2007;107(10):4048–77.

[35] Muhich CL, Ehrhart BD, Al-Shankiti I, Ward BJ, Musgrave CB, Weimer AW. A review and perspective of efficient hydrogen generation via solar thermal water splitting. Wiley Interdisciplinary Reviews: Energy and Environment 2016;5(3):261–87.

[36] Allendorf MD, Diver RB, Siegel NP, Miller JE. Two-step water splitting using mixed-metal ferrites: thermodynamic analysis and characterization of synthesized materials. Energy & Fuels 2008;22(6):4115–24.

[37] Meng Q-L, Lee C, Ishihara T, Kaneko H, Tamaura Y. Reactivity of CeO_2-based ceramics for solar hydrogen production via a two-step water-splitting cycle with concentrated solar energy. International Journal of Hydrogen Energy 2011;36(21):13435−41.
[38] Le Gal A, Abanades S, Flamant G. CO_2 and H_2O splitting for thermochemical production of solar fuels using nonstoichiometric ceria and ceria/zirconia solid solutions. Energy & Fuels 2011;25(10):4836−45.
[39] Le Gal A, Bion N, Le Mercier T, Harle V. Reactivity of doped ceria-based mixed oxides for solar thermochemical hydrogen generation via two-step water-splitting cycles. Energy & Fuels 2013;27(10):6068−78.
[40] Scheffe J, Welte M, Steinfeld A. Reduction of cerium dioxide in an aerosol tubular reactor for the thermal dissociation of CO_2 and H_2O two-step solar thermochemical cycle. Industrial & Engineering Chemistry Research 2014;53(6):2175−82.
[41] Furler P, Scheffe J, Marxer D, Gorbar M, Bonk A, Vogt U, Steinfeld A. Thermochemical CO_2 splitting via redox cycling of ceria reticulated foam structures with dual-scale porosities. Physical Chemistry Chemical Physics 2014;16:10503−11.
[42] Venstrom LJ, De Smith RM, Hao Y, Haile SM, Davidson JH. Efficient splitting of CO_2 in an isothermal redox cycle based on ceria. Energy & Fuels 2014;28(4):2732−42.
[43] Venstrom LJ, Petkovich N, Rudisill S, Stein A, Davidson JH. The effects of morphology on the oxidation of ceria by water and carbon dioxide. Journal of Solar Energy Engineering 2012;134(1):011005/1−8.
[44] McDaniel AH, Miller EC, Arifin D, Ambrosini A, Coker EN, O'Hayre R, Chueh WC, Tong J. Sr- and Mn-doped $LaAlO_{3-\delta}$ for solar thermochemical H_2 and CO production. Energy & Environmental Science 2013;6(8):2424.
[45] Demont A, Abanades S. High redox activity of Sr-substituted lanthanum manganite perovskites for two-step thermochemical dissociation of CO_2. RSC Advances 2014;4: 54885−91.
[46] Demont A, Abanades S, Beche E. Investigation of perovskite structures as oxygen-exchange redox materials for hydrogen production from thermochemical two-step water-splitting cycles. The Journal of Physical Chemistry C 2014;118(24):12682−92.
[47] Yang C-K, Yamazaki Y, Aydin A, Haile SM. Thermodynamic and kinetic assessments of strontium-doped lanthanum manganite perovskites for two-step thermochemical water splitting. Journal of Materials Chemistry A 2014;2:13612−23.
[48] Jiang Q, Tong J, Zhou G, Jiang Z, Li Z, Li C. Thermochemical CO_2 splitting reaction with supported $La_xA_{1-x}Fe_yB_{1-y}O_3$ (A=Sr, Ce, B=Co, Mn; $0\leq x, y\leq 1$) perovskite oxides. Solar Energy 2014;103:425−37.
[49] Demont A, Abanades S. Solar thermochemical conversion of CO_2 into fuel via two-step redox cycling of non-stoichiometric Mn-containing perovskite oxides. Journal of Materials Chemistry A 2015;3:3536−46.
[50] Dey S, Naidu BS, Rao CNR. $Ln_{0.5}A_{0.5}MnO_3$ (Ln=Lanthanide, A=Ca, Sr) perovskites exhibiting remarkable performance in the thermochemical generation of CO and H_2 from CO_2 and H_2O. Chemistry - A European Journal 2015;21:7077−81.
[51] Bork AH, Kubicek M, Struzik M, Rupp JLM. Perovskite $La_{0.6}Sr_{0.4}Cr_{1-x}Co_xO_{3-\delta}$ solid solutions for solar-thermochemical fuel production: strategies to lower the operation temperature. Journal of Materials Chemistry A 2015;3(30):15546−57.
[52] Gálvez ME, Jacot R, Scheffe J, Cooper T, Patzke G, Steinfeld A. Physico-chemical changes in Ca, Sr and Al-doped La−Mn−O perovskites upon thermochemical splitting of CO_2 via redox cycling. Physical Chemistry Chemical Physics 2015;17:6629−34.

[53] Meredig B, Wolverton C. First-principles thermodynamic framework for the evaluation of thermochemical H_2O- or CO_2-splitting materials. Physical Review B - Condensed Matter and Materials Physics 2009;80. 245119/1−8.

[54] Deml AM, Stevanović V, Muhich CL, Musgrave CB, O'Hayre R. Oxide enthalpy of formation and band gap energy as accurate descriptors of oxygen vacancy formation energetics. Energy & Environmental Science 2014;7:1996−2004.

[55] Muhich CL, Aston VJ, Trottier RM, Weimer AW, Musgrave CB. First-principles analysis of cation diffusion in mixed metal ferrite spinels. Chemistry of Materials 2016;28(1): 214−26.

[56] Haussener S, Coray P, Lipinski W, Wyss P, Steinfeld A. Tomography-based heat and mass transfer characterization of reticulate porous ceramics for high-temperature processing. Journal of Heat Transfer 2010;132(2):023305.

[57] Keene DJ, Lipiński W, Davidson JH. The effects of morphology on the thermal reduction of nonstoichiometric ceria. Chemical Engineering Science 2014;111:213−43.

[58] Wheeler VM, Randrianalisoa J, Tamma K, Lipinski W. Spectral radiative properties of three-dimensionally ordered macroporous ceria particles. Journal of Quantitative Spectroscopy and Radiative Transfer 2014;143:63−72.

[59] Randrianalisoa J, Lipinski W. Effect of pore-level geometry on far-field radiative properties of three-dimensionally ordered macroporous ceria particle. Applied Optics 2014; 53(7):1290−7.

[60] Rudisill SG, Venstrom LJ, Petkovich ND, Quan T, Hein N, Boman DB, Davidson JH, Stein A. Enhanced oxidation kinetics in thermochemical cycling of CeO_2 through templated porosity. The Journal of Physical Chemistry C 2013;117(4):1692−700.

[61] Gao, X., Vidal, A., Bayon Sandoval, A., Bader, R., Hinkley, J., Lipiński, W., and Tricoli, A., Efficient ceria nanostructures for enhanced solar fuel production via high-temperature thermochemical redox cycles. Journal of Materials Chemistry A 2016; 4:9614−24.

[62] Suter S, Steinfeld A, Haussener S. Pore-level engineering of macroporous media for increased performance of solar-driven thermochemical fuel processing. International Journal of Heat and Mass Transfer 2014;78:688−98.

[63] Steinfeld A, Sanders S, Palumbo R. Design aspects of solar thermochemical engineering—a case study: two-step water-splitting cycle using the Fe_3O_4/FeO redox system. Solar Energy 1999;65(1):43−53.

[64] Loutzenhiser PG, Steinfeld A. Solar syngas production from CO_2 and H_2O in a two-step thermochemical cycle via Zn/ZnO redox reactions: thermodynamic cycle analysis. International Journal of Hydrogen Energy 2011;36(19):12141−7.

[65] Lapp J, Davidson JH, Lipiński W. Efficiency of two-step solar thermochemical non-stoichiometric redox cycles with heat recovery. Energy 2012;37(1):591−600.

[66] Hao Y, Yang C-K, Haile SM. High-temperature isothermal chemical cycling for solar-driven fuel production. Physical Chemistry Chemical Physics 2013;15:17084−92.

[67] Bader R, Venstrom LJ, Davidson JH, Lipiński W. Thermodynamic analysis of isothermal redox cycling of ceria for solar fuel production. Energy & Fuels 2013;27(9):5533−44.

[68] Ermanoski I. Cascading pressure thermal reduction for efficient solar fuel production. International Journal of Hydrogen Energy 2014;39(25):13114−7.

[69] Felinks J, Brendelberger S, Roeb M, Sattler C, Pitz-Paal R. Heat recovery concept for thermochemical processes using a solid heat transfer medium. Applied Thermal Engineering 2014;73(1):1004−11.

[70] Brendelberger S, Sattler C. Concept analysis of an indirect particle-based redox process for solar-driven H_2O/CO_2 splitting. Solar Energy 2015;113:158−70.

[71] Bulfin B, Call F, Lange M, Sattler C, Pitz-Paal R, Shvets IV. Thermodynamics of CeO_2 thermochemical fuel production. Energy & Fuels 2014;29(2):1001−9.

[72] Roeb M, Sattler C, Klüser R, Monnerie N, de Oliveira L, Konstandopoulos AG, Agrafiotis C, Zaspalis VT, Nalbandian L, Steele A, Stobbe P. Solar hydrogen production by a two-step cycle based on mixed iron oxides. Journal of Solar Energy Engineering 2006;128(2):125−33.

[73] Roeb M, Säck J-P, Rietbrock P, Prahl C, Schreiber H, Neises M, de Oliveira L, Graf D, Ebert M, Reinalter W, Meyer-Grünefeldt M, Sattler C, Lopez A, Vidal A, Elsberg A, Stobbe P, Jones D, Steele A, Lorentzou S, Pagkoura C, Zygogianni A, Agrafiotis C, Konstandopoulos AG. Test operation of a 100 kW pilot plant for solar hydrogen production from water on a solar tower. Solar Energy 2011;85(4):634−44.

[74] Diver RB, Miller JE, Allendorf MD, Siegel NP, Hogan RE. Solar thermochemical water-splitting ferrite-cycle heat engines. Journal of Solar Energy Engineering 2008;130(4): 041001/1−8.

[75] Gokon N, Mataga T, Kondo N, Kodama T. Thermochemical two-step water splitting by internally circulating fluidized bed of $NiFe_2O_4$ particles: successive reaction of thermal-reduction and water-decomposition steps. International Journal of Hydrogen Energy 2011;36(8):4757−67.

[76] Loutzenhiser PG, Meier A, Steinfeld A. Review of the two-step H_2O/CO_2-splitting solar thermochemical cycle based on Zn/ZnO redox reactions. Materials 2010;3(11):4922−38.

[77] Villasmil W, Brkic M, Wuillemin D, Meier A, Steinfeld A. Pilot scale demonstration of a 100-kW_{th} solar thermochemical plant for the thermal dissociation of ZnO. Journal of Solar Energy Engineering 2014;136:011017/1−11.

[78] Koepf E, Advani SG, Steinfeld A, Prasad AK. A novel beam-down, gravity-fed, solar thermochemical receiver/reactor for direct solid particle decomposition: design, modeling, and experimentation. International Journal of Hydrogen Energy 2012;37(22): 16871−87.

[79] Melchior T, Piatkowski N, Steinfeld A. H_2 production by steam-quenching of Zn vapor in a hot-wall aerosol flow reactor. Chemical Engineering Science 2009;64:1095−101.

[80] Bader R, Bala Chandran R, Sedler S, De Smith R, Venstrom LJ, Banerjee A, Chase TR, Davidson JH, Lipiński W. Design of a solar reactor to split CO_2 via isothermal redox cycling of ceria. Journal of Solar Energy Engineering 2015;137:031007/1−10.

[81] Ermanoski I, Siegel NP, Stechel EB. A new reactor concept for efficient solar-thermochemical fuel production. Journal of Solar Energy Engineering 2013;135(3): 031002/1−10.

[82] Martinek J, Viger R, Weimer AW. Transient simulation of a tubular packed bed solar receiver for hydrogen generation via metal oxide thermochemical cycles. Solar Energy 2014;105:613−31.

[83] Houaijia A, Sattler C, Roeb M, Lange M, Breuer S, Säck JP. Analysis and improvement of a high-efficiency solar cavity reactor design for a two-step thermochemical cycle for solar hydrogen production from water. Solar Energy 2013;97:26−38.

[84] Perkins C, Lichty PR, Weimer AW. Thermal ZnO dissociation in a rapid aerosol reactor as part of a solar hydrogen production cycle. International Journal of Hydrogen Energy 2008;33(2):499−510.

[85] Melchior T, Perkins C, Weimer AW, Steinfeld A. A cavity-receiver containing a tubular absorber for high-temperature thermochemical processing using concentrated solar energy. International Journal of Thermal Sciences 2008;47(11):1496−503.

[86] Kaneko H, Miura T, Fuse A, Ishihara H, Taku S, Fukuzumi H, Naganuma Y, Tamaura Y. Rotary-type solar reactor for solar hydrogen production with two-step water splitting-process. Energy Fuels 2007;21(4):2287–93.

[87] Miller JE, Allendorf MD, Ambrosini A, Chen KS, Coker EN, Dedrick DE, Diver RB, Hogan RE, Ermanoski I, Johnson TA, Kellogg GL, McDaniel AH, Siegel NP, Staiger CL, Ellen B. Reimagining liquid transportation fuels : sunshine to petrol. Sandia Report 2012. SAND2012–0307.

[88] Lapp J, Davidson JH, Lipiński W. Heat transfer analysis of a solid-solid heat recuperation system for solar-driven nonstoichiometric redox cycles. Journal of Solar Energy Engineering 2013;135:031004/1–11.

[89] Agrafiotis C, Roeb M, Konstandopoulos AG, Nalbandian L, Zaspalis VT, Sattler C, Stobbe P, Steele AM. Solar water splitting for hydrogen production with monolithic reactors. Solar Energy 2005;79(4):409–21.

[90] Haueter P, Moeller S, Palumbo R, Steinfeld A. The production of zinc by thermal dissociation of zinc oxide - solar chemical reactor design. Solar Energy 1999;67(1–3): 161–7.

[91] Schunk LO, Haeberling P, Wepf S, Wuillemin D, Meier A, Steinfeld A. A receiver-reactor for the solar thermal dissociation of zinc oxide. Journal of Solar Energy Engineering 2008;130(2):021009/1–6.

[92] Schunk LO, Lipiński W, Steinfeld A. Heat transfer model of a solar receiver-reactor for the thermal dissociation of ZnO—Experimental validation at 10 kW and scale-up to 1 MW. Chemical Engineering Journal 2009;150(2–3):502–8.

[93] Hathaway BJ, Bala Chandran R, Sedler SJ, Thomas D, Gladen A, Chase TR, Davidson JH. Effect of flow rates on operation of a solar thermochemical reactor for splitting CO_2 via the isothermal ceria redox cycle. Journal of Solar Energy Engineering 2015;138(1). 011007/1–12.

[94] O'Brien JE, McKellar MG, Stoots CM, Herring JS, Hawkes GL. Parametric study of large-scale production of syngas via high-temperature co-electrolysis. International Journal of Hydrogen Energy 2009;34(9):4216–26.

[95] Herring JS, O'Brien JE, Stoots CM, Hawkes GL, Hartvigsen JJ, Shahnam M. Progress in high-temperature electrolysis for hydrogen production using planar SOFC technology. International Journal of Hydrogen Energy 2007;32(4):440–50.

[96] Abraham BM, Schreiner F. General principles underlying chemical cycles which thermally decompose water into the elements. Industrial & Engineering Chemistry Fundamentals 1974;13(4):305–10.

[97] Thomey D, De Oliveira L, Säck JP, Roeb M, Sattler C. Development and test of a solar reactor for decomposition of sulphuric acid in thermochemical hydrogen production. International Journal of Hydrogen Energy 2012;37(21):16615–22.

[98] Bayer Botero N, Thomey D, Guerra Niehoff A, Roeb M, Sattler C, Pitz-Paal R. Modelling and scaling analysis of a solar reactor for sulphuric acid cracking in a hybrid sulphur cycle process for thermochemical hydrogen production. International Journal of Hydrogen Energy 2016;41(19):8008–19.

[99] Krenzke PT, Davidson JH. Thermodynamic analysis of syngas production via the solar thermochemical cerium oxide redox cycle with methane-driven reduction. Energy & Fuels 2014;28(6):4088–95.

[100] Chueh WC, Haile SM. A thermochemical study of ceria: exploiting an old material for new modes of energy conversion and CO_2 mitigation. Philosophical Transactions. Series A, Mathematical, Physical, and Engineering Sciences 2010;368(1923): 3269–94.

[101] von Zedtwitz P, Petrasch J, Trommer D, Steinfeld A. Hydrogen production via the solar thermal decarbonization of fossil fuels. Solar Energy 2006;80(10):1333–7.
[102] Piatkowski N, Wieckert C, Weimer AW, Steinfeld A. Solar-driven gasification of carbonaceous feedstock—a review. Energy & Environmental Science 2011;4(1): 73–82.
[103] Kodama T. High-temperature solar chemistry for converting solar heat to chemical fuels. Progress in Energy and Combustion Science 2003;29(6):567–97.
[104] von Zedtwitz P, Steinfeld A. The solar thermal gasification of coal - energy conversion efficiency and CO_2 mitigation potential. Energy 2003;28(5):441–56.
[105] Ng YC, Lipiński W. Thermodynamic analyses of solar thermal gasification of coal for hybrid solar-fossil power and fuel production. Energy 2012;44(1):720–31.
[106] Müller R, von Zedtwitz P, Wokaun A, Steinfeld A. Kinetic investigation on steam gasification of charcoal under direct high-flux irradiation. Chemical Engineering Science 2003;58(22):5111–9.
[107] Trommer D, Steinfeld A. Kinetic modeling for the combined pyrolysis and steam gasification of petroleum coke and experimental determination of the rate constants by dynamic thermogravimetry in the 500–1520 K range. Energy & Fuels 2006;20(3): 1250–8.
[108] Trommer D, Noembrini F, Fasciana M, Rodriguez D, Morales A, Romero M, Steinfeld A. Hydrogen production by steam-gasification of petroleum coke using concentrated solar power - I. Thermodynamic and kinetic analyses. International Journal of Hydrogen Energy 2005;30(6):605–18.
[109] Lipiński W, Davidson JH, Haussener S, Klausner JF, Mehdizadeh AM, Petrasch J, Steinfeld A, Venstrom L. Review of heat transfer research for solar thermochemical applications. Journal of Thermal Science and Engineering Applications 2013;5(2): 021005.
[110] von Zedtwitz P, Lipiński W, Steinfeld A. Numerical and experimental study of gas–particle radiative heat exchange in a fluidized-bed reactor for steam-gasification of coal. Chemical Engineering Science 2007;62(1–2):599–607.
[111] Melchior T, Perkins C, Lichty P, Weimer AW, Steinfeld A. Solar-driven biochar gasification in a particle-flow reactor. Chemical Engineering and Processing 2009;48(8): 1279–87.
[112] Z'Graggen A, Haueter P, Trommer D, Romero M, de Jesus JC, Steinfeld A. Hydrogen production by steam-gasification of petroleum coke using concentrated solar power—II reactor design, testing, and modeling. International Journal of Hydrogen Energy 2006; 31(6):797–811.
[113] Piatkowski N, Wieckert C, Steinfeld A. Experimental investigation of a packed-bed solar reactor for the steam-gasification of carbonaceous feedstocks. Fuel Processing Technology 2009;90(3):360–6.
[114] Hathaway BJ, Davidson JH, Kittelson DB. Solar gasification of biomass: kinetics of pyrolysis and steam gasification in molten salt. Journal of Solar Energy Engineering 2011;133(2):021011/1–9.
[115] Chen J, Lu Y, Guo L, Zhang X, Xiao P. Hydrogen production by biomass gasification in supercritical water using concentrated solar energy: system development and proof of concept. International Journal of Hydrogen Energy 2010;35(13):7134–41.
[116] Maag G, Lipiński W, Steinfeld A. Particle-gas reacting flow under concentrated solar irradiation. International Journal of Heat and Mass Transfer 2009;52(21–22): 4997–5004.

[117] Maag G, Zanganeh G, Steinfeld A. Solar thermal cracking of methane in a particle-flow reactor for the co-production of hydrogen and carbon. International Journal of Hydrogen Energy 2009;34(18):7676–85.
[118] Kogan A, Kogan M, Barak S. Production of hydrogen and carbon by solar thermal methane splitting. III. Fluidization, entrainment and seeding powder particles into a volumetric solar receiver. International Journal of Hydrogen Energy 2005;30(1):35–43.
[119] Abanades S, Flamant G. Experimental study and modeling of a high-temperature solar chemical reactor for hydrogen production from methane cracking. International Journal of Hydrogen Energy 2007;32(10–11):1508–15.
[120] Dahl JK, Buechler KJ, Weimer AW, Lewandowski A, Bingham C. Solar-thermal dissociation of methane in a fluid-wall aerosol flow reactor. International Journal of Hydrogen Energy 2004;29(7):725–36.
[121] Wyss J, Martinek J, Kerins M, Dahl JK, Weimer A, Lewandowski A, Bingham C. Rapid solar-thermal decarbonization of methane in a fluid-wall aerosol flow reactor: fundamentals and application. International Journal of Chemical Reactor Engineering 2007; 5(1):A69.
[122] Rodat S, Abanades S, Flamant G. Experimental evaluation of indirect heating tubular reactors for solar methane pyrolysis. International Journal of Chemical Reactor Engineering 2010;8(1):A25.
[123] Möller S, Kaucic D, Sattler C. Hydrogen production by solar reforming of natural gas: a comparison study of two possible process configurations. Journal of Solar Energy Engineering 2006;128:16–23.
[124] Wörner A, Tamme R. CO_2 reforming of methane in a solar driven volumetric receiver–reactor. Catalysis Today 1998;46(2–3):165–74.
[125] Tamme R, Buck R, Epstein M, Fisher U, Sugarmen C. Solar upgrading of fuels for generation of electricity. Journal of Solar Energy Engineering 2001;123(2):160–3.
[126] Ozalp N, Shilapuram V. Step-by-step methodology of developing a solar reactor for emission-free generation of hydrogen. International Journal of Hydrogen Energy 2010; 35(10):4484–95.
[127] Agrafiotis C, Von Storch H, Roeb M, Sattler C. Solar thermal reforming of methane feedstocks for hydrogen and syngas production—A review. Renew Sustain Energy Rev 2014;29:656–82.
[128] Meier A, Bonaldi E, Cella GM, Lipinski W, Wuillemin D. Solar chemical reactor technology for industrial production of lime. Solar Energy 2006;80(10):1355–62.
[129] Meier A, Bonaldi E, Cella GM, Lipinski W, Wuillemin D, Palumbo R. Design and experimental investigation of a horizontal rotary reactor for the solar thermal production of lime. Energy 2004;29:811–21.
[130] Nikulshina V, Halmann M, Steinfeld A. Coproduction of syngas and lime by combined $CaCO_3$-calcination and CH_4-reforming using a particle-flow reactor driven by concentrated solar radiation. Energy & Fuels 2009;23(12):6207–12.
[131] Matthews L, Lipinski W. Thermodynamic analysis of solar thermochemical CO_2 capture via carbonation/calcination cycle with heat recovery. Energy 2012;45:900–7.
[132] Reich L, Melmoth L, Gresham R, Simon T, Lipiński W. Design of a solar thermochemical reactor for calcium oxide based carbon dioxide capture. In: Proceedings of the ASME 2015 Power & Energy Conference, San Diego, PowerEnergy 2015–49507; 2015.
[133] Reich L, Bader R, Simon TW, Lipiński W. Thermal transport model of a packed-bed reactor for solar thermochemical CO_2 capture. Special Topics & Reviews in Porous Media–An International Journal 2015;6(2):197–209.

[134] Flamant G, Hernandez D, Bonet C. Experimental aspects of the thermochemical conversion of solar energy decarbonation of $CaCO_3$. Solar Energy 1980;24:385–95.
[135] Nikulshina V, Gebald C, Steinfeld A. CO_2 capture from atmospheric air via consecutive CaO-carbonation and $CaCO_3$-calcination cycles in a fluidized-bed solar reactor. Chemical Engineering Journal 2009;146(2):244–8.
[136] Imhof A. The cyclone reactor - an atmospheric open solar reactor. Solar Energy Materials 1991;24:733–41.
[137] Steinfeld A, Imhof A, Mischler D. Experimental investigation of an atmospheric-open cyclone solar reactor for solid-gas thermochemical reactions. Journal of Solar Energy Engineering 1992;114:171–4.
[138] Kruesi M, Galvez ME, Halmann M, Steinfeld A. Solar aluminum production by vacuum carbothermal reduction of alumina-thermodynamic and experimental analyses. Metallurgical and Materials Transactions B 2011;42(1):254–60.
[139] Steinfeld A. Solar thermochemical production of hydrogen—a review. Solar Energy 2005;78(5):603–15.
[140] Steinfeld A, Meier A. Solar fuels and materials. Encyclopedia of Energy. Amsterdam: Elsevier; 2004.
[141] Halmann M, Frei A, Steinfeld A. Vacuum carbothermic reduction of Al_2O_3, BeO, MgO-CaO, TiO_2, ZrO_2, HfO_2+ZrO_2, SiO_2, SiO_2+Fe_2O_3, and GeO_2 to the metals. A thermodynamic study. Mineral Processing and Extractive Metallurgy Review 2011; 32(4):247–66.
[142] Loutzenhiser PG, Tuerk O, Steinfeld A. Production of Si by vacuum carbothermal reduction of SiO_2 using concentrated solar energy. JOM 2010;62(9):49–54.
[143] Levêque G, Abanades S. Investigation of thermal and carbothermal reduction of volatile oxides (ZnO, SnO_2, GeO_2, and MgO) via solar-driven vacuum thermogravimetry for thermochemical production of solar fuels. Thermochim Acta 2015;605:86–94.
[144] Chubukov BA, Palumbo AW, Rowe SC, Hischier I, Groehn AJ, Weimer AW. Pressure dependent kinetics of magnesium oxide carbothermal reduction. Thermochim Acta 2016; 636:23–32.
[145] Halmann M, Frei A, Steinfeld A. Magnesium production by the pidgeon process involving dolomite calcination and MgO silicothermic reduction: thermodynamic and environmental analyses. Industrial & Engineering Chemistry Research 2008;47(7):2146–54.
[146] Sheline W, Matthews L, Lindeke N, Duncan S, Palumbo R. An exploratory study of the solar thermal electrolytic production of Mg from MgO. Energy 2013;51: 163–70.
[147] Halmann M, Epstein M, Steinfeld A. Carbothermic reduction of alumina by natural gas to aluminum and syngas: a thermodynamic study. Mineral Processing and Extractive Metallurgy Review 2012;33(5):352–61.
[148] Halmann M, Epstein M, Steinfeld A. Vacuum carbothermic reduction of bauxite components: a thermodynamic study. Mineral Processing and Extractive Metallurgy Review 2012;33(3):190–203.
[149] Halmann M, Frei A, Steinfeld A. Carbothermal reduction of alumina: thermochemical equilibrium calculations and experimental investigation. Energy 2007;32(12): 2420–7.
[150] Gálvez ME, Frei A, Meier F, Steinfeld A. Production of AlN by carbothermal and methanothermal reduction of Al_2O_3 in a N_2 flow using concentrated thermal radiation. Industrial & Engineering Chemistry Research 2009;48:528–33.
[151] Commodity summary for zinc. [Online]. Available: www.usgs.gov.

[152] Fletcher EA, Noring JE. High temperature solar electrothermal processing - zinc from zinc oxide. Energy 1983;8(4):247−54.

[153] Steinfeld A, Frei A, Kuhn P, Wuillemin D. Solar thermal production of zinc and syngas via combined ZnO-reduction and CH_4-reforming processes. International Journal of Hydrogen Energy 1995;20(10):793−804.

[154] Steinfeld A, Kuhn P, Reller A, Palumbo R, Murray J, Tamaura Y. Solar processed metals as clean energy carriers and water-splitters. International Journal of Hydrogen Energy 1998;23:767−74.

[155] Steinfeld A. Solar hydrogen production via a two-step water-splitting thermochemical cycle based on Zn/ZnO redox reactions. International Journal of Hydrogen Energy 2002; 27(6):611−9.

[156] Abanades S, Charvin P, Flamant G. Design and simulation of a solar chemical reactor for the thermal reduction of metal oxides: case study of zinc oxide dissociation. Chemical Engineering Science 2007;62(22):6323−33.

[157] Schroeder R, Matthews L, Leatzow D, Kndratko J, Will J, Duncan S, Sheline W, Lindeke N, Palumbo RD, Neves P. Solar thermal electrolytic process for the production of Zn from ZnO: the electrolysis of ZnO from 1275−1500 K. Journal of Solar Energy Engineering 2011;133(4):041013/1−11.

[158] Steinfeld A, Palumbo R. Solar thermochemical process technology. Encyclopedia of Physical Science and Technology 2001;15:237−56.

[159] Gstoehl D, Brambilla A, Schunk LO, Steinfeld A. A quenching apparatus for the gaseous products of the solar thermal dissociation of ZnO. Journal of Materials Science 2008; 43(14):4729−36.

[160] Müller R, Steinfeld A. H_2O-splitting thermochemical cycle based on ZnO/Zn-redox: quenching the effluents from the ZnO dissociation. Chemical Engineering Science 2008; 63(1):217−27.

[161] Schunk LO, Lipinski W, Steinfeld A. Ablative heat transfer in a shrinking packed bed of ZnO undergoing solar thermal dissociation. AIChE Journal 2009;55(7): 1659−66.

[162] Dombrovsky L, Schunk L, Lipiński W, Steinfeld A. An ablation model for the thermal decomposition of porous zinc oxide layer heated by concentrated solar radiation. International Journal of Heat and Mass Transfer 2009;52(11−12):2444−52.

[163] Villasmil W, Meier A, Steinfeld A. Dynamic modeling of a solar reactor for zinc oxide thermal dissociation and experimental validation using IR thermography. Journal of Solar Energy Engineering 2014;136(1):010901/1−11.

[164] Müller R, Lipiński W, Steinfeld A. Transient heat transfer in a directly-irradiated solar chemical reactor for the thermal dissociation of ZnO. Applied Thermal Engineering 2008; 28(5−6):524−31.

[165] Möller S, Palumbo R. The development of a solar chemical reactor for the direct thermal dissociation of zinc oxide. Journal of Solar Energy Engineering 2001;123(2):83.

[166] Adinberg R, Epstein M. Experimental study of solar reactors for carboreduction of zinc oxide. Energy 2004;29(5−6):757−69.

[167] Osinga T, Frommherz U, Steinfeld A, Wieckert C. Experimental investigation of the solar carbothermic reduction of ZnO using a two-cavity solar reactor. Journal of Solar Energy Engineering 2004;126(1):633−7.

[168] Kräupl S, Frommherz U, Wieckert C. Solar carbothermic reduction of ZnO in a two-cavity reactor: laboratory experiments for a reactor scale-up. Journal of Solar Energy Engineering 2005;128(1):8−15.

[169] Wieckert C, Frommherz U, Kräupl S, Guillot E, Olalde G, Epstein M, Santen S, Osinga T, Steinfeld A. A 300 kW solar chemical pilot plant for the carbothermic production of zinc. Journal of Solar Energy Engineering 2007;129(2):190−6.

[170] Epstein M, Olalde G, Santen S, Steinfeld A, Wieckert C. Towards the industrial solar carbothermal production of zinc. Journal of Solar Energy Engineering 2008;130: 014505/1−4.

[171] Weidenkaff A, Reller A, Steinfeld A. Solar production of zinc from the zinc silicate ore willemite. Journal of Solar Energy Engineering 2001;123:98−101.

[172] Tzouganatos N, Matter R, Wieckert C, Antrekowitsch J, Gamroth M, Steinfeld A. Thermal recycling of Waelz oxide using concentrated solar energy. JOM 2013;65(12): 1733−43.

[173] Schaffner B, Hoffelner W, Sun H, Steinfeld A. Recycling of hazardous solid waste material using high-temperature solar process heat. 1. Thermodynamic analysis. Environmental Science & Technology 2000;34(19):4177−84.

[174] Schaffner B, Meier A, Wuillemin D, Hoffelner W, Steinfeld A. Recycling of hazardous solid waste material using high-temperature solar process heat. 2. Reactor design and experimentation. Environmental Science & Technology 2003;37(1):165−70.

[175] Myers WR, Fishel WP. The preparation and hydrolysis of manganese carbide (Mn_3C). Journal of the American Chemical Society 1945;67(11):1962−4.

[176] Seetharaman S. Fundamentals of Metallurgy. Woodhead Publishing Ltd. 2005.

[177] Michalsky R, Pfromm PH. Chromium as reactant for solar thermochemical synthesis of ammonia from steam, nitrogen, and biomass at atmospheric pressure. Solar Energy 2011; 85(11):2642−54.

[178] Gálvez ME, Halmann M, Steinfeld A. Ammonia production via a two-step Al_2O_3/AlN thermochemical cycle. 1. Thermodynamic, environmental, and economic analyses. Industrial & Engineering Chemistry Research 2007;46:2042−6.

[179] Gálvez ME, Frei A, Halmann M, Steinfeld A. Ammonia production via a two-step Al_2O_3/ AlN thermochemical cycle. 2. Kinetic analysis. Industrial & Engineering Chemistry Research 2007;46:2047−53.

[180] Gálvez ME, Hischier I, Frei A, Steinfeld A. Ammonia production via a two-step Al_2O_3/ AlN thermochemical cycle. 3. Influence of the carbon reducing agent and cyclability. Industrial & Engineering Chemistry Research 2008;47:2231−7.

[181] Michalsky R, Parman BJ, Amanor-Boadu V, Pfromm PH. Solar thermochemical production of ammonia from water, air and sunlight: thermodynamic and economic analyses. Energy 2012;42(1):251−60.

[182] Michalsky R, Pfromm PH, Steinfeld A. Rational design of metal nitride redox materials for solar-driven ammonia synthesis. Interface Focus 2015;5(3):20140084/1−10.

[183] Wagener P, Barcikowski S, Bärsch N. Herstellung von Nanopartikeln und -materialien durch Laserablation in Flüssigkeiten. Photonik 2011;1:50−3.

[184] Zhu S, Lu YF, Hong MH. Laser ablation of solid substrates in a water-confined environment. Applied Physics Letters 2001;79(9):1396−8.

[185] Liu P, Cai W, Fang M, Li Z, Zeng H, Hu J, Luo X, Jing W. Room temperature synthesized rutile TiO_2 nanoparticles induced by laser ablation in liquid and their photocatalytic activity. Nanotechnology 2009;20(28):285707/1−6.

[186] Kroto H, Heath JR, O'Brien SC, Curl R, Smalley RE. C_{60}: Buckminsterfullerene. Nature 1985;318:162−3.

[187] Krätschmer W, Lamb LD, Fostiropoulos K, Huffman DR. Solid C_{60}: a new form of carbon. Nature 1990;347:354−8.

[188] Haufler RE, Conceicao J, Chibante LPF, Chai Y, Byrne NE, Flanagan S, Haley MM, O'Brien SC, Pan C, Xiao Z, Billups E, Ciufolini MA, Hauge RH, Margrave JL, Wilson LJ, Curl RF, Smalley RE. Efficient production of C_{60} (Buckminsterfullerene), $C_{60}H_{36}$, and the solvated buckide ion. The Journal of Physical Chemistry 1990;94(24):8634–6.
[189] Flamant G, Luxembourg D, Robert JF, Laplaze D. Optimizing fullerene synthesis in a 50 kW solar reactor. Solar Energy 2004;77(1):73–80.
[190] Chibante LPF, Thess A, Alford JM, Diener MD, Smalley RE. Solar generation of the fullerenes. The Journal of Physical Chemistry 1993;97(34):8696–700.
[191] Fields CL, Pitts JR, Hale MJ, Bingham C, Lewandowski a, King DE. Formation of fullerenes in highly concentrated solar flux. The Journal of Physical Chemistry 1993; 97(34):8701–2.
[192] Laplaze D, Bernier P, Flamant G, Lebrun M, Brunelle A, Della-Negra S. Solar energy: application to the production of fullerenes. Journal of Physics B: Atomic, Molecular and Optical Physics 1996;29(21):4943–54.
[193] Guillard T, Flamant G, Robert J-F, Rivoire B, Giral J, Laplaze D. Scale up of a solar reactor for fullerene and nanotube synthesis. Journal of Solar Energy Engineering 2002; 124(1):22–7.
[194] Flamant G, Bijeire M, Luxembourg D. Modeling of a solar reactor for single-wall nanotube. Journal of Solar Energy Engineering 2006;128:24–9.
[195] Albu-Yaron A, Arad T, Levy M, Popovitz-Biro R, Tenne R, Gordon JM, Feuermann D, Katz EA, Jansen M, Mühle C. Synthesis of fullerene-like Cs_2O nanoparticles by concentrated sunlight. Advanced Materials 2006;18(22):2993–6.
[196] Wiesel I, Arbel H, Albu-Yaron A, Popovitz-Biro R, Gordon JM, Feuermann D, Tenne R. Synthesis of WS_2 and MoS_2 fullerene-like nanoparticles from solid precursors. Nano Research 2009;2(5):416–24.
[197] Levy M, Albu-Yaron A, Tenne R, Feuermann D, Katz EA, Babai D, Gordon JM. Synthesis of inorganic fullerene-like nanostructures by concentrated solar and artificial light. Israel Journal of Chemistry 2010;50(4):417–25.
[198] Lu H-B, Chan BCY, Wang X, Chua HT, Raston CL, Albu-Yaron A, Levy M, Popowitz-Biro R, Tenne R, Feuermann D, Gordon JM. High-yield synthesis of silicon carbide nanowires by solar and lamp ablation. Nanotechnology 2013;24(33):335603/1–7.
[199] Brontvein O, Stroppa DG, Popovitz-Biro R, Albu-Yaron A, Levy M, Feuerman D, Houben L, Tenne R, Gordon JM. New high-temperature Pb-catalyzed synthesis of inorganic nanotubes. Journal of the American Chemical Society 2012;134(39): 16379–86.
[200] Feuermann D, Gordon JM, Huleihil M. "Solar fiber-optic mini-dish concentrators: first experimental results and field experience. Solar Energy 2002;72(6):459–72.
[201] Feuermann D, Gordon JM. Solar surgery: remote fiber optic irradiation with highly concentrated sunlight in lieu of lasers. Optical Engineering 1998;37(10):2760–7.
[202] Feuermann D, Gordon JM. Solar fiber-optic mini-dishes: a new approach to the efficient collection of sunlight. Solar Energy 1999;65(3):159–70.
[203] Feuermann D, Gordon JM, Huleihil M. Light leakage in optical fibers: experimental results, modeling and the consequences for solar concentrators. Solar Energy 2002;72(3): 195–204.
[204] Liang D, Fraser Monteiro L, Ribau Teixeira M, Fraser Monteiro M, Collares-Pereira M. Fiber-optic solar energy transmission and concentration. Solar Energy Materials, Sol Cells 1998;54(1–4):323–31.
[205] International Energy Agency. [Online]. Available: www.iea.org.

[206] Kalogirou S. The potential of solar industrial process heat applications. Applied Energy 2003;76(4):337−61.
[207] IT Power. Renewable energy options for Australian industrial gas users. ARENA Background Technical Report 2015.
[208] Zanganeh G, Pedretti A, Zavattoni S, Barbato M, Steinfeld A. Packed-bed thermal storage for concentrated solar power − Pilot-scale demonstration and industrial-scale design. Solar Energy 2012;86:3084−98.
[209] Müller-Steinhagen H. Applications of solar heat for temperatures ranging from 50−2000°C. In: 5th European thermal-sciences conference, The Netherlands; 2008.
[210] Duffie JA, Beckman WA. Active solar collectors and their applications. Hoboken: John Wiley & Sons; 2006.
[211] Mekhilef S, Saidur R, Safari A. A review on solar energy use in industries. Renew Sustain Energy Rev 2011;15(4):1777−90.
[212] Luque AL, Andreev VM, editors. Concentrator photovoltaics. Berlin: Springer-Verlag; 2007.
[213] Kraemer D, Poudel B, Feng H-P, Caylor JC, Yu B, Yan X, Ma Y, Wang X, Wang D, Muto A, McEnaney K, Chiesa M, Ren Z, Chen G. High-performance flat-panel solar thermoelectric generators with high thermal concentration. Nature Materials 2011;10(7): 532−8.
[214] Chow TT. A review on photovoltaic/thermal hybrid solar technology. Appl Energy 2010; 87(2):365−79.
[215] Li C, Goswami Y, Stefanakos E. Solar assisted sea water desalination: a review. Renew Sustain Energy Rev 2013;19:136−63.
[216] Legrini O, Oliveros E, Braun AM. Photochemical processes for water treatment. Chemical Reviews 1993;93(2):671−98.
[217] Chong MN, Jin B, Chow CWK, Saint C. Recent developments in photocatalytic water treatment technology: a review. Water Research 2010;44(10):2997−3027.
[218] Cuce E, Cuce PM. A comprehensive review on solar cookers. Appl Energy 2013;102: 1399−421.
[219] Mayhoub MS. Innovative daylighting systems' challenges: a critical study. Energy Build 2014;80:394−405.
[220] Arashi H, Kaneda Y. Solar-pumped laser and its second harmonic generation. Solar Energy 1993;50(5):447−51.
[221] Walter MG, Warren EL, McKone JR, Boettcher SW, Mi Q, Santori EA, Lewis NS. Solar water splitting cells. Chemical Reviews 2010;110(11):6446−73.

Index

'*Note*: Page numbers followed by "f" indicate figures, "t" indicate tables.'

I